Nanoscale Science and Engineering Forum

Core Programming Topic at the 2011 AIChE Annual Meeting

Minneapolis, Minnesota, USA
16-21 October 2011

ISBN: 978-1-61839-741-6

Printed from e-media with permission by:

Curran Associates, Inc.
57 Morehouse Lane
Red Hook, NY 12571

CURRAN ASSOCIATES INC.
proceedings
.com

Some format issues inherent in the e-media version may also appear in this print version.

Printed by Curran Associates, Inc. (2012)

For permission requests, please contact AIChE
at the address below.

AIChE
3 Park Avenue
New York, NY 10016-5991

Phone: (203) 702-7660
Fax: (203) 775-5177

www.aiche.org

Additional copies of this publication are available from:

Curran Associates, Inc.
57 Morehouse Lane
Red Hook, NY 12571 USA
Phone: 845-758-0400
Fax: 845-758-2634
Email: curran@proceedings.com
Web: www.proceedings.com

TABLE OF CONTENTS

(665e) Metallic Nanowire Growth In Solution: The Effect of Mechanical Disturbance

Chao Wang, Materials Science Division, Argonne National Laboratory, Argonne, IL

The effect of mechanical disturbance on 1D nanocrystal growth in solution phase isinvestigated by controlled growth of Au nanowires with and without stirring. While a static growth leads tostraight, single-crystal Au nanowires, the mechanic disturbance by stirring tends to bend the nanowire growth, yielding nanowire kinks abundant in various types of crystal defects including dislocations, twin boundariesand grain boundaries. Mechanical modeling and analysis is introduced to elucidate the nanowire growthmechanisms in these two conditions. This work provides fundamental understanding of crystal defectformation at nanoscale which could be applied to guide the development of advanced nanomaterials withshape control and unique mechanical properties.

See more of this Session: Nanowires I: Synthesis and Modeling
See more of this Group/Topical: Nanoscale Science and Engineering Forum

(609c) Solar Cells Based On Sensitized Nanowire Arrays

Jason B. Baxter, Department of Chemical and Biological Engineering, Drexel University, Philadelphia, PA

Solar cells can provide an abundant, clean, and sustainable source of electricity, but high costs have limited their implementation. The use of sensitized nanostructured architectures may enable both low-cost processing and high efficiency by decoupling the functions of light harvesting and charge transport into different materials. Sensitized solar cells consist of a bicontinuous interpenetrating network of electron- and hole-transporting materials with an interfacial absorber layer. In this architecture, interfacial recombination is the dominant loss process, so controlling the interfacial chemistry during deposition is critical. I will present our group's work on the use of ZnO nanowire arrays sensitized by thin CdSe coatings to produce extremely thin absorber (ETA) solar cells. Nanowire arrays can provide direct pathways for electron transport as well as sufficient surface area for sensitization. Progress in sensitized nanowire-based solar cells will be presented, as well as some of the challenges for improving their efficiencies. Overcoming these challenges will require detailed fundamental understanding and control of charge transfer in the materials and interfaces within the cells.

See more of this Session: Semiconductor Nanowires and Nanotubes for Next Generation Photovoltaics - From Fundamentals to Devices
See more of this Group/Topical: Topical 5: Nanomaterials for Energy Applications

(470b) Advanced Nanostructures for Thermoelectric Energy Harvesting

Yue Wu, School of Chemical Engineering, Purdue University, West Lafayette, IN

Substantial efforts have been devoted to use nanostructured materials for thermoelectric energy harvesting and solid-state cooling in the past decade. However, the majority of the research is still limited in lab scale due to the incapability to mass-produce well-defined nanostructured materials with low yet industrial-compatible process. In addition, a lot of widely used thermoelectric materials contain toxic and expensive elements that prevent the large-scale deployment of the thermoelectric devices. In this presentation, we will discuss our research on the development of mass production of molecular scale nanostructures of chalcogenides and metal oxides, as well as their heterostructures, for the manufacture of thermoelectric generators operating at different temperature ranges. Particularly, we will talk about our approach to discover and investigate the non-toxic and abundant nanostructured materials to achieve an environmentally friendly process. Our preliminary research indicated that thermoelectric figure of merit (ZT) close to 2 can be achieved in the molecular scale nanowires of certain chalcogenide due to significantly enhanced power factor and reduced thermal conductivity as a result of quantum confinement.

See more of this Session: Nanomaterials for Thermal-to-Electric Conversion I
See more of this Group/Topical: Topical 5: Nanomaterials for Energy Applications

(532e) Biodegradation of Carbazole by Microbial Cells Coated with Magnetite Nanoparticles

Jianmin Xing, National Key Lab of Biochemical Engineering, Institute of Process Engineering, Chinese Academy of Sciences, Beijing, China

Microbes can remove nitrogen from carbazole through a series of enzyme-catalyzed reactions. Microbial cells of *Klebsiella* sp. LSSE-H2 were coated with magnetic Fe_3O_4 nanoparticles, which were synthesized by a coprecipitation method followed by modification with ammonium oleate. The average size of the magnetic particles was found to be in the range from 10 to 15 nm. Transmission electron microscopy (TEM) cross section analysis of the cells showed further that the Fe_3O_4 nanoparticles were strongly absorbed by the surfaces of the cells and coated the cells. The coated cells had superparamagnetic properties. The coated cells not only had the same activity as free cells but could also be reused. Biodenitrogenation rate was increased with the addition of Tween 80. The activity decreased less than 23% after three recycles.

See more of this Session: Green and Renewable Energy Aspects of Nanotechnology
See more of this Group/Topical: Topical C: Environmental Aspects, Applications, and Implications of Nanomaterials and Nanotechnology

BIO-REFINERY BASED ON INDIAN PAPER INDUSTRY WASTES

Sanjay Tyagi*, Narayan Chandra Mishra, and A.K.Ray
Indian Institute of Technology, Saharanpur Campus, Saharanpur, India
*Central Pulp and paper Research Institute, Saharanpur, India

ABSTRACT

Paper Industry is one of the major industries which discharge some wastes that can produce many valuable by-products. These are black liquor, steam and electrical energy, wood/ bark wastes, sawdust/ bamboo dust, solid wastes like lime mud, grits, dregs, ETP sludges and silica. These wastes can be converted to many value added chemicals, such as activated carbon from black liquor, sawdust, bamboo dust, broken chips and partly pulverized bark.

Black liquor /sulphite liquor produces ligno sulphones from rice straw, wheat straw, bagasse, bamboo, sarkanda and hardwood. It has diversified uses e.g. as a base for industrial detergents(washing powder and liquid soaps), oil-well drilling mud, cement and concrete additives, dispersants for a variety of chemicals, emulsifiers and stabilizers, brightening agents for electrolytic refining during electroplating, binders and adhesives for pellets and briquettes chemicals such as clay, calcium carbonate and titanium di-oxide resin ingredients, lignin-phenol-formaldehyde mixtures as adhesives for ply wood and other wood based panels, rubber additives, adsorbent/desorbent, micronutrient for agricultural applications, asphalt emulsions. Besides black liquor can also be converted to lignin-poly isocyanate foam, lignin reinforced polymers, lignin adducts, chlorobrominated lignins as effective fire retarding agents, composites of alkali lignin and biologically active agents such as fungicides, insecticides, nematocides, herbicides etc. ,activated carbon by coking or gasification of black liquor, hydrolytic degradation of ligno-cellulosic waste to sugars such as xylose and xylitol used as food stuff as sweeteners, synthetic fuel using solar furnace, silica from straw black liquor, hydrogen gas by electrolysis, vanillin, DMSO, methanol, acetate, formate, actate, saccharinic acids, lactones, turpentine and tall oil, alkali lignin, guaiacol, pyrocatechol, acetoguaiacone, biogas(methane) etc.

Cellulosic wastes can also produce cellulosic bio-ethanol using thermophilic bacterial species like Clostridium thermocellum. In the similar manner, the sludge can produce glucose by treatment with cellulase enzymes. Syngas from gasification of pith of some nonwood based plants can also be produced.

From straw soda black liquor, silica can be precipitated for industrial applications such as amorphous silica (for high quality synthetic pigments as paper fillers which can replace soap stone and TiO_2), filler, retention aid and as flocculating agent, for production of ceramics. The later can be employed for household china, fine earthen ware, concrete block, electric porcelain, silica light weight brick, clay brick, calcium silicate thermal insulating materials, reinforcement in rubber, in paint manufacture, adsorbent, grease thickener, insecticide and cosmetic industry.

An attempt has been made in this paper to review the existing status of Indian industry to examine up to what extent they are able to exploit the potential of aforesaid cellulose and lignin by-products. Further exploratory investigations were made to look into the feasibility of their manufacture in plant practice in India for its sustainability.

INTRODUCTION:

Pulp and paper Industry is the World's largest non-food biomass collection system. India is the 15th largest paper producer in the World. The installed capacity of Indian paper industry is 12 million tonnes and 11 million tonnes is in operation, however annual production is around 10 million tonnes with per capita consumption of 10.5 million tonnes. Now the Indian paper industry is growing at 9.5%CAGR per annum along with booming economy with GDP growth rate 8.6 % .It is projected a demand of paper and paper board in the country will be 15 million tonnes by the year 2015 and 22 million tonnes by the year 2020[1]. On the utilization of recovered paper side, India imports around 4.5-5 million tonnes of waste paper as the recovery rate at India rests at meager value of 20-22% [1]. In the coming decade Asia will be major hub for paper production and trade, particularly in a time when drastic paper production cut is taking place in Europe (minus 0.3%),North America (minus 0.5%) and other parts of the World**[2]**.As a major economy of Asia, compounded with the booming economy of India and increasing demand of paper and paper board, India will provide consistent growth for decades to come. Together India and China are estimated to make up 59% of global incremental paper and paper board demand during 2009-2020. At the same time , North America and Western Europe is forecast to decline from 46% to 38% of global production by 2020.However, Indian manufacturers have to be more conscious on the emerging opportunities and growing market size as stringent competitions are ahead for them which have to be met by adopting clean technologies and quality products [2].

Pulp and paper industry is capital-intensive industry, intensive in terms of energy, chemical, water, labor and raw materials concomitant with high pollution load generation. Besides it demands various large expensive pieces of equipments for its manufacture to produce quality products. Because of its capital intensiveness, investment in the paper industry is always a long-term proposition. The technology of operations is also very much intricate. Today's plant personnel need the concept of production integrated with energy-environmental-productivity-quality issues. The discharges from paper industry include solid waste, liquid effluent and gaseous emissions. There are regulatory limits enforced on the discharge of these pollutants by the regulatory authorities and the industry is expected to treat its effluents to meet the standards. In some of the states zero effluent concepts are being implemented in waste paper based mills. The new process must induct lower energy, water consumption along with little or no impact on environment. In order to follow this strategy the technology has been continuously passing through many revolutionary changes over the last decades to achieve low manufacturing costs, to earn profits derived from the ability of the producer and to face international competition for quality standards in the global market.

PAPER MILL BASED BIOREFINERY:

The pulp and paper industry is at the forefront of the concept of bio-refinery since it is already a biorefinery -1st generation of Biorefinery –enabling sustainable production of pulp/ paper enhanced by the coproduction of bioenergy and biofuels and bioproducts, through matured technological processes that maximizes the value chain of pulpwood **[3]**. However, the decreasing profitability of standard pulp and paper mills and the world's demand for a non-fossil Carbon Economy resulted a new biorefinery platform based upon complex technology concepts with higher revenue potential [3]. Biorefinery is an overall concept of an integrated and diversified processing plant where biomass feedstocks are converted into a wide range of valuable products, much likewise to petroleum refineries. Integrated birefinery is a processing facility that extracts carbohydrates, oils, lignin, and other materials from biomass, converts them into fuels, high value chemicals and other materials, with a near zero waste draw [4]. Paper Industry is one of the major industries which discharges some wastes as given below that can produce many valuable by-products [5-26].These are black liquor, steam and electrical energy, tall oil, wood/ bark wastes, sawdust/ bamboo dust, solid wastes like lime mud, grits, dregs, ETP sludge and silica. These wastes can be converted to many value added chemicals, such as activated carbon from black liquor, sawdust, bamboo dust, broken chips and partly pulverized bark, turpentine and tall oil. Important chemicals like methanol, acetate, formate, lactate, saccharinic acids, lactones, alkali lignin, guaiacol, pyrocatechol, acetoguaiacone can also be prepared. These are discussed as under:

Solid waste from big paper mills[12]

(Avg. data based on data collected from 8 paper mills)

Organic Waste: (T/T of Paper)

1.	Raw Material Preparation (Wood /bamboo dust)	0.1
2.	Pulp mill rejects	0.01
3.	Paper machine cleaner rejects	0.1
4.	ETP Sludge	0.2
	Sub-Total	0.41

Inorganic Waste: (T/T of Paper)

1.	Ash (Cinder + Fly ash) from boilers	0.80
2.	Lime sludge and grits (Recovery + Bleach liquor preparation)	0.70
	Sub-Total	1.50
	Total	1.91

Utilization of wastes from a paper industry can have enough potential for a perfect biorefinery in India. Production of many value added products can be explored. There are many Indian industries engaged in establishing biorefineries for producing multiple products. Some are mentioned below:

UTILIZATION OF WASTES FOR BIOREFINERY IN INDIA

1. Utilization of wood/nonwood wastes:

Large amounts of ligno cellulosic waste are generated through forestry and agricultural practices , pulp and paper industries, timber and agro-industries.

1.1 Production of bioethanol:

Cellulosic wastes, solid wastes like wood dust, bagasse pith, waste paper and even sludge can produce cellulosic bioethanol [8]. Alcohol can be produced from cellulosic wastes using thermophilic anaerobic bacteria species like Clostridium thermocellum at 60°C and pH of 7.5 under control anaerobic conditions [14]. Bioethanol from bagasse pith has been produced in the laboratory using thermophilic ethanologen, yeast strain at 50°C through sachharification both chemically and enzymetically (using hydrolytic enzymes like celluloses, laccases and xylanase) followed by fermentation of the hydrolysate. It is reported that almost 80% glucose (30 g/l)of enzyme hydrolysate was fermented to ethanol(9.5gm/l) within 40 h and thereafter some of the oligosachharides also converted to ethanol bringing the final ethanol concentration 11.5 gm/l in the broth [6,7]. Naithani et al. [9] describes utilization of Lantana Camara and pine needle as lignocellulosic wastes for ethanol production with a yeast culture of Saccharomyces Cerevisiae NCIM-3570 with final pH to 4.5 at a temperature of 30°C. The acid hydrolysate of hemicelluloses extracted bagasse with the yeast S.Cerevisiae for fermentation can produce ethanol after concentration and supplementation can produce 8.3% alcohol with fermentation efficiency of 83.3% compared to 7.8% for molasses medium of 28°Bx with FE of the order of 83.3%[25].

2. Utilization of fiber rich Sludge:

In the similar manner, the sludge can produce glucose by treatment with cellulase enzymes [10]. Treatment of cellulosic sludge with cellulase enzymes rapidly shortens the fiber length and then dissolves the fibers in to glucose. The glucose can then be sent to an ethanol plant for fermentation to ethanol. The economics are based on sludge containing 50 tonnes per day of cellulose. The enzyme cost is projected at $6,000 per day. Water treatment cost is $1,000 per day (including capital and O &M) assuming reverse osmosis will be needed. Thus the total costs are $7,100. The value of the glucose produced (assuming 70% conversion from fiber) is $13,000 per day. The value of sludge disposal avoidance is taken at $3,000 per day for a total benefit of $16,000 per day. Addition of c-PAM accelerates the rate of fiber hydrolysis. The other costs considered for investment are due to some equipment to handle the residual solids,acid and power. The cost-benefits have been found favorable [10]

3. Biomass Gassification:

Syngas from gasification of pith of some nonwood based plants such as bagasse can be produced [11]. Biomass thermo-conversion for energy production from bagasse has been attempted by Umrao et.al. [11] for pyrolysis in a free fall reactor followed by gasification in a quartz reactor to obtain a more efficient and complete conversion of the bagasse as well as production of a gas with high heating value or with a high H_2 + CO (Syngas) content to be used as a fuel for chemical industry. Pyrolysis of bagasse is done on a fluidized bed at high temperature due to which charcoal is formed and the volatile components are released. The charcoal is sent to the two stage gasifier. As a result solid ashes, and a gas mixture which is composed of CO_2,CO,H_2, water vapour, pyrolysis product , tar and hydrocarbons. This process saves one liter of liquid fuel by using 4-5 kg of bagasse.

4. **Bio-gas production [21, 24]**

Primary and secondary (biological) sludges generated in pulp and paper industry can be digested anaerobically to produce bio-gas through acid fermentation followed by conversion to methane and carbon di oxide by methane forming bacteria. Benefits of anaerobic digestion are:

- Production of Bio-gas which can be utilized as a fuel
- Reduction of organic solids (and stabilization)
- Sludge can be used as a soil conditioner or fertilizer.
- Sludge can be treated without pre-dewatering.

Typical values of bio-gas production and organic solid waste reduction using anaerobic digestion of pulp and paper mill sludges are:

Reduction of organic solids : 45-75%)

Bio-gas produced : 0.5-0.7 m^3/kg organic solids

The process details for a typical 30TPD bagasse based soda pulp mill in India producing 0.45 - 0.50 m^3/kg COD gas (75% CH_4 + 25% CO_2) with calorific value of 5400 kcals/m^3 is available. The process removes 90% BOD, and 70% COD at temperature range 35-37°C with saving of 4000kWh /day power and 5 tons per month nutrients [24].

5. **Particle board**: Pith + Lignin can make particle board.
6. **Other important value added products:**

Many valuable chemicals can be manufactured from paper industry wastes. Majority of the processes for chemical manufacture from paper industry wastes are patented [22].These are excellently reviewed by a number of investigators [19-23]. Some of the patented processes are highlighted below:

6.1 Activated carbon:

Ligno-cellulosic material such as saw dust, bamboo dust, broken chips, and partly pulverized bark is reported to have been converted into activated carbon.

Method I: The material was first treated with inorganic acid to remove sugars at low temperature range. Acid extracted material was then heated at **450°C for nine hours and** for one hour at 800°C. While steam was being injected, the tar was removed as condensate in this high temperature range. About 22 % yield of activated carbon was reported.

Method II:

Continuous production of activated carbon can be possible by carbonization of cellulosic material impregnated with activating agent such as zinc chloride. The process involved continuous supply of activating agent-impregnated material into carbonizing zone containing previously carbonized cellulosic material at such constant flow rate as to allow the impregnated material to reside for fifteen minutes in the carbonizing zone. The impregnated material was carbonized with the carbonized cellulosic material heated to 350°C by counter currently supplied hot gases and which formed an agitated fluidized bed by being fluidized with hot gas stream and being stirred by a mechanical means. The total weight of the product was two third of the original weight of the wood material supplied.

6.2 Chemicals from Bark:

Bark along with other refuse creates a disposal problem for paper mills using hard wood as raw material. Utilization of bark can be developed on following lines.

I. As a source of fuel for steam generation
II. As a source of chemicals, particularly tannins and tannin like compounds, which would be recovered by water/acid/alkali extraction of bark, prior to burning.
III. Marginal utilization as soil conditioner, extender for thermosetting resins' oil well mud additive, reinforcing and stabilizing agent for polypropylene moulding compounds.

6.3 Hydrolytic Degradation of Lignocellulosic waste to sugars, Xylose and Xylitol:

During acid hydrolysis of lignocellulosic biomass a number of chemicals are produced. These are sugars (Xylose), aliphatic acids (acetic, formic and levulinic), furan derivatives, furfural and 5-hexamethyl furfural(HMF) and phenolic compounds. Hydrolytic degradation of lignocellulosic waste to sugars such as xylose and xylitol used as food stuff as sweeteners can be potential chemicals. Hardwood and agricultural residues are rich in pentosans which mainly contain xylan. Waste and refuse generated from hard woods/straw/bagasse etc. while preparing these materials for pulping, could be degraded hydrolytically to D (+) Xylose and Xylitol, which are used in food stuff as sweeteners.

Hydrolysis of waste was carried out in first stage with sodium hydroxide of 4% concentration which removes acetic acid chemically bonded in the wood as sodium acetate. The liquid extract of this stage can be re-used after it has been brought to the alkali concentration to be used.

In the second stage of processing the residue from the first stage containing pentosan and cellolignin was further hydrolyzed with mineral acid e.g. sulfuric acid. When working without pressure, 1.5 to 6 % strength of H_2SO_4, and a ratio of liquid to solid of 3 to 6 parts by volume was preferred. Temperature was preferably 50 to 125°C and time was 2-4 hours. After completion of the second stage, the batch was filtered; the filtrate could be worked to Xylose or directly to Xylitol.The insoluble residue from the second stage could next be used to obtain organosolve lignin which was thermoplastic and very reactive; the material, this lignin preparation could be used as a base material for dyestuffs and pesticides. The residue remained after removal of lignin, on treatment with sodium chlorite followed by sodium hydroxide could be converted to practically white cellulose which could be used for industrial applications. Thus all wood waste could be converted to useful chemicals.

6.4 Synthetic Fuel using Solar Furnace:

A process of producing gases such as CO_2, CO, CH_4 and H_2 from pyrolysis and gasification of solid organic waste contained within a reactor has been patented [22]. The process comprises solar heating of working fluid (steam, CO_2 or a mixture of gases) to 600°C to 700°C by means of solar top furnace. The working fluid was contained in the lower section of the reactor situated at the focus of the solar top furnace. The heated working fluid was injected under pressure into porous reactant bed on which the catalyst ($NaHCO_3$) - mixed organic waste was situated. The catalyst-mixed organic waste was pyrolyzed causing the formation of CO_2, CO, CH_4 and H_2 gases and char residue.The hydrogen produced could be utilized to replace natural gas in a hydrogen economy, to make fertilizer or as a chemical feed stock. It is claimed that process did not require a source of pure oxygen, and provided an attractive method for using solar energy.

7.0 Basic chemicals from wastes

Wood waste has a complex carbohydrate lignin structure requiring severe conditions to reduce it to such basic chemicals as Methanol, Ethanol, Furfural and Phenols. Chemical production costs are directly related to scale and complexity of operation **[22].** Utilization of pine needles and bagasse as bed material in solid state fermentation for production of lactic acid by Lactobacillus strains has been an active research in IITRoorkee[26].

8.0 Cellulose nano crystals and nanocomposites:

Cellulose nanocrystals have been largely applied as reinforcing fillers in the preparation of nano composites materials with improved mechanical and barrier properties.

9.0 Utilization of black liquor and sulphite waste liquor

Pulp and Paper industry has a very huge lignin potential. According to rough estimate for every ton of bleached pulp produced, about 0.5 ton of lignin is either wasted to sewer or used as a low value fuel. If the means could be found to isolate it from spent pulping liquor and to convert it to usable chemicals, the financial returns might be sufficient to support separation and processing lignin preparations from spent liquor, thus helping abatement of stream-pollution. In many ways Kraft lignin is a versatile raw material. It can be easily isolated by acidification, preferably carried out with carbon dioxide (stack gas), whereby phenolic group are liberated and lignin precipitates. The precipitated lignin can be removed by filtration at elevated temperature and purified further by re-dissolution in water and re-precipitation by sulphuric acid.

9.1 Lignin Isolation and separation:

Lignin compounds from sulphate black liquor could be precipitated by bubbling CO_2 into it at about 60°C. The precipitated lignin was separated by filtration and purified by dissolving in dry hexane and re-precipitating in ether. Alkali lignin could be isolated, from soda process black liquor by neutralizing it with CO_2 from stack gases to pH 8.3 to 8.5; about 40% of the lignin was reported[9] to have been precipitated. The mix was heated to 85°C – 90°C to coagulate the lignin, was filtered. Residual lignin could be further precipitated by addition of sulfuric/hydrochloric acid. Isolated lignin was washed first with cold water and then with dilute sulfuric acid and dried on double drum drier.

The slimy, gelatinous and hard to filter form of the precipitated lignin from black liquor could be changed to a crystalline, easily filterable and quick settling precipitate by mixing small amount of water-immiscible solvent such as chloroform, methelyene bromide, l-nitropropane or the like, with lignin containing liquid such as Kraft/soda black liquor. The chloroform, or other agent, may then be recovered from the filtered liquid by vacuum- or steam distillation.

Chemical modification of Kraft lignin drastically changes its properties and product obtained are of commercial value. Common modifications can be summarized as follows:

1. Formation of ethers by reaction with alkyl/aryl halides
2. Formation of esters by, reaction with carboxylic acid
3. Formation of polymeric condensation product with phenols, aldehydes and amines
4. Conversion to sparingly soluble phenol derivatives in reaction with various salts of heavy metals
5. Sulphonation to Kraft lignosulphonates
6. Nitration carried out with nitric acid
7. Chlorination with aqueous chlorine
8. Oxidation, with air or oxygen produces derivatives with changed solubility and melting point
9. Demethylation of lignin can be accomplished by acidic and alkaline treatment and results in the formation of catechol group increasing the reactivity in condensate e.g. with formaldehyde.

9.2 Lignin derivatives:

Lignin thus isolated could be converted to large surface area. Lignin is useful in drug formulations for adsorbing acids in the digestive system, due to their extremely high surface area and their acid-alkali resistance. Other suggested uses include carriers for adhesives, catalyst substrate, reinforcement for plastics and carriers for insecticides and herbicides.

A method to produce biodegradable uncolored lignin solutions was described wherein aqueous lignin solutions such as pulp mill effluent were treated with polyvalent metal cations. The amount of cations should be sufficient to react with lignin, and the aromatics to produce a water insoluble precipitate which was separated from remaining liquor. The precipitate was subsequently acidified with acid solution and was allowed to stand for sufficient time to permit digestion. The precipitate thus formed was separated by centrifugation; and was treated with alkali to produce aqueous basic solution. This solution was found to be effective in controlling Scale formation in boiler water systems, when added to these systems in amount of 5-50 ppm. Thus method not only recovers valuable product from effluent but also reduces the amount of non-biodegradable and toxic materials prior to discharge. In addition color of the final discharge is also reduced.

Sulpho methylated alkali lignins mixed with aqueous ferrous salt and dichromate and spray dried to a powdered product was reported to be useful as drilling mud thinner.

9.3 Lignin Reinforced Polymers:

A method for forming lignin reinforced polymers was developed by co-precipitating lignin/latex slurry and gas drying the slurry to leave a lignin reinforced polymer free of inorganic salts. In a process, alkali lignin was dissolved in ammonium hydroxides; and the ammonium lignate thus formed was then modified by adding hexamethylenetetramine to the lignate and heating to 90°-110°C for two: hours. Thus treated lignin was mixed with latex to give emulsion, which was precipitated by CO_2. When the precipitate was dried at 300°F or below, inorganic salts formed in precipitation, sublimed leaving lignin/latex polymer.

9.4 Lignin- polyisocyanate/polyurethane foam:

These foams are useful as plastics and adhesives. Lignin-polyisocyanate foam can be prepared from polyethylene glycol solution of lignin when the later is treated with toluene di-isocyanate at 115°C which afforded foam having good compressive strength and 74% recovery. Polyurethane type foam from lignin preparations and organic isocyanates are also reported.

9.5 Lignin as dispersants:

Method: preparation of ozone treated lignin from Kraft black liquor by treating lignin with oxygen containing 3% ozone at 50°C for two hours; and spray drying the resulting solution has been reported. Oxidized lignin was found to be useful as dispersant for a variety of inorganic materials such as Clay, Cement, Calcium Carbonate, and Titanium Dioxide.

9.6 Lignin Adducts as surfactants:

Lignin adducts with 3-chloro-2-hydroxypropane-l-sulphonate, chloromethane sulphonate, acrolein and the like compounds are also reported to be useful as surfactants in dyestuff compositions.

9.7 Adhesive from lignin: Lignin-phenol-formaldehyde mixtures as adhesives are used for ply wood and other wood based panels:

Method: An adhesive from Kraft or Soda black liquor was prepared by reacting the black liquor (dry matter more than 25%) with formaldehyde and combining the product with separately prepared phenol-formaldehyde adhesive.

9.8 Lignin as coagulants and flocculants:

Lignin preparations from pulping liquor chemically modified by reaction with compounds such as formaldehyde, triazines, alkylamines were found to be useful as coagulants in treatment of sewage and other wastes. Cationic water soluble lignin amines prepared by reacting Kraft/Soda lignins with the reaction product of epichlorohydrin and a tertiary amine were suitable for a variety of uses such as flocculating agents, coagulants, retention aids, dispersing agents, asphalt emulsifiers and emulsion stabilizers.

9.9 Fire retarding agents:

Chlorobrominated lignins were highly effective as fire retarding agents. In one method, the inflammable material was coated or soaked with an aqueous slurry or alkaline solution of chlorobrominated lignin followed by drying. In addition chlorobrominated lignins could be combined with bonding compositions used for the production of bonded articles from particulate material, such as fibers, chips, granules, saw dust etc. eg. in the formation of chip board, particle board and the like. Chlorobromination was affected by adding chlorine and bromine concurrently or in two step operation. After completion the reaction, the chlorobrominated lignin was filtered, washed and dried.

9.10 Composites

Composites of alkali lignin and biologically active agents such as fungicides, insecticides, nematocides, herbicides and the like are useful as controlled release pesticides. These composites were prepared by intermixing ammonium salt of lignin with active agent under conditions of high shear mixing to form stable emulsion; acidifying the emulsion with mineral acid at pH 4-5 to precipitate the active agent/lignin emulsion with the active agent being physically absorbed within the lignin and drying the precipitate if desired.

9.11 Coke and activated carbon from black liquor

Process for activated carbon by coking or gasification of black liquor: In a process black liquor was heated in absence of free oxygen to an elevated temperature (450°-625^oF) and pressure to prevent vaporization of the water, for a period sufficient to produce granular coke. The coke was separated from the aqueous phase to yield an aqueous effluent of low COD. The coke produced might be burned to supply heat, or used as a soil conditioner or carbonized to produce high grade coke for metallurgical purposes or used to produce activated carbon.

Organic values of black liquor have been converted into heat energy and activated carbon by a method which eliminates the recovery boiler from pulping liquor regeneration systems. Black liquor containing 16% solids or even lower or at any solids content up to 65% was hydropyrolyzed under very high pressure at 550°-650°F. The organic matter separates as a water-insoluble precipitate along with inorganic constituents such as Na_2CO_3, and minor amounts of Na_2SO_4, Na_2S. The organic matter could be utilized to produce heat energy or activated carbon which itself could be used for purifying effluents from various section of paper mill.

10.0 By-products from lignin

Generally by-products from lignin can be categorized as follows:

1. Low molecular weight compounds
2. Polymeric products

Table 1 gives a detailed picture of by-products (both commercial and non-commercial) obtained from Kraft cooking of wood using different methods of isolation. It can be seen from the table that the main by-products of Kraft cook are the degradation products of carbohydrates and lignin. The most consistent efforts on dry distillation of the alkaline black liquor powdered together with excess lime or alkali were made to obtain methanol, acetone, methylethylketone, higher ketones and other components of light and heavy oils in appreciable quantity (60-70 kg/ton of pulp each of the low molecular and high molecular compounds).

Low molecular weight compounds

The most practical approach to synthesis low molecular weight compounds from lignin is to heat the black liquor for 10-20 minutes at 250-290°C together with small amount of alkali and sodium sulphide. Table 2 covers the chemicals obtained during this process with their corresponding yield in kg/PTP. Under optimum conditions (291°C, 20.4 % of organic substance, 3.2 % Na_2S and 1.6 % of the NaOH based on total liquor) a maximum yield of 33 % of the organic products can be obtained as ether soluble phenols.

Of all chemicals listed in Table 2 the following chemicals are most important.

Use of lignin and lignin derivatives: Lignin Sulphonates and Kraft Lignin, Vanillin, Dimethyl sulphonates, Dimethyl Sulphoxide, Furfural, Methyl Mercaptan, Phenolics,thermosetting resins and graft polymers including polyurethanes, polyesters, polyamines and epoxies[13].

1. Vanillin
2. Furfural
3. Organosulphur compounds
    a) Dimethyl sulphide (DMS)
    b) Methyl mercaptans (MM)
    c) Dimethyl sulphoxide (DMSO)
    d) Dimethyl sulphone ($DMSO_2$)
4. Low molecular weight Phenolic compounds

10.1 Vanillin

Vanillin has a wide range of industrial application. Vanillin is the most widely flavoring agent,, commonly known as vanilla, due to its pleasant aroma and its property of enhancing other flavors.. It is universally used as flavoring agent with a property to mask other odors. Vanillin also finds use as a reductant and as material absorbing ultraviolet light. It can be incorporated into various plastics such as regenerated cellulose, polyvinyl ethers and polyamides to protect them against ultraviolet radiation. Ointments for protection against sunburn on the skin can be prepared with vanillin as active agent. Polymeric materials prepared from vanillin and vanillic acid have potential as fibres and films.

Generally vanillin is synthesized from Guaiacol or Eugenol.

It can be produced from lignin through oxidation of spent liquor with suitable oxidising agent or by extraction by a suitable solvent. There are a few methods for manufacturing Vanillin.

Method I: Catalytic oxidation under pressure in pressure vessel by prior raising pH of black liquor from 9.5 to 13 by sodium hydroxide and then heating for about 2 hours followed by extraction with chloroform. The aqueous layer is acidified with sulphuric acid to bring down the pH to 3.0 and the filtrate is again extracted with chloroform.

Commercial process based on alkaline oxidation of lignin with oxidising agent as cupric oxide (2% wt/wt.), various nitrobenzene compounds, mercuric oxide etc. have reported a better yield of vanillin from lignosulphonates. For example cupric hydroxide oxidation of lignin gives 22 % of vanillin at 160°C and 12.5 % at 102°C. In comparison a maximum yield of 23 % is obtained from nitrobenzene oxidation under optimum conditions of 1 hr at 180°C. Table No.2 gives the product obtained by alkaline oxidation of lignin. While vanillin is the most commonly reported simple compound obtained by alkaline oxidation of lignin, there are a number of other products that are formed which may eventually prove to be of commercial interest.

Method II: The black liquor is dissolved in chloroform and then extracted with saturated sodium bisulphate solution.

Purification of vanillin is an essential step when synthesized from lignin. Most practical way to purify it is to distillate under reduced pressure or co-distillation with saturated hydrocarbon.

10.2 Organo sulphur compounds from lignin

The most important products are DMS and DMSO.

DMSO is used as a catalyst and organic solvent for dyes and other chemical reactions. Since DMSO is an excellent solvent for many carbohydrate materials, it has found application in preparation of sugar derivatives. DMSO also has found utility as a material to protect cellulose tissue from freeze damage. In this case, it has been shown to penetrate-the cell walls better than glycerin and thus prevents excessive concentration of electrolytes in the cell during freezing. It has also been used successfully to protect red blood carpusals against freezing. Widespread interest in DMCO as a solvent for carrying drugs across cell membrane and as a drug itself have been developed.

Depending upon the source of lignin whether softwood or hardwood, the methoxyl contents of Kraft lignin varies from 14-20 %. Since methyl group represents about half of the weight of the methoxy group, it is theoretically possible to recover 7-10 % of lignin calculated as methyl radicals.

Synthesis of DMS

Method of production of DMS and MM is by heating Kraft spent liquor with sodium sulphide at 200°C and to purify the product obtained. The yields of DMS are about 3 % and MM 0.2-3 per cent on lignin solids.

Commercial process consist of heating concentrated spent liquor with elemental sulphur to a temperature of 200-250°C. The DMS formed is flashed from the reactor along with some water, MM and H_2S and collected by means of a condenser. The crude DMS is scrubbed by passing it through Sodium hydroxide and distilled.

DMS itself does not find any application but is a precursor for its halogenated and oxidized derivatives.

Synthesis of DMSO

This is a compound obtained from lignin with a variety of applications in chemical and pharmaceutical industries. This can be commercially obtained by oxidizing DMS by nitrogen-dioxide under optimum conditions. The reaction can be summarized as

$$3Me_2S + 2HNO_3 \rightarrow 3Me_2SO + 2NO + H_2O$$

The application of DMSO has extended in many fields. It has been used as a outstanding medium for promoting organic reactions and as a reactant solvent itself. The following points outline some of the most potential application of DMSO.

10.3 Polymeric compounds from lignin:

Lignosulphonates:

Bulk of the materials are imported from Russia and China **[19]**

This class of lignin polymers has attracted a great deal of interest because of their diversified uses. They can also be utilized as the base for-industrial detergents such as washing powders and liquid soaps.

10.3.1 Synthesis of lignosulphonates

Lignosulphonates can easily be prepared from purified Kraft lignin by sulphonation with solutions of alkali sulphites at elevated temperature and pressure. Sulphite spent liquor contains lignosulphonates which are formed during pulping reactions. Practical approach to isolate and purify this compound is to remove most sulphite and sulphate by lime addition upto pH of 10.5, more lime is added to the filtrate to give basic calcium lignosulphonate which precipitates in the pH interval of 10.5 - 12.2. More than half the lignosulphonate is thereby removed and separated by filtration. By the addition of sulphuric acid and magnesium and sodium lignosulphonates can be obtained.

10.3.2 Application of Lignosulphonates:

Black liquor /sulphite liquor produces lignosulphones from rice straw, wheat straw, bagasse, bamboo, sarkanda etc. and also from hardwood. Fundamental studies on the colloidal properties of purified lignosulphonates demonstrated their specific value for emulsification, defloccuation, chelation, adsorption and adhesion uses. It has diversified industrial applications; some major ones are as under:

i. Oil well drilling muds: In the drilling of oil wells, water and dispersants are continuously fed into the hole, to form, together with suspended fine clay, called as 'drilling mud'. Oil in water emulsions in a drilling fluid has been in use since the 1940's. Lignin sulphonates are valuable for stabilizing the oil in water emulsions which give the mud improved properties in control of water loss, reduction of torque on the drill stem, increase of bit life and general improvement of bore hole condition.

ii. Cement and concrete additives: Lignosulphonates are used as grinding aids in Portland cement and concrete. Their function is to reduce the agglomeration of the-ground particles, to reduce surface-tension at the solid-air interference of the particles-and to keep the surface of the grinding media clean & free. It reduces settling time; water reduction and these can also be used as air entraining agents in concretes. The entrained air gives increased resistance to spalling and cracking of the finished concrete. The dispersive action of ligno-sulphonates is useful in mixing, and placing concrete. Incorporation of 0.07 - 0.15 kg of lignosulphonates-concrete admixture per sack of cement will allow water reduction up to 20%. This results in better compressive strength and durable concrete. In oil well cementing it thus controls settling time and strength.

iii. Dispersants for variety of Chemicals: Lignin sulphonate dispersants usually have a molecular weight in the range of 10,000 - 40,000. The colloidal lignin coats the surface imparting a negative charge to the particle. The similarly charged particles now repel one another and a good dispersion is obtained. This is used as dispersants in refractories, tiles, absorbants, cosmetics, crayons, medicines, paint, paper coatings, tooth powder and dyes. It can also be base for industrial detergents (washing powder and liquid soaps

iv. Ore floatation: Grinding aids, refining of ores as floatation agent, improves filtering properties in Pb-Zn system.

v. Emulsifiers and stabilizers: Lignin emulsifiers are quite competitive with other commercial emulsifiers stabilizing oil-in-water emulsions. Emulsions formed are very stable even under frozen or boiling stage. The stabilization of fire fighting foams that results from the interaction of sodium bicarbonate solution and alum is achieved by an addition of sodium salt of Kraft lignin. Better results can be obtained by using lignosulphonates.

vi. Electrolytic Refining: Lignosulphonates have been successfully used as 'Brightening' agent for the plated surface preventing pitting and deposition of rough surface in electroplating.The electrolytic refining of copper , zinc etc. are some examples.

vii. Binders and adhesives: Lignosulphonate enhances the properties of the specific products e.g., pellets, briquettes. Large tonnage of spent liquor solids are used as binders for pellets and briquettes. 2% calcium lignin sulphonate in iron ore gives a pellet ten times as hard as that made with bentonite binder. The most familiar adhesive product is linoleum cement containing principally lignin sulphonates and clay. These cements are also used as adhesive for ceiling tiles, road surfacing, hard durable surface and laying of dust, chemicals such as clay, calcium carbonate and titanium di-oxide resin ingredients,

viii. Resin ingredients: Combination of lignin sulphonates, with low molecular weight urea and phenol formaldehyde resins have received attention as adhesive or binding agent for fibrous materials such as plywood, particle and chip board and paper laminates. Many good quality glues can be produced by incorporating lignin sulphonates with polyvinyl acetate or alcohol.

ix. Rubber additives: As extender, modified reinforcing pigment. Many types of lattices such as neutral nitrile and neoprene are compatible with lignin either in the acid or sodium salt form. Mixture of rubber with Kraft lignin is also suitable for use as rubber to glass adhesives.

x. Protein precipitants: Forms insoluble complex with proteins.

xi. Tanning agents: Forms more soluble tanning solution which facilitates recovery procedures.

xii. Sequestering agents: Ca, Ni, Sn, Zn & Al can be sequestered by lignosulphonates above pH 7.

xiii. Miscellaneous uses of lignosulphonates:

Other uses of lignosulphonates are lime plastering medium, storage battery plates-as expander and component of separator., lime plaster-use of less water with improved settling properties, crystal growth inhibitor, Ingot mold wash, and many more.

11.0 Other uses of black liquor: **Hydrogen gas by electrolysis,** Gasification of black liquor, Biogas from black liquor, Preparation of rubber additives, adsorbent/desorbent, micronutrient for agricultural applications, asphalt emulsions,

12. Utilization of lime sludge:

As lime sludge contain high silica content, so its burning is not considered economical. However, the following approach can be considered.

a. Burning to get low grade lime
b. In the cement industry for partial substitution of lime stones
c. As a conditioner for soil especially acidic one.
d. Burning after desilication
e. Lime mud comprising calcium carbonate was found to be useful in removing colour from pulping liquid waste effluents. In a process[32] effluents were treated with lime mud followed by clarification, separation of effluents from classifiers to remove colour bodies.
f. Puzzulanna Cement

13.0 Silica recovery from from straw black liquor:

Amorphous silica (for high quality synthetic pigments as paper fillers which can replace soap stone and TiO_2), filler, retention aid and as flocculating agent, for production of ceramics. The later can be employed for household china, fine earthen ware, concrete block, electric porcelain, silica light weight brick, clay brick, calcium silicate thermal insulating materials, reinforcement in rubber, in paint manufacture, adsorbent, grease thickener, insecticide and cosmetic industry.

Silica from boiler Fly ash, and Sludge are also used for manufacturing light weight bricks.

14. Industry Status in India:

- **Mission New Energy :** Uses agri-residues using enzymatic hydrolysis process produces 0.07Ml/year of cellulosic ethanol[3]
- **Utilization of bamboo dust:** Gasfication, Syn Gas by Hindustan Paper Corporation(HPC), **Nogaon,** Assam
- **Pith + Lignin to particle board**: Kailash Paper Mill , Awagwanpur, Moradabad, India
- **Biomethanation Process**: Satia Paper Mills, Pudumjee Paper Mills
- **Biomass cogeneration:** ABC paper mill

- **Sludge and Boiler Fly Ash to light weight Bricks**: ITC, Bhadrachalam, Rainbow Papers, Ahmedabad.
- **Corrugated roof sheets and Diesel from waste plastics in recycled paper:** Rainbow Papers, Damanganga
- **Producer gas plant using wood chips :** Andhra Pradesh Paper Mill (APPM), now International Paper, USA
- **Biogas, cogeneration and biomass gasification:**TNPL
- **PAFC :** BHEL, India[15].
- **Installation of biomethanation plant in pulp and paper industry with UASB reactor:** United Envirotechier
- **Biomass gassifier :** W2E projects and Systems
- **Lignosulphonates:** ABC paper mill, SailaKhhurd, Punjab

15. Status in Research Institutions

ETP sludge, pith and paper to Bio ethanol: Indian Institute of petroleum (IIP, Dehradun), Central pulp and paper research institute (CPPRI), Cellulose and paper branch,Forest research institute(FRI), National Sugar Institute , Kanpur,Basant Dada Patil Sugar Institute, Pune.

Cellulosic waste to alcohol: College of technology,Osmania University, Hyderabad, National Sugar Institute, Govt. of India, Indian Institute of Petroleum, Dehradun, Vasant Dada Sugar Institute, Pune, IITRoorkee.

Utilization of lime sludge,Fly ash :Central building research Institute, National thermal power corporation(NTPC)

Lignin grafted copolymers:Regional Research Laboratory(RRL at Bhubaneswar,Jammu, Jorhat)

Biogas –fuel cell based co-generation: Tata Energy Research Institute (TERI)

Production of lactic acid from waste sugarcane bagasse derived cellulose,cellulase production with Tricoderma citriviridae on solid bed ,use of acid hydrolysates of pine needles,and bagasse for lactic acid production using various strains such as Lactobacillus delbrueckii or lactobacillus pentosus have been explored in Department of paper technology, IITRoorkee( ), National sugar Institute , Kanpur, H.B.T.I., Kanpur

Ministry of New and Renewable Energy(MNRF) is keen to provide technical and financial assistance for setting up of energy generating plant based on biogas, cogeneration and biomass gasification

The technical parameters of a typical HRAD-PAFC (high rate anaerobic digester with phosphoric acid fuel cell) cogeneration system for a 50 TPD paper mill have been estimated by

TERI and development of a prototype PAFC have been explored by Bharat Heavy Elecricals Ltd.[15].

Table 1: By-products (both commercial & non-commercial) of the kraft cook

Compound	**Origin**	**Quantity kg per ton of pulp**
Methanol	Methoxyl groups of lignin & possibly glucuronoxylan	5
Acetate	Acetyl groups of the xylan as well as degraded carbohydrates	100-200
Formiate	Degradation of carbohydrates	40
Lactate	Degradation of carbohydrates	100
Saccharinic acids & lactones	Degradation of carbohydrates	250
Aliphatic sulfur compounds (methyl mercaptan, sulfide, disulfide)	Methoxyl groups of lignin and possibly glucuronoxylan	1
Turpentine	Terpenes	8-10
Tall oil (fatty & resin acid salts, phytosterols, etc.)	Wood resin components	20-100
Alkali lignin	Lignin	400-600

Table 2: By-products of the Kraft cook and from subsequent pressure heating of the black liquor with alkali and sulfide at 250-285°C (48 atm)

Before Pressure Heating	**kg ptp**	**After Pressure Heating**	**kg ptp**
Tall Oil	50	Acetic acid	60
Turpentine	17	Formic acid	60
Acetic acid	40	Methyl sulfide	40
Formic acid	40	Methyl mercaptan	8

Methanol	5	Pyrocatechol	30
Vanillin	2	Methyl catechol	11
Acetoguaiacone	2	Ethyl catechol	9
Guaiacol	2	Other ether-soluble phenols	150
Pyrocatechol	1	Homoprotocatechuic, protocatechuic and other phenol carboxylic acids, perhaps	150
Other phenols	42	Ether soluble acids of carbohydrate origin	150
Phenol carbonic acids	45	Demethylated lignin of novalac properties	100
Non-volatiles and lactic acids	145	Butanol-soluble lactones	35
TOTAL	**391**	**TOTAL**	**803**

CONCLUSIONS

Paper industry wastes can be properly utilized to produce value added products in order to act as second generation biorefinery. Lignin in Black Liquors has excellent potential for reuse in Fine Chemical Manufacture. The Indian Paper Industry is not yet ready for such a drastic transformation, since it will jeopardize the Chemical Recovery Processes. The Paper Industry is totally dependent upon the lignin black liquors as a source of energy fuel in Recovery Boilers. Serious attempts have to be made for medium and small scale manufacturing process. Some initiatives have been taken by Indian pulp and paper industry in recent years. A few research institutions have started research activities for establishing biorefinery based on paper/ sugar mill wastes.

It is necessary to determine the quantum of the wood waste/ agri-residue facility for the maximum size consistent with availability of raw material. A single product facility appears not to be competitive. In order to be economic utilization of the wood waste, multiple product facility is essential.

REFERENCES

1. Mukundan, P.G., Editorial, Inpaper International, Vol.12, No.2, Jan.-March 2010, Vol.13, No.1, Oct.-No.2010, www.inpaper.com.
2. Inpaper India, Vol.13(3), 14(2), March-April,2011, pp.24, 30.

3. Gabriel D.A., Sousa, Biorefinery development pathways: A survey for the pulp and paper industry, Proceedings of XXI Encontro Nacional da TECNICELPA/VI CIADICYP 2010, 12-15 Outubro 2010-Lisboa , Portugal.
4. Florbela Carvalheiro, Luis C.Duarte and Francisco M Girio, Hemicellulose biorefineries: a review on biomass pretreatments, JSIR, Vol.67, November, 2008, pp.849-864.
5. 56th Chemical weekly, July 05, 2011.
6. Thakur Vasanta V, Diwakar Pandey, R.K.Jain, R.M.Mathur, Second generation biofuels: Bioethanol from lignocellulosic materials, Inpaper India, Vol. 14(2), March-April, 2011, pp.4-12.
7. Jain, R.K., Thakur Vasanta V., Pandey Diwakar, Adhikari, Dixit A.K. and Mathur R.M., Bio ethanol from Bagasse Pith-A lignocellulosic waste biomass from paper / sugar Industry, IPPTA J., Vol.23, No.1, Jan-March, 2011, 169-173.
8. Mark E. Reno, Monique STREFF, Andrew Hird and Anjan Ray, Conversion of biomass to fuel as an environmental impact reduction opportunity for the pulp and paper industry, Inpaper International, Vol.13, No.3, April-June, 2011, pp.18-20.
9. Naithani Sanjay, Ganesh Joshi and Alok Kumar Dubey, Detoxification of Acid Catalyzed Lignocellulosic Hydrolyzate to enhance Fermentation Efficiency for Ethanol Production, INPAPER INDIA, Vol.14(2), March –April 2011, pp.14-22.
10. Banerjee Sujit, Energy from Pulp Mill Sludge,INPAPER INTERNATIONAL, Vol.13, No.1, Oct.-Nov. 2010, pp.10-12.
11. Umrao S., Pankaj Kumar and Sadhana Sachan, Method of producing Syngas from gasification of bagasse, Proceedings on "Advances in Chemical Engineering"-MacMillan, Advanced Research series February 27-28, 2009, Patiala, pp. 445-450.
12. Dhingra H.K. and Mohendru V.K., Status of Solid Waste Pollution From Indian Paper Industry and Future Prospects of its Management, IPPTA Vol.8,No.1, March 1996, pp.49-55.
13. Sahoo N, Rohella R.S.,Choudhury S.,Paul S.C.,and Chakravortty, V., Preparation of Value-Added Products From Black Liquor,IPPTA, Vol.8(3), September, 1996,pp.45-50.
14. Padmaja A, Venkateshwar S,Seenayya G and Bhagavanth Rao M, Evaluation of cellulosic substrates in the conversion of biomass to ethanol,Indian Chemical Engr., Section A . Vol.37 No.3, Jul-Sep(1995),pp.140-144.
15. Kurup ,G. Rajendra Chawla S.K.,Mathur A., Biogas –fuel cell based cogeneration in the pulp and paper Industry, IPPTA convention issue ,1992,pp.25-30.
16. Singh S.P., Recent developments in characterization and utilization of black liquor lignins from small pulp and paper mills, IPPTA, Vol.2 (3), 1990, pp.11-15.
17. Adsul, M.G., Verma A.J. and Ghokle, D.V., Lactic acid production from waste sugarcane bagasse derived cellulose, Green chemistry, 9, 58-62, 2007.
18. Guerra, G.,Casado, M.R.L.G., Arguelles,M.I., Sanchez M.I.,Manzano, A.M., and Guzman,T., Cellulase production with sugarcane straw by Trichoderma citrinoviridae on solid bed, sugar Tech.,8(1),30-35.

19. Managaonkar , N.D. and Kulkarni, Y.S.,Preparation and utilization of lignin by-products and lignosulphonates obtained from wood black liquor,IPPTA convention Issue,1987 pp.40-52
20. Ramaswamy, V. B iotechnology –applications in waste utilization and pollution abatement, ibid, pp.1-13.
21. Jain Sailendra, Environmental protection in Paper Industry, ibid, pp.53-61.
22. Dhake, J.D., Utilization of pulp and paper industry waste for chemicals,ibid, pp.19-29.
23. Trivedi, M.K., Lignin recovery from alkaline pulping of bagasse, ibid, pp.30-35.
24. NIEM (UNEP) Project Under Phase II- Activity III, October, 1990, Institute of Paper Technology (University of Roorkee), Saharanpur-247001 (India), pp.64-65, 66-67.
25. Lehri Madhu, Vartika Saxena and P.K. Agarwal, Preliminary studies on ethanolic fermentation of acid hydrolysates of hemicelluloses-extracted bagasse,pp.G587-G593, , Proceedings of the 57th Annual Convention of STAI, New Delhi, 13th -15th October,1995
26. Manoj Kumar Ghosh, Uttam K. Ghosh, Utilization of pine needles as bed material in solid state fermentation for production of lactic acid by Lactobacillus strains, Bio Resources, Vol.6, No.2,2011, pp.1556-1575.

Enhanced Cardiomyocyte Function on Poly-Lactic-co-Glycolic Acid: Carbon Nanofiber Composites under Electrical Stimulation

David A. Stout, Brown University, Providence, RI
Dr. Thomas J. Webster, Brown University, Providence, RI

Introduction

Cardiomyocytes (a specialized contractile muscle cell that generates part of the myocardium tissue of the heart) depends on continuous conductivity to function [1, 2]. However, such conductivity may break down during heart disease or malfunction. For instance, a myocardial infarction, also known as a heart attack, usually occurs because a major blood vessel supplying the heart's left ventricle is suddenly blocked by an obstruction, such as a blood clot [3-5]. During myocardial infarction, part of the cardiac muscle, or myocardium, is deprived of blood and therefore oxygen, which destroys cardiomyocytes leaving dead tissue [3-5] as well as denervation of the myocardium [6-8]. In particular, nerve damage to cardiac tissue can result in nerve sprouting in the left ventricle [6, 9] and development of arrhythmias [10]. Scarred cardiac muscle results in heart failure for millions of heart attack survivors worldwide. In 2009, an estimated 785,000 Americans had a new coronary attack and about 470,000 had a recurrent heart attack leading to a coronary event [11].

In recent years, various techniques have been developed to promote cardiomyocyte growth around dead tissue after a myocardial infarction. Each approach has their own advantages [12], as well as disadvantages [12], but in general, all of the above can be divided into two groups: a) conductive cardiovascular patches (using scaffolding and 3D printing techniques usually with polypyrrole) and b) non-conductive cardiovascular patches (mostly involving direct cell injection, scaffolding, and injectable scaffolds). One can also observe that cardiovascular biomaterials can be either based on biodegradable or on non-biodegradable materials. Within this matrix of conductive vs. non-conductive and biodegradable vs. non-biodegradable material lie the most commonly studied materials and techniques used for promoting heart health. However, one area that has been largely omitted to date is the exploration of nanotechnology (or materials with one dimension less than 200nm) in cardiovascular applications. Numerous articles suggest that using nanotechnology can specifically promote cell functions on a variety of materials, ranging from titanium to silicon [13-15] due to optimal surface chemistry, wettability which control protein adsorption onto the surface. While some degree of nanostructured features may promote tissue growth, it is also known that certain topographies can hinder cell activity [16-18]. For example, when comparing nano to micro diamond features, Yang et al. showed that osteoblast (or bone forming cell) adhesion and proliferation increased on the nano-rough topography [16]. On the contrary, using a 1718 gene microarray Dalby et al. were able to suggest that nano-rough topography greater than 40 nm decreased human fibroblast cell adhesion [17, 18].

In recent research, it was demonstrated that the use of nano-materials can promote the growth of cardiomyocytes when compared to the use of conventional or micro-structured materials. One such nano-material consisted of poly-lactic-co-glycolic acid (PLGA) with carbon nanofibers (CNFs) where it was shown that cardiomyocytes proliferated faster on the composites compared to a pure PLGA film after 5 days.

The object of this study was to better understand the cytocompatibility properties of this composite material for myocardial applications by using electrical stimulation which mimics that of the heart. For this reason, an *in vitro* continuous electrical stimulation model was used to determine cardiomyocyte functions on PLGA: CNF composite materials.

Experiments

Fabrication

Purified carbon nanofibers (CNFs) (99.9% by weight %, Catalytic Materials, MA) with diameters of 60, 100, and 200 nanometers were sonicated in 20 ml of chloroform at 20W for 30 minutes. Two pellets of poly-lactic-co-glycolic acid or PLGA (50:50 PLA:PGA wt.) (Polysciences Cat #23986) were diluted in a 50 ml flask with 30 ml of tetrahydrofuran and were sonicated in a water bath below 30 ^{o}C for thirty minutes.

After the PLGA and CNF solutions were prepared, various PLGA:CNF weight percent ratios were created (100:0, 75:25, 50:50, 25:75, 0:100) by adding the appropriate amount of CNF to PLGA in 20 ml disposable scintillation vials. When the appropriate ratios were reached, each composite material was sonicated at 10W for 20 minutes. 1 ml of the appropriate PLGA:CNF solution was placed onto a glass substrate and then placed into an oven below 50 ^{o}C for 15 minutes. Each composite film was then vacuum dried at -20 inches of Hg gauge pressure for 48 hours to allow the THF and chloroform to evaporate.

Composite Characteristics

A Hitachi 2700 scanning electron microscope was used to characterize the surface of the PLGA:CNF samples (Fig. 1). The electrical resistance of the PLGA:CNF samples was determined using a multi-meter (HP 34401A) by connecting two probes, via alligator clip connectors, of the meter to opposite ends of the sample. The sample was tested dry at room temperature and at opposite ends of the sample, which were 22mm apart. Three measurements were taken for each sample.

X-Ray diffraction (Bunker AXS: D8 Focus) was used with settings at 0.5° per 2 minutes between 2Θ = 10-35° to characterize the crystallinity of the composites. Also, Raman spectroscopy results were obtained and recorded using an argon ion laser attached with a spectrophotometer (Acton Research Corporation, model AM505F). A low laser power of 7mW was applied to avoid any local surface heating. To acquire the spectrum, an incident light with a wavelength of 514.5 nm was used in a wavenumber region of 200–3700 cm^{-1}. The scattered radiation was collected at 180° (backscattered geometry) to the incoming beam and detected using a CCD cooled to -120°C. The spectral resolution of the Raman experimental set up was better than 1 cm^{-1} and the total integration time was 3 minutes.

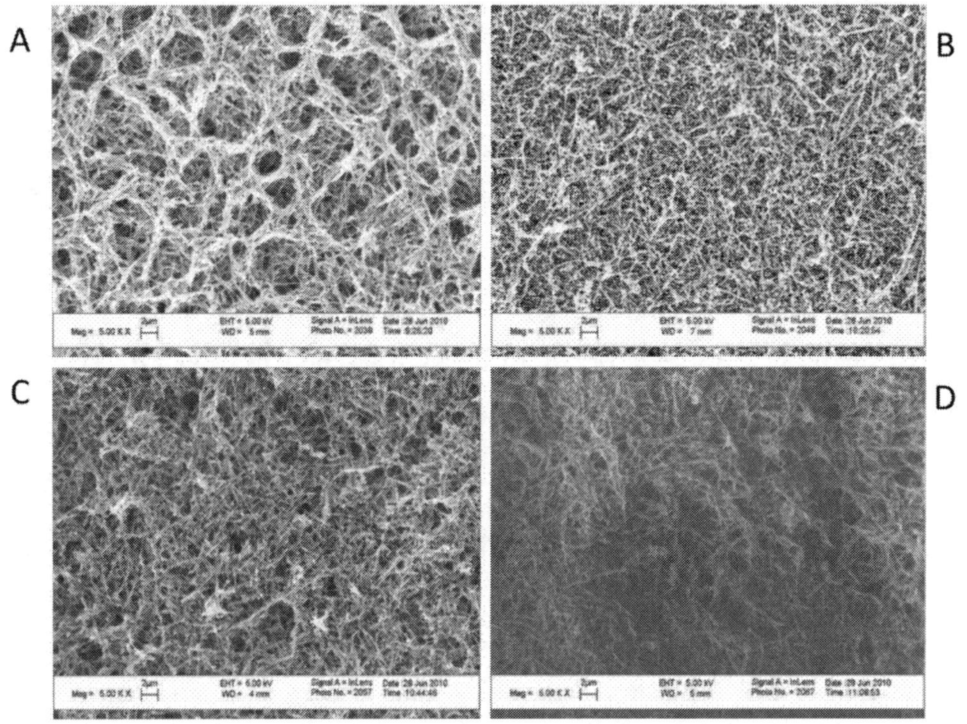

Figure 1: SEM images at 5K magnification showing the different PLGA:CNF (wt:wt) material ratios of 75:25 (A), 50:50 (B), 25:75 (C), 0:100 (D). Scale bar = 2 µm. SEM image analysis showed that more CNFs were observed at higher CNF ratio samples and the material width, with glass substrate, was 15.01 ± 1.22 µm (n = 4, mean ± s.d.).

Cytocompatibility

All samples and controls were sterilized using ultraviolet light for 24 hours prior to cell seeding. Human cardiomyocytes (Celprogen, Cat #36044-15) were seeded in human cardiomyocyte complete stem cell culture growth media with serum (Celprogen, Cat #M36044-15S) at a cell concentration of 10 x 10^4 cells/cm^2 for the cell adhesion assay and 1.5 x 10^4 cell/cm^2 for the cell proliferation assay. Cells were seeded into 8-well rectangle nunc cell plates on PLGA:CNF samples, and 22 mm diameter microscope cover slips as controls. Samples were incubated and were continuously stimulated (rectangular, 2 nm, $5\frac{V}{cm}$, 1 Hz) for 24, 72, and 120 hours under standard incubation conditions (at 5% CO_2 95% humidified air and 37^0C, changing the media every other day) (Fig. 2). MTT assay was used and compared to determine cell proliferation.

Troponin-I ELISA biomarker assays were completed in conjunction with the proliferation assay. This was done to make sure that the cardiomyocytes had the appropriate protein complexes that are specific to cardiomyocyte cell characteristics and did not differentiate throughout the experiment or loose phenotypes.

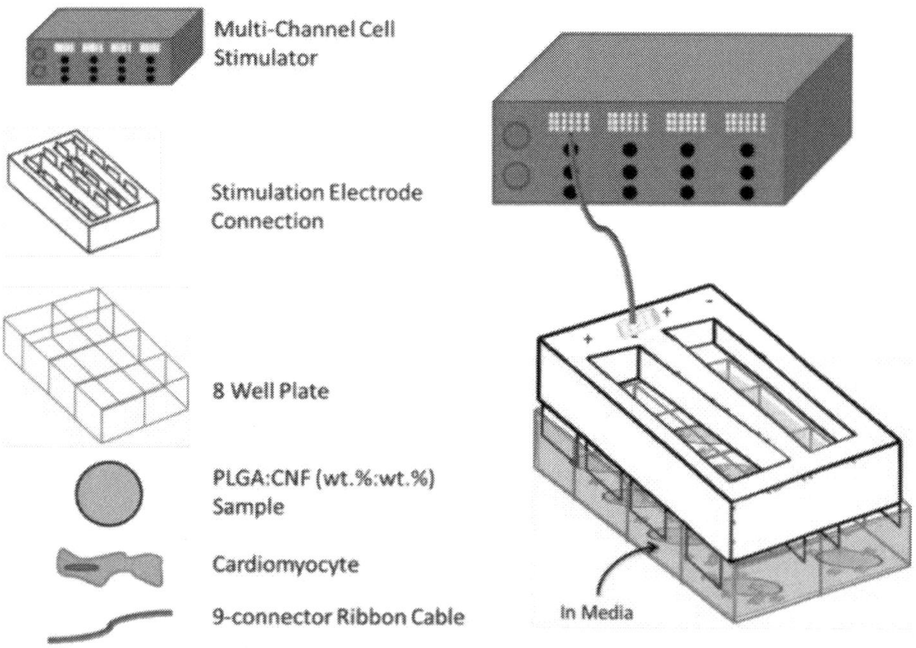

Fig. 2: Schematic of in vitro electrical stimulation apparatus.

Statistical Analysis

All proliferation assays were performed at least in triplicate and results were compared to a controlled glass surface at 24, 72, and 120 hours. Data were plotted as mean ± standard error of the mean and statistical analyses were performed. When data were compared, ANOVA software and a student T-test were used. A p-value of < 0.05 was considered to be significant.

Results

Material Properties

Scanning electron microscopy showed that the CNFs were uniformly dispersed within the PLGA material and more CNFs were observed at higher CNF ratio samples. It was shown that increasing the CNF weight ratio in PLGA increased conductivity of the sample. Comparing the same sample weight ratio with different CNF sizes, it was shown that decreasing the CNF size increased sample conductivity (Fig. 3). Also, XRD spectra obtained from the as-synthesized PLGA:CNF composites are shown in Fig. 2. As expected an extremely broad and flat peak was recorded for the PLGA matrix, confirming its amorphous nature. However, with addition of CNF in PLGA, the characteristic X-ray peak at $2\Theta = 26.5°$ evolved into a rather strong and sharp peak in PLGA:CNF composites for the 25:75 [PLGA:CNF (wt:wt)] ratio (Fig 4). The curves were fitted to the characteristic peak by Gaussian distribution. Clearly, the intensity increased as full width at half maximum (FWHM) decreased as CNFs were added to.

Complimentary evidence of the confirmation of the crystallinity was obtained using Raman spectroscopy and the results are provided in. Two sets of Raman bands are clearly visible in the composites with one set of Raman shift at 173 and 278 cm^{-1} recorded in the presence of CNF. These Raman bands are characteristic of C-C bonds [19]. The Raman bands at 1345 cm^{-1} and 1570 cm^{-1} arose particularly due to C=C bending or bonding [19]. Another observation was the slight shift of these Raman bands in higher CNF containing PLGA composites or with 100% CNF.

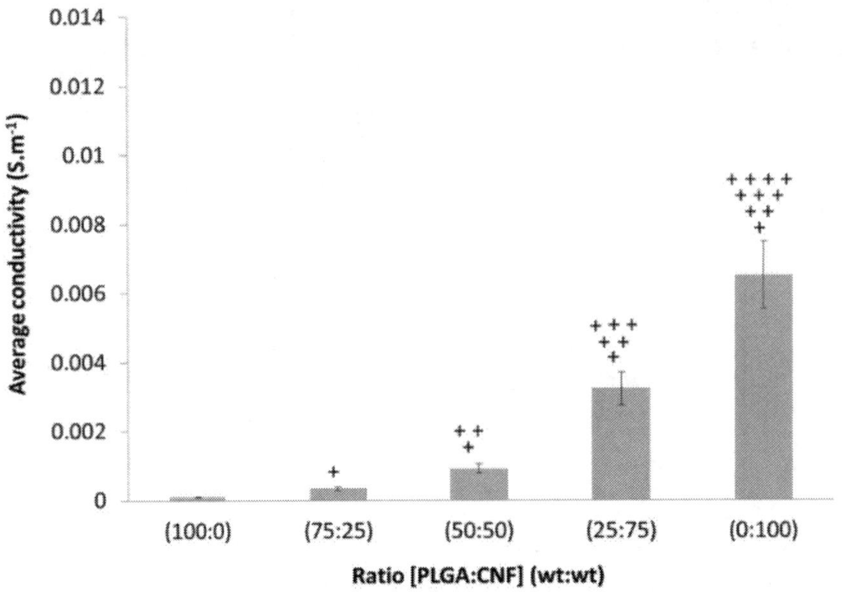

Figure 3: Comparative conductivity analysis of different PLGA:CNF (wt:wt) composite ratios with d. Data = mean ± SEM; N =3 + $p < 0.05$ compared to the 100 nm diameter CNF 100:0 (PLGA:CNF) ratio. ++ $p < 0.05$ compared to the 100 nm diameter CNF 75:25 (PLGA:CNF) ratio. +++ $p < 0.05$ compared to the 200 nm diameter CNF 50:50 (PLGA:CNF) ratio. ++++ $p < 0.05$ compared to the 100 nm diameter CNF 25:75 (PLGA:CNF) ratio.

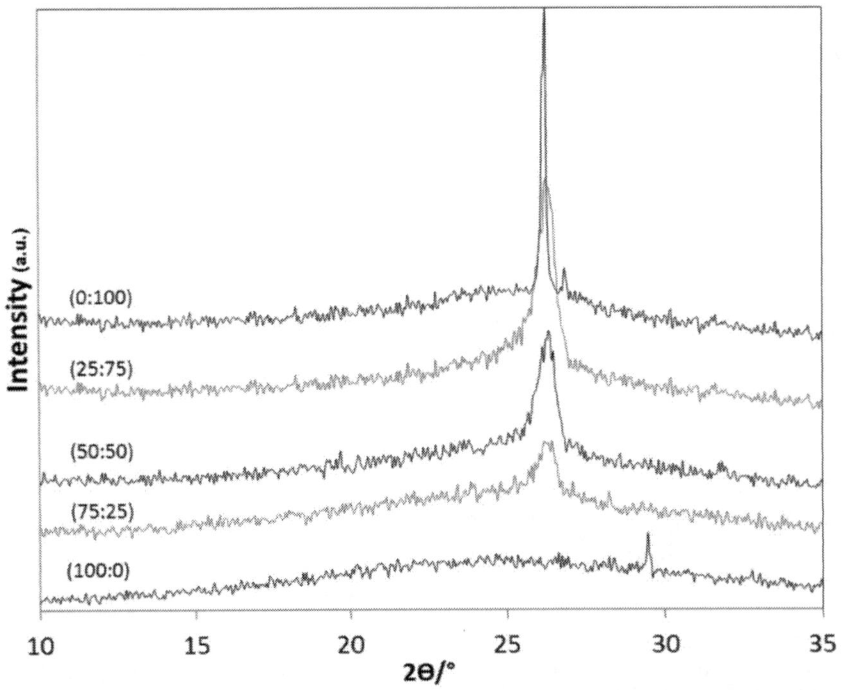

Fig. 4: XRD analysis of the 100 nm diameter PLGA:CNF composites. Measurements were completed between 10-35 degrees at a speed of 0.5 degrees per 2 minutes. Increasing the CNF weight ratio, Increased the intensity and slenderness of the peak. This is due to more crystallinity of the composites, therefore the 0:100 ratio [PLGA:CNF (wt:wt)] had the most slender and intense peak.

Cytocompatibility

Results indicated that PLGA:CNF composites promoted human cardiomyocyte proliferation better when electrically stimulated. General trends were seen between each ratio

(Fig. 5) and support published results [20]. For 100nm diameter CNFs, it was determined that the 50:50 ratio [PLGA:CNF; (wt:wt)] had the highest cardiomyocyte density all three time points, whereas the lowest density was on the 100:0 ratio [PLGA:CNF; (wt:wt)].

Troponin-I assay indicated that cardiomyocytes had a functional Troponin Complex throughout the cytocompatible assays. All CNF experiments for all time points showed a slight increase of Troponin-I indicating more cell proliferation and production. It is also good to note that Troponin-I assays presented the same trends seen in the proliferation cytocompatibility assays.

Discussion

Many materials have been developed to promote cardiomyocyte function. Recently, it has been shown that composite materials composed of different matrices (conductive vs. non-conductive and biodegradable vs. non-biodegradable) can stabilize heart tissue to decrease the consequences of a myocardial infarction and such approaches include but are not limited to: culturing patient tissue [21-30], direct cell injection [31-36], biomaterial scaffolds (materials ranging from elastin to 3D PLGA fibers) [37-43], 3D printing [44-46], and injectable scaffolds using fibrin based materials [47-49]. This present work assessed a brand new material, PLGA:CNF composites, for enhancing conductivity and cytocompatibility properties necessary for cardiovascular applications.

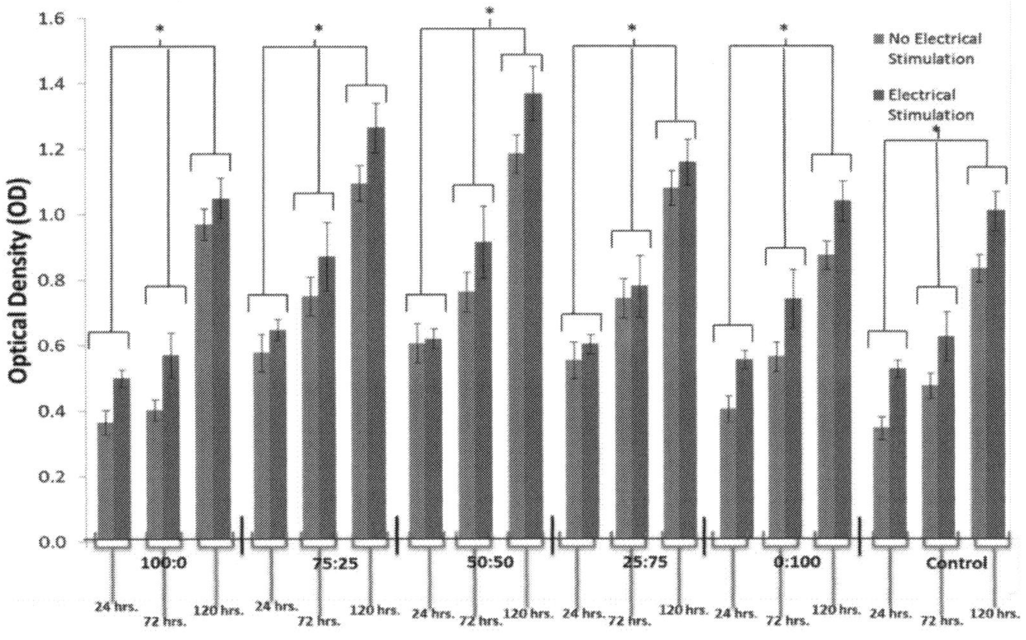

Fig. 5: Comparative electrical stimulation analysis of different PLGA:CNF (wt:wt) composite ratios with d. Data = mean ± SEM; N =3. * $p < 0.05$ compared to its non-electrically stimulated counterpart as well to other PLGA:CNF (wt:wt) ratios.

Results showed improved cardiomyocyte functions on electro-stimulated composites than those samples that were not electro-stimulated.

It is worth speculating why greater cardiomyocyte density on electro-stimulated composited were observed. Pedrotty et al. showed that numerous cardiac cell functions (including adhesion, proliferation, and migration) may be modulated by electrical stimulation [50], hence requiring the use of a conductive material in cardiac applications. Also, Mihardjo et al. demonstrated that enhanced myocardial repair following an ischemic injury could be achieved using conductive polymers, such as polypyrrole [51]. The conductivity values measured in the present work were lower than that of the heart tissue (ranging from 0.16 S/m longitudinally to 0.005 S/m transversely) [52], but future techniques (such as CNF alignment) may increase the anisotropic conductivity [53] to match that of heart tissue.

Conclusion

In this study, it was shown that adding CNFs to PLGA improved that conductivity and compatibility properties of PLGA for myocardial tissue engineering applications. Different PLGA:CNF ratios were created to improve conductivity and cytocompatibility properties with human cardiomyocytes. It was further shown that increasing CNF weight ratio increased composite conductivity. Also, when then the samples were in vitro electrically stimulated, increased cell proliferation were observed. In this manner, the results of this study suggest that CNFs should be further studied in cardiac tissue engineering applications.

Acknowledgements

The authors would like to thank the National Science Foundation Graduate Student Fellowship Program (NSF #069063) and the Hermann Foundation for funding.

References

1. M. Radisic, M.V. Sefton, Cardiac Tissue, in: Principles of Regenerative Medicine (Second edition), Academic Press, San Diego, 2011, pp. 877-909.
2. J.A. Armour, Autonomic Nervous System: Cardiovascular Control, in: Larry R. Squire (Ed.), Encyclopedia of Neuroscience, Academic Press, Oxford, 2009, pp. 849-854.
3. D. Locca, C. Bucciarelli-Ducci, G. Ferrante, A. La Manna, N.G. Keenan, A. Grasso, P. Barlis, F. Del Furia, S.K. Prasad, J.C. Kaski, D.J. Pennell, C. Di Mario, New Universal Definition of Myocardial Infarction: Applicable After Complex Percutaneous Coronary Interventions? 3 (2010) 950-958.
4. K. Thygesen, J.S. Alpert, H.D. White, Universal Definition of Myocardial Infarction 50 (2007) 2173-2195.
5. A.K. Jain, C. Knight, Myocardial infarction - new definitions and implications 38 (2010) 421-423.
6. M. Barber, T. Mueller, B. Davies, R. Gill, D. Zipes, Interruption of sympathetic and vagal-mediated afferent responses by transmural myocardial infarction, Circulation 72 (1985) 623-631.
7. E.F. Hirsch, M. Jellinik, T. Cooper, G.C. Kaiser, H.C. Barner, A.M. Borghard-Erdle, The Innervation of the Vertebrate Heart (1970).
8. Y. Oh, A.Y. Jong, D.T. Kim, H. Li, C. Wang, A. Zemljic-Harpf, R.S. Ross, M.C. Fishbein, P. Chen, L.S. Chen, Spatial distribution of nerve sprouting after myocardial infarction in mice, Heart Rhythm 3 (2006) 728-736.
9. R. Vracko, D. Thorning, R. Frederickson, Fate of nerve fibers in necrotic, healing, and healed rat marycardium, Laboratory Investigation 63 (1990) 490-501.

10. W.C. Randall, M.P. Kaye, G.R. Hageman, H.K. Jacobs, D.E. Euler, W. Wehrmacher, Cardiac dysrhythmias in the conscious dog after surgically induced autonomic imbalance, Am. J. Cardiol. 38 (1976) 178-183.
11. WRITING GROUP MEMBERS, Committee,for the American Heart Association Statistics, Stroke Statistics Subcommittee, Heart Disease and Stroke Statistics--2009 Update: A Report From the American Heart Association Statistics Committee and Stroke Statistics Subcommittee 119 (2009) e21-181.
12. S. Cohen, J. Leor, REBUILDING BROKEN HEARTS, Scientific American 291 (2004) 45-51.
13. D. Khang, J. Lu, C. Yao, K.M. Haberstroh, T.J. Webster, The role of nanometer and sub-micron surface features on vascular and bone cell adhesion on titanium 29 (2008) 970-983.
14. R.L. Price, M.C. Waid, K.M. Haberstroh, T.J. Webster, Selective bone cell adhesion on formulations containing carbon nanofibers 24 (2003) 1877-1887.
15. A. Ranella, M. Barberoglou, S. Bakogianni, C. Fotakis, E. Stratakis, Tuning cell adhesion by controlling the roughness and wettability of 3D micro/nano silicon structures 6 (2010) 2711-2720.
16. L. Yang, B.W. Sheldon, T.J. Webster, The impact of diamond nanocrystallinity on osteoblast functions 30 (2009) 3458-3465.
17. M.J. Dalby, D. Pasqui, S. Affrossman, Cell response to nano-islands produced by polymer demixing: a brief review, Nanobiotechnology, IEE Proceedings - 151 (2004) 53-61.
18. M.J. Dalby, D. Giannaras, M.O. Riehle, N. Gadegaard, S. Affrossman, A.S.G. Curtis, Rapid fibroblast adhesion to 27 nm high polymer demixed nano-topography, Biomaterials 25 (2004) 77-83.
19. S. Aslan, C.Z. Loebick, S. Kang, M. Elimelech, L.D. Pfefferle, P.R. Van Tassel, Antimicrobial biomaterials based on carbon nanotubes dispersed in poly(lactic-co-glycolic acid), Nanoscale 2010.
20. D.A. Stout, B. Basu, T.J. Webster, Cytocompatible Poly-Lactic-co-Glycolic Acid: Carbon Nanofiber Composite Analysis for Cardiovascular Applications, IEEE 2011 Northeast Bioengineering Conference 11 (2011).
21. L. Bengtsson, K. Radegran, A. Haegerstrand, In vitro endothelialization of commercially available heart valve bioprostheses with cultured adult human cells 7 (1993) 393-398.
22. C.R. Horres, J.F. Aiton, M. Lieberman, E.A. Johnson, Electrogenic transport in tissue cultured heart cells 11 (1979) 1201-1202, IN31, 1203-1205.
23. H.J. Jongsma, L. Tsjernina, J. de Bruijne, The establishment of regular beating in populations of pacemaker heart cells. A study with tissue-cultured rat heart cells 15 (1983) 123-133.
24. E.C. Martinez, T. Kofidis, Adult stem cells for cardiac tissue engineering In Press, Corrected Proof.
25. S. Ramaswamy, D. Gottlieb, G.C. Engelmayr Jr, E. Aikawa, D.E. Schmidt, D.M. Gaitan-Leon, V.L. Sales, J.E. Mayer Jr, M.S. Sacks, The role of organ level conditioning on the promotion of engineered heart valve tissue development in-vitro using mesenchymal stem cells 31 (2010) 1114-1125.
26. D. Schmidt, A. Mol, S. Neuenschwander, C. Breymann, M. GÃ¶ssi, G. Zund, M. Turina, S.P. Hoerstrup, Living patches engineered from human umbilical cord derived fibroblasts and endothelial progenitor cells 27 (2005) 795-800.
27. A. Wada, H. Yoneda, N. Shibata, Y. Inui, H. Fushimi, K. Takemura, S. Onishi, Tissue-cultured heart cells from the cardiomyopathic hamster 8 (1976) 619-620.

28. H. Wei, C. Chen, W. Lee, I. Chiu, S. Hwang, W. Lin, C. Huang, Y. Yeh, Y. Chang, H. Sung, Bioengineered cardiac patch constructed from multilayered mesenchymal stem cells for myocardial repair 29 (2008) 3547-3556.
29. C. Yang, R. Sodian, P. Fu, C. LÃ¼ders, T. Lemke, J. Du, M. HÃ¼bler, Y. Weng, R. Meyer, R. Hetzer, In Vitro Fabrication of a Tissue Engineered Human Cardiovascular Patch for Future Use in Cardiovascular Surgery 81 (2006) 57-63.
30. K. Zitta, B. Brandt, A. Wuensch, P. Meybohm, B. Bein, M. Steinfath, J. Scholz, M. Albrecht, Interleukin-1[beta] regulates cell proliferation and activity of extracellular matrix remodelling enzymes in cultured primary pig heart cells 399 (2010) 542-547.
31. N. Borenstein, V. Chetboul, P. Bruneval, M. Hekmati, R. Tissier, L. Behr, G. Derumeaux, D. Montarras, Non-cultured cell transplantation in an ovine model of non-ischemic heart failure 31 (2007) 444-451.
32. C. Chen, Y. Chang, C. Wang, C. Huang, C. Huang, Y. Yeh, S. Hwang, H. Sung, Construction and characterization of fragmented mesenchymal-stem-cell sheets for intramuscular injection 28 (2007) 4643-4651.
33. S.C. Li, L. Wang, H. Jiang, J. Acevedo, A.C. Chang, W.G. Loudon, Stem cell engineering for treatment of heart diseases: Potentials and challenges 33 (2009) 255-267.
34. E.C. Perin, H.F. Dohmann, R. Borejevic, S.A. Silva, A.L. Sousa, G.V. Silva, J.A. Assad, L. Belem, A.C. Carvalho, R. Esporcatte, W.K. Vaughn, H.J. Dohmann, J.T. Willerson, 814-2 Time course of improvement following stem cell injections in humans with heart failure 43 (2004) A187-A187.
35. Y. Shintani, S. Fukushima, A. Varela-Carver, J. Lee, S.R. Coppen, K. Takahashi, S.W. Brouilette, K. Yashiro, C.M.N. Terracciano, M.H. Yacoub, K. Suzuki, Donor cell-type specific paracrine effects of cell transplantation for post-infarction heart failure 47 (2009) 288-295.
36. L. Zhang, T.J. Webster, Nanotechnology and nanomaterials: Promises for improved tissue regeneration, Nano Today 4 (2009) 66-80.
37. A. Callegari, S. Bollini, L. Iop, A. Chiavegato, G. Torregrossa, M. Pozzobon, G. Gerosa, P. De Coppi, N. Elvassore, S. Sartore, Neovascularization induced by porous collagen scaffold implanted on intact and cryoinjured rat hearts 28 (2007) 5449-5461.
38. J. Hu, X. Sun, H. Ma, C. Xie, Y.E. Chen, P.X. Ma, Porous nanofibrous PLLA scaffolds for vascular tissue engineering 31 (2010) 7971-7977.
39. Q. Lu, K. Ganesan, D.T. Simionescu, N.R. Vyavahare, Novel porous aortic elastin and collagen scaffolds for tissue engineering 25 (2004) 5227-5237.
40. M. Siepe, M. Giraud, M. Pavlovic, C. Receputo, F. Beyersdorf, P. MenaschÃ©, T. Carrel, H.T. Tevaearai, Myoblast-seeded biodegradable scaffolds to prevent post-myocardial infarction evolution toward heart failure 132 (2006) 124-131.
41. P.M. Taylor, A.E.G. Cass, M.H. Yacoub, Extracellular matrix scaffolds for tissue engineering heart valves 21 (2006) 219-225.
42. Y. Wang, L.B. Wong, H. Mao, Creation of a long-lifespan ciliated epithelial tissue structure using a 3D collagen scaffold 31 (2010) 848-853.
43. W.Y. Yeong, N. Sudarmadji, H.Y. Yu, C.K. Chua, K.F. Leong, S.S. Venkatraman, Y.C.F. Boey, L.P. Tan, Porous polycaprolactone scaffold for cardiac tissue engineering fabricated by selective laser sintering 6 (2010) 2028-2034.
44. X. Cui, T. Boland, Human microvasculature fabrication using thermal inkjet printing technology 30 (2009) 6221-6227.

45. V. Mironov, T. Boland, T. Trusk, G. Forgacs, R.R. Markwald, Organ printing: computer-aided jet-based 3D tissue engineering 21 (2003) 157-161.
46. T. Xu, C.A. Gregory, P. Molnar, X. Cui, S. Jalota, S.B. Bhaduri, T. Boland, Viability and electrophysiology of neural cell structures generated by the inkjet printing method 27 (2006) 3580-3588.
47. G. Morales, M.I. Barrena, J.M.G.d. Salazar, C. Merino, D. RodrÃguez, Conductive CNF-reinforced hybrid composites by injection moulding 92 (2010) 1416-1422.
48. K.L. Christman, A.J. Vardanian, Q. Fang, R.E. Sievers, H.H. Fok, R.J. Lee, Injectable Fibrin Scaffold Improves Cell Transplant Survival, Reduces Infarct Expansion, and Induces Neovasculature Formation in Ischemic Myocardium 44 (2004) 654-660.
49. H. Wang, J. Zhou, Z. Liu, C. Wang, Injectable cardiac tissue engineering for the treatment of myocardial infarction, Journal of Cellular and Molecular Medicine 14 (2010) 1044-1055.
50. D.M. Pedrotty, J. Koh, B.H. Davis, D.A. Taylor, P. Wolf, L.E. Niklason, Engineering skeletal myoblasts: roles of three-dimensional culture and electrical stimulation 288 (2005) H1620-1626.
51. S.S. Mihardja, R.E. Sievers, R.J. Lee, The effect of polypyrrole on arteriogenesis in an acute rat infarct model, Biomaterials 29 (2008) 4205-4210.
52. K.C. Roberts-Thomson, P.M. Kistler, P. Sanders, J.B. Morton, H.M. Haqqani, I. Stevenson, J.K. Vohra, P.B. Sparks, J.M. Kalman, Fractionated atrial electrograms during sinus rhythm: Relationship to age, voltage, and conduction velocity 6 (2009) 587-591.
53. S. Bal, Experimental study of mechanical and electrical properties of carbon nanofiber/epoxy composites 31 (2010) 2406-2413.

(470a) Silicon-Based Bulk Nanostructured Thermoelectric Generators

Akram Boukai, Materials Science and Engineering, University of Michigan, Ann Arbor, MI

State-of-the-art thermoelectric generators are typically based on exotic alloys and have low device efficiencies. Therefore, research targeting new discoveries in earth abundant thermoelectric materials that increase device efficiency are needed. This talk will present recent progress on bulk nanostructured thermoelectric materials that are silicon-based. These include functional silicon nanowire devices and silicide/silicon eutectics. These bulk nanostructured thermoelectric devices exploit dramatic reductions of thermal conductivity due to phonon scattering at disordered interfaces, which is expected to increase ZT. Thermoelectric properties, structural characterization, and device efficiencies will be presented.

See more of this Session: Nanomaterials for Thermal-to-Electric Conversion I
See more of this Group/Topical: Topical 5: Nanomaterials for Energy Applications

(246f) Self Assembly of POSS and Sorbitol and Their Effects On the Reinforcement of Polypropylene Spun Fiber

Sayantan Roy and Sadhan C. Jana, Department of Polymer Engineering, University of Akron, Akron, OH

Abstract:

Thisstudy developed fundamental understanding of nanocomposite formation by "bottom-up"self-assembly of inorganic-organic hybrid molecules (POSS) in isotactic polypropylene(iPP) matrix. A sorbitol nucleating agent (DBS) compatible with POSS withsilanol (Si-OH)functionalities promoted dispersion of POSS molecules andprovided templates for self-assembly into cylindrical nanoparticles in the spunfibers. It was found that at the optimum combination of phenyl-POSS and DBS, iPPcomposites can be spun into fibers at higher draw-ratios compared to unfilled polymerand provided superior mechanical properties. Rheological studies revealedphenyl silanol-POSS prevented fibrillation of DBS. The interactions between DBSand silanol POSS with various functionalities were investigated using infra-redspectroscopy (FT-IR), differential scanning calorimetry (DSC), wide-angle X-raydiffraction (WAXD), nuclear magnetic resonance spectroscopy (NMR), massspectrometry (MS), and their synergistic morphology by scanning electronmicroscopy (SEM). The mixtures of phenyl silanol-POSS with DBS when heated atthe processing temperature of iPP (200°C) formed a low viscous transparentliquid. The cylindrical shaped nanoparticles as found in spun fibers werebelieved to have resulted from the stretching of this liquid complex. Nochemical reaction occurred between DBS and POSS as identified from the solidstate NMR and complexes of different molecular ratios were identified from massspectrometry study. It was inferred that a combination of hydrogen bondinginteractions and intermolecular π-π stacking between the phenyl ringsplayed a major role in the formation of supramolecular complexes. Higher orientationof polymer chains at higher draw-ratios was identified to be the possible causeof mechanical strength improvement.

1. Introduction

Our study focused on dispersion of polar, crystalline polyhedraloligomeric silsesquioxane (POSS) molecules of 1-2 nm size in non-polar polyolefins to obtainlarge enhancements in mechanical properties. The incompatible POSS molecules do not dispersewell and remain as agglomerates of 10-20 mm size in polyolefins.[1] Such large size POSS particles act as defects anddeteriorate mechanical properties.[2] In this context, POSS with non-polar side groupsdisperses well in polyolefins, but do not contribute much to the properties ofthe molded composites due to greater plasticization effects.[2, 3] So the scientific challenge of the work is to find a dispersionaid that can produce nano-scale dispersion of an incompatible POSS in non-polar polyolefins andhaving a subsequent improvement in mechanical properties.

Thisresearch exploited the fundamentals of nanocomposite formation by a "bottom-up"self-assembly approach from dispersions of POSS molecules in polyolefins. Thesecomposite materials had the processing ease of unfilled polymers and weresuitable for manufacturing of articles with micro- and nano-scale features byhigh-speed fiber spinning. The fundamentals developed in this work offeredsuperior alternatives to most "top-down" polymer nanocomposites, where ordersof magnitude increase in viscosity over that of the host polymers is a norm andachieving nanoscale dispersion is a challenge. The degree of nanofiller/polymerand nanofiller/nanofiller interactions was governed by the choice of polymer andnature of the side groups of POSS.[1, 4, 5]

In our work, non-reactive POSS containingmultiple silanol groups were dispersed in iPP with the aid of sorbitol-typenucleating agent, di-benzylidene sorbitol (DBS). Theunderlying chemistry between DBS and POSSramified some unique molecular structures in molten PP, which in turn led tonanoparticle formation in the final stages of fiber spinning and providedmechanical property improvement.

2. Materials and methods

Variousgrades of POSS were chosen tetra-silanol phenyl-POSS (tetra-POSS),tri-silanol phenyl-POSS (tri-POSS), tri-silanol cyclopentyl-POSS (cyclo-POSS),and tri-silanol isobutyl POSS (iso- POSS). These were obtained from HybridPlastics. The sorbitol derivative chosen for this study wasdi(benzylidene)sorbitol (DBS) and was obtained from Milliken Chemicals. Forcomposite preparation, isotactic polypropylene (iPP), P4G2Z-159, was obtainedfrom Flint Hill Resources (MFI 1.95 g/10 min, T_m 165°C). The materials were mixed in a batch mixer at 200°C for 5minutes to obtain compounds. In some compounds the nucleating agent DBS wasnot used to determine the potential of nanoparticle formation by POSS molecules alone. Anoptimum ratio of DBS and POSS were determined using rheological study for furtherprocessing. The optimized polymer composites were subsequently melt-spun into monofilamentfibers using a capillary rheometer (Instron) and a take-up device. Thevalues of tensile modulus, tensile strength and yield strength of the fibers were determined. The fibers were characterized by DSC andWAXD and morphology by SEM.

The synergistic interactions between POSS and DBS moleculeswere studied using FT-IR, DSC, and MS and any possible reaction by solid stateNMR. DBS/POSSself-assembled structures interdependent of iPP were observed by SEM.

3. Results and discussion

The POSS molecules with multiple silanol groups were selected sothat non-covalent hydrogen bonding interactions can occur with the freehydroxyl groups of DBS which was identified by FT-IR. DBS is known toform 3-D fibrillar networks in PP melt.[6] However, the supramolecular complexes between phenyl silanol-POSS and DBS suppressed such fibrillar network formation as revealed from oscillatory shear rheology.[7] For cyclo- and iso-butyl POSS, the DBS fibrillation was still visible which suggests insufficient interactions between non-phenyl POSS and DBS. This trend was interpreted by invoking the possibility of π-π stacking between the phenyl rings. It was confirmed by molecular dynamics simulation that DBS/tri-POSS and DBS/tetra-POSS molecules first engaged in π-π stacking and then formed hydrogen bonding leading to the formation of mutlimolecular complexes, which in turn subdued DBS fibrillation completely. Such π-π stacking was not possible in the case of cyclo-POSS and iso-POSS with DBS system. In corroboration with the rheology data, the optimum combinations of DBS with tri- and tetra-POSS in PP also produced lowest crystallinity. In this case, the nucleating function of DBS[8] was arrested due to participation of DBS in DBS-POSS multi-molecular complexes. Insufficient interaction of cyclo-POSS and iso-POSS with DBS also reflected in no such changes in crystallinity with increasing POSS content.

Tri-POSSand tetra-POSS formed crystalline lamella structures through self-aggregation viaintermolecular hydrogen bonding between the silanol POSS molecules. In the presenceof DBS, the phenyl silanol-POSS formed much smaller crystalline aggregates anddistributed uniformly into the DBS network (Figure 1). On the other hand, cyclo-POSS and iso-POSS retained their circular and lamellar domain aggregates respectively, when mixed with DBS. The nature of assembly of cyclo-POSS and iso-POSS indicates greater POSS-POSS affinity,unlike phenyl silanol-POSS, which showed greater affinity towards DBS.

Theinteractions towards the formation of supramolecular complexes between POSS andsorbitol derivatives at various molecular ratios were also systematicallyinvestigated using WAXD, NMR and MS. Nochemical reaction occurred between them as identified from the solid state ^{29}Siand ^{13}C NMR and supramolecular complexes of different molecularratios were identified from MS - ElectrosprayIonization study. Only complexes with 1:1 molecular ratio wereidentified between cyclo-POSS and DBS, whereas higher molecular ratio complexeswere observed for tri-POSS and tetra-POSS with sorbitol derivatives.

Finefibers were obtained with DBS/tri-POSS and DBS/tetra-POSS compounds. PPcomposite formulations with 2-5 wt% of POSS were spun into fibers with close to40% reduction in diameter. These fibers offered 60-80% increase in modulus,50-60% increase in tensile strength, and 100% increase in yield strengthcompared to unfilled iPP. On the other hand, in spun fibers, the domain size of POSS particles were found to be cylindrical with close to 100 nm in diameterand 200-300 nm in length (Figure 2). The complex materials formed between DBS and phenyl silanol-POSS were liquids with low viscosity which enhanced the melt-spun fiber to be drawn at very high draw-ratio, where the liquid droplets stretched and deformedto form the cylindrical nanoparticles. Maximum tensile properties were identifieddue to increased draw-ratio of the fibers, which in turn produced finer fibers.

4. Conclusions

The following conclusions can be drawn from the resultsobtained in this study.

PP/DBS/POSS ternary blend system provided the best reinforcementof PP at its optimum combination with maximum draw-ratio of spun fiber.

The spun fiber allowed formation of self-assembled cylindrical nanoparticlesof POSS/DBS complex.

A combination of hydrogen bonding interactions and intermolecularπ-π stacking between the phenyl rings played a major role in theformation of supramolecular complexes between DBS and POSS molecules.

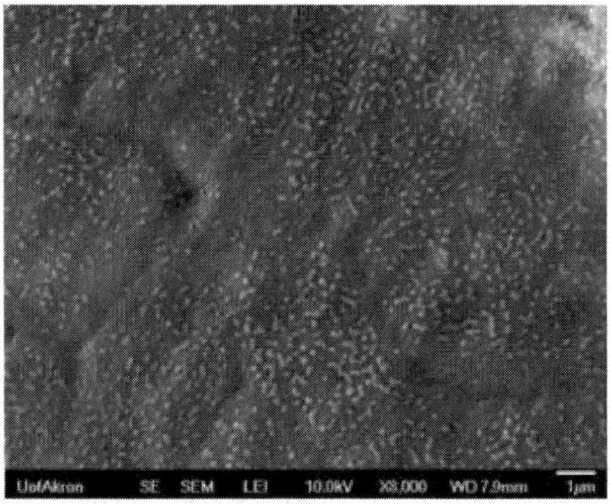

Figure 1. Uniformly dispersed POSSnanoparticles on DBS fibrils (solution cast)

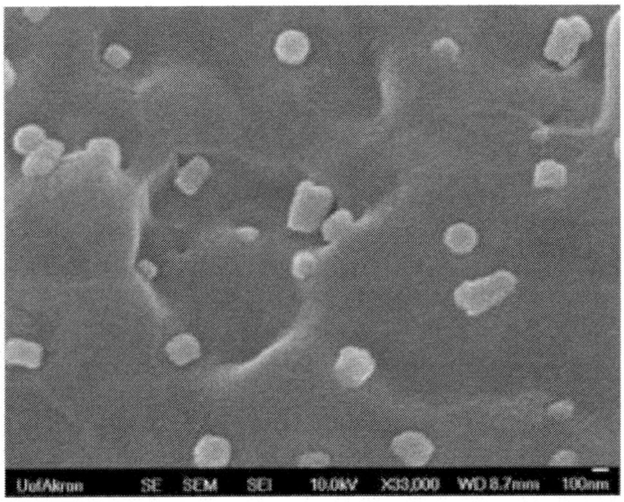

Figure 2. Cylindricalnanoparticle formation in iPP spun fiber

5.Acknowledgements:

Financialsupport from National Science Foundation in the form of award CMMI-0727231 isgratefully acknowledged.

6.>References:

1. Fina A, Tabuani D, Frache A, andCamino G. Polymer 2005;46(19):7855-7866.

2. Baldi F, Bignotti F, Fina A, TabuaniD, and Ricco T. Journal of Applied Polymer Science 2007;105(2):935-943.

3. Misra R, Fu BX, and Morgan SE.Journal of Polymer Science, Part B Polymer Physics 2007;45(17):2441-2455.

4. Jeon HG, Mather PT, and Haddad TS.Polymer International 2000;49(5):453-457.

5. Zheng L, Farris RJ, and CoughlinEB. Macromolecules 2001;34(23):8034-8039.

6. Kristiansen M,Werner M, Tervoort T, Smith P, Blomenhofer M, and Schmidt H-W. Macromolecules2003;36(14):5150-5156.

7. Perilla JE, Lee B-J, and Jana SC.Journal of Rheology (Melville, NY, United States) 2010;54(4):761-779.

8. Thierry A, Fillon B, Straupe C,Lotz B, and Wittmann JC. Progress in Colloid & Polymer Science1992;87(Solidif. Processes Polym.):28-31.

See more of this Session: Nanoscale Structure In Polymers I
See more of this Group/Topical: Materials Engineering and Sciences Division

(394a) Integrating Nanocellulose Production with Biofuel

Junyong Zhu, USDA Forest Products Laboratory, Madison, WI

This study demonstrated the feasibility of integrating the production of nano-fibrillated cellulose (NFC), a potentially highly valuable biomaterial, with sugar/biofuel (ethanol) from wood fibers. Commercial cellulase enzymes were used to fractionate the less recalcitrant amorphous cellulose from a bleached kraft eucalyptus pulp resulting in a highly crystalline and recalcitrant cellulose (RC). The RC is difficult to hydrolyze to sugars but very suitable for producing biobased nanomaterials through mechanical homogenization. A range of fractionation yields of RC from 10–70% can be achieved by varying fractionation duration and enzyme dosage. The crystallinity of the RC was found to be as much as 24% higher than that of the original bleached pulp. The cellulase fractionation process facilitated mechanical homogenization by significantly reducing the degree of polymerization (DP) to about 400 and the length of the fibers to about 200 mm. The hydrolyzed sugars were found to be easily converted to ethanol through yeast fermentation with excellent efficiency of 91%. Films made from nano-fibrillated cellulose were found to be optically transparent with opacity as low as 12%. The NFC films were mechanically strong and stiff with tensile strengths and moduli approximately 10 and 6 times higher than film made from fibers that had not been nano-fibrillated.

See more of this Session: Nanocellulose-Based Materials and Composites
See more of this Group/Topical: Forest and Plant Bioproducts Division

(600a) Coating Process Regimes In Particulate Film Production by Forced Convection-Assisted Drag-Out

Damien D. Brewer, **Satish Kumar** and Michael Tsapatsis, Department of Chemical Engineering and Materials Science, University of Minnesota, Minneapolis, MN

Operating conditions for deposition of monolayer and bilayer particulate coatings from aqueous silica dispersions are identified in the context of a drag-out operation assisted by forced convection. Dry film thickness, uniformity, and morphology are assessed within an operating window parameterized by the capillary number and silica dispersion weight fraction. Three film deposition regimes with respect to the capillary number are observed: convective film deposition at low process rates, film entrainment at moderate process rates, and a thin-film transition regime at intermediate process rates. Locally ordered particulate films of variable layering thickness, including discontinuous submonolayer or mixed submonolayer and monolayer, mixed monolayer and bilayer, and multilayer are dominant under convective deposition conditions. A map of morphologies is presented within the capillary number – weight fraction operating window, where monolayer and mixed monolayer/bilayer films are demonstrated in the thin-film transition regime at intermediate dispersion weight fraction. These operating maps are constructed using a model silica dispersion, and are therefore relevant to particulate coatings of other inorganic materials.

See more of this Session: Nanostructured Thin Films
See more of this Group/Topical: Materials Engineering and Sciences Division

(321d) Characterization of Oxygen Scavenging Films Based On Butadiene-Containing Polymers

Kevin K. Tung, Richard H. Li, Benny D. Freeman and Don R. Paul, Department of Chemical Engineering, The University of Texas at Austin, Austin, TX

Butadiene-containing polymers, such as styrene-butadiene-styrene block copolymers and 1,4 polybutadiene are of use as oxygen scavenging polymers (OSP) in barrier applications. As part of an evaluation of this use in such applications, films of these polymers were oxidized catalytically at four catalyst loadings and various thicknesses. Oxidation led to formation of crust layers at the film outer surfaces. Gas permeabilities of the crust layers are 2-3 orders of magnitude lower than those of neat, un-oxidized polymers. The diffusion-controlled oxidation creates a transient moving boundary that separates the oxidized and un-oxidized regions. Maximum oxygen uptake was measured to understand the kinetics and oxidation capacities of OSP systems and extracted from thin film experiments.

The thickness of the oxidized regions at long oxidation times is called the critical thickness, and it reaches a maximum at a certain catalyst loading. Films that are thinner than the critical thicknesses will oxidize fully and homogeneously. Results from three methods of analysis: 1) Thickness Model Analysis, 2) Inverse Thickness Analysis, and 3) SEM Imaging Analysis, all indicate a similar critical thickness value in each OSP system. Critical thicknesses appear related to the values of maximum uptake. Critical thicknesses set the upper OSP thickness limit in future diffusion-controlled oxidation systems, such as multi-layered and blend films.

See more of this Session: Nanoscale Structure In Polymers II
See more of this Group/Topical: Materials Engineering and Sciences Division

(662a) Engineering of Nanocrystals for Clean Energy Applications

Michael Z. Hu, Energy & Transportation Sciences Division, Oak Ridge National Laboratory, Oak Ridge, TN

Nanocrystals such as quantum dots are important for electronic/optoelectronic energy technologies. In the course of developing energy devices or energy process technologies, chemical synthesis and process engineering play significant roles in controllably achieving designed tailoring of materials nanostructures, as well as drastically improving the process production yield and nanomaterials properties. Distinct from other nanotechnology applications, energy processes or devices (such as large-dimension panels for solar cells or for solid state lighting) could consume large quantities of high-quality, but low-cost nanomaterials (such as nanocrystals, films, or membranes). Thus, new engineering principles must be developed to enable high performance/cost ratio and the nanoscience-to-nanomanufacturing transition in order to meet the energy demands. Energy efficiency performance is determined by the high quality (e.g., well controlled fine nanostructures) of the produced materials. The speaker will focus on chemical/molecular process engineering perspectives by presenting a few examples of energy nanomaterials being developed at the Oak Ridge National Laboratory, including quantum-dot nanocrystals for solid-state lighting and biofermentation-produced CIGS nanoparticles for thin film solar cells.

See more of this Session: Nanocrystal Science and Technologies
See more of this Group/Topical: Materials Engineering and Sciences Division

(385g) Synthesis and Characterization of Core-Shell Fluorescent Nanoparticles with Magnetic and Plasmon Properties, for Bio-Distribution Studies and Application In Photo-Ablation Cancer Therapy

Angel A. Galvis, Juan R. Reyes and Watson L. Vargas, Chemical Engineering, Universidad de los Andes, Bogota, D.C, Colombia

The development of bimetallic/bifunctional nanoparticles that exhibit a core-shell architecture has been the focus of attention in recent years due to their unique properties in "anti-cancer therapy'' and other potentially attractive applications. This study presents results on the synthesis and potential application in photo-ablation cancer therapy of a set of core-shell bifunctional gold-coated Silica spheres with both magnetic and plasmonic properties. Naked magnetite-maghemite Fe_3O_4-Fe_2O_3 nanoparticles are covered with a thin shell of silica using the well-known method developed by Stöber with some modifications. The silica-modified (magnetite-maghemite) nanoparticles are then functionalized with Silane (APES) monolayers which are subsequently covered with a thin solid shell using Au nanoparticles as seeds to template the growth of the shell. Also SiO_2@Au, and Fe_3O_4-Fe_2O_3@Au have been synthesized with the purpose of studying their plasmon properties; in the SiO_2@Au particles the silica core has been prepared ••by a novel procedure using amino-acids as catalysts. The particles are characterized by X-ray diffraction(XRD), Dynamic Light Scattering (DLS), Zeta potential, Scanning Electron microscopy (SEM), Transmission electron microscopy (TEM), Atomic force microscopy (AFM), Fourier transform infrared spectroscopy (FT-IR), and UV-VIS-NIR spectroscopy. Preliminary results on bio-distribution *(in-vitro* assays) and photo-thermal ablation activity (on a prototypical tissue) will also be presented.

See more of this Session: Magnetic Nanoparticles In Biotechnology and Medicine
See more of this Group/Topical: Nanoscale Science and Engineering Forum

(346f) Application of Core-Shell Fluorescent Nanoparticles with Magnetic and Plasmon Properties In Biotransport Studies and Photo-Ablation Cancer Therapy

Angel A. Galvis, Juan R. Reyes, Ruben D. Vargas, Oscar A. Alvarez and Watson L. Vargas, Chemical Engineering, Universidad de los Andes, Bogota, D.C, Colombia

Bimetallic/bifunctional nanoparticles that exhibit a core-shell with precise control of size, morphology, surface chemistry, and assembly process of each component have been the focus of attention in recent years. These functional nanocomposite nanospheres possess a core of silica-protected magnetite particles and *in situ* growth of a thin solid shell using Au nanoparticles as seeds to template the growth of the outer layer, such nanoparticles provide, fluorescence, magnetic and plasmon functionalities. These unique nanostructures are a highly efficient platform for several applications. In this work two applications of such materials will be discussed: transport in prototypical tissues and photo-ablation cancer therapy.

For tissue bio-distribution assays we have used fluorescent nanocomposite nanopheres with a monolayer of rhodamine encapsulated into the core of SiO_2. A diffusion cell with pig skin as prototype tissue has been implemented to conduct the transport experiments. A model based on the concepts of anomalous diffusion and the analytic and numerical of solution of the time fractional diffusion equation (TFDE) has been implemented, for determining the sub-diffusion parameters that describe the transport in such complex media.

For the purpose of improving bio-distribution of the particles, and the "anti-cancer" drugs that they could potentially transport, we have studied the change in the diffusion, cytotoxicity, hemolytic activity, and genotoxicity; in emulsions of type (O/W) with dispersed Fe_3O_4- Fe_2O_3@SiO_2@Au nanoparticles. Finally a preliminary study of the phothermal ablation activity of the Fe_3O_4-Fe_2O_3@SiO_2@Au nanoparticles, will also be discussed.

See more of this Session: Applications of Magnetic Nanoparticles
See more of this Group/Topical: Nanoscale Science and Engineering Forum

(472f) Iron Chelation by Polyamidoamine Dendrimers: A Second Order Kinetic Model for Metal-Amine Complexation

Michael R. Mankbadi and John N. Kuhn, Chemical and Biomedical Engineering, University of South Florida, Tampa, FL

Iron Chelation by Polyamidoamine Dendrimers: A Second Order Kinetic Model for Metal-Amine Complexation

Michael R. Mankbadi and John N. Kuhn
Department of Chemical & Biomedical Engineering, University of South Florida, 4202 East Fowler Avenue, ENB 118, Tampa, FL 33620

To whom correspondence should be addressed. jnkuhn@usf.edu & (813) 974-6498

ABSTRACT

This study presents a kinetic model of the chelation of iron ions by generation 4 hydroxyl-terminated polyamidoamine (PAMAM) with ethylenediamine core (G4-OH). The coordination processes of iron ions from ferric chloride, $FeCl_3$, and ferrous bromide, $FeBr_2$, to G4-OH dendrimers were analyzed using ultraviolet-visible (UV-Vis) spectroscopy, proton nuclear magnetic resonance (^{1}H-NMR) spectroscopy, and liquid chromatography-mass spectrometry (LC-MS). In the visible region, a charge-transfer was observed when the dendrimer was added to a ferric chloride solution. This phenomenon is a ligand-to-metal charge-transfer (LMCT) between the free electron group of the dendrimer's internal amines and the dehalogenated iron ion that takes 2 hours to complete at room temperature. Analysis of potential rate laws and diffusion effects led to a second-order kinetic model for this reaction. By measuring the rate coefficients as a function of temperature (22 °C to 37 °C), an apparent activation energy of 41.5 kJ/mol was obtained using the Arrhenius method. The results of this study will fuel research of PAMAM dendrimers for environmental, pharmaceutical, and materials applications.

See more of this Session: Nanotechnology for Biotechnology and Pharmaceuticals I
See more of this Group/Topical: Nanoscale Science and Engineering Forum

(649b) Enabling Energy Applications with Nanocarbons

Randy L. Vander Wal, Energy and Mineral Engineering & The EMS Energy Institute, Pennsylvania State University, University Park, PA

Introduction

Energy and Nanomaterials are intimately related. Most energy processes occur at the interface. Nanomaterials, when used as interfacial modifiers have the potential to significantly alter the energy landscape. Although carbon nanomaterials have been explored for a suite of energy applications, to-date the approach has been based on replacing bulk materials, not as interfacial modifiers. The work presented here will highlight this latter approach by showing results for a suite of energy applications from the author's own work.

Experimental

Experimental details will be presented at the presentation with regards to the synthesis and utilization of the nanomaterials. Our applications include energy conversion, generation, storage, efficiency, conservation and control.

Results and Discussion

Nanoscale materials are redefining the relation between material composition, size and properties. Chemical properties (e.g. reactivity) and physical properties (e.g. surface area) become a strong function of size at the nanoscale. Applications illustrated include the following selected examples from the author's work: catalysis, composite materials, energy storage, sensors, thermal management, and tribology. The key concept is that the nanomaterials serve as interfacial modifiers. Since most energy related processes are dominated by interfacial reactions, nanomaterials have the potential to dramatically affect energy conversion rates and magnitudes. Examples follow.

Catalysis is central to particulate and NO_x after-treatment systems. A prime example of reducing materials to nanoscale and realizing new properties is catalysis by nanoscale gold. Au nanoparticles supported on oxides such as CeO_2, TiO_2 and Fe_2O_3 offer ambient temperature oxidation of CO, volatile organic compounds (VOCs) and potentially exhaust hydrocarbons.

Lightweight materials will reduce weight significantly yielding substantial benefits in fuel efficiency with reduced emissions. Composites with substantial gains in Young's modulus, tensile strength, and

EM shielding may be realized in polymeric composites using carbon nanotubes as an interfacial modifier rather than bulk filler [1]. The purpose of the foil is to serve as a gas impermeable barrier layer within the polymer composite.

Advances in energy storage include batteries, ultra-capacitors. These can support lighting, appliances, a starter, cooling fans, transmission and hydraulic systems, fuel and air handling systems and ultimately enable hybrid systems. Towards these goals, substantial gains in Li ion battery cathode and anode materials have been realized using carbon nanofibers, coating processes and including elements such as tin and silicon. Modifications of the CNT surfaces increases the Li ion capacity beyond the theoretical limit of normal graphite. As an illustration, CNTs were directly fabricated for this end-application use upon ultra-fine SS mesh. Significantly no harvesting, purification or processing (using binders) was required. The CNTs could be used directly as synthesized [2].

With regards to overall system integration and control, sensors will play a prominent role [3]. With ultrahigh surface exposure relative to bulk material, nanoscale materials are exceedingly sensitive to gas adsorption. Exploitation of nanoscale properties will lead to new NO_x sensors and in-cylinder oxygen sensors. Examples include catalyst coated metal oxide semiconductors capable of ambient temperature operation.

Thermal management will benefit from nanofluids. Nanofluids can increase thermal conductivity and reduce radiator and heat exchanger size. Carbon-based nanofluids using nano-onions and carbon nanotubes have increased water conductivity by ~ 20% [4]..

Lubrication is critical to many engine components and powertrain systems. Nanolubricants can bridge the gap between fluid and solid materials [5]. As additives with liquids or greases, synergistic properties may be realized, particularly in boundary-phase lubrication. Improved coating formulations and properties can reduce or eliminate fretting and pitting. Results with nanocarbons show superior performance relative to graphite, diamond like carbon (DLC) and even Teflon.

Conclusions

Nanoscale materials are redefining the relation between material composition, size and properties. Their applications to energy processes include selected examples from the author's work. Selected technologies include catalysis, composite materials, energy storage, sensors, thermal management, and tribology. Interfacial modification serves to affect changes in energy conversion rates, and transfer magnitudes. This appears to be a rather universal concept and suggests R&D directions for nano-engineering of interfaces.

Acknowledgements

Funding through The Penn State Institutes for Energy and the Environment (PSIEE) and the Pennsylvania Keystone Innovation Starter Kit (KISK) is gratefully acknowledged.

References

(1) Vander Wal, R. L., and Hall, L. J., Advanced Engineering Materials **2004,** 6, 48-52.

(2) Luo, Y., Vander Wal, R. L., and Scherson, D. A., Electrochemical and Solid State Letters **2003,** 6, A56-A58.Vander Wal, R. L., Mozes, S. D., and Pushkarev, Nanotechnology ***2009***, 20, 105702-10.

(3) Hunter, G. W., Vander Wal, R. L., Liu, C. C., Xu, J. C., and Berger, G. M., Sensor and Actuators B ***2009***, 138, 113-119.

(4) Vander Wal, R. L., Mozes, S. D., and Pushkarev, Nanotechnology ***2009***, 20, 105702-10.

(5) Street, K. W., Miyoshi, K., and Vander Wal, R. L., "Application of Carbon Based Nano-particles to Aeronautics and Space Lubrication", in *Superlubricity*, Chapter 19, pp 311-340. Edited by Drs. Jean-Michel Martin and Ali Erdemir, Elsevier, Amsterdam (2007). (also NASA/TM-2007-214473).

See more of this Session: Graphene and Carbon Nanotubes: Applications
See more of this Group/Topical: Nanoscale Science and Engineering Forum

(615c) Aqueous Only Route Toward Graphene From Graphite Oxide

Ken-Hsuan Liao, Anudha Mittal, Shameek Bose, Christopher Leighton, K. Andre Mkhoyan and Christopher W. Macosko, Chemical Engineering and Materials Science, University of Minnesota in Twin Cities, Minneapolis, MN

We report a new, simple, hydrazine-free, high-yield method for producing single-layer graphene sheets. Graphene sheets were formed from graphite oxide by reduction with simple deionized water at 95°C under atmospheric pressure. Over 65% of the sheets are single graphene layers; the average sheet diameter is 300 nm. We speculate that dehydration of GO is the main mechanism for oxygen reduction and transformation of C-C bonds from sp^3 to sp^2. The reduction appears to occur in large uniform interconnected oxygen-free patches so that despite the presence of residual oxygen the sp^2 carbon bonds formed on the sheets are sufficient to provide electronic properties comparable to reduced graphene sheets obtained using other methods.

See more of this Session: Synthesis of Graphene and Carbon Nanotubes: Kinetics, Mechanisms and Reactor Design
See more of this Group/Topical: Nanoscale Science and Engineering Forum

(420ac) Immobilized Ionic Liquids In Layered Compounds Via Mechanochemical Intercalation As Catalysts for CO2 Fixation

Hang Hu[1], Jarett C. Martin[2], Min Xiao[1], Cara S. Southworth[2], Yuezhong Meng[1] and **Luyi Sun**[2], (1) School of Physics and Engineering, Sun Yat-sen University, Guangzhou, China, (2)Department of Chemistry and Biochemistry, Texas State University-San Marcos, San Marcos, TX

A facile mechanochemical intercalation approach was adopted to immobilize 1-butyl-3-methylimidazolium chloride (BMIMCl) in layered α-zirconium phosphate (ZrP). The intercalation compounds with expanded layer distance were confirmed by X-ray diffraction. Thermogravimetric characterization revealed that partial BMIMCl was intercalated into the ZrP galleries, while partial was adsorbed on the layer surface. The immobilized BMIMCl was evaluated for catalysis evaluation of the coupling reaction of CO_2 and propylene oxide to synthesize propylene carbonate. The immobilized BMIMCl exhibited close reactivity as free BMIMCl. Overall, the mechanochemical approach proves to be effective to immobilize ionic liquids in layered compounds, and thus may expand the applications of ionic liquids, and meanwhile improve catalyst separation and recycling.

See more of this Session: Poster Session: Nanoscale Science and Engineering
See more of this Group/Topical: Nanoscale Science and Engineering Forum

(472d) Photo-Activated Antibacterial and Antiviral Activity of Porphyrin-Conjugated Multiwalled Carbon Nanotubes

Indrani Banerjee, Marc P. Douaisi, Dhananjoy Mondal and Ravi S. Kane, Chemical and Biological Engineering, Rensselaer Polytechnic Institute, Troy, NY

We describe the design of porphyrin-conjugated multi-walled carbon nanotubes (MWNT-P) that show significant activity against *S. aureus* and influenza viruses. *S. aureus* infections can lead to life-threatening situations, especially when caused by antibiotic-resistant strains. Influenza spreads around the world in seasonal epidemics, resulting in thousands of deaths every year, up to millions in some pandemic years. To design antibacterial and antiviral conjugates, we attached Protoporphyrin IX (PPIX) to acid-functionalized multiwalled carbon nanotubes (MWNTs). We selected carbon nanotubes as scaffolds because of their ease of recovery from solution via simple filtration, and their ability to form large area coatings. In the presence of visible light, MWNT-P deactivated *S. aureus* and also significantly reduced the ability of Influenza A virus to infect mammalian cells. Films made of MWNT-P also exhibited potent biocidal activity against *S. aureus*. This work thus provides a facile approach to incorporate porphyrins into coatings and makes them re-usable which may lead to their more widespread use as antibacterial and antiviral agents/coatings.

See more of this Session: Nanotechnology for Biotechnology and Pharmaceuticals I
See more of this Group/Topical: Nanoscale Science and Engineering Forum

(385i) In Situ Synthesis of Amino-Functionalized Magnetite/Silica Nanoparticles In Microemulsion and Their Application for Immobilization of Catalase

Wei Zhang, **Shaoyi Jia**, Songhai Wu and Yong Liu, School of chemical engineering and technology, Tianjin University, Tianjin, China

Abstract

Magnetite/Silica (Fe_3O_4/S_iO_2) nanoparticles with amino-functionalized were synthesized in a simple one-pot process. The synthetic reaction took place in a water-in-oil microemulsion system, which was composed of cyclohexane, Alkylphenols polyoxyethylene (OP-10), n – butanol and water. This method was based on a three-step process, involving: (i) synthesis of the Fe_3O_4 nanoparticles by adding the microemulsion containing the basic precipitant of ammonia into another microemulsion containing Fe^{2+} and Fe^{3+} with atmospheric nitrogen under ambient conditions; (ii) formation of the Fe_3O_4/S_iO_2 nanocomposites via direct decomposition of tetraethyl orthosilicate (TEOS) in the reverse micelles nanoreactors under the presence of freshly synthesized Fe_3O_4 nanoparticles; (iii) functionalization of the Fe_3O_4/S_iO_2 nanocomposites with amine groups by hydrolysis of 3-aminopropyl- trimethoxysilane (APTES) in situ of the above microemulsion system. The structure composition and the surface functional groups of the product were characterized by transmission electron microscopy (TEM), X-ray diffractometer (XRD) and Fourier transform infrared analysis (FT-IR). The magnetic behavior was measured by Vibrating sample magnetometer (VSM). It was found that the synthesized product was embedded structure of Fe_3O_4/S_iO_2 nanocomposites with a diameter range from 10 to 20 nm. These magnetic silica nanocomposites with a functional surface had excellent superparamagnetic properties which implied potentially promising applications in biomedicine areas. Accordingly, the amino-functionalized Fe_3O_4/S_iO_2 nanocomposites were used as magnetic separation carriers for immobilization of catalase (CAT). Activity of the immobilized CAT was examined and contrasted with the free CAT by the catalysis of hydrogen peroxide in aqueous buffer. The results showed that the immobilized CAT obtained through covalent attachment exhibited higher resistance to pH and thermal stability than the free CAT. In addition, the immobilized CAT still retained a high activity after 10 repeated batches reaction.

SURFACE AND AGING CHARACTERIATICS OF PAPER COATED WITH NANO SIZE ZINC OXIDE PIGMENT

Sanjay Tyagi. Renu Tyagi* , B P Thapliyal* R M Mathur* and A K Ray***

**Central Pulp and Paper Research Institute, Saharanpur – 247001*
*** Indian Institute of Technology, (Saharanpur Campus), INDIA*

ABSTRACT

Quality of paper is improved on pigment coating in terms of surface, optical properties and printability characteristics. One of the major parameter for aqueous pigment coating is the type of pigment used and also its particle size and distribution. Zinc oxide is one of the pigment s employed for special coating application. It is offlet nano size particles are playing prominent roles in several areas ranging from medicine to electronics. In paper making there is no exception. Few specialty grades of paper are already using nano size calcium carbonates both as filler and also as coating pigments. Application of water based barrier coating in paper and board in food packaging industry have proven successful. It is also important due to substantial smaller size particles (nano range) compared to macro size solid particles (in micron range) can provide better functional properties of paper in some specialized applications. However, ZnO is present in various forms nano size ZnO particles can enhance the properties significantly to improve optical and surface properties and specially the effect of aging. In this present therefore an attempt has been made to study the potential technological applications of nano zinc oxide particles on paper surface made from hardwood furnish and bamboo using 100% nano ZnO and with admixture of coating grade China clay using laboratory rod coater. All the required properties were evaluated as per ISO standard and compared with the same base paper coated with micro size ZnO pigment particles. From the detailed study it is observed that paper coated with nano ZnO particles imparts better properties as compared to bulk ZnO in terms of brightness, whiteness, gloss, smoothness and printability. Aging properties were also thoroughly studied under artificial condition of varying temperature (50% – 120%)and humidity (10% - 90%). And also UV light, paper coated with nano ZnO showed better resistance to brightness reversion and UV protecting potential.

INTRODUCTION

Paper is pigment coated to improve the quality of paper in terms of surface, optical and printability characteristics. Paper coating ingredients which generally contains pigments, binders and additives is applied to the paper surface to improve the above properties of the paper. One of the major parameter for aqueous pigment coating is the type of pigment used and also its particle size distribution [(1,2,3)]. A commonly used pigment includes china clay, calcium carbonate, silica, zeolite, titanium dioxide, alumina trihydrate and polystyrene [(2)]. Zinc oxide is one of the pigments employed for special coating application [(4)]. Presently all above mentioned conventional pigments are available in the market have their particle size in the range of 1 – 10 micron [(5,6)]. The size of these particles is too big for the preparation of coating pigments to be used efficiently in coating machines. Evenness and uniformity in surface smoothness and thickness of coating layer which is one of the two prime requirements for producing good multi color print, generally lacks with conventional micron sized coating pigments [7,8,9]. Pore structure and contact angle wettability effect print quality by preventing ink spreading. In order to prevent wicking and feathering, it is important that the thickness of coating layer be homogeneous to a scale of few microns in depth which also helps in the absorption of successive droplets of ink at high delivery rates and any water present[9,10].

Application of nano size Zinc oxide in papermaking

While papermaking is normally a statistical process, nano technology makes use of the classic process where the position of every atom is specified and controlled. Yet despite this apparent discrepancy, between two technologies, the paper industry has been practicing nano particles over the conventional pigments long before it was marketed as latest technology trend [5,6].

Nano sized calcium carbonate has been used by paper industry since long as filler due to the advantage of high opacity, high brightness and whiteness [11]. At present, it is also used in special paper products such as baby diapers, sanitary napkins, etc.

Zinc oxide is also one of the pigments finding its use for special coating applications[4]. It is offlet nano size particles are playing prominent role in several areas ranging from medicine to electronics [12]. ZnO is essential for rubber and tier industry and also used as natural rubber, synthetic rubber and latex vulcanization active agents, reinforcing agent and coloring agents. In paint industry, ZnO in addition to color strength and hiding power outside but also a light coating of preservative and agents [13]. In medical and health and food industries, ZnO has a Sida, bleeding, hygienic convergence of functions. In electronics industry, ZnO varistor is the main raw material is magnetic, optical material such as the main additive. Prepared using ZnO varistor, not only has low sintering temperature, but also improve the performance of porous sensitive resistor such as the flux, the non linear coefficient [14,15].

In the present study, an attempt has been made to study the potential technological applications of nano zinc oxide particles on the paper surface made from the pulp of hardwood furnish and bamboo using 100% nano ZnO and with the admixture of coating grade china clay by taking trials on laboratory rod coater. The Zinc oxide soluble starch nano composite was prepared by wet chemical method using Zinc nitrate and sodium hydroxide as precursors and soluble starch as stabilizing and compared with the same base paper coated with micro size ZnO pigment particles.

EXPERIMENTAL

Coating formulation

The coating trials were carried out using Laboratory rod coater. The coated sheets were dried immediately in laboratory oven at the temperature of 105 C for maximum 2 minutes. The following two coating formulations were used for coating trials.

Ingredient	Composition I	Composition II
Pigment		
• Zinc Oxide (Nano or Ordinary)	500 parts	1000 parts
• China clay	500 parts	
Binder		

• Oxidized Starch	90 parts	90 parts
• Styrene Butadiene Latex (SBR Latex)	60 parts	60 parts
Dispersant		
• Anionic polymer	5 parts	5 parts
	Total solids maintained- 65±2%	

Preparation of coating colour

The preparation method for the coating slip is as under-

(a) Starch cooking

The oxidized starch solution was prepared separately by cooking under continuous agitation in hot water at temperature 90-95 ^{0}C, for about 15 minutes, i.e. upto the point when the whole solution became fully gelatinized.

(b) Pigment preparation

Pigment slurry was made in laboratory high-shear coating mixer. The dispersant was firstly added to desired quantity of water. When homogeneous solution was formed, weighed quantities of pigments were added, under constant stirring. A thick paste thus resulted, which became uniform at high shear. Time of agitation was, however kept at about 15 minutes.

(c) Addition of binders to pigments

The freshly prepared starch solution and synthetic binder SBR latex (in required amount) were added to the pigment paste in laboratory high-shear coating mixer, under constant agitation. The mixture was kept under constant agitation at high shear for about 30 minutes to ensure uniform mixing and dispersion of all the ingredients. Finally the slurry was taken out of the mixer.

(d) Screening

The coating slurry was next screened on a 300-mesh screen to eliminate foreign matters, coagulated coating and particles of dried coating slurry etc. Thus a uniform coating slurry; free from dirt, was ready for coating.

Calendering

Coated sheets after dried in laboratory oven were calendered using the laboratory calender with softnip mode, at 40 ^{0}C temperature & 30 bar pressure. The sheets were passed twice through the softnip.

Evaluation of optical characteristics of the pigments

The two pigments- Zinc Oxide (ordinary) and Zinc Oxide (nano particles), which were used in coating trials, were evaluated also for following optical characteristics; as an extension of the detailed study.

- Brightness (%)
- Whiteness (%)
- Yellowness (%)

Evaluation of base paper and coated paper

Base paper as well the coated & calendared papers were conditioned at 27±1 ^{0}C temperature and 65±2 % relative humidity prior to testing of following properties as per standard methods mentioned below.

Property	Standard methods
Grammage (g/m^2)	ISO 534
Brightness (% ISO)	ISO 2470

Gloss (%)	Tappi 480 om-99
Whiteness (%)	ISO 2470
PPS Roughness (micron)	ISO 8791/4
Picking velocity (cm/s)	ISO 3783
Print density (g/cm^3)	Using IGT printability tester Model AIC2-5
Print uniformity index	Using IGT printability tester Model AIC2-5
Stain length (mm), oil absorbency	Using IGT printability tester Model AIC2-5

RESULTS & DISCUSSION

Optical characteristics of base paper

The base paper used in the coating trials, was procured from commercial paper mill. The base paper was made from the pulp containing furnish of hardwood pulp and bamboo. Base paper was evaluated for different optical & printing properties for comparison purpose with coated & calendered paper. The result is tabulated in Table-I.

Optical characteristics of pigments

The optical characteristics of the pigments used in coating are given in Table-II. By comparing the values of the properties obtained for nano Zinc Oxide and ordinary Zinc Oxide, it was found that the starch-encapsulated nano Zinc Oxide particles had slightly lower brightness, whiteness and was relatively more yellowish in colour as compared to ordinary Zinc Oxide particles. This is probably due to the preparation process.

Characteristics of Coated & Calendered Paper samples

The coated paper samples, after softnip calendering, were evaluated for

(a) Optical characteristics
(b) Surface characteristics
(c) Printing characteristics
(d) Aging characteristics

(a) Optical characteristics

The effect of nano size zinc oxide particles on brightness of coated and calendered paper is given in Figure 1. Results show that nano zinc oxide gives better brightness as compared to conventional zinc oxide in both cases. i.e. when used alone and when used in combination with coating grade china clay. The same trend is true with whiteness value of coated paper (Fig. 3).

The value gloss of paper coated with nano ZnO also improved as compared to paper coated with conventional ZnO (Fig. 2). However, the improvement is low as compared to brightness and whiteness values of base paper, there is a fall in their values after coating with ZnO and China clay. The reduction in values is more in case when china clay is used in coating formulation. This is probably due to the lower brightness of china clay and binder additives which results in overall reduction of optical properties of coated paper.

(b) Surface characteristics

PPS Roughness

The surface smoothness of coated paper which is an important property in deciding the print performance of paper, improved with the application of nano ZnO. The PPS roughness in case of nano ZnO is relatively lower than conventional ZnO. The same trend was obtained when nano ZnO is used with china clay. A smooth surface was obtained when china clay is used in coating furnish. This is probably due to particle shape. China clay being platelet shape flattened more during the calendaring operation in compared to ZnO which is spherical in shape (Fig. 4).

(c) Printing characteristics

Picking velocity

Picking velocity of paper coated with nano ZnO was found to be higher as compared to conventional ZnO (Fig. 5) in both cases when used alone and in combination with china clay. This indicates that paper coated with nano ZnO can run easily on high speed printing presses or at the same printing speed, less binder is required in coating formulations with nano ZnO. The excess amount of binder always has negative effect on the printing characteristics.

Print Mottle (Print Non-Uniformity)

Print non uniformity leads to print mottle which is uneven appearance of colour density in a print. This is due to the surface factors to absorb ink evenly due to variation in base paper formation or coat weight. The non uniformity printing index is lower for paper coated with nano ZnO as compared to paper coated with conventional ZnO (Fig. 8).

The variation in sheet matrix formation formed by coating layer of conventional ZnO (due to non uniform size of ZnO particles) probably the reason for this. Paper coated with ZnO alone showed lower number as compared to paper coated with a mixture of china clay and ZnO since particle size of china clay is largely varied. The sheet matrix variation can lead to print mottle non uniform compression during calendering, resulting in variation in coating layer densification. Sheets that contain binder migration, coat weight variation, inferior coating profile and heterogeneous pigment blends can all cause mottle formation on the printed sheet.

Print density

Print density is affected by ink setting/ink holdout. For good density the vehicle of printing ink should stay on the top of paper together with ink pigment. The more the dried ink is retained on the outer surface of the paper, the higher will be the ink density, accordingly, the lower will be the ink demand. Paper with large pores in coating will have more penetration of printing ink leading the lower density. This is observed in case of papers coated with conventional ZnO coating (Fig. 7). The same is true with china clay in the coating formulation. This indicates that more ink will be required to print coated papers with conventional ZnO coatings.

Oil absorbency

Oil absorbency which is an indication of rate of penetration of oil was found to be more in the case of paper coated with nano size ZnO alone and in combination with china clay. (Fig 6)

(d) Effect of Aging

The effect of aging on the optical properties of coated and calendered paper is shown in the graph (Fig. 9-11) and tabulated in table (4-6). The results show that nano zinc oxide gives excellent UV protecting strength in both cases when used alone and when used in combination with coating grade china clay. The reduction in brightness and whiteness is minimum in the case of nano zinc oxide. The lower value of indicated the higher yellowness, which incase after aging maximum when china clay is used.

CONCLUSIONS

The use of nano zinc oxide as a coating pigment improves the paper properties in terms of optical, surface and printability as compared to conventional zinc oxide. This indicates that use of nano ZnO can be a good substitute for the replacement of conventional ZnO in the case of speciality grade papers. This opens a new scope for the use of nano Zinc oxide as a coating pigment for speciality grade of papers. The application of nano ZnO also exhibits excellent UV protection characteristics when tested at different condition of humidity and temperature. Paper represents good aging resistance properties which increases the longevity of paper. This opens a new scope of manufacturing paper with much longer life which lasts for longer years paper manufactured under connectional methods

The use of nano Zinc oxide as a coating pigment results in the paper with better properties as compared to conventional ZnO in terms of brightness, whiteness, gloss, smoothness, UV protection and printability characteristics i.e. picking velocity, print uniformity, print density, and oil absorbency .

Refrences:

1. 'Y.V.Sood, S.Tyagi. R.Tyagi, P.C.Pande, R.Tandon & T.K.Roy', 'Quality of indigenous coated paper and ways to improve it', *'J.IPPTA, 2008, 20 (4), pp'*
2. 'Coating pigments: from size and shape to nanotech', *'www.thefreelibrary.com/coating+pigments:+from+size+and+shape+to+nanotech.....'*
3. 'Surface characteristics of coating layers formed by coating pigments with different particle sizes and size distribution' *'J. Coat. Tech. and Res'*
4. 'N. Vatansever & S. Polat', 'Effect of zinc oxide type on aging properties of styrene butadiene rubber compounds', *'www.sciencedirect.com, 2009'*
5. Roxare, 'Nanotechnology in paper making', *'http://pulpandpapercircle.com/2008/07/nanotechnology –in-paper-making'*
6. 'J.Shen, Z.Song, X.Ojan, F.Yang & F.Kong', 'Nano fillers for papermaking wet end applications', *'www.bioresources.com, 2010, 5(3)'*
7. G. Chinga, 'Detailed characterization of paper surface structure for gloss assessment', *'JPPS, 2004, 30 (8), pp 222-227'*
8. G. Ström, S. E. Research, A. Englund & M. Karathanasis, 'Effect of coating structure on print gloss after sheet fed offset printing', *'J. Nord. Pulp & Paper, 2003, 18 (1), pp108-115'*
9. 'G. Chinga-Carrasco, M. Axelsson, Ø. Eriksen & S. Svensson', 'Structural characteristics of pore networks affecting print through', *'JPPS, 2008, 34 (1), pp 13-22'*
10. 'D.W. Donigian, J. N. Ishley & K.J. Wise, 'Coating pore structure and offset printed gloss', *'J.TAPPI, 1997, 80 (5), pp 163-172'*
11. T. Nypelö, M. Österberg & J. Laine, 'Ultra thin coating of paper by tailor made nanoparticles', *'TAPPI International Conf. on Nanotechnology for the Forest Product Industry, 2010'*
12. 'V. Bianchard & P. Blanchet', 'Color stability for wood products during use: Effects of inorganic nanoparticles', *'www.bioresources.com, 2011, 6(2)'*
13. 'B. Wang, N. Han, D. Meng, R. Yue, J. Yan & Y. Chen', 'Filter paper- templated preparation of ZnO thin films and examination of their gas sensing properties', *'www.sciencedirect.com, 2011'*
14. 'V. Prasad, A. J. Shaikh, A. A. Kathe, D. K. Bisoyi, A.K. Verma, N. Vigneshwaran', 'Functional behavior of paper coated with zinc oxide-soluble starch nanocomposites', *'J. Mat. Proc. Tech., 2010, 210 (14), pp 1962-1967'*
15. 'Surface characteristics of coating layers formed by coating pigments with different particle sizes and size distribution' *'J. Coat. Tech. and Res.'*

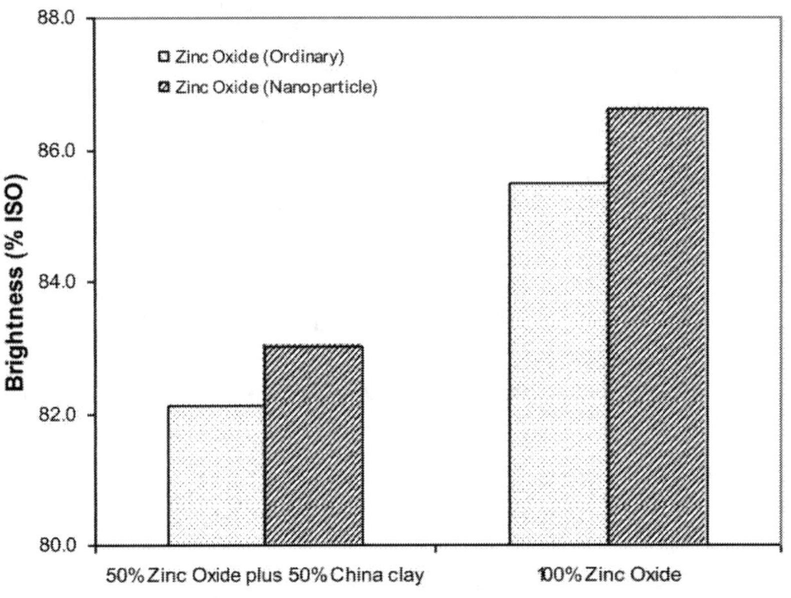

Fig. 1: Effect of Zinc Oxide nanoparticles on brightness of coated & calendered paper

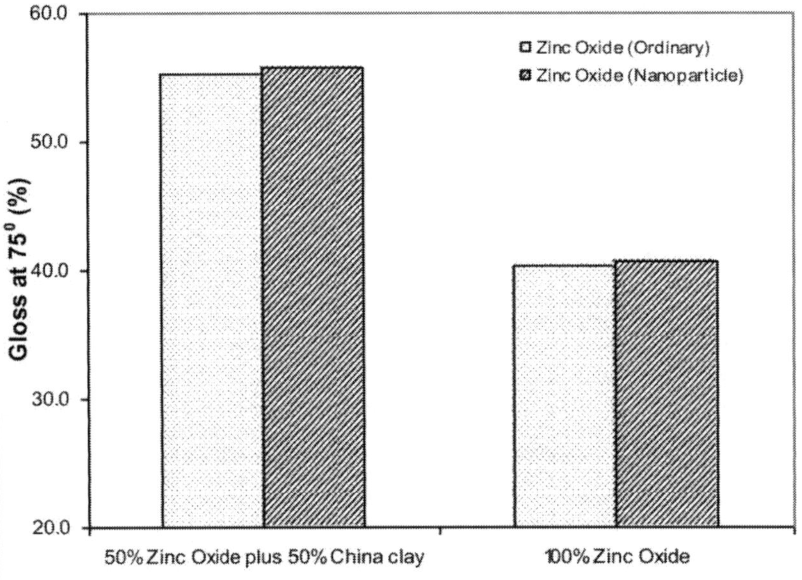

Fig. 2: Effect of Zinc Oxide nanoparticles on gloss of coated & calendered paper

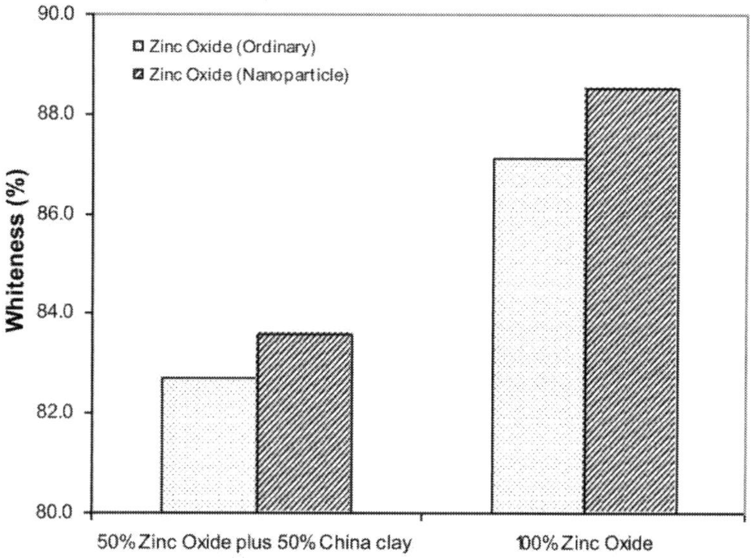

Fig. 3: Effect of Zinc Oxide nanoparticles on whiteness of coated & calendered paper

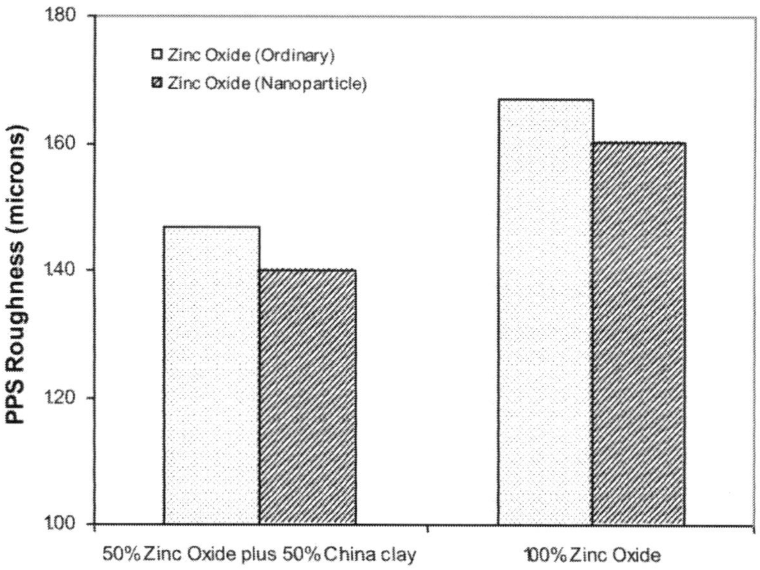

Fig. 4: Effect of Zinc Oxide nanoparticles on the roughness of coated & calendered paper

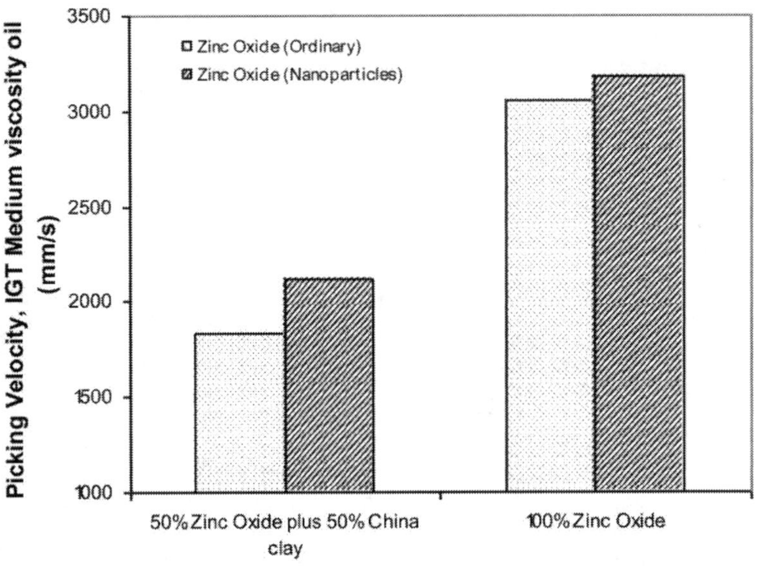

Fig. 5: Effect of Zinc Oxide nanoparticles on picking velocity of coated & calendered paper

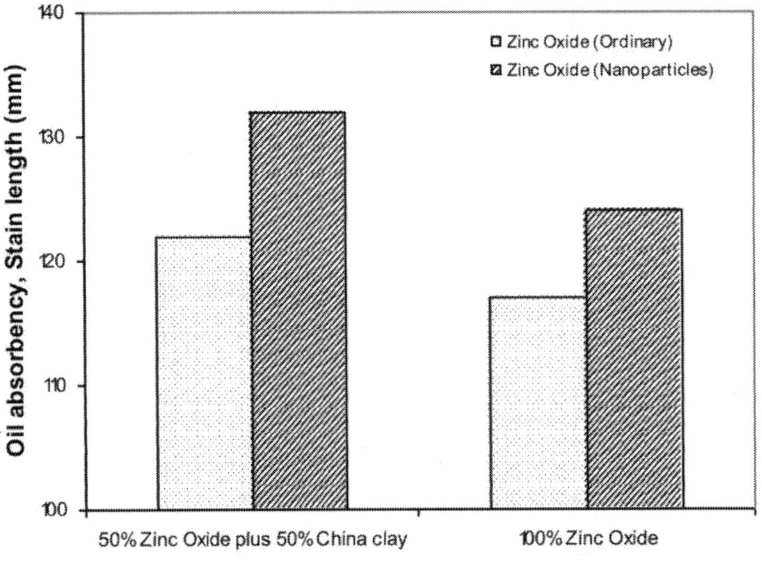

Fig. 6: Effect of Zinc Oxide nanoparticles on oil absorbency of coated & calendered paper

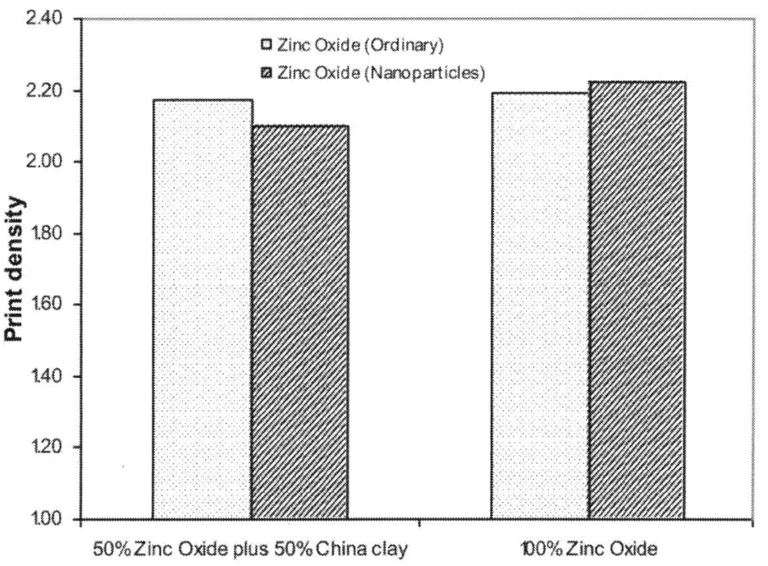

Fig. 7: Effect of Zinc Oxide nanoparticles on print density of coated & calendered paper

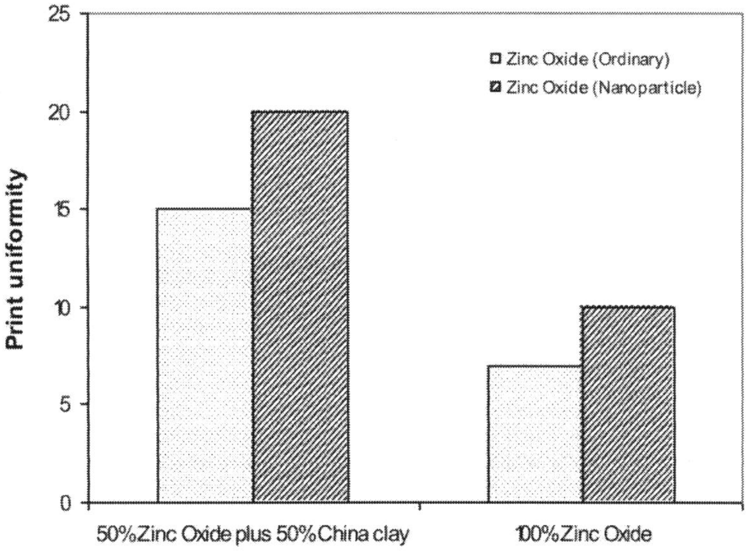

Fig. 8: Effect Zinc Oxide nanoparticles on print uniformity of coated & calendered paper

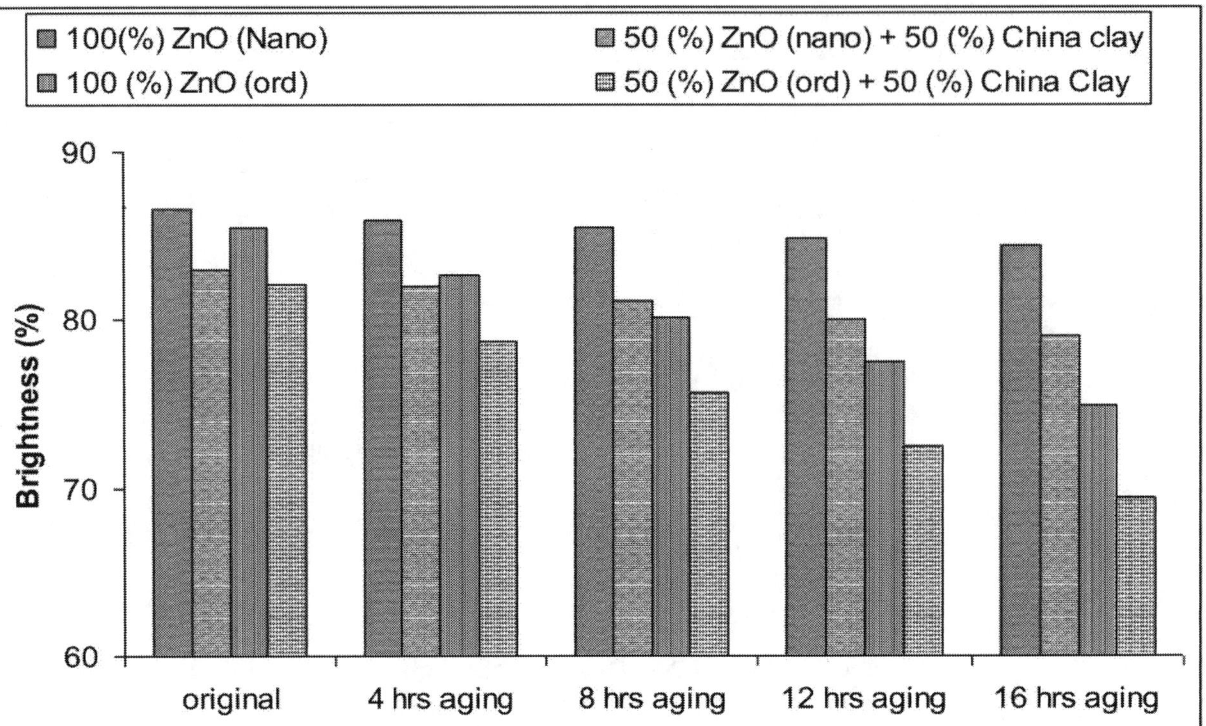

Fig. 9: Effect of zinc oxide nano particle on brightness reversion of coated and calendared paper

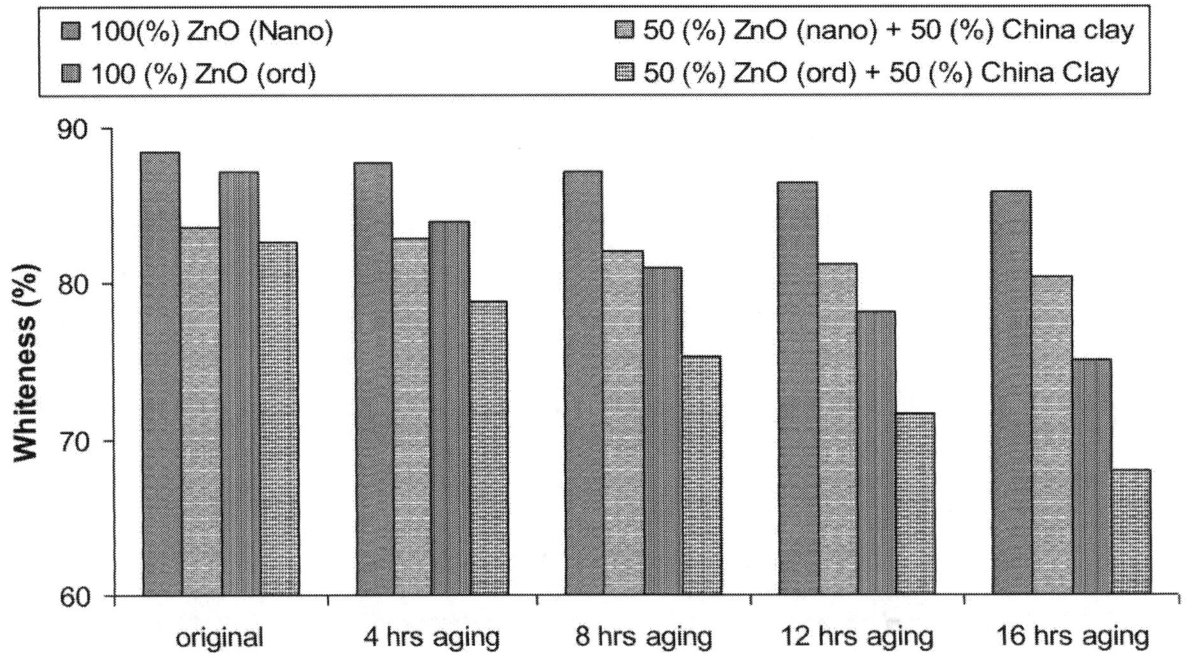

Fig. 10: Effect of zinc oxide nano particles on whiteness reversion of coated and calendared paper

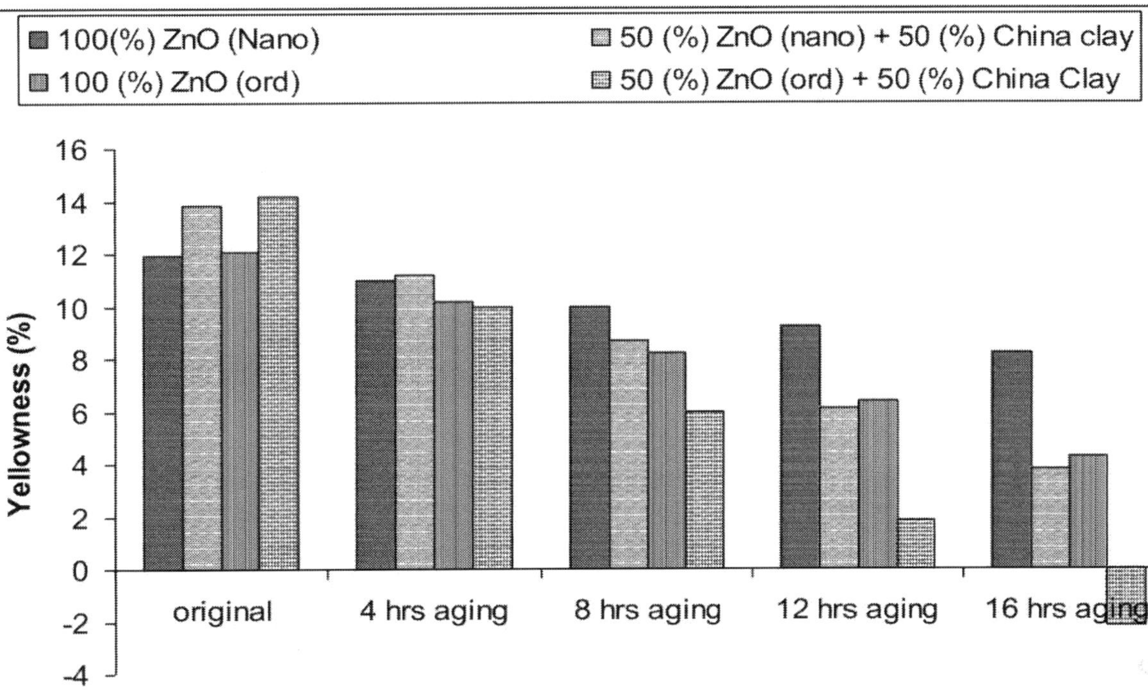

Fig.11: Effect of zinc oxide nano particle on yellowness reversion of coated and calendared paper

Table-I: Optical & printing characteristics of base paper used for coating

Property	Values obtained for the sample
Brightness (%)	88.5
Opacity (%)	80.0
Yellowness (%)	-5.13
Whiteness (%)	124.2
Gloss at 75^0 (%)	6.9
Porosity, Bendtsen (ml/min)	593

Table-II: Optical & printing characteristics of Zinc Oxide pigments used in coating

Property	Zinc Oxide (Ordinary)	Zinc Oxide (Nanoparticle)
Brightness (%)	97.8	96.2
Whiteness (%)	93.9	91.2
Yellowness (%)	1.64	1.70

Table-III: Different optical & printing characteristics of coated & calendared paper

Property	Pigment composition for coating			
	50% Zinc Oxide (Ord.) plus 50% China clay	50% Zinc Oxide (Nano) plus 50% China clay	100% Zinc Oxide (Ord.)	100% Zinc Oxide (Nano)
Coat weight (g/m^2)	12.7	12.6	12.8	12.4
Brightness (%)	82.1	83.0	85.5	86.6
Whiteness (%)	82.7	83.6	87.1	88.5
Gloss at 75°(%)	55.2	55.8	40.4	40.6
PPS Roughness (microns)	1.47	1.40	1.67	1.60
Oil absorbency, IGT Stain length (mm)	122	132	117	124

Picking velocity, IGT Medium viscosity oil (mm/s)	1830	2120	3060	3180
Picking Viscosity Velocity Product (VVP), IGT(Poise.m/s)	805	933	1346	1399
Print density	2.17	2.10	2.19	2.22
Print uniformity	15	20	7	10

Table-IV: Brightness reversion of coated and calendared paper after aging

Pigment Composition	Brightness (%)				
	Original	4 hrs aging	8 hrs aging	12 hrs aging	16 hrs aging
100(%) ZnO (nano)	86.6	86	85.5	84.9	84.4
50(%) ZnO (nano) + 50 (%) China Clay	83	82	81.1	80	79
100 (%) ZnO (ord)	85.5	82.7	80.2	77.5	74.9
50 (%) ZnO (ord) + 50 (%) China Clay	82.1	78.7	75.7	72.5	69.4

Table-V: Whiteness reversion of coated and calendared paper after aging

Pigment Composition	Whiteness (%)				
	Original	4 hrs aging	8 hrs aging	12 hrs aging	16 hrs aging
100 (%) ZnO (nano)	88.5	87.7	87.2	86.5	85.9
50 (%) ZnO (nano) + 50 (%) China Clay	83.6	82.9	82.1	81.2	80.4
100 (%) ZnO (ord)	87.1	84	81	78.1	75.1
50 (%) ZnO (ord) + 50 (%) China Clay	82.7	78.8	75.3	71.6	68

Table-VI: Yellowness reversion of coated and calendared paper after aging

Pigment Composition	Whiteness (%)				
	Original	4 hrs aging	8 hrs aging	12 hrs aging	16 hrs aging
100 (%) ZnO (nano)	11.9	11	10	9.2	8.2
50 (%) ZnO (nano) + 50 (%) China Clay	13.86	11.2	8.7	6.1	3.8
100 (%) ZnO (ord)	12.1	10.2	8.2	6.4	4.3
50 (%) ZnO (ord) + 50 (%) China Clay	14.2	10	6	1.8	-2.2

(599a) Harnessing Self-Assembly to Design Functionalized Nanotube-Lipid Hybrid Structures

Meenakshi Dutt[1], Olga Kuksenok[2], Michael J. Nayhouse[2], Steven R. Little[2] and Anna C. Balazs[2], (1) Chemical and Petroleum Engineering, University of Pittsburgh, Pittsburgh, PA, (2)Chemical Engineering, University of Pittsburgh, Pittsburgh, PA

Via Dissipative Particle Dynamics (DPD) approach, we study the design and creation of functionalized amphiphilic nanotube-lipid hybrid structures. Individual lipids are composed of a hydrophilic head group and two hydrophobic tails. Each bare nanotube encompasses an ABA architecture, with a hydrophobic shaft (B) and two hydrophilic ends (A). To allow controlled transport through the nanotube, we also introduce hydrophilic hairs at one or both ends of the tube.Our earlier investigations on nanotube-lipid bilayer interactions (Nanoscale, 2011, 3, 240) demonstrated that bare and single-end hairy nanotubes spontaneously penetrate and assume a trans-membrane position in the bilayer; this process is found to critically depend upon the membrane tension. On the other hand, the double-end hairy nanotubes are not unable spontaneously self-organize into the bilayer, and require the formation of a stable pore for its insertion. Based upon our earlier findings, we use two different approaches to generate the hybrid structures. (1) For the double-end hairy nanotube, we use the self-assembly of the amphiphilic lipids and the nanotubes in a hydrophilic solvent to create equilibrium hybrid structures such as a vesicle or a bicelle. The formation of a specific structure depends upon the concentrations of each component (M. Dutt et al. submitted). (2) For the bare and single-end hairy nanotubes, we add a nanotube into a solvent bath containing a pre-assembled vesicle and observe its spontaneous insertion into the vesicle bilayer to assume a transmembrane position. We sequentially add the nanotubes one at a time after the previous nanotube has been inserted (M. Dutt et al. submitted). In both these approaches, the nanotubes in the hybrid vesicles are found to self-organize in the bilayer into smectic-, tripod- or tetrapod-like structures. We also show that the nanotubes insertion and clustering within the vesicle strongly affects the vesicle shape in cases of a sufficiently large number of tubes. Ultimately, these nanotube-lipid systems can be used for making hybrid controlled release vesicles.

(532a) Ultra-Rapid Elimination of Biofilms On Metal and Ceramic Surfaces Via the Combustion of a Nanoenergetic Coating

Shramik Sengupta[1], Rajagopalan Thiruvengadathan[2], Byung-Doo Lee[1], Brandon M. Smith[1], Sachidevi Puttaswamy[1], Keshab Gangopadhyay[3] and Shubhra Gangopadhyay[3], (1)Biological Engineering, University of Missouri, Columbia, MO, (2)University of Missouri Columbia, Columbia, MO, (3) Electrical and Computer Engineering, University of Missouri, Columbia, MO

In Heat Exchangers and Cooling-Water Systems, which are an integral part of a wide variety of industrial processes, the presence of biofilms can reduce the effective heat transfer coefficient by as much as 50%. Biofilms consist of microbials colonies embedded on the surface of the equipment, and protected by a layer of extra-cellular polymeric substances secreted by the same microbes.

Removal of the biofilms requires that the equipment be taken "offline" and subjected to chemical and physical treatments that not only take a long time, but also require the use of toxic chemicals. Reducing the down-time for the equipment and the development of processes that do not require the use and disposal of toxic chemicals, is thus highly desirable from both economic and environmental points of view.

We present here a novel process that uses nano-energetic materials to eliminate biofilms in just a few seconds. The process involves spray-coating of an approximately 100 μm thick layer of nano-energetic material on top of the biofilm, allowing it to dry, and igniting the dried coating to incinerate the biofilm. The nanoenergetic material used is a two component mixture consisting of aluminum nanoparticles (fuel) dispersed in a THV220A (fluoropolymer oxidizer) matrix. Upon ignition, the Al nanoparticles react with THV 220A exothermically, producing high temperatures (> 2500K) for an extremely brief period (~100mS) that destroys the biofilm underneath, along with the bacteria harbored within. However, due to the small mass of nanoparticles present (~1 mg/cm^2 of coating), the total *amount* of heat produced is low (~0.1 KJ/cm^2) and the underlying metal or ceramic surface does not suffer observable damages. Since the byproducts of the process consist only of carbonaceous residue (that can be easily washed away) and gaseous combustion products, the process is not extremely fast, but also environmentally friendly.

See more of this Session: Green and Renewable Energy Aspects of Nanotechnology
See more of this Group/Topical: Topical C: Environmental Aspects, Applications, and Implications of Nanomaterials and Nanotechnology

(374b) From Embedded to Supported Metal/Oxide Nanomaterials: Thermal Behavior and Modelling of Structural Evolution At Elevated Temperatures

Stephanie B. Bubenhofer, Wendelin J. Stark and Robert N. Grass, Institute for Chemical and Bioengineering, ETH Zurich, Zurich, Switzerland

The relation between structure and function is of major importance at material interfaces which often excel in terms of chemical or physical effects. Therefore, multi-phase nanocomposites (e.g. core-shell, janus-shaped and carrier-supported nanoparticles) attract attention due to their possibility to contact two materials at the nanometer scale profiting from their highly increased interaction area.

Especially at metal/oxide interfaces, accounting for a majority of heterogeneous catalysts, the structure-function relationship has been studied intensively. High temperature, dry aerosol processes (flame spray synthesis) are common for building such composites. But the formation of key structural elements in such multi-component systems, depending on coagulation-order and solid-solid diffusion, is not fully understood.

Here, we present the synthesis of silica coated crystalline Pd nanoparticles (Pd in silica) in a single step flame spray pyrolysis process. The nanometer scale transformation of this core-shell material to nanocrystalline Pd supported on the amorphous silica matrix (Pd on silica) was achieved at temperatures exceeding 700°C. A physical, population balance based aggregation/diffusion model gives an understanding of the influence of temperature, matrix viscosity, particle sizes and concentrations on this transformation process. It enables morphological predictions (core-shell vs. carrier-supported particles) in nanocomposite formation processes such as flame spray synthesis. It further allows to understand the performance of multi-phase materials during thermal cycling in catalysis or fuel cells.

Figure 1. Morphological transformation of noble metal encapsulated in oxide (core-shell, left TEM image) nanoparticles to supported on oxide particles (right TEM image).

Reference :

S. Bubenhofer et al., J. Phys. Chem. C, 115(4), 1269-76 (2011).

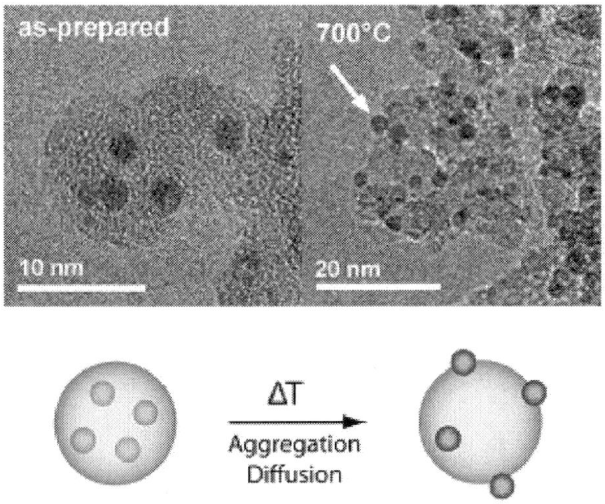

See more of this Session: Functional Nanoparticles and Nanocoatings on Particles I
See more of this Group/Topical: Particle Technology Forum

(217f) Forensic Analysis of Nano-Material Environmental Regulation

Coby A. Scher and Dennis L. Caputo, Quest Consulting, Inc., Bellaire, TX

Forensic Analysis of Nanomaterial Environmental Regulation

Coby A. Scher – Quest Consulting, Inc. (Presenter)

& Dennis L. Caputo – Quest Consulting, Inc.

Some considerable effort has recently been displayed by manufacturers, researchers, regulatory bodies, professional organizations, NGOs, and other interested parties addressing how nanomaterials and their manufacture should be environmentally regulated. In the USA, everyone including EPA, FDA, ABA, universities and a host of other entities (including companies) are trying to voice an opinion. The executive branch of the federal government with congress's approval and funding has created an investigatory group examining the regulatory potential. Internationally, the EU and ISO are investigating the subject of "nano-regulation". There is a body of thought within science and engineering termed forensics. In fact, the pure definition of forensic according to Webster's Collegiate Dictionary, 10th edition, is "1; belonging to, used in, or suitable to courts of judicature or to public discussion and debate 2: Argumentative, rhetorical 3: Relating to or dealing with the application of scientific knowledge to legal problems." All three parts of the definition are applicable to the issue at hand – the application of forensics to the environmental regulation of nanomaterials. It is the authors' contention that use of environmental forensics to encompass scientific investigative methods will provide the answer to what level of regulation is required. Forensic environmental science and engineering is a state of mind; an analytical approach to solving this problem. And it need not be complicated, just complete and scientifically defensible. This paper will update the current governmental movement toward environmental regulation of nanomaterials and apply a forensic approach to solving the problem of nanomaterial environmental regulation.

See more of this Session: Environmental Health and Safety (EHS) Concerns of Nanomaterials
See more of this Group/Topical: Topical C: Environmental Aspects, Applications, and Implications of Nanomaterials and Nanotechnology

(615g) Formation, Structure, and Poisoning of Catalyst Nanoparticles During Growth of Single-Walled Carbon Nanotubes

Juan C. Burgos and Perla B. Balbuena, Chemical Engineering, Texas A&M University, College Station, TX

Single-walled carbon nanotubes (SWCNTs) are generally synthesized by chemical vapor deposition (CVD) techniques through catalytic decomposition of hydrocarbons on metallic nanoparticles. An appropriate selection of CVD process variables has been suggested to be the key of preferential growth of SWCNTs over other kind of carbon species such as amorphous carbon, graphite, and graphene. Temperature, pressure, rate of addition of carbon precursor species and composition of the precursor gas phase are some of the CVD parameters to be adjusted in order to gain SWCNTs selectivity. However, the metallic catalyst nature and geometry are perhaps the most determinant variables, not only on the preferential growth of SWCNTs, but also on the definition of SWCNTs diameter and chirality. In this study, reactive molecular dynamics simulations are carried out with the objective of developing a thorough analysis of the metallic catalyst behavior during SWCNT synthesis, identifying the critical parameters that assure long catalytic life avoiding graphitic poisoning. We have shown that low metal-carbon interactions, high temperatures, and small catalyst particles tend to favor the cap lift-off over graphitic encapsulation.[1-2] Here we discuss how a substrate strongly attracting metal atoms may define the catalyst particle structure and the exposed facets for carbon deposition. Moreover, our simulations show that the metal-substrate strength influences the quality of the growing nanotube and may lead to formation of multiple-walled carbon nanotubes[3] on large diameter catalyst particles, in agreement with experimental results.[4] Finally, we present preliminary results on the clustering process of catalytic nanoparticles during the initial stages of the SWCNT synthesis process, where we observe simultaneous metal atoms agglomeration and catalytic activity. In addition, cluster sintering may occur once carbon structures are already formed on the nanoparticle surface, which may induce sudden diameter/chirality changes during growth.

References

(1) Ribas, M. A.; Ding, F.; Balbuena, P. B.; Yakobson, B. I. *The Journal of Chemical Physics* **2009**, *131*, 224501.

(2) Burgos, J. C.; Reyna, H.; Yakobson, B. I.; Balbuena, P. B. *The Journal of Physical Chemistry C* **2010**, *114*, 6952.

(3) Burgos, J. C.; Jones, E.; Balbuena, P. B. *The Journal of Physical Chemistry C* **2011**, *115*, 7668.

(4) Yamada, T.; Namai, T.; Hata, K.; Futaba, D. N.; Mizuno, K.; Fan, J.; Yudasaka, M.; Yumura, M.; Iijima, S. *Nat Nano* **2006**, *1*, 131.

See more of this Session: Synthesis of Graphene and Carbon Nanotubes: Kinetics, Mechanisms and Reactor Design
See more of this Group/Topical: Nanoscale Science and Engineering Forum

(176c) Controllable Fabrication of Clustered Quantum Dots for Time-Correlated Hyperspectral Studies

Rajasekhar Anumolu[1], Hyeong Gon Kang[2], Matthew L. Clarke[2], Jeeseong Hwang[2] and Leonard F. Pease III[3], (1)Chemical Engineering, University of Utah, Salt Lake City, UT, (2)National Institute of Standards and Technology, (3)Chemical Engineering and Pharmaceutics & Pharmaceutical Chemistry, University of Utah, Salt Lake City, UT

Optical properties of clustered core(CdSe)-Shell(ZnS) quantum dots (QDs) are investigated. Monomers, dimers, and trimers of QDs are generated through a self-induced droplet assembly using an electrospray differential mobility analyzer. A fixed number of QDs in the clusters is achieved through controlled electrostatic deposition. QD clusters composed of both homoclusters and heteroclusters are generated. Multimodal time-correlated hyperspectral studies of these QD clusters are carried out, simultaneously measuring the dynamic photoluminescence (PL) intermittencies, PL life times, and spectral diffusion revealing the unique properties of interacting QDs in QD clusters under photooxidation. Controlled assembly of a fixed number of QDs into single clusters and analysis of their dynamic PL properties leads to better understanding of the excitonic charge transfer (photonic to electronic energy transfer) during photooxidation in single and clustered QDs. The clustered QDs will provide unique optical characteristics which will be beneficial for applications in energy devices such as solar cells and devices with a long-term photostability. Analysis of homoclusters and heteroclusters reveals a distinct variation in their optical properties which will lead to interesting aspects for QD based energy devices.

See more of this Session: Nanostructured Materials for Electronic and Photonic Applications
See more of this Group/Topical: Materials Engineering and Sciences Division

(298e) Solution-Phase Adsorption of 1-Pyrenebutyric Acid Using Single-Wall Carbon Nanotubes

Reginald E. Rogers Jr., Travis I. Bardsley, Steven J. Weinstein and Brian J. Landi, Chemical & Biomedical Engineering, Rochester Institute of Technology, Rochester, NY

The time-dependent adsorption of aqueous 1-pyrenebutyric acid (PBA) onto single-wall carbon nanotubes (SWCNT) is studied theoretically and experimentally. Optical absorption spectroscopy was used to measure determine the amount of PBA adsorbed to the nanotubes as a function of time. Short time (≤ 3 hours) adsorption data obtained at various temperatures exhibit a linear dependence of adsorbed PBA with time, where the slopes of these curves also vary linearly with initial bulk concentration. Equilibrium analysis after 78 days is fit well by a Langmuir isotherm, and suggests favorable adsorption of PBA onto SWCNTs at low solution concentrations with a maximum adsorption of 0.27 mg PBA/mg SWCNT. A mass transport model was developed to describe the batch adsorption process, and agrees favorably with experimental results. The model suggests that the mass transfer process from the bulk to the SWCNTs is adsorption-limited, which enables an interpretation that the adsorption process is first order in kinetics. The Arrhenius activation energy can be directly related to the mass transfer coefficient and is found to be 20.3 kJ/mol further suggesting that the PBA binds to the SWCNT by physisorption. The model and experiments together provide a semi-empirical approach to streamline the study of adsorption behavior of organic molecules onto carbon nanotubes, which may be applicable to other systems of interest.

See more of this Session: Graphene and Carbon Nanotubes: Absorption and Transport Processes
See more of this Group/Topical: Nanoscale Science and Engineering Forum

(394c) Investigation of Mass Transport Properties of Microfibrillated Cellulose (MFC) Films

Matteo Minelli[1], **Marco Giacinti Baschetti**[2], Ferruccio Doghieri[2], Mikael Ankerfors[3], Tom Lindstrm[3], David Plackett[4] and Istvn Sir[4], (1)Department of Chemical Engineering, Mining and Environment Technology, University of Bologna, Bologna, Italy, (2)Department of Chemical Engineering, Mining and Environmental technology, University of Bologna, Bologna, Italy, (3)Innventia AB, Stockholm, Sweden, (4)Solar Energy Programme, Ris National Laboratory for Sustainable Energy, Technical University of Denmark, Roskilde, Denmark

Investigation of masstransport properties of microfibrillated cellulose (MFC) films

M.Minelli, M. Giacinti Baschetti, F. Doghieri, M. Ankerfors, T. Lindstrm, I.Sir, D. Plackett

Cellulose-based nanocomposites are now emerging as promising newmaterials suitable for a variety of applications. One source for such newnanocomposites is microfibrillated cellulose (MFC), produced by delamination ofcellulosic fibers in high-pressure homogenizers. MFC is an organic andbiodegradable reinforcement in polymer nanocomposites because of its highaspect ratio, good mechanical properties and ability to form networks: promisingproperties that may make this material suitable for use in a number ofindustrial fields and products.

The structure and transport properties of a four different filmsbased on two different generations of microfibrillated cellulose (MFC), aloneor in combination with glycerol as plasticizer, were investigated throughFE-SEM analysis and sorption or permeation experiments. FE-SEM revealed the existenceof complex structures in the different samples. A porous, closely packed fibernetwork, more homogeneous in the samples containing glycerol, wascharacteristic of the surface of MFC films; while film cross-sections presenteda dense layered structure with no evidence of porosity. Water vapor sorptionexperiments confirmed the hydrophilic character of these cellulosic materialsand showed a dual effect of glycerol, which reduced the water uptake at lowwater activity while enhancing it at high relative humidity, as showed inFigure 1. The observed trends of water sorption can be described considering thatboth physical adsorption on fiber surfaces and absorption in the celluloseamorphous phase takes place and therefore a dual-model sorption model wasemployed.

The water diffusivity (in the panel in Figure 1) in dry samples wasremarkably low for a porous material (10^{-11}-10^{-12}cm^2/s), confirming the existence of complex structures below thefilm surface; however, when water is present in the system, *D* rapidlyincreases with an exponential trend. Diffusivity is also definitely affected byplasticization, being higher for glycerol-containing samples.

Similar behavior was observed in permeation experiments. Dry MFCfilms showed excellent oxygen barrier properties, comparable with those ofpolymers usually considered suitable for ultra-high barrier applications.Indeed, when dry, the cellulose network presents a particularly compact andstiff structure, causing the observed barrier effect. Furthermore, although MFCfilms are characterized by a certain porosity, FE-SEM images of sample surfacessuggests that the pores are not substantially interconnected

When the water content in the membrane is raised, however, adramatic decrease in these properties was observed (Figure 2).

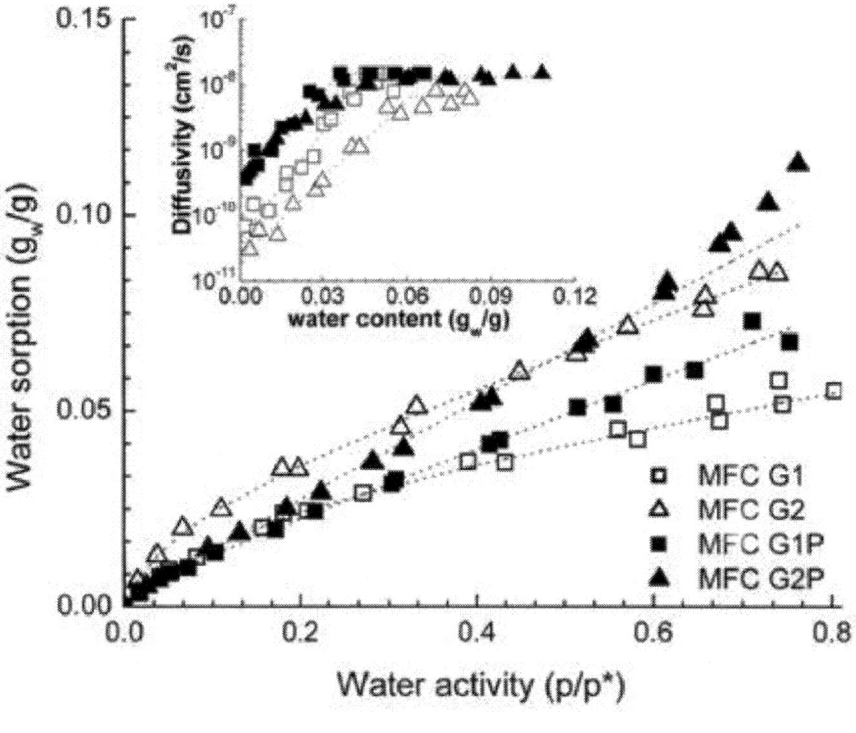

Figure 1. Water vapor sorption isothermsin MFC samples at 35°C, dotted lines are given by the best fitting of the data withthe dual-mode model equation. In the panel: water diffusivity at 35°C in the four different MFC samples as function of the average water concentration.

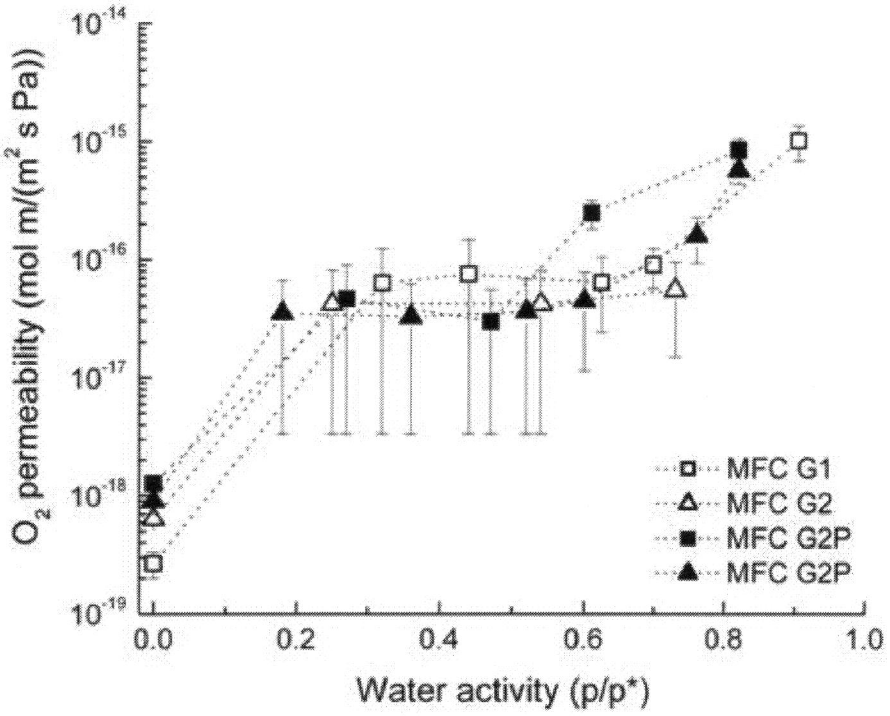

Figure 2. Oxygen permeability in thefour MFC samples at 35°C as a function of water activity.

See more of this Session: Nanocellulose-Based Materials and Composites
See more of this Group/Topical: Forest and Plant Bioproducts Division

(543d) Magnetic Resonance Imaging of Drug Release From 3D Poly(propylene fumarate) Scaffolds

Jonghoon Choi[1], Kyobum Kim[2], Taeho Kim[3], Guanshu Liu[4], Taeghwan Hyeon[3], Michael T. McMahon[4], Jeff W.M. Bulte[5], John P. Fisher[6] and Assaf A. Gilad[5], (1)Department of Chemical Engineering, The David H. Koch Institute for Integrative Cancer Research, Massachusetts Institute of Technology, Cambridge, MA, (2)Department of Bioengineering, Rice University, Houston, TX, (3) National Creative Research Initiative Center for Oxide Nanocrystalline Materials & School of Chemical and Biological Engineering, Seoul National University, 151-744 Seoul, South Korea, (4)F.M. Kirby Research Center for Functional Brain Imaging, Kennedy Krieger Institute, Baltimore, MD, (5)Russell H. Morgan Dept. of Radiology and Radiological Science, Johns Hopkins University, Baltimore, MD, (6) Fischell Department of Bioengineering, University of Maryland, College Park, MD

Implantable, three-dimensional (3D) polymer scaffolds have been drawing attention from various disciplines because of their unique properties. Efficient drug delivery, with a controlled drug-release manner at the targeted site, is one of the major applications for polymer scaffolds in biomedical research. The use of 3D poly(propylene fumarate) (PPF) scaffolds as a drug-release matrix is relatively new, but is a promising application in several aspects. First of all, the highly porous structure of PPF scaffolds provides sizable drug-loading spaces. Furthermore, a photo-crosslinked PPF scaffold is able to retain its initial porosity and mechanical properties for 18-32 weeks *in vitro*, and its degradability can be controlled by adjusting fabrication parameters, such as PPF molecular weight and photoinitiator content. Rapid improvements in magnetic resonance imaging (MRI) instrumentation and techniques have led to increased spatial resolution (up to 50-100 μm for rodent in vivo imaging). In addition, a variety of novel nanoparticles designed for MRI can enhance the MRI contrast even further, making it possible to image cellular and molecular events non-invasively and co-register these events with 3D anatomical structures. Here we report on the use of porous PPF scaffolds loaded with doxorubicin (DOX)-coated iron oxide and manganese oxide nanoparticles as a vehicle for sustained anti-cancer drug release. Using nanoparticles as a drug carrier contributes to more efficient loading of drug molecules onto the PPF scaffold. It also allows monitoring the release of drug-nanoparticle complexes from the PPF scaffold surface via changes in MRI contrast and absorption spectra in a media containing scaffold pieces. In addition, as a proof-of-concept experiment, it was demonstrated that the diamagnetic CEST contrast agent, protamine sulfate (PS), can also be used directly, without nanoparticle carriers, to monitor release from the PPF scaffolds by MRI. This report about drug-delivering 3D PPF scaffolds, with a sustained release rate and bimodal imaging (fluorescence and magnetic resonance) capabilities, suggests this new system may potentially be used in various biomedical applications.

See more of this Session: Nanotechnology for Biotechnology and Pharmaceuticals II
See more of this Group/Topical: Nanoscale Science and Engineering Forum

(217d) Human Blood Biocompatibility of Silver Nanoparticles

Jonghoon Choi[1], Vytas Reipa[2], Victoria M. Hitchins[3], Peter L. Goering[3] and Richard A. Malinauskas[3], (1)Department of Chemical Engineering, The David H. Koch Institute for Integrative Cancer Research, Massachusetts Institute of Technology, Cambridge, MA, (2)Biochemical Science, National Institute of Standards and Technology, Gaithersburg, MD, (3)Center for Devices and Radiological Health, Food and Drug Administration, Silver Spring, MD

Silver nanoparticles are increasingly being used in medical devices for their antimicrobial properties. Present toxicological techniques may not be sufficient to fully determine their safety for use in medical devices. In this talk, *in vitro* assessment of the hemolytic properties of silver particles in human blood will be discussed. In addition to the observed hemolysis, the possible mechanisms of red blood cell damage were explored in regard to the physical and chemical properties of silver particles. Four silver particle (two nano-sized, two micro-sized) preparations were dispersed in water and characterized using multiple analytical methods. Particle size and agglomeration were dependent on the suspension media. Under similar conditions to the hemolysis assay, i.e., with particles added to phosphate buffered saline and plasma, the size of the nanoparticles increased compared to particles suspended in water. Nano-sized particles exhibited higher hemolysis than micro-sized particles at mass concentrations greater than 220 ug/ml. While significant particle size changes occurred due to interaction with media components, the higher level of hemolysis observed with nanoparticles compared to micro-sized particles may be related to greater surface area, increased silver ion release, and direct nanoparticle interactions with red blood cells. These *in vitro* hemolysis results suggest that blood-contacting medical devices containing silver nanoparticles should be carefully evaluated for their hemolytic potential.

See more of this Session: Environmental Health and Safety (EHS) Concerns of Nanomaterials
See more of this Group/Topical: Topical C: Environmental Aspects, Applications, and Implications of Nanomaterials and Nanotechnology

(746e) Enzyme-Free Method for Amplifying Detection of Cytokines Released by Single Immune Cells Ex Vivo

Jonghoon Choi, Kerry Routenberg Love, Yuan Gong, Todd M. Gierahn and J. Christopher Love, Department of Chemical Engineering, The David H. Koch Institute for Integrative Cancer Research, Massachusetts Institute of Technology, Cambridge, MA

The human immune system comprises a diverse set of cells that contribute to a productive immunological response. Characterizing the functional behaviors for individual cells is critical for gaining knowledge on how the various subsets of cells of the immune system operate. Current bioanalytical technologies are challenged to resolve the diverse functional activities and identities of individual subsets of cells in part because of limited signal associated with events of interest, and the number of parameters concurrently measured in most assays. In this talk, we will present the integration of microengraving—a soft lithographic method for printing arrays of secreted proteins from thousands of single cells—with a new DNA-based strategy for amplifying signals from immunoassays and protein microarrays. Combination of this DNA-based method for enzyme-free amplification with microengraving extends the sensitivity of the microengraving for detecting proteins such as human cytokines and chemokines released by individual, activated cells. This method for amplification lowers the limits of detection for detecting secretions from single cells without requiring enzymes. The advantages of the method are that it is easy to integrate with standard protocols for processing microarrays, and it is extensible to increase the degree of multiplexed characterization. Our approach is potentially suited for use in a high-throughput, multiplexed, and directly *ex vivo* screening of immune cells of interest that may benefit early detection of disease and designing of immunotherapies.

See more of this Session: Nanotechnology and Nanobiotechnology for Sensors
See more of this Group/Topical: Nanoscale Science and Engineering Forum

(217e) Nanotechnology - What's So Big about the Small Stuff?

William C. Looney, Environment, AECOM, Milwaukee, WI

Nanotechnology - What's So Big about the Small Stuff?

Nanotechnology is "the Next Big Thing." It is a truly internationalphenomenon that "will have a major impact on the health, wealth and security ofthe world's people that will be at least as significant in this century asantibiotics, the integrated circuit, and manmade polymers." [1]

The market for nanotechnology-based products is expected to reach $2.5trillion by 2015, up from $147 billion in 2007.[2]

Nanotechnologypresents the promise of a diverse array of manufactured goods and products thatincorporate improved and innovative properties. Studies have determined that, due to their unique properties, nanomaterials (NM) have unpredictableproperties in biological systems and that the health implications from NMexposures are not yet known. Although numerous productsincorporating nanotechnology are here today, specific regulatory drivers for NM are still emerging. Companies producing or using NM must addressthe potential hazards of NM in the workplace to comply with existing workplace standards under the Occupational Healthand Safety Act. For example, the GeneralDuty Clause requires that employers provide each employee a place of employmentthat is free from recognized hazards that are causing or are likely to causedeath or serious physical harm and the Hazard Communication Standard (29 CFR1900.1200) provides workers with the right to know the hazards that they areexposed to in the workplace. Uncertaintyabout adverse health effects from exposures to NM and which work practices toadopt today should not impede efforts by companies to develop worker safetyprograms that minimize potential health effects. Such an approach can reduce the likelihood ofworker exposures to NM and the financial costs of subsequent liability claimsand litigation.

This presentation will provide an overview ofnanotechnology basics, an inventory of potentially applicable regulations andguidelines and the benefits of a proactive approach.

[1]National Science and Technology CouncilCommittee on Technology

[2]*Lux Populi Newsletter*, February 22, 2010, LuxResearch

Contact Information:

William C. Looney

Senior Program Manager

Director AECOM Nanotechnology Initiative

AECOM Environment

D 414.577.1362   C 262.893.0658

bill.looney@aecom.com

AECOM

11425 W. Lake Park Drive, Suite 100

Milwaukee, WI 53224

T 414.359.3030 F 414.359.0822

www.aecom.com

See more of this Session: Environmental Health and Safety (EHS) Concerns of Nanomaterials
See more of this Group/Topical: Topical C: Environmental Aspects, Applications, and Implications of Nanomaterials and Nanotechnology

(570c) Polymeric Prodrug Micelles for Gene-Directed Enzyme Prodrug Therapy

Alicia Jane Sawdon and Ching-An Peng, Chemical Engineering, Michigan Technological University, Houghton, MI

Acyclovir (ACV) is a guanosine-based prodrug used in anti-herpes virus therapy. Since ACV can be converted into its toxic form ACV-triphosphate by herpes simplex type-1 thymidine kinase enzyme (HSV-Tk), ACV has potential for cancer cell treatment via gene-directed enzyme prodrug therapy (GDEPT). This study showed that ACV can be used as an initiator for the ring-opening polymerization of ε-caprolactone (ε-CL) with the aid of tin(II) 2-ethylhexanoate ($Sn(Oct)_2$). Unlike commonly used grafting methods (i.e., conjugating ACV to biocompatible polymers), this innovative synthetic method simplified the process of making ACV-poly(ε-caprolactone) (ACV-PCL). The kinetics of the reaction was investigated by measuring the molecular weights of ACV-PCL at various time intervals via gel-permeation chromatography (GPC). The hydrophobic ACV-PCL polymer was then grafted with hydrophilic chitosan to form amphiphilic copolymers for the preparation of stable micellar nanoparticles. The characteristics of these polymers were determined by GPC, proton nuclear magnetic resonance (^{1}H NMR) and Fourier transform infrared spectroscopy (FTIR). ^{1}H NMR analysis demonstrated that the ring-opening polymerization of ε-CL was caused by the hydroxyl group of ACV. FTIR spectra reveal characteristic absorption peaks associated with PCL and ACV are simultaneously present in ACV-PCL polymers. The average sizes of the micelles were approximately 210 nm and their zeta potentials were positive due to the chitosan segment deployed on the outer layer of the polymeric micelles. The prepared polymeric prodrug micelles were used to challenge colon HT-29 cells with HSV-TK gene expression and showed toxic effects on cells by the MTT assay.

See more of this Session: Bionanotechnology for Gene and Drug Delivery III
See more of this Group/Topical: Nanoscale Science and Engineering Forum

(121e) Fundamental Mechanisms of DNA Self-Assembly

Daniel M. Hinckley, Gordon S. Freeman and Juan J. de Pablo, Department of Chemical and Biological Engineering, University of Wisconsin-Madison, Madison, WI

DNA is becoming increasingly more common as a building block for controlling the assembly of nanostructured materials. The predictable self-assembly of small DNA oligomers attached to building blocks such as nanoparticles leads to the reversible formation of structures with long-range order. Despite the proven effectiveness of DNA-directed self-assembly, the mechanisms underlying the hybridization of such DNA oligomers have, until recently, been largely unknown, even for naked strands in the bulk. Prior work using Transition Path Sampling and a coarse-grained DNA model shed some insight into possible key intermediates in the pathways of bulk hybridization [1] and surface hybridization [2]. Recently, the same model was used with Forward Flux Sampling and Langevin dynamics to determine the effect of salt concentration, supercooling, and oligomer length on the rate of bulk hybridization [3]. A general mechanism for DNA oligomer bulk hybridization was also proposed. In the present work, we expand on our previous results to examine the effect of geometry, as well as the effect of salt concentration, degree of supercooling, and sequence length and composition on hybridization rate. So doing allows us to construct a more complete description of the mechanism of DNA oligomer hybridization. This new understanding provides design principles that can be used by experimentalists to rationally design non-equilibrium kinetically-limited systems.

1. Sambriski, E. J., Schwartz, D. C., and J. J. de Pablo. "Uncovering Pathways in DNA Oligonucleotide Hybridization via Transition State Analysis." *PNAS*, 106, No. 43 (October 2009): 18125-30.

2. Hoefert, M. J., Sambriski, E. J., and J. J. de Pablo. "Molecular Pathways in DNA-DNA Hybridization of Surface-bound Oligonucleotides." *Soft Matter* 7, No. 2 (2011): 560.

3. Hinckley, D. M., Freeman, G. S., and J. J. De Pablo. A Mechanism of DNA Oligonucleotide Renaturation. *In preparation.*

See more of this Session: Self and Directed Assembly at the Nanoscale
See more of this Group/Topical: Nanoscale Science and Engineering Forum

(292c) Oxidation of Biomolecules by Emerging Inorganic Nanoparticles

Antonia Luna-Velasco[1], Reyes Sierra-Alvarez[2] and Jim A. Field[2], (1)Chemical and Environmental Engineering, University of Arizona, Tucson, AZ, (2)Chemical & Environmental Engineering, University of Arizona, Tucson, AZ

Oxidationof Biomolecules by Emerging Inorganic Nanoparticles

Antonia Luna-Velasco, Reyes Sierra-Alvarez, Jim A.Field

Departmentof Chemical and Environmental Engineering

Universityof Arizona, P.O. Box 210011, Tucson, Arizona 85721, USA

In recent years, inorganicnanoparticles (NPs) are finding increasingapplication in manufacturing and consumer products. The unusual properties ofthese emerging pollutants (e.g., very small size, high surface area andreactivity) has lead to growing public and regulatory concern about theirpotential toxicity. An important mechanism suggested for nanotoxicityis oxidative stress caused by reactive oxygen species (ROS). Testing thechemical generation of ROS by the interaction of NPs with biomoleculescould be a helpful tool for nanotoxicity screening. Inthis study, a rapid screening assay was developed to determine the chemicalgeneration of ROS by NPs through their interaction with biomoleculessusceptible to oxidation such as, L-3,4-dihydroxyphemylalanine(L-dopa) and 1,2-dihydroxybenzene (catechol). The assay utilized the dye dichlorofluorescin(DCFH), which is oxidized to its highly fluorescent product dichlorofluorescein(DCF) by ROS. Additionally, oxidation of proteins by NPs was evaluated using anELISA assay (based on detecting carbonyl groups in amino acid side chains).

Theinteraction of nanosized CeO_2 with L-dopa orcatechol significantly enhanced the oxidation of theROS indicator dye. Nanoscale CeO_2, $Fe_2O_3$and Fe^0 enhanced ROS production during the autooxidationof L-dopa by more than four-fold in reactions that were dependent on O_2. Mn_2O_3 NPs oxidized DCFHin the absence of O_2 or L-dopa, suggesting a direct oxidation of thedye. Mn_2O_3 NPs also significantly oxidized the protein bovineserum albumin indicating their potential reactivity with cell components. ROSgeneration by NPs (CeO_2, Mn_2O_3 and Fe^0)in aerobic aqueous suspensions was confirmed by electron paramagnetic resonance(EPR) signature for hydroxyl radicals with spin traps. The results taken as awhole indicate that NPs can generate ROS via chemical reactions with mediumcomponents and biomolecules susceptible to oxidation,such as L-dopa. The rapid assay with L-dopa and DCFH is a method proposed toscreen for chemical ROS production by NPs.

(570d) Combinatorial Synthesis of Chemically Diverse Core-Shell Nanoparticles for siRNA Delivery

Daniel J. Siegwart, Lutz Nuhn, Matthias Leiendecker, Kathryn A. Whitehead, Gaurav Sahay, Hao Cheng, Minglin Ma, Shan Jiang, Arturo Vegas, Robert Langer and Daniel G. Anderson, Department of Chemical Engineering, Massachusetts Institute of Technology, Cambridge, MA

Chemically diverse nanoparticles form the basis of a number of emerging applications in engineering, including drug and gene delivery vehicles. However, it is difficult to predict the optimal chemical and physical properties for delivery of a specific drug or biomolecule. A key challenge in nanoparticle development is the need to perform distinct synthesis, characterization, and formulation steps before performance can be evaluated. Combinatorial, high-throughput (HTP) methods can aid in the discovery of new nanoparticles for delivery, while simultaneously elucidating important design parameters of a given system for a targeted application.

We will present a "modular" approach that applied combinatorial Controlled Radical Polymerization (CRP)[1-2] and ring-opening polymerization (ROP) for the formation, characterization, and screening of more than 1,800 core cross-linked nanoparticles for siRNA delivery.[3] Combinatorial methods have driven small molecule drug discovery,[4] but have been less explored for the discovery of polymers[5] for specific applications in part due to the inherent challenge in synthesizing and characterizing such complex materials. Advances made in the controlled synthesis of polymers (particularly CRP and functional ROP methods) and in robotic technology motivated us to examine physical and chemical properties in relation to performance.

We aimed to prepare a library of structurally distinct nanoparticles comprised of cationic cross-linked cores and variable shells with precise control over the particle size, chemical composition, and architecture. siRNA can complex to or inside of the core, while the hairy arms provide a protective shield. To achieve the desired structure, we employed a method of cross-linking block copolymers prepared by Reversible Addition-Fragmentation chain Transfer (RAFT) polymerization and ROP of functional polyesters, building on previously published reports.[6-10] For the cross-linking, we sought an efficient and versatile reaction that would be permit library formation. We chose the ring-opening reaction of amines with epoxides[9,10] due to the availability of a large number of amines that can incorporate cationic charge into the core, and the efficiency and high rate of catalyst-free reaction.

Analysis revealed structure-function relationships and beneficial design guidelines, including a higher reactive block weight fraction, stoichiometric equivalence between epoxides and amines, and short hydrophilic shells. Cross-linkers optimally possess tertiary dimethylamine or piperazine groups, and potential buffering capacity. Covalent cholesterol attachment allowed for transfection *in vivo* to liver hepatocytes and blood immune cells.

To examine the effect of block length on nanoparticle formation, a series of block copolymers were synthesized by RAFT that varied in weight % of oligoethylene oxide methacrylate (OEOMA) and glycidyl methacrylate (GMA) blocks. After cross-linking with different amines, optimal particle size (10-100 nm) was obtained for most block copolymers, except for when the GMA block was less than 5 wt.% or greater than 70 wt.%. In addition, ROP of allylglycolide from monomethoxy PEG initiators was conducted to prepare biodegradable nanoparticles. Oxidation to epoxides was performed using m-chloroperoxybenzoic acid (mCPBA).

We found that the block length, concentration, temperature, and choice of cross-linker all play roles in the formation of the hairy nanoparticles. AFM images of nanoparticles formed from the cross-linking of $P(GMA_{39}$- b- $(OEO_5MA)_{41})$ with the linear aliphatic 1,10-diaminodecane show uniform particles with a diameter of 34 nm. Polyamines were also able to efficiently cross-link the block copolymers to form the hairy nanoparticle shape. The reaction of $P((OEO_5MA)_{25}$-b- $GMA_{10})$ with pentaethylenehexamine formed particles with a diameter of 21 nm as measured by DLS and 25.4 nm by AFM. To verify the core-shell structure, transmission TEM was employed. An examination of these particles revealed a higher electron density in the core, compared with the shell, indicating the core cross-linked morphology. Culturing fluorescently labeled particles with HeLa cells showed that they could be efficiently internalized.

After completion of preliminary studies, the synthesis were scaled up by reacting 19 different blocks copolymers with more than 100 amines in a combinatorial manner on a Symyx fluid handling robot to yield >1,800 structurally distinct core-shell nanoparticles. We characterized these nanoparticles using HTP GPC. We then screened them for siRNA complexation, siRNA delivery, and pDNA delivery *in vitro*. Many nanoparticles were able to complex tightly to siRNA and effectively deliver siRNA *in vitro*, silencing more than 80% of luciferase expression. The best performing nanoparticles were tested *in vivo* and were capable of silencing more than 50% of Factor VII in mice. In a separate experiments, these siRNA-loaded nanoparticles silenced >30% of CD45 protein expression in the blood monocyte/ macrophage (CD11b+ cells) population.

We believe that the development of this library of core-shell nanoparticles represents an important step forward for the HTP synthesis of polymers. We aimed to move beyond simple linear polymers by exploring CRP and ROP for the formation of complex architectures with great chemical diversity in the core and shell. This one pot synthetic method enables parallel generation of large libraries of chemically distinct core-shell materials. Notably, a number of materials were identified with both *in vitro* and *in vivo* utility and the common structural features of these materials suggest certain design criteria for creating future intracellular delivery agents. These trends, full synthetic details and characterization, and *in vivo* biodistribution and knockdown data will be presented.

1. Braunecker, W.; Matyjaszewski, K. *Prog. Poly. Sci.* **2007**, *32*, 93-146.
2. Moad, G.; Rizzardo, E.; Thang, S. *Aust. J. Chem.* **2005**, *58*, 379-410.
3. Whitehead, K.; Langer, R.; Anderson, D. *Nat. Rev. Drug. Disc.* **2009**, *8*, 129-138.
4. Geysen, H.; Schoenen, F.; Wagner, D.; Wagner, R. *Nat. Rev. Drug. Disc.* **2003**, *2*, 222-230.
5. Brocchini, S.; James, K.; Tangpasuthadol, V.; Kohn, J. *J. Am. Chem. Soc.* **1997**, *119*, 4553-4554.
6. Chen, D; Peng, H; Jiang, M. *Macromolecules* **2003**, *36*, 2576-2578.
7. Gao, H.; Matyjaszewski, K. *Prog. Poly. Sci.* ***2009,*** *34*, 317-350.
8. O'Reilly, R.; Hawker, C.; Wooley, K. *Chem. Soc. Rev.* **2006**, *35*, 1068-1083.
9. Tsarevsky, N.; Bencherif, S.; Matyjaszewski, K. *Macromolecules* **2007**, *40*, 4439-4445.
10. Van der Ende, A.; Kravitz, E.J.; Harth, E. *J. Am. Chem. Soc.* **2008**, *130*, 8706-8713.

See more of this Session: Bionanotechnology for Gene and Drug Delivery III
See more of this Group/Topical: Nanoscale Science and Engineering Forum

(395a) Template Directed Synthesis and Characterization of Tunable Mesoporous Polymer Resins

Manasa Sridhar, Krishna Reddy Gunugunuri, Panagiotis Smirniotis and Neville G. Pinto, Chemical Engineering Program, School of Energy, Environmental, Biological and Medicinal Engineering, University of Cincinnati, Cincinnati, OH

Ordered organic porous materials possess tremendous potential to serve as radically advanced adsorbents in the field of bio-separations. These biocompatible organic mesoporous polymer replicas are expected to offer unique selectivities based both on size and surface-biomolecule affinity, ultrahigh capacity, reversible adsorption and rapid adsorption/desorption kinetics. In our current study we have synthesized a broad spectrum of ordered mesoporous phenol formaldehyde polymer resins by a facile nanocasting method with siliceous Mesocellular Foams (MCFs) as hard templates. A systematic investigation into the textural, structural and framework properties has been carried out through BET, SAXS, TEM, ^{13}C NMR and FT-IR studies. Faithful replication of pore structure in the resultant mesoporous organic polymer materials by optimizing the resol precursor to template ratio and the thermopolymerization temperature has been achieved. Highly cross-linked phenolic resin frameworks with high surface areas (up to 1200 m^2/g) and large pore volumes (up to 1. 3 cm^3/g) have been synthesized. Well-defined gradation of pore characteristics in the replicas has been established by controlling the morphology of the inorganic hard template used. These interesting findings will be discussed in detail during the presentation.

See more of this Session: Nanoscale Structure In Polymers III
See more of this Group/Topical: Materials Engineering and Sciences Division

Effect of Shear Stress on Cytotoxicity of Silica Nanoparticles

Donghyuk Kim, Yu-Shen Lin, and Christy L. Haynes

In this work, nanotoxicity in the bloodstream was modeled as the cytotoxicity of sub-50 nm mesoporous silica nanoparticles to human endothelial cells was investigated under microfluidic flow conditions. Compared to traditional *in vitro* cytotoxicity assays performed under static conditions, unmodified mesoporous silica nanoparticles show higher and shear stress-dependent toxicity to endothelial cells under flow conditions. Interestingly, even under flow conditions, highly organo-modified mesoporous silica nanoparticles show no significant toxicity to endothelial cells. This paper clearly demonstrates that shear stress is an important factor to be considered in *in vitro* nanotoxicology assessments and provides a simple device for pursuing this consideration.

(172a) High Density Memory Devices Using Self-Assembled Gold Nanoparticle Arrays As Floating Gates

Girish Muralidharan[1], Navakanta Bhat[2] and Venugopal Santhanam[1], (1)Department of Chemical Engineering, Indian Institute of Science, Bangalore, India, (2)Department of Electrical Communication Engineering and Centre for Excellence in Nanoelectronics, Indian Institute of Science, Bangalore

Floating gate memory devices find applications in many portable electronic devices such as mobile phones, cameras, laptops etc. Floating gate memory devices consist of field effect transistor with dual gate architecture, namely, floating gate and control gate for its memory characteristics. Depending on the presence or absence of electrons in floating gate, memory state can be read as 0 or 1. Programming or erasing (P/E) operation is performed by controlling the flow of electrons from or to the substrate through the tunneling oxide. Until now, miniaturization was accomplished by the reduction of feature sizes with concomitant reduction of tunneling oxide thickness. However, with the feature size reaching sub-25 nm, further device scaling is hampered by: (1) significant cell-to-cell interference as the spacing between the floating gates has now become comparable to memory cell size, and (2) non-scalability of tunneling oxide thickness due to unreliable charge retention after several program/erase cycles. Use of nanoparticles as charge-trapping nodes will aid in both minimizing interference and enhancing reliability through compartmentalization of charges. Metal nanoparticles, in particular, gold nanoparticle arrays as floating gate will result in realizing superior endurance and retention characteristics, as it offers deeper potential well in comparison to its semiconductor counterparts. Thus, manufacture of next generation floating gate memory devices entails process development for fabricating large scale (wafer level), high density (~ 10^{12} particle per cm^{-2}) ordered gold nanoparticle arrays.

Here, we present fabrication and electrical characterization of metal-oxide-semiconductor (MOS) capacitor using self-assembled gold nanoparticle arrays as floating gates. We will also show a simple and cost-effective approach for fabricating wafer scale (~ 3 inch), close-packed arrays of gold nanoparticle by drop-casting gold colloids containing small amounts of excess ligand molecules on water surface. We will present the device charging, endurance and retention characteristics using both capacitance-voltage and electrostatic force microscopic measurements. Salient features of the MOS capacitor include: (a) less device-to-device variability over cm scale (b) reduced power consumption (c) good endurance characteristics and (d) reliable charge retention characteristics. This approach can be readily integrated into existing fabricating facilities for manufacturing next generation non-volatile memory devices.

See more of this Session: Nanoelectronic Materials
See more of this Group/Topical: Nanoscale Science and Engineering Forum

(746a) Longitudinal Plasmon Resonance of Gold Nanorods for Glucose Sensing

Sushil M. Pachpinde, Chemical Engineering Department, Michigan Technological University, Houghton, MI and Ching-An Peng, Chemical Engineering, Michigan Technological University, Houghton, MI

Gold nanorods are known for their unique surface plasmon resonance, a weak transverse band and a strong longitudinal band. In this study the strong longitudinal band was harnessed to detect glucose concentrations. First, cetyltrimethylammonium bromide (CTAB) stabilized gold nanorods modified by having negatively charged dextran sulfate form a coat electrostatically around the positively charged CTAB bilayer. The decorated gold nanorods were able to form clusters in the presence of concanavalin A (ConA), and thereby revealing apparent red shift of longitudinal plasmon band when compared to the band of original gold nanorods. The shift was found proportional to ConA concentrations. The cluster of ConA treated gold nanorods were demonstrated by the TEM micrographs. When ConA treated gold nanorods were mixed with glucose containing solution, the peak of longitudinal plasmon band shifts back towards original peak of the gold nanorods coated with dextran sulfate. The peak back shifting of longitudinal plasmon band is strongly affected by the concentration of glucose, therefore making the ConA associated gold nanorods a potential sensor system.

See more of this Session: Nanotechnology and Nanobiotechnology for Sensors
See more of this Group/Topical: Nanoscale Science and Engineering Forum

(321a) Melt and Solid-State Structures of Semicrystalline Linear ABC "Block-Random" Copolymers

Bryan S. Beckingham and Richard A. Register, Chemical and Biological Engineering, Princeton University, Princeton, NJ

The solid-state structure ofsemicrystalline block copolymers is set either by block incompatibility or bycrystallization of one or more blocks. A variety of solid-state morphologiesmay be observed depending on the block interaction strength, ranging fromspherulitic to confined crystallization within preexisting microphase-separateddomains[1]. We aim to explore crystallization from both homogeneousand microphase-separated melts and to characterize the resulting solid-statestructure for linear ABC "block-random" copolymers that incorporate asemicrystalline polyethylene endblock.

"Block-random"copolymers-represented generally by the structure (A_xB_{1-x})-(A_yB_{1-y}),where each of the two blocks is a random copolymer of monomers A and B, simplywith different fractions of A (x and y) present a convenient and usefulvariation on the typical block copolymer architecture, as the interblockinteractions and physical properties can be tuned continuously, via x and y,through the random block's composition. The ability to tune the effectiveinteraction parameter between the blocks continuously, allows for theorder-disorder transition temperature (T_{ODT}) to be tunedindependently of molecular weight using only two monomers, via the difference[2]between x and y. This flexibility makes block-random copolymers a versatileplatform for the exploration of polymer phase behavior and structure-propertyrelationships.

In previouswork we showed that in a cyclohexane/triethylamine mixture, narrow-distributioncopolymers of styrene (S) and isoprene (I) of any desired composition, with nomeasurable down-chain gradient could be synthesized. These random copolymers (S*r*I)have been successfully incorporated into well-defined symmetric blockcopolymers (I-S*r*I diblocks) and subsequent isoprene-selectivehydrogenation yields thermally stable hI-S*r*hI diblocks, whichself-assemble into well-defined lamellar morphologies with sharply-definedorder-disorder transitions, whose temperatures scale predictably with diblockmolecular weight. The use of S*r*hI in lieu of a styrene homopolymer blockeffectively dilutes the unfavorable contacts between the two blocks in thehomogeneous phase, reducing the effective χ; this reductionwas generally consistent with the mean-field °Copolymer equation[3].By studying these block-random diblocks we were able to extract solubilityparameters for both hI and S*r*hI. By comparison with the saturatedhydrocarbon polymers studied extensively by Graessley and coworkers,[4,5]we find that our determined solubility parameter for S*r*hI matched wellwith that calculated from homopolymer solubility parameters, for hI and S,using a volume-fraction-weighted average and, interestingly, that thedetermined value of dS*r*hI is close to that forpolyethylene,[5] suggesting potential miscibility between polyethyleneand S*r*I which would facilitate the formation of homogeneous melts. Here,we extend these hI-S*r*hI studies by adding a semicrystalline endblock,polyethylene (E), via hydrogenation of high-1,4-polybutadiene (B), to formlinear ABC (E-hI-S*r*hI) and ACB (E-S*r*hI-hI) "block-random"copolymers. Such ABC triblock copolymers are more complex than more typical ABdiblock copolymers, with two additional independent interaction parameters andtwo independent volume fractions.

We synthesize linear triblockcopolymers, B-I-S*r*I and B-S*r*I-I, via organolithium-initiatedanionic polymerization in cyclohexane. After polymerization of the butadieneblock, triethylamine is added facilitating the random copolymerization ofstyrene and isoprene while also increasing the vinyl content of the isopreneblock. Selective hydrogenation of the diene units with a Ni-Al catalyst yieldsa semicrystalline polyethylene endblock. For E-hI-S*r*hI triblock-randomcopolymer with block molecular weights of 30-14-14 kg/mol, small-angle x-rayscattering has revealed the formation of a well-ordered three-domain lamellarmelt from which crystallization of the hydrogenated polybutadiene(polyethylene) block proceeds. Currently this polymer is being studied afterorientation, compression in a channel die[6] above T_m(E),to further examine the solid state structure, to confirm confinement ofcrystallization into the melt morphology, and to reveal the orientation of thecrystal stems relative to the pre-existing lamellar interfaces formed byself-assembly in the melt. We are currently synthesizing, hydrogenating andcharacterizing additional E-hI-S*r*hI and E-S*r*hI-hI triblock-randomcopolymers to explore the resulting morphologies.

This work was generously supported by the National ScienceFoundation, Polymers Program (DMR-1003942).

References

(1) Loo,Y. L.; Register, R. A.; Ryan, A. J. *Macromolecules*, **2002,** *35*,2365-2374.

(2) Smith,S. D.; Ashraf, A.; Satkowski, M. M.; Spontak, R. J. *Polym. Prepr. (Am. Chem.Soc., Div. Polym. Chem.)* **1994**, *35(1)*, 651-652.

(3) Graessley,W. W. *Polymeric Liquids & Networks: Structure and Properties*;Garland Science: New York, 2004, pp. 341-407.

(4) Graessley,W.W.; Krishnamoorti, R.; Reichart, G. C.; Balsara, N. P.; Fetters, L. J.;Lohse, D. J. *Macromolecules* **1995**, *28*, 1260-1270.

(5) Krishnamoorti,R.; Graessley, W. W.; Dee, G. T.; Walsh, D. J.; Fetters, L. J.; Lohse, D. J. *Macromolecules***1996**, *29*, 367-376.

(6) Lee,H. H.; Register, R. A.; Hajduk, D. A.; Gruner, S. M; *Polymer Engineering andScience*, **1996,** *36*, 1414-1424.

(420i) Gold Nanoparticle-Based Biosensor for Colorimetric Detection of Helicobacter Pylori In Water

Mayra del Pilar Quiroz, Laura M. Calle, Juan R. Reyes, Angel A. Galvis and Watson L. Vargas, Department of Chemical Engineering, Universidad de los Andes, Bogot, Colombia

Bottom–up nanotechnology approaches offer a wide array of nanoparticles (NPs) with special interest in biosensing. Gold nanoparticles (Au-NPs) have been the focus of attention in recent years due to their unique size and shape-dependent properties (including, optical, electrical, magnetic, catalytic, mechanical, and chemical) which make them interesting for biological applications. Therefore, biosensing platforms based on the application of Au-NPs are of great interest and are actively being pursued.

In this study, a colorimetric biosensor has been developed to give a rapid response for the presence or absence of *Helicobacter pylori* in water based on differences in electrostatic properties of single- and double-stranded oligonucleotides (ssDNA and dsDNA), no functionalization of the gold, the probe, or the target DNA is involved. The biosensor operates in two steps; the first step involves hybridization of a DNA probe with a complementary ssDNA sequence of the target pathogen, and second; a visual detection from colloidal gold particle aggregation for a positive result due to surface plasmon resonance which is greatly affected by agglomeration. DNA probe, target DNA and colloid amounts were optimized in order to improve the colorimetric readout.

Target DNA was extracted from a *Helicobacter Pylori* culture and denatured to obtain ssDNA, a lyophilized oligonucleotide complementary sequence was purchased and used as received, salt solution of given ionic strength was used to screen the repulsive interactions of Au-NPs and induced the aggregation. Colloidal gold was synthesized by means of citrate reduction of $HAuCl_4$ (Turkevich method). Gold nanoparticles were characterized by Scanning Electron microscopy (SEM), Ultraviolet-visible spectroscopy, Fourier transform infrared spectroscopy (FTIR), X-ray diffraction (XRD) and Atomic force microscopy (AFM). This simple colorimetric hybridization assay has also been integrated into a disposable microfluidic device for simple widespread use. The results show that this relatively simple assay is surprisingly robust in its ability to detect the presence of the target pathogen.

See more of this Session: Poster Session: Nanoscale Science and Engineering
See more of this Group/Topical: Nanoscale Science and Engineering Forum

Reduced-order model development for multiphase flow through patterned, orthotropic, and structured porous materials

Scott A. Roberts and P. Randall Schunk, Sandia National Laboratories, Albuquerque, NM

Flow through thin porous media occurs in many industrial and consumer processes. One example of this is the nano-imprint lithography process, where picoliter size drops are squeezed under a porous template. This template may have regions of closed pores, where liquid will be pulled into the template but not spread, and open pores such as channels, where liquid can spread through the template, often in specifically engineered directions. These pores are structured, with length scales on the order of tens of nanometers, so capillary forces are incredibly strong, influencing the bulk flow under the template and the deformation of the template itself.

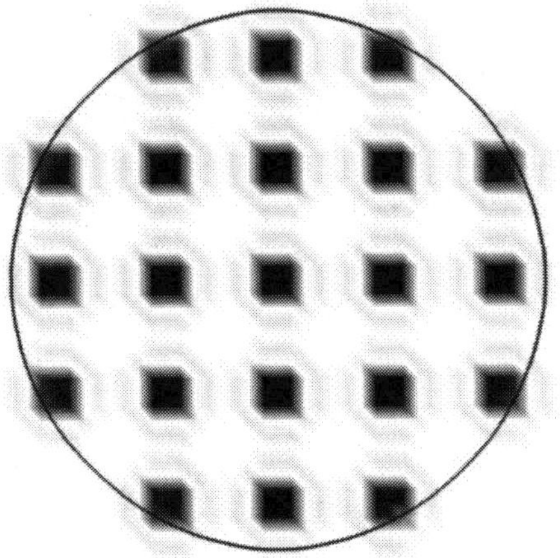

Figure 1: Illustration of a liquid drop spreading under a template. Dark regions represent areas of the template saturated with liquid

In this presentation we detail a mesoscale model for the porous and multiphase bulk flows, including fluid-structural interactions. Bulk flow is modeled with multiphase lubrication hydrodynamics using a level-set field to track fluid-air interfaces. The individual features of the porous material are coarse-grained and modeled by an analytical structure representation. The porous flow model is based on Darcy flow, but with the order reduced by integrating through the thickness of the porous material, taking into account the representation of the structure. The coupled problem is solved using the finite element method, with 2-D shell elements for the bulk and porous flows and 3-D continuum elements for the structural deformation. This model is used to investigate model applications, including air entrapment and fill rate in nano-imprint lithography processes.

* Sandia is a multiprogram laboratory operated by Sandia Corporation, a Lockheed Martin Company, for the United States Department of Energy's National Nuclear Security Administration under Contract DE-AC04-94AL85000.

(420c) Influences of External Magnetic and Gravity Fields On Self-Assembly Lattice Structures Composed of Cubic Hematite Particles

Masayuki Aoshima, Yuki Satoh and Akira Satoh, Machine Intelligence & System Engineering, Akita Prefectural University, Akita, Japan

We have carried out Monte Carlo simulations in order to investigate influences of external magnetic and gravity fields on the self-assembled lattice structures of a monolayer composed of cubic hematite particles. We emploied the hematite particle model which is a cubic rigid body, and has a magnetic moment along to the diagonal direction at the particle center. The energies of interaction between particle and both the magnetic and in-plane gravity fields, and also the energy of magnetic dipole-dipole interaction are taken into account in the simulation. The obtained results are summarized as follows: Particles are integrated near the bottom wall and form ordered lattice structures under relatively weak gravity field. For an uniform magnetic field is applied in an opposite direction of the gravity field, oblique and square lattice structures can be formed depending on the barnce between energies of the magnetic and gravity fields. For the magnetic field direction along to the upper-right (45° from the *x*-axis), well-ordered oblique lattice structures are formed and they have different lattice constant from the first-mentioned oblique lattice for the magnetic field along the y-axis direction. It suggest that relative angle between the magnetic and the gravity fields is also important for the lattice structures.

See more of this Session: Poster Session: Nanoscale Science and Engineering
See more of this Group/Topical: Nanoscale Science and Engineering Forum

(744c) Conductive and Shape Reformable Polymeric Nanofibers Entrapped with QDs As a Material for Enzyme Stabilization

Ee Taek Hwang[1], Rameshwar Tatavarty[1], Hyun Lee[1], Jee-Woong Park[1], Jungbae Kim[2] and Man Bock Gu[1], (1)College of Life Sciences and Biotechnology, Korea University, Seoul, South Korea, (2) Department of Chemical and Biological Engineering, Korea University, Seoul, South Korea

In this study, we report the structure and shape of new polymeric nanofibers entrapped with QDs based on its unique properties, along with its very successful implementation as an enzyme carrier. Electrospinning was employed to obtain uniformly distributed Cadmium Selenide (CdSe)/Zinc Sulfide (ZnS) core shell quantum dot (QD) encapsulated nanofibers in polymeric mixture of polystyrene (PS) and polystyrene-co-maleic anhydride (PSMA) copolymer. The introduction of QDs into the PS–PSMA nanofibers was found to bring electrical conductivity on them. In addition, uniformly distributed QDs within the polymer matrix seem to induce the high degree of compactness and shape rigidity in the nanofibers, and resulted in the efficient enzyme immobilization. Therefore, after immobilization, the esterase enzymes coated on the QDs–nanofibers showed good stability with no loss of enzyme activity even after being recycled for more than ten times. The present study demonstrates the successful application of shape reformable polymeric nanofibers entrapped with QDs for the first time, and so it is anticipated that this technology can be employed in various enzyme applications due to its facile shape control feature in the fabrication of the nanofibers for enzyme reactor, and applied for electrical conductance or gas based sensing applications.

See more of this Session: Nanoscale Science and Engineering In Biomolecular Catalysis II
See more of this Group/Topical: Nanoscale Science and Engineering Forum

(615e) Chiral-Selective Growth of Single-Walled Carbon Nanotubes by Lattice Matching Criterion

Debosruti Dutta[1], Wei-Hung Chiang[2], R. Mohan Sankaran[2] and Venkat R. Bhethanabotla[1], (1) Department of Chemical and Biomedical Engineering, University of South Florida, Tampa, FL, (2) Chemical Engineering, Case Western Reserve University, Cleveland, OH

Chiral-selective growth of SWCNTs on bimetallic nanoparticles using lattice matching criterion

The synthesis of CNTs with predefined chirality and, therefore, functionality, remains one of the ultimate goals of nanoscience because of its potential impact on a wide-range of emerging technologies including nanoelectronics, renewable energy, nanocomposites, sensor and medical. Recently, we showed that changing the composition of Ni_xFe_{1-x} bimetallic nanoparticles via a novel microplasma synthesis route resulted in chiral-selective growth of specific nanotube chiralities. In this work, we used density functional theory calculations to explain and predict how the catalyst structure determines the chirality of as-synthesized SWCNTs. Tuning composition will perturb the crystal structure, and affect the lattice matching of the nanoparticle with certain chiralities of the SWCNT, promoting chiral-selective growth. To understand the effect of compositional tuning and perturbation to the crystal structure of a bimetallic catalyst on SWCNT chirality, we studied the energetics of representative chiralities such as the (8,4), (6,5), (9,4), (8,6), (7,5) and (10,3) nanotubes, which are predominantly seen in our experimental results. We observe that a small increase in the average Ni-Ni bond length resulted in the increased stability of the (8,4) and (6,5) nanotube caps adsorbed on the Ni (111) surface whereas the (9,4) and (8,6) caps were destabilized. Remarkably, the (8,4) and (6,5) nanotubes are predominatly observed when the CVD growth took place on the $Ni_{0.27}Fe_{0.73}$ bimetallic nanoparticle which have a distorted fcc structure with a slightly increased Ni bond length. The findings strengthen our belief that chiral-selective growth of SWCNTs can be obtained by controlling the constituent atoms and their composition that results in a change in the lattice parameter of the bimetallic nanoparticle.

See more of this Session: Synthesis of Graphene and Carbon Nanotubes: Kinetics, Mechanisms and Reactor Design
See more of this Group/Topical: Nanoscale Science and Engineering Forum

(615a) High-Throughput Production of Graphene Nanostructures (nanoribbons and quantum dots) with Controlled Dimensions and Smooth Edge Terminations

Nihar Mohanty and **Vikas Berry**, Chemical Engineering, Kansas State University, Manhattan, KS

Graphene nanostructures (GNSs): narrow strips (*nanoribbons: GNRs*) or nano sheets (*quantum dots: GQDs*) of graphene with dimensions less than 100 nm, have attracted attention of experimentalists and theorists alike owing to their exceptional properties which include widely tunable bandgap and relativistic quantum particle nature of charge carriers leading to potential application in electronics, optics, spintronics and quantum computation. Even though numerous theoretical studies have been carried out for characterizing the GNSs, experimental studies have lagged behind. The current strategies of fabrication of GNSs, via the top-down lithographic, sonochemical and chemical techniques or via the bottom up chemical condensation and high temperature sublimation based synthesis are plagued by either low throughput or they produce GNSs with limited control on sizes and / or edge crystallographies. The sensitive dependence of the intrinsic electronic and optical properties of the GNSs on their spatial dimensions and the edge crystallography mandates a high throughput production strategy with a high degree of control on the size and the edge crystallography for most practical applications. In this talk we would demonstrate a novel nanotome (ultra-microtome) based fabrication of GNSs with controlled spatial dimensions and smooth lattice terminations from the diamond knife-based cleavage of HOPG blocks. With an overall efficiency of ~ 80 %, we achieved production of ~ 10^8 GNRs / hour / HOPG / nanotome (and ~ 10^5 GQDs / hour / HOPG / nanotome) which equals to or surpasses most current GNR (or GQD) production strategies. The average Raman $I_{D/G}$ ratios for our GNSs (for ~ 15 nm wide GNRs ≈ 0.25 – 0.28; for ~ 30 X 30 nm GQDs ≈ 2) are one of the lowest reported till date for similar structures indicating smoother edge terminations. We will present a detailed structural, chemical, electrical and optical characterization of the as-produced GNSs.

Further we will demonstrate the following:

(1) the employment of our narrow, pristine GNRs for fabricating flexible, semiconducting thin-films with the highest mobilities reported till date (~ 20 cm^2/VS) and high bandgaps (~ 35 meV for ~ 15 nm wide GNR-based thin-film);

(2) the facile tunability of the optical properties (optical bandgaps and photoluminescence emission spectra) of the GQDs via dimension / shape control.

(531g) Impermeable Graphenic Wrapping of Bacteria

Nihar Mohanty[1], Ashvin Nagaraja[2], Monica Fahrenholtz[1], Daniel L. Boyle[3] and **Vikas Berry**[1], (1) Chemical Engineering, Kansas State University, Manhattan, KS, (2)Electrical Engineering, Kansas State University, Manhattan, KS, (3)Division of Biology, Kansas State University, Manhattan, KS

Transmission electron microscopy (TEM) of hygroscopic, permeable, and electron-absorbing biological cells has been an important challenge due to the volumetric-shrinkage, electrostatic charging, and structural degradation of cells under high vacuum and fixed electron beam. In this talk we would demonstrate that the bacterial cells can be encased within a graphenic chamber to preserve their dimensional and topological characteristics under high vacuum (10-5 Torr) and beam current (150 A/cm2). The strongly-repelling p-clouds in the interstitial sites of graphene's lattice reduces the graphene-encased-cell's permeability from 7.6 - 20 nm/s to 0 nm/s. The C-C bond flexibility enables conformal, hermetic encasement of cells. Additionally, graphene's high Young's modulus retains cell's structural integrity under TEM conditions, while its high electrical and thermal conductivity significantly abates electrostatic-charging. This graphenic encasement approach will facilitate real-time TEM imaging of fluidic samples and potentially biochemical activity.

See more of this Session: Graphene and Carbon Nanotube Based Devices
See more of this Group/Topical: Materials Engineering and Sciences Division

(470c) Molecular and Hybrid Solution Processible Thermoelectrics

Rachel A. Segalman[1], Shannon Yee[2], Boris Russ[2], Nelson Coates[2] and Jeffrey Urban[3], (1)Department of Chemical and Biomolecular Engineering, University of California, Berkeley, Berkeley, CA, (2)University of California at Berkeley, Berkeley, CA, (3)Lawrence Berkeley National Laboratories, Berkeley, CA

Thermoelectric materials for energy generation have severaladvantages over conventional power cycles including lack of moving parts,silent operation, miniaturizability, and CO2 freeconversion of heat to electricity. Excellentthermoelectric efficiency requires a combination of high thermopower(S, V/K), high electrical conductivity (σ, S/cm), and low thermalconductivity (κ, W/mK). To date the bestmaterials available have been inorganic compounds with relatively low earthabundance and highly complex, vacuum processing routes (and hence greaterexpense), such as Bi_2Te_3. Molecular materials andhybrid organic-inorganics bring the promise ofinexpensive, solution processible, mechanicallydurable devices. While highlyconductive polymers are now common place, theygenerally demonstrate low thermopower. Our work onmolecular scale junctions demonstrates that nanostructuringof organics allows them to act as thermionic filters between inorganicjunctions which can lead to enhanced thermoelectric properties. We have takeninspiration from this fundamental understanding to design material systems inwhich we combine a high electrical conductivity, low thermal conductivitypolymer with a nanoparticle that contributes high thermopower. Additionally, the work functions of the two materials are well-aligned which introduces the possibility of thermionicfiltering at the interface and an additional boost to the power factor. The combination of these effects results in a new hybrid,solution processible material with a thermoelectricfigure of merit within an order of magnitude of the Bi_2Te_3. In this talk, I will discuss both the use ofthermoelectric measurements to gain insight to molecular junctions and how thisinsight translates to design principles for polymer and hybrid thermoelectrics.

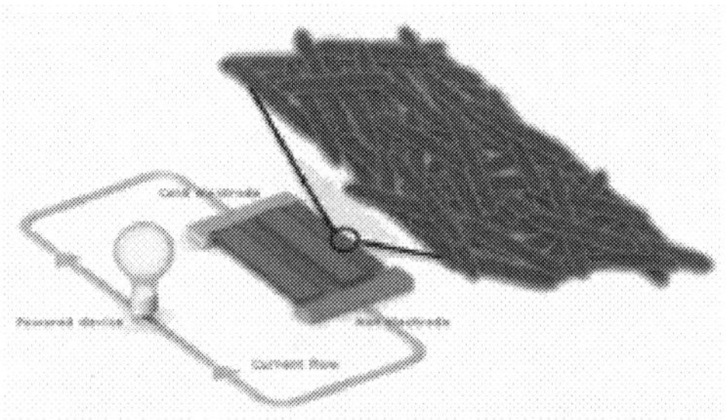

See more of this Session: Nanomaterials for Thermal-to-Electric Conversion I
See more of this Group/Topical: Topical 5: Nanomaterials for Energy Applications

(298d) Heat Transfer Through a Dual-Walled Carbon Nanotube

Khoa N. D. Bui, Huong Nguyen, Alberto Striolo and Dimitrios V. Papavassiliou, School of Chemical Biological and Materials Engineering, The University of Oklahoma, Norman, OK

Molecular Dynamics simulations are used to investigate the resistance to heat transfer between the walls of dual-walled carbon nanotubes (DWNTs). Recent computational results have indicated that the carbon nanotube to carbon nanotube thermal boundary resistance (TBR) can be as high or higher than the TBR between a nanotube and the surrounding matrix (e.g., epoxy or octane) in these case of composites [1-4]. This generates the question of whether heat transfer in multi-walled carbon nanotubes occurs through only one nanotube (most likely the outer nanotube) only or not. We discuss here results for DWNTs composed with the same chirality nanotubes and with different chirality between the inner and the outer tubes.

We conclude that, for the same intertube distance, the dual-walled with different chirality between the inner and the outer tube conducts heat better in the radial direction than the one with the same chirality. In addition, the Kapitza resistance appears to be strongly dependent on the direction of heat flux as it travels across the interface.

References:

[1] Bryning, M. B. Milkie, D. E. Islam, M. F. Kikkawa, J. M. Yodh, A. G., Applied Physics Letters, 87 (16), 2005.

[2] Shenogin, Sergei Xue, Liping Ozisik, Rahmi Keblinski, Pawel Cahill, David G., J. Applied Physics, 95 (12), 2004.

[3] Khoa Bui, Hai M. Duong, Alberto Striolo, and Dimitrios V. Papavassiliou, J. Phys. Chem. C, 115 (10), 2011.

[4] Khoa Bui, Brian P. Grady and Dimitrios V. Papavassiliou, Chemical Physics Letters, doi:10.1016/j.cplett.2011.04.005, 2011.

See more of this Session: Graphene and Carbon Nanotubes: Absorption and Transport Processes
See more of this Group/Topical: Nanoscale Science and Engineering Forum

(346a) Core-Shell Nanomagnets for Magnetic Chemistry and Precious Metal Recovery: Long-Term Stability In Acids and Organic Solvents

Christoph M. Schumacher, Robert N. Grass, Alexander Schaetz and Wendelin J. Stark, Institute for Chemical- and Bioengineering, ETH Zurich, Zurich, Switzerland

Core-shellnanomagnets for magnetic chemistry and precious metalrecovery: Long-term stability in acids and organic solvents

ChristophM. Schumacher, Robert. N. Grass, Alexander Schtz,

*Wendelin J. Stark**

**Institute for Chemical- and Bioengineering,*

Department of Chemistry and applied Biosciences,

ETH Zurich,HCI E 107, Wolfgang-Pauli-Strasse 10,

8093 Zurich,Switzerland, email: wstark@ethz.ch

The useof magnetic nanoparticles has gained increasing attention in a multitude ofapplication areas over the past years. Comparably high and readily availablespecific surfaces in combination with the ease of separation make these kindsof materials promising candidates as precious metal extraction agents (Rossier,2009) and in magnetic chemistry where a reagent is linked to a nanomagnet, enabling fast separation (Grass, 2007). The chemicalstability in conjunction with a high saturation magnetization is decisiveparameter for a broad range of practicability.

In thefollowing work we examine the long-term stability of core - shell cobalt -carbon and ironcarbide - carbon nanoparticles underdistinct conditions such as strong acidity and alkalinity compared to other wellestablished magnetic nanomaterials, such as ironoxide nanoparticles and silica coated iron oxides. Therefore we treat thecorresponding materials under the mentioned conditions for several weeks withconstant monitoring of the stable fractions. In addition, the extraction ofseveral precious metal species under relatively harsh conditions isinvestigated concerning kinetics, selectivity and the influence of the carbon shellproperties.

Thiscontribution will further show how ultra-stable nanomagnetsallow an extension of magnetic separation from its present broad use in biochemistryto catalysis (Schaetz, 2010).

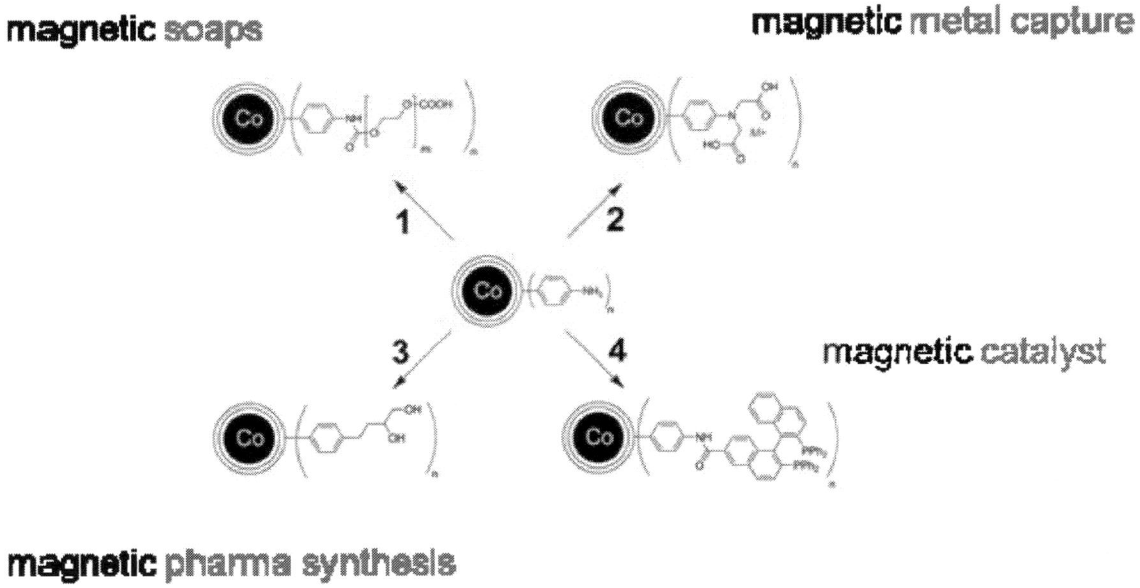

Figure 1. The use of chemically stable nanomagnetsenables to tag classical molecules and enables their separation after usewithin seconds. This is of particular interest for recycling of precious reagents,noble metal extraction and in fine chemical synthesis.

References

M.Rossier, F.M. Koehler, E.K. Athanassiou, R.N. Grass, B. Aeschlimann,D. Gnther, W.J. Stark, Gold adsorption on the carbon surface of C/Conanoparticles allows magnetic extraction from extremely diluted aqueoussolutions, ***J. Mater. Chem.,19, 8239-43 (2009).***

R.N.Grass, E.K. Athanassiou, W.J. Stark, Covalently functionalized cobaltnanoparticles as a platform for magnetic separations in organic synthesis, Angew. ***Chem. Int.Ed.***, 46, 4909-12 (**2007**).

A. Schaetz, O. Reiser, W. J. Stark, Nanoparticles asSemi-Heterogeneous Catalyst Supports, ***Chem. Eur.J.***, 16(30), 8950-67 (**2010**).

A. Schaetz, R.N. Grass, W.J. Stark, MagnetischeChemie: Winzig und mit metallischem Kern, ***Nachrichtenaus der Chemie***, 58, 857 (**2010**).

See more of this Session: Applications of Magnetic Nanoparticles
See more of this Group/Topical: Nanoscale Science and Engineering Forum

(586e) Magnetic Catch & Release: Reversible Organic Contaminant Adsorption and Enrichment From Water

Roland Fuhrer, Inge K. Herrmann, Evagelos K. Athanassiou, Robert N. Grass and Wendelin J. Stark, Institute for Chemical- and Bioengineering, ETH Zurich, 8093 Zurich, Switzerland

There is an increasingneed for fast and efficient extraction methods to remove low concentrated (ppblevels) organic compounds from water with high adsorbent recyclability. Surface-modified magnetic nanoparticles can be used in extraction processesas they readily disperse in common solvents and combine high saturationmagnetization with excellent accessibility. Reversible and recyclableadsorption and desorption through solvent changes and magnetic separationprovide technically attractive alternatives to classical solvent extraction. Carbon-coatedcobalt nanoparticles with a thin polymer layer were tagged withβ-cyclodextrin. The resulting material reversibly adsorbed organiccontaminants in water within minutes. Isolation of the immobilized inclusioncomplex was easily carried out within seconds by magnetic separation due to thestrong magnetization of the nanomagnets (metal core). The trapped moleculeswere fully and rapidly recovered by filling the cyclodextrin cavity with amicrobiologically well accepted substitute, e.g. benzyl alcohol.Phenolphthalein was used as a model compound for organic contaminants such aspolychlorinated dibenzodioxins (PCDDs) or bisphenol A (BPA). The nanomagnetscan be regenerated under mild conditions and reused in the next cycle at fullefficiency (16 recycles tested). Tagging cyclodextrins with magnetic, stable nanoparticles(carbon shell, polymer layer) makes them magnetoresponsive and may lead to anew generation of adsorbents in separation, contaminant enrichment or drugdelivery. Experiments at ultralow concentrations (160 ppb) underline the highpotential of cyclodextrin modified nanomagnets as a fast, recyclable extractionmethod for organic contaminants in large water streams or as an enrichment toolfor analytics.

Reference:

1) R. Fuhrer, I.K. Herrmann, E.K.Athanassiou, R.N. Grass, W.J. Stark, **Langmuir**,2011, 27(5), 1924-9

Figure:

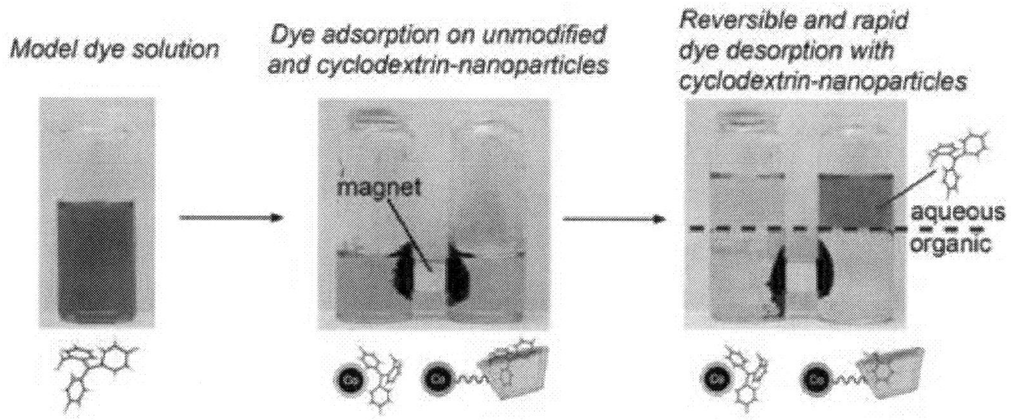

See more of this Session: Environmental Applications of Nanotechnology and Nanomaterials
See more of this Group/Topical: Topical C: Environmental Aspects, Applications, and Implications of Nanomaterials and Nanotechnology

(175b) Electrospinning Nafion Nanofibers

Jason Ballengee and Peter N. Pintauro, Chemical and Biomolecular Engineering, Vanderbilt University, Nashville, TN

Nafion® nanofiber electrospinning has received recent attention for applications in electrochemical/ electromechanical devices including proton exchange membrane fuel cells, sensors, and polymer-based actuators.[1-3] Unfortunately, Nafion is notoriously difficult to electrospin. Nafion does not form a true solution in solvent, but rather a micellar dispersion. Such dispersions lack the required chain entanglements for polymer electrospinning and can only be electrosprayed as discrete droplets. Thus, a carrier polymer such as poly(ethylene oxide) (PEO), poly(acrylic acid), or poly(vinyl alcohol) must be added to Nafion solutions to provide chain entanglements and permit electrospinning of nanofibers. Previously, researchers have electrospun Nafion with high concentrations of carrier polymer (15-25 wt% carrier polymer). A low concentration of carrier is desirable for many end uses, particularly for fuel cell applications since one does not want to contaminate Nafion and compromise its proton conductivity properties by partial dilution with an inert carrier material. In this talk we will present the results from our recent studies demonstrating Nafion electrospinning with low (1 wt%) carrier polymer concentrations.[4]

The effects of relative humidity, electrospinning voltage, polymer solution flow rate, carrier polymer molecular weight, and solvent were investigated for their effect on the fiber diameter of Nafion/PEO nanofibers and the presence of bead or bead-on-fiber structures. Relative humidity during was found to have a profound effect on Nafion electrospinning. Electrospinning fibers into air at a high relative humidity increased the rate of Nafion precipitation, thus leading to the presence of bead-on-fiber structures and an increase in the diameter of electrospun nanofibers. In general, higher voltages and lower flow rates led to fibers of smaller diameter. Fiber diameter was a strong function of the carrier polymer's molecular weight; higher molecular weights led to more viscous solutions and larger electrospun fiber diameters. Furthermore, at carrier polymer high molecular weight, ribbon structures (i. e. flattened fibers) appeared in electrospun mats. Ribbons form due to rapid precipitation of polymer at the fiber surface and subsequent collapse of the resulting tube-like structure upon contact with the electrospinning collector plate. Solvent volatility also effects fiber diameter, where electrospinning with a more volatile solvent produced fibers of larger diameter.

Over the course of the electrospinning experiments, experimental conditions were identified to electrospin well-formed mats of Nafion nanofibers where the average fiber diameter could be controlled between 300 and 900 nm. These conditions will be presented and discussed.

1. Lee, K. M.; Choi, J.; Wycisk, R.; Pintauro, P. N.; Mather, P., Nafion nanofiber membranes. *ECS Transactions*: 2009; Vol. 25, pp 1451-1458.

2. Nah, C.; Lee, Y. S.; Cho, B. H.; Yu, H. C.; Akle, B.; Leo, D. J., Preparation and properties of nanofibrous Nafion mats for ionic polymer metal composites. *Composites Science and Technology* **2008,** 68, (14), 2960-2964.

3. Zhou, C. S.; Liu, Z.; Dai, J. Y.; Xiao, D., Electrospun Ru(bpy)(3)(2+)-doped nafion nanofibers for electrochemiluminescence sensing. *Analyst* **2010,** 135, (5), 1004-1009.

4. Ballengee, J. B.; Pintauro, P. N., Morphological control of electrospun Nafion nanofiber mats. *Journal of the Electrochemical Society* **2011,** 158, (5), B568-B572.

See more of this Session: Nanoscale Structure In Polymers I: Nanostructured Polymeric Materials
See more of this Group/Topical: Nanoscale Science and Engineering Forum

(649g) Nanostructured Polymeric Systems

Arianna Watters, Materials Science and Engineering, Drexel University, Philadelphia, PA and Giuseppe R. Palmese, Chemical and Biological Engineering, Drexel University, Philadelphia, PA

Room temperature ionic liquids offer a new way to homogeneously disperse SWNTs in polymer systems while simultaneously initiating epoxy polymerization. The small particle size of SWNT minimizes disruption of the polymer matrix, while high surface area increases the interactions between the polymer and the SWNTs. The resulting composite materials show improved modulus, fracture toughness, thermal and electrical conductivity while maintaining a glass transition temperature comparable to the neat epoxy. This non-volatile solvent process creates nanocomposites with simplified processing and improved interfacial interactions. The nanocomposite enables control of desired material properties including mechanical, thermal and electrical by varying parameters such as particle size, surface area of the nanoparticle and the strength of the interfaces.

See more of this Session: Graphene and Carbon Nanotubes: Applications
See more of this Group/Topical: Nanoscale Science and Engineering Forum

(374e) Single-Walled Aluminosilicate Nanotubes with Organic-Modified Interiors

Dun-Yen Kang, Ji Zang, Christopher W. Jones and Sankar Nair, School of Chemical & Biomolecular Engineering, Georgia Institute of Technology, Atlanta, GA

This talk describes a methodology for modifying the interior of single-walled metal oxide (aluminosilicate) nanotubes by covalently immobilizing organic functional entities on the interior surface of the nanotube structure. Characterization of the modified nanotubes by a range of solid-state characterization techniques - including nitrogen physisorption, thermogravimetric analysis, transmission electron microscopy (TEM), powder X-ray diffraction (XRD), and solid-state NMR - strongly indicates that the organic entities are immobilized on the inner surface of the nanotubes by reaction with the silanol groups on the interior wall of the nanotube. The resulting organic-modified SWNTs show higher hydrophobicity than bare nanotubes based upon water adsorption measurements. Furthermore, a mechanistic understanding of water adsorption in the modified SWNTs is developed, by interpretation of the water adsorption data with a multilayer adsorption model. The degree of interior surface silanol substitution is estimated, with up to 35% of the silanols being substituted through the present modification chemistry. This methodology of immobilizing various functional entities at the inner wall of aluminosilicate nanotubes opens up a range of previously inaccessible "molecular recognition"-based applications for nanotube materials in areas such as catalysis, molecular encapsulation, sensing, and separation.

See more of this Session: Functional Nanoparticles and Nanocoatings on Particles I
See more of this Group/Topical: Particle Technology Forum

(462b) Enabling In Situ Real-Time Characterization of Interfaces with Quartz Crystal Microbalance with Dissipation Monitoring

Mark A. Poggi, Q-Sense and Farfield, Biolin Scientific, Linthicum Heights, MD, Archana Jaiswal, Q-Sense, Biolin Scientific, Linthicum Heights, MD and Matthew Dixon, Biolin Scientific, Linthicum Heights, MD

There is a growing need for new technologies to quantitatively measure the surface properties of biological and material interfaces. One technique in particular, the Quartz Crystal Microbalance with Dissipation Monitoring (QCM-D), fulfills the need for monitoring real-time dynamic adsorption and desorption phenomena. Capable of operating in liquids and in real-time, QCM-D provides a powerful approach to analyze the *in situ* thickness, structural, and viscoelastic properties of molecules and materials at interfaces. This presentation will focus on the application of QCM-D to monitoring self assembly processes. Examples to be discussed will be recent research done in the areas of: protein deposition or fouling, polyelectrolyte multilayer build-up, nanoparticle/nanotube assembly, lipid membrane fuel cell interfaces. The QCM-D technique has allowed scientists to achieve a more fundamental understanding of how dynamic materials behave on the molecular scale and current research in several of these areas will be highlighted.

See more of this Session: Life Cycle Studies Associated with Nanomaterials and Nano-Enhanced Products
See more of this Group/Topical: Topical C: Environmental Aspects, Applications, and Implications of Nanomaterials and Nanotechnology

(292d) Demonstration of Nanoparticle-Bound Polymer Biodegradation and Resulting Nanoparticle Destabilization

Teresa L. Kirschling[1], Patricia L. Golas[2], Krzysztof Matyjaszewski[2], Kelvin B. Gregory[3], Gregory V. Lowry[3] and **Robert D. Tilton**[4], (1)Department of Chemical Engineering, Carnegie Mellon University, Pittsburgh, PA, (2)Department of Chemistry, Carnegie Mellon University, Pittsburgh, PA, (3) Department of Civil and Environmental Engineering, Carnegie Mellon University, Pittsburgh, PA, (4) Department of Chemical Engineering and Department of Biomedical Engineering, Carnegie Mellon University, Pittsburgh, PA

Many engineered nanoparticles are modified by adsorption or grafting of polymers onto their surfaces in order to impart colloidal stability in suspension, to promote adhesion with matrix materials, to increase biocompatibility, or to impart some particular biomolecular targeting capability. By controlling nanoparticle flocculation and surface deposition, surface-bound polymer layers strongly influence nanoparticle fate, transport, and subsequent biological impact in environmental systems. Biodegradation is one potential route to the breakdown of surface-bound polymers, but it is unknown whether such nanoparticle-bound polymers are bioavailable or if the surface confinement restricts microbial access to the chains so that they become recalcitrant to biodegradation. Most engineered nanoparticles are in a size range that cannot be internalized by bacteria, so the bound chains must be accessed extracellularly. Here we demonstrate, for the first time, that a polymer brush bound to nanomaterials can be bioavailable to microbes. Model core/shell nanomaterials with covalently attached polyethylene oxide (PEO) chains were prepared for this study. These were synthesized to obtain a nanomaterial in which biodegradation was the only available coating breakdown mechanism. Taking the form of a densely crosslinked bottlebrush copolymer with a hydrophobic core, these nanomaterials consisted of poly(ethylene oxide) methacrylate macromonomers crosslinked by hydrophobic divinyl benzene. PEO is a common constituent of consumer products, and PEO-based polymers are common, effective steric stabilizers for nanoparticles and other colloids. Accordingly, this an appropriate model system to represent a class of nanoparticle coatings that is likely to be introduced to the environment. The nanoparticles were approximately 30 nm in hydrodynamic radius and were colloidally stable in the culture media used in the biodegradation experiments. Microbial samples isolated from river water were enriched by supplying PEO homopolymer as the sole carbon source. The PEO-homopolymer degrading microbial enrichment cultures were transferred to fresh media, then supplied with PEO brush-bound nanoparticles as the sole carbon source. Biomass and CO_2 production measurements indicated that the particle-bound PEO brush was indeed bioavailable to these microbes. DNA sequencing indicated that the mixed culture responsible for degradation included *Novosphingobium sp., Pseudomonas sp.*, and *Hydrogenophaga sp.* Brush biodegradation resulted in nanoparticle flocculation. Control experiments demonstrated that byproducts of microbial metabolism or cell debris did not cause the observed nanoparticle flocculation. This points to the need to understand biologically mediated transformations of nanoparticle coatings in order to understand the fate and transport of nanoparticles in the environment.

See more of this Session: Environmental Implications of Nanomaterials: Biological Interactions
See more of this Group/Topical: Topical C: Environmental Aspects, Applications, and Implications of Nanomaterials and Nanotechnology

(531c) Structural Stability of Transparent Conducting Films Assembled From Single-Wall Carbon Nanotubes Purified by Electronic Type

John M. Harris[1], Jeffrey A. Fagan[2], Steven D. Hudson[3], Christopher M. Stafford[2] and **Erik K. Hobbie**[1], (1)NDSU, Fargo, ND, (2)Polymers Division, National Institute of Standards and Technology, Gaithersburg, MD, (3)NIST, Gaithersburg, MD

Strain-induced structural changes in thin polymer-supported membranes of purified single-wall carbon nanotubes (SWCNTs) are evaluated through the wrinkling instabilities that develop under uniaxial and isotropic compression. Nanotubes that have been separated by electronic type using density-gradient ultracentrifugation are assembled as surfactant-free films on prestrained polydimethylsiloxane (PDMS) substrates, and the strain response is measured using a broad range of techniques. The small-strain behavior is inferred from kinetic changes in the wrinkling topography during the slow drying of pre-swelled polymer supports. Our measurements reveal a remarkable degree of strain softening that couples strongly to the sheet resistance of the membranes through microscale anisotropy. By depositing comparable films on quartz, we use optical (UV-Vis-NIR) absorption spectroscopy to compute the London dispersion spectra of the purified materials, and from these we compute the attractive part of the van der Waals potential between nanotubes of identical electronic type. We find significant differences in the strength of the contact potential between metallic vs. semiconducting nanotubes, which in turn are evident in the modulus and yield strain of the films. Supported by the NSF through CMMI-0969155.

See more of this Session: Graphene and Carbon Nanotube Based Devices
See more of this Group/Topical: Materials Engineering and Sciences Division

(105b) Preparation of Monodisperse Silicon Nanocrystals Through Density-Gradient Ultracentrifugation In Organic Solvents

Joseph B. Miller[1], Austin R. Van Sickle[1], Rebecca Anthony[2], Uwe R. Kortshagen[3] and Erik K. Hobbie[1], (1)NDSU, Fargo, ND, (2)Department of Mechanical Engineering, University of Minnesota, Twin-Cities, Minneapolis, MN, (3)Mechanical Engineering, University of Minnesota, Minneapolis, MN

Monodisperse colloidal suspensions of ligand-stabilized silicon nanocrystals, synthesized through a nonthermal low-pressure plasma reaction, are prepared through density-gradient ultracentrifugation in mixed organic solvents. Density-gradient profiles of mixed chloroform and m-xylene are used to control the settling speed of the nanoparticles and hence optimize their transient separation by size along the depth of polyoxymethylene ultracentrifuge tubes. The mean size and polydispersity of the extracted fractions are characterized through photoluminescence spectroscopy and transmission electron microscopy, and the self-assembly of fractions into close-packed crystal lattices is achieved using an immiscible two-fluid evaporation scheme. The photophysical properties of the purified nanocrystal lattices are compared with those of the starting materials, and the influence of atmospheric oxygen and temperature on the stability of nanocrystal ensemble photoluminescence is measured. Supported by the DOE through DE-FG36-08GO88160.

See more of this Session: Nanofabrication and Nanoscale Processing
See more of this Group/Topical: Nanoscale Science and Engineering Forum

(755b) Targeted Bioresorbable Polymersomes for the Delivery of Cisplatin

Matthew A. Petersen[1], Marc A. Hillmyer[2] and Efrosini Kokkoli[1], (1)Chemical Engineering and Materials Science, University of Minnesota, Minneapolis, MN, (2)Chemistry, University of Minnesota, Minneapolis, MN

Chemotherapy today is often plagued by major side effects due to anticancer drugs damaging healthy tissues in addition to cancer cells. These side effects can often be mitigated by encapsulating chemotherapeutics in a targeted drug delivery system. One of the most promising systems is the polymersome, a self-assembled nanoscale aggregate of amphiphilic polymers that has the potential to emerge as a long-circulating, robust carrier capable of encapsulating both hydrophilic and hydrophobic payloads.

We have developed a nontoxic, polymersome-forming, bioresorbable polymer that can efficiently and site-selectively tether targeting peptides under mild conditions (water at pH 7.4) without the need for potentially toxic catalysts. Here we characterize delivery and release using these polymersomes to human colon cancer cells overexpressing the $\alpha_5\beta_1$ integrin. We demonstrate that delivery efficacy using these carriers can be markedly improved over commonly-used RGD targeting peptides by targeting these receptors with the next-generation targeting peptide PR_b. Further, delivery is demonstrated to be specific to cells overexpressing the integrin with minimal delivery observed in a cell line that does not overexpress the receptor. Finally, we encapsulate the anticancer drug cisplatin and demonstrate that PR_b targeting can greatly improve delivery efficacy to human colon cancer cells with commensurate gains in cisplatin-induced cell death. These results demonstrate that our polymersomes may be an attractive route to minimizing side effects associated with current cisplatin chemotherapy.

See more of this Session: Self-Assembled Biomaterials
See more of this Group/Topical: Nanoscale Science and Engineering Forum

(420j) Enzyme Inhibition by Inorganic Nanoparticles of Different Shapes

Sang-Ho Cha, Chemical Engineering, University of Michigan, Ann Arbor, Ann Arbor, MI and Nicholas Kotov, Chemical Engineering, University of Michigan, Ann Arbor, MI

Although there are a few reports about the effect of inorganic nanoparticles on the catalytic activity of enzyme, to the best of our knowledge, the systematic study about the direct effect for the shapes of inorganic nanoparticles on the structure and activity of enzyme has never been established before. On the basis of these considerations, herein, we carried out an investigation on the activity change in HRP upon the addition of different shaped ZnO nanomaterials. The relative activity of HRP decreases with the increasing the concentrations of ZnO nanopyramids (nPY) and nanoplates (nPL), while the enzyme activity is found to be approximately invariant for all concentrations of spherical ZnO nanoparticles (nPA). In the case of ZnO nPA, the notable effect on the activity of HRP was not observed even in the concentrations exceeded 3.0 uM. In any concentrations of ZnO nanomaterials, ZnO nPY show much higher inhibition effect on HRP than that mediated by ZnO nPL. For example, with 1.25 uM concentration of ZnO nPY, the activity of HRP was reduced by ~ 75 % of the original HRP activity, whereas the similar concentration of ZnO nPL leads to only ~ 21 % enzyme activity lost. When the concentration of ZnO nPY increased to ~ 3.0 uM, the activity of HRP was almost lost. However, in the case of ZnO nPL, the HRP still had about 30 % of enzyme activity.

See more of this Session: Poster Session: Nanoscale Science and Engineering
See more of this Group/Topical: Nanoscale Science and Engineering Forum

(704e) Design of Nanoparticle-Based Platforms for Multi-Enzyme Co-Localization

Feng Jia, Department of Chemical and Biological Engineering, Iowa State University, Ames, IA, Surya Mallapragada, Department of Chemical and Biological Engineering and Ames Laboratory, Iowa State University, Ames, IA and Balaji Narasimhan, Chemical and Biological Engineering, Iowa State University, Ames, IA

In this research, we have designed nanoparticle-based platforms to co-localize multi-enzyme complexes with reactive intermediates. An example of these multi-enzyme systems is anthocyanidin synthase (ANS) and anthocyanidin reductase (ANR), where the two enzymes need to be co-localized in order to rapidly transport the highly reactive intermediate cyanidin from ANS to ANR to maximize the production of flavon-3-ols, a powerful anti-oxidant, which are beneficial for plant and human health. As a first step, glucose oxidase (GOX) and horseradish peroxidase (HRP) were used as model enzymes to demonstrate co-localization of multiple enzymes on polymeric nanoparticles. Hydrogen peroxide formed in the first reaction catalyzed by GOX is consumed by HRP in the second reaction in the presence of N-acetyl-3,7-dihydroxyphenoxazine (Amplex Red), to yield a specific fluorescent product, resorufin. To optimize the enzyme attachment, covalent binding, streptavidin-biotin coupling and polyethylene glycol (PEG) spacer linkage were evaluated and compared by immobilizing HRP on carboxylated polystyrene nanoparticles. The blank, functionalized, and enzyme-attached nanoparticles were characterized by scanning electron microscopy (SEM), X-ray photoelectron spectroscopy (XPS), and dynamic light scattering (DLS). The enzyme kinetics were evaluated before and after the attachment. The results showed that the PEG spacer linkage and streptavidin-biotin coupling were comparable, and both were superior to covalent binding in terms of maintaining the stability and activity of HRP. Furthermore, different spacer lengths of PEG were investigated. The GOX and HRP were co-localized on the nanoparticles through PEG spacer linkage and the two step reaction kinetics were evaluated and compared with single-enzyme complexes and free enzymes. Based on the optimization carried out with the model system, ANS and ANR were co-localized on the nanoparticles through appropriate PEG spacer attachments. This study demonstrates a simple and effective way for multiple enzyme co-localization to mimic the efficient multi-enzyme complexes that widely exist in Nature.

See more of this Session: Nanoscale Science and Engineering In Biomolecular Catalysis I
See more of this Group/Topical: Nanoscale Science and Engineering Forum

(298g) Dynamical Heterogeneity In the Ionic Liquid [BMIM+][PF6-] Confined In a Graphitic Slit Pore

Ramesh Singh, Joshua D. Monk and Francisco R. Hung, Cain Department of Chemical Engineering, Louisiana State University, Baton Rouge, LA

Ionic liquids (ILs) have attracted extensive attention in recent years due to the potential application as electrolytes in electrochemical double-layer capacitors (EDLCs) and other electrochemical devices. Compared with traditional solutions of electrolytes, ILs offer many advantages, such as good chemical and thermal stability, negligible vapor pressure, nonflammability, high ionic conductivity, and wide electrochemical window. The dynamics of the confined ions is one important factor that determines the internal resistance in an EDLC, which ultimately affects its specific power (i.e., how quickly an EDLC can deliver energy).

In this work we have used molecular simulation to investigate the relaxation dynamics of the IL [BMIM $^{+}$][PF_6^-] under confinement in a slit-like graphitic pore in the temperature range of 300-450K. The dynamics of the confined IL are analyzed in detail by measuring mean squared displacements (MSDs) in different directions, van Hove correlation functions and intermediate scattering functions. The latest are then fitted to modified Kohlrausch-Williams-Watts (KWW) functions to attempt to quantify the timescales associated with the α and β relaxation processes. All these properties, when combined with measurements of different structural properties (local density profiles, radial distribution functions, etc.), can be used to detect and characterize spatial heterogeneities in the dynamics of the IL. Our results suggest that the local relaxation dynamics of the ions depend strongly on their position inside the pore. Close to the pore walls, the typical relaxation times are larger than those for the ions in the center of the pore. Cations have faster relaxation times than anions everywhere in the system. Important spatial heterogeneities in the dynamics of the confined IL are observed, and those are compared against results observed for the same IL in bulk systems at the same temperatures.

See more of this Session: Graphene and Carbon Nanotubes: Absorption and Transport Processes
See more of this Group/Topical: Nanoscale Science and Engineering Forum

(321c) Polystyrene Nanoparticles As a Model System to Investigate 3-Dimensional Confinement Effects On the Glass Transition Temperature

Chuan Zhang, Yunlong Guo and Rodney D. Priestley, Chemical and Biological Engineering, Princeton University, Princeton, NJ

When confined to the nanoscale, the glass transition temperature (T_g) of polymers can deviate substantially from the bulk, in what is commonly referred to as the T_g-confinement effect. Due to ease of processing, most studies have focused on the size-dependent T_g of thin films, while few have extended investigations to other geometries. As polymers confined to higher geometrical dimensions become the enabling material in technologies ranging from drug delivery to plastic electronics, a greater understanding of size effects on the T_g is warranted. Here, we investigate the effect of soft and hard three-dimensional confinement on the T_g of polymer nanoparticles. Via modulated differential scanning calorimetry, we show that the T_g can decrease significantly with size for aqueous suspended polymer nanoparticles, i.e., the case of soft confinement. Furthermore, capping of the nanoparticles with a silica shell leads to a size-invariant T_g, i.e., the case of hard confinement. Lastly, using a unique flash nanoprecipitation method to generate polymer nanoparticles of the same size but different molecular weights, we examine the impact of polymer molecular weight on the T_g of 3-dimensional confined polystyrene nanoparticles.

See more of this Session: Nanoscale Structure In Polymers II
See more of this Group/Topical: Materials Engineering and Sciences Division

(600b) High-Performance Randomly Oriented Zeolite Membranes Using Brittle Seeds and Rapid Thermal Processing

Won Cheol Yoo[1], Jared A. Stoeger[1], Pyung-Soo Lee[1], Andreas Stein[2] and Michael Tsapatsis[1], (1) Department of Chemical Engineering and Materials Science, University of Minnesota, Minneapolis, MN, (2)Department of Chemistry, University of Minnesota, Minneapolis, MN

To exhibit high flux and high selectivity for energy efficient membrane separation processes, thin and defect-free zeolite membranes on a porous support must be fabricated. Experimental efforts have been directed to the confined synthesis of zeolite particles to be used as a seed layer for the secondary hydrothermal growth of membranes, primarily due to controllable size and good monodispersity of the seed particles. Here, we present a facile method to fabricate MFI zeolite membranes with high flux (ca. 2.6×10^{-7} $molm^{-2}s^{-1}Pa^{-1}$) and high separation factors (SF, 123–139 and in one case as high as 335) for xylene isomer separation by optimizing seed deposition and calcination processes. Zeolite nanoparticles prepared by aforementioned confined syntheses were deposited as randomly oriented seed layers on porous supports by rubbing and leveling methods to obtain thin zeolite films (400–500 nm) after secondary hydrothermal growth. Rapid thermal processing (RTP) was used in a single calcination step that substantially reduced processing times and energy usage and improved the separation efficiency of the membranes.

See more of this Session: Nanostructured Thin Films
See more of this Group/Topical: Materials Engineering and Sciences Division

(266c) Nature-Inspired Hybrid Membranes for Molecular Separations

Silo L. Meoto and Marc-Olivier Coppens, Chemical & Biological Engineering, Rensselaer Polytechnic Institute, Troy, NY

A mesoporous hybrid membrane has been synthesized for size-exclusive separation of molecules. This membrane is composed of columnar silica structures containing mesopores of a controlled diameter, grown within the channels of an anodic alumina template, which are oriented perpendicularly to the external membrane surface. The mesopores in the silica columns are formed by soft templating, using surfactants, which form micelles, around which the silica grows; removal of the surfactant leads to mesopores.

The fabrication of mesoporous membranes using anodic alumina as a macroporous template facilitates the orientation of the pores, and the control over other geometrical parameters, such as wall thickness, length and diameter. To this end extensive research on the template-directed process using macroporous membranes has been carried out. The nanocomposite formed inside the alumina pores is an assembly of silica and surfactant, with a column diameter equal to the columnar alumina pore diameter. The surfactant-filled nanochannels are oriented parallel to the walls of the columnar alumina pore. For further application as a molecular separator, after removal of the surfactant, it is desirable to produce hybrid membranes in which the mesoporous material completely fills the anodic alumina membrane (AAM) channels to minimize loss of size selective separation and transport of molecules. Additionally, a stronger interaction between the mesoporous material and the alumina channel wall is desired, as shrinkage of the mesostructure is observed in calcined samples, leaving a gap between silica fibers and alumina channel walls. It is also essential to obtain permeable mesostructures such as a hexagonal columnar phase.

In this study, the AAM was loaded with precursor solutions having different surfactant-to-silica ratios and left to dry under ambient condition and controlled humidity. The silica nanocomposites are grown by sol-gel chemistry of silica in confinement and other methods, such as evaporation-induced self-assembly, and electrophoretic deposition. Temperature was found to play a key role in phase formation and, when used, aspiration pressure influenced the type of mesostructure obtained. By functionalizing the mesoporous silica inside the AAM channels, chemical groups could be incorporated, guided by the structure and function of certain biological pores.

Mesoporous hybrid membranes were characterized by scanning electron microscopy (SEM), small-angle X-ray scattering, and nitrogen adsorption and desorption. These provided evidence of pore filling, the nature of structural order and accessible mesoporosity. Functionalized silica nanostructures were characterized by high-resolution transmission electron microscopy (TEM) and size-exclusion permeation experiments. This allowed evaluation of the effective filling and thickness of the mesoporous silica columns.

See more of this Session: Templated Assembly of Inorganic Nanomaterials I
See more of this Group/Topical: Nanoscale Science and Engineering Forum

(105c) Formation of Organosilane "Nano-Beaker" Arrays by Particle Lithography

Sunxi Wang, Daniel Sobczynski and Guangzhao Mao, Chemical Engineering and Materials Science, Wayne State University, Detroit, MI

The ability to make nano-arrays enables high-throughput screening and nanofluidics. This paper describes a simple method to make nano-arrays of organosilanes on oxidized silicon wafers based on latex particle lithography. Particle lithography was carried out on 300 and 900 nm polystyrene latex particle array templates by chemical vapor deposition (CVD) of octadecyltrichlorosilane (OTS). OTS reacts with the substrate in the water menisci in between latex particles to form a ring pattern. In order to use the ring pattern as "beakers" to contain droplets and particles, it is necessary to increase the OTS ring height beyond its monolayer thickness of 2.6 nm. We discovered CVD conditions that favor OTS bulk polymerization resulting in 14 nm thick rings for 900 nm particles and 9 nm thick rings for 300 nm particles. Mechanisms for the "nano-beaker" array formation were investigated. In addition, the use of OTS nano-arrays for distributing and containing solutions was demonstrated with hydrophobic compounds and nanoparticles. It is estimated that a 1×1 cm^2 wafer containing ordered "nano-beaker" arrays made with 900 nm particle lithography holds more than 100 million droplets with droplet volume less than 3×10^{-10} nL. The OTS nano-arrays are stable against common solvents, which allow repeated use in high-throughput crystallization and confinement screening experiments.

See more of this Session: Nanofabrication and Nanoscale Processing
See more of this Group/Topical: Nanoscale Science and Engineering Forum

(346b) Temperature and Magnetic Field-Responsive Nanoparticles Based On a Superparamagnetic Magnetite Core and a Poly(N-isopropylacrylamide) Shell

Sharavanan Balasubramaniam[1], Nikorn Pothayee[1], Yinnian Lin[2], Mike House[3], Robert Woodward[3], Timothy G. St. Pierre[3], Judy S. Riffle[2] and Richey M. Davis[4], (1)Macromolecules and Interfaces Institute, Virginia Tech, Blacksburg, VA, (2)Chemistry, Virginia Tech, Blacksburg, VA, (3)Physics, The University of Western Australia, Crawley, Australia, (4)Chemical Engineering, Virginia Tech, Blacksburg, VA

Multistimuli-responsive nanostructures have interesting existing and potential *in vivo* applications. We report on the temperature-induced aggregation and relaxometric behavior of magnetic nanostructures comprised of a magnetite core and a poly(N-isopropylacrylamide) (PNIPAM) shell. Discrete magnetite nanoparticles were synthesized by a high-temperature decomposition method using Iron (III) acetylacetonate as the precursor. Controlled synthesis of bis(phosphonate)-terminated PNIPAM was achieved by atom transfer radical polymerization. Ligand exchange between a dispersion of oleic acid-coated magnetite nanoparticles and the PNIPAM solution afforded highly water-dispersible magnetite-PNIPAM nanoparticles (D_I = 38 nm) with good colloidal stability in physiological media. The hydrodynamic sizes of the core-shell nanoparticles were predicted with a density distribution model using the measured sizes of the magnetite cores (~ 7.8 nm) from transmission electron microscopy. Good agreement between the measured and predicted hydrodynamic sizes suggested that the complexes were primarily discrete, non-agglomerated nanoparticles under ambient conditions. SQuID magnetometry confirmed that the nanoparticles were fully superparamagnetic at 300 K with a saturation magnetization of ~ 55 emu g^{-1}.

The PNIPAM shell around the magnetite core afforded thermosensitive properties to the nanoparticles. When heated above ~ 33.5 °C, the intensity-weighted diameter increased rapidly, signaling aggregation induced by the lower critical solution temperature (LCST) of PNIPAM. At the transition temperature, the PNIPAM steric chains lose their water of hydration to become mutually attractive and drive aggregation of the magnetite nanoparticles. Steady-state fluorescence experiments on Nile red-labelled magnetite-PNIPAM nanoparticles at 25 °C (< LCST) and 35 °C (> LCST) showed fluorescence quenching above the LCST because of the aggregation of nanoparticles and concomitant dye aggregation. NMR relaxometry performed on aqueous dispersions of the magnetite-PNIPAM nanoparticles revealed dramatic shortening of transverse relaxation times (T_2) of the water protons and an increase in the relaxometric ratio (R_2/R_1) above the LCST transition of PNIPAM. We envisage that temperature and magnetic field sensitivity offer potential for the design of 'smart' multifunctional magnetic nanocarriers for biomedical applications such as site-specific drug delivery and magnetic resonance imaging.

See more of this Session: Applications of Magnetic Nanoparticles
See more of this Group/Topical: Nanoscale Science and Engineering Forum

(472b) Block Ionomer Complexes Containing Gentamicin to Kill Intracellular Brucella Melitensis

N. Pothayee[1], N. Jain[2], T.P. Vadala[1], R. Mejia[1], L.M. Johnson[1], N. Sriranganathan[2], J.S. Riffle[1] and **R. M. Davis**[3], (1)Macromolecules and Interfaces Institute, Virginia Tech, Blacksburg, VA, (2)VA-MD Regional School of Veterinary Medicine, Virginia Tech, Blacksburg, VA, (3)Chemical Engineering, Virginia Tech, Blacksburg, VA

Brucellosis, caused by intracellular *Brucella spp.*, is the most common disease that can be passed from infected animals to humans. It is difficult to eliminate as the pathogen resides partially within phagocytic host cells and the recommended polar antibiotics for treatment do not enter these cells efficiently. Core-shell block ionomer complexes (BICs) carrying antibiotics in their cores were designed to transport these drugs into cells. Polyether-polyacrylate copolymers were complexed with cationic aminoglycoside antibiotics to form BICs with hydrodynamic diameters of 170-340 nm in water. An anionic poly (acrylate-*b*-ethylene oxide-*b*-propylene oxide-*b*-ethylene oxide-*b*-acrylate) copolymer blended with a poly(ethylene oxide-*b*-acrylate) diblock was complexed with the cationic drug gentamicin in PBS to form complexes containing up to 42 wt% of the antibiotic. The poly(propylene oxide) provided hydrophobic interactions that enhanced complex stability in aqueous media while the hydrophilic blocks generated steric interactions to disperse the MBICs. These core-shell BICs were fabricated in a multi-inlet-vortex mixer that involved mixing solutions of the cationic drug and anionic block copolymers under turbulent flow conditions so that the mixing time was rapid (<50 ms). This promoted reproducible and uniform complex formation. The drug release profiles in PBS showed two distinctly different regimes - a "burst" phase within the first ~10 hours followed by slower release over the next 40-60 hours. *In vitro* efficacy of the BICs to reduce intracellular *Brucella* was studied in murine macrophage-like cells. Significant reductions of 2.78 and 2.85 logs were obtained, while free gentamicin reduced the bacteria by only 0.75 logs. This suggests that the BICs are efficient transporters of polar antibiotics into phagocytic cells. No cell toxicity or nitric oxide production was detected on incubation of the macrophages with a range of BIC concentrations. These BICs were, however, ineffective in significantly reducing *Brucella melitensis* in the spleens and livers of infected mice, either alone or in combination with doxycycline.

See more of this Session: Nanotechnology for Biotechnology and Pharmaceuticals I
See more of this Group/Topical: Nanoscale Science and Engineering Forum

(266d) A Tetrahedral Lattice Model for Simulating the Formation of Ordered Nanoporous Materials

Lin Jin[1], Scott Auerbach[2] and Peter A. Monson[1], (1)Chemical Engineering, University of Massachusetts, Amherst, MA, (2)Chemistry, University of Massachusetts, Amherst, MA

Ordered nanoporous materials are of great scientific and technological importance. Understanding how such materials form and how to control and tailor the pore geometries will help facilitate advanced applications of these materials. Molecular modeling of syntheses of porous materials is challenging because large system sizes are needed to have representative sampling of the processes, and long time scales are required to observe the formation of ordered structures. To address this need, we have developed a relatively simple silicon-oxygen lattice model where each silicic acid molecule is represented as a tetrahedron on a body centered cubic lattice with one silicon atom at the center and hydroxyl groups at the corners [J. Chem. Phys., **134**, 134703 (2011)]. The low-coordination lattice model captures the basic mechanism of silica polymerization and provides structural evolution information during the polymerization processes in dynamic Monte Carlo simulations. The simplicity of this lattice model makes it possible to study the polymerization process to higher degree of condensation and larger system sizes than has been possible with previous atomistic models. Using templating procedures at different length scales the model can be used to study the directed assembly of both ordered microporous crystalline materials and ordered mesoporous materials. Although this low-coordination lattice model may not represent real zeolite morphologies because of the lattice restrictions, it is capable of generating a variety of architectural motifs including layered materials, crystalline microporous materials and chalcogenide zeolite analogs. Combining silica polymerization with surfactant self-assembly using Larson's lattice model of surfactants, we may also model the formation of ordered mesoporous materials with realistic atomic structure of the silica within these materials.

See more of this Session: Templated Assembly of Inorganic Nanomaterials I
See more of this Group/Topical: Nanoscale Science and Engineering Forum

(246b) Self Assembly of Soft Matter Quasicrystals and Their Approximants

Christopher R. Iacovella[1], Aaron S. Keys[2] and Sharon C. Glotzer[2], (1)Chemical and Biomolecular Engineering, Vanderbilt University, Nashville, TN, (2)Department of Chemical Engineering, University of Michigan, Ann Arbor, MI

The discovery of soft-matter quasicrystals and related periodic approximants in dendrimer [1-2] and block copolymer systems [3-4] hints at a unique thermodynamic mechanism that gives rise to their stability. In the past, specific interaction potentials have been contrived to stabilize quasicrystals and their approximants in computer simulations. However, such complex interactions are seemingly incompatible with the known properties of such soft-matter systems. Here, we use molecular simulation to demonstrate an alternative approach for assembling dodecagonal quasicrystals and their approximants based solely on particle functionalization and shape [5]. These effects mimic the characteristics of the specialized potentials, thereby replacing complex energetic interactions with simpler-to-achieve bonded and excluded-volume interactions. Our model consists of spherical building blocks functionalized with mobile surface entities to encourage the formation of structures with low surface contact area, including non-close-packed and polytetrahedral structures. The building blocks also posses shape polydispersity, where a subset of the building blocks deviate from the ideal spherical shape, which discourages the formation of close-packed crystals. We show that a model system possessing both mobile surface entities and shape polydispersity consistently assembles complex quasicrystal-like structures. We also report the spontaneous assembly of quasicrystal approximants in two micelle-forming systems of tethered nanoparticle building blocks that possess these key features. We argue that this mechanism can be widely exploited to assemble quasicrystals and approximants on the nano and microscales, and may further elucidate the recent formation of soft matter quasicrystals in experiment [1-4].

[1] G. Ungar, et al., Giant supramolecular liquid crystal lattice, *Science* 299 (2003)
[2] X. Zeng, et al., Supramolecular dendritic liquid quasicrystals, *Nature* 428, (2004)
[3] S. Lee, M.J. Bluemle, F.S. Bates, Discovery of a frank-kasper sigma phase in sphere-forming block copolymer melts, *Science*, 330 (2010)
[4] S. Fischer, et al. Colloidal quasicrystals with 12-fold and 18-fold diffraction symmetry, *Proc. Natl. Acad. Sci.*, 108, (2011)
[5] C.R. Iacovella, A.S. Keys, S.C. Glotzer, Self assembly of soft matter quasicrystals and their approximants, preprint, http://arxiv.org/abs/1102.5589

See more of this Session: Nanoscale Structure In Polymers I
See more of this Group/Topical: Materials Engineering and Sciences Division

(420k) Electron Beam Induced Radiation Damage In Nafion and the Lifetime of Fuel Cells

Qianping He, Department of Chemical and Biomolecular Engineering, University of Tennessee, Knoxville, Knoxville, TN, David C. Joy, Materials Science and Engineering, University of Tennessee, Knoxville, Knoxville, TN and David J. Keffer, Chemical and Biomolecular Engineering, University of Tennessee Knoxville, Knoxville, TN

Polymer electrolyte membrane fuel cells(PEMFCs) are a potentially viable, clean, and renewable source of energy. There are numerous technical challenges that must be overcome in order toachieve significant market penetration. The interface between theelectrode and electrolyte remains one of the areas of greatest ignorance. A molecular-level understanding of fuel cell structure will be very helpful tounderstand the electron/proton transportation mechanism at the electrode/electrolyteinterface, see Fig. 1. SEM and TEM are two powerful tools commonly used to gainmolecular level information about the electrode/electrolyte interface structure. However, Schneider [1], has reported that Nafion (used as electrolyte in thePEMFC) is susceptible to radiolysis during SEM imaging (Fig. 2.),and suggested that such damage was dependent on the microstructure of thesubstance being viewed. Because beam induced damage could have majorimplications for transport phenomena occurring at the electrolyte/electrodeinterface of PEMFC and could possibly suggest one of the possible mechanisms offuel cell aging, we have investigated the response of Nafion to electron beamirradiation.

In the work described here, high angleannular dark field (HAADF) STEM images of our interface samples were recordedas a function of the integrated beam dose using a Zeiss MERLIN SEM operated inscanning transmission (STEM) mode. The magnitude of the HAADF signal islinearly proportional to the mass thickness of the specimen [2]. So thickerregions of the sample or areas with higher density will appear bright, whilst ahole through the sample in the beam path will appear darker. To makemeasurements the probe beam, containing a known beam current, is scanned at TVrates in a square raster a few micrometers in size across the Nafion sample fortime periods varying from a few seconds up to of the order of one minute. Animage of the exposed area is then recorded and, during the photo-recording, thebeam is blanked for one or two seconds to provide a zero-signal ("black level")reference. The recorded STEM image is then analyzed by a histogram whichindentifies both the zero signal baseline value and the signal level in theirradiated region (Fig. 3.), which permits the brightness of the irradiatedarea to properly determined. This procedure is then repeated, as required, toincrease the deposited beam dose while simultaneously measuring the change inSTEM image brightness. The relative change in sample thickness with irradiationcan then be found by plotting the signal intensity as a function of the beamdose deposited. An absolute measurement of the damage rate versus dose can be producedby using a mass standard, such as ferritin, to calibrate the sensitivity of theHAADF signal under the given experimental conditions.

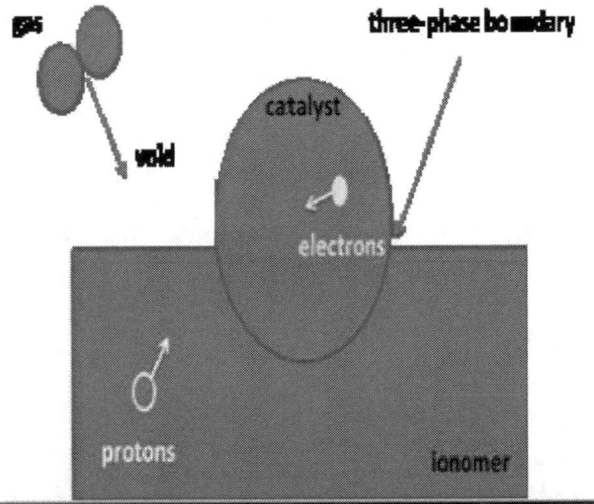

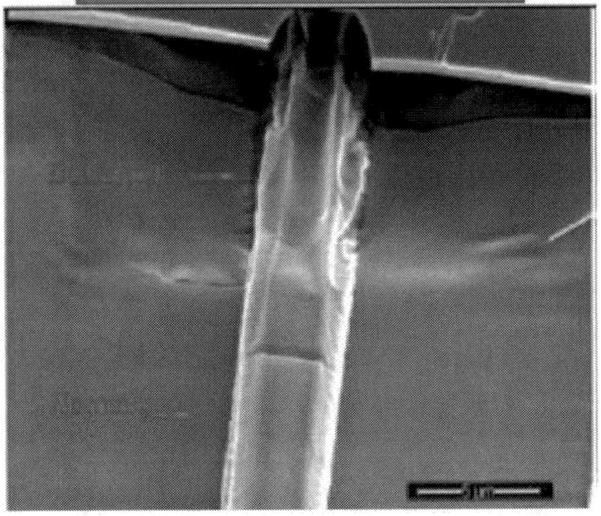

FIG. 1. An illustration of the three phase boundary at which the electrochemical reaction takes place in a PEMFC.

FIG. 2. A 5000X image of both the damaged and undamaged portion of a feature in the Nafion sample dry pressed at 140 ℃ [1]

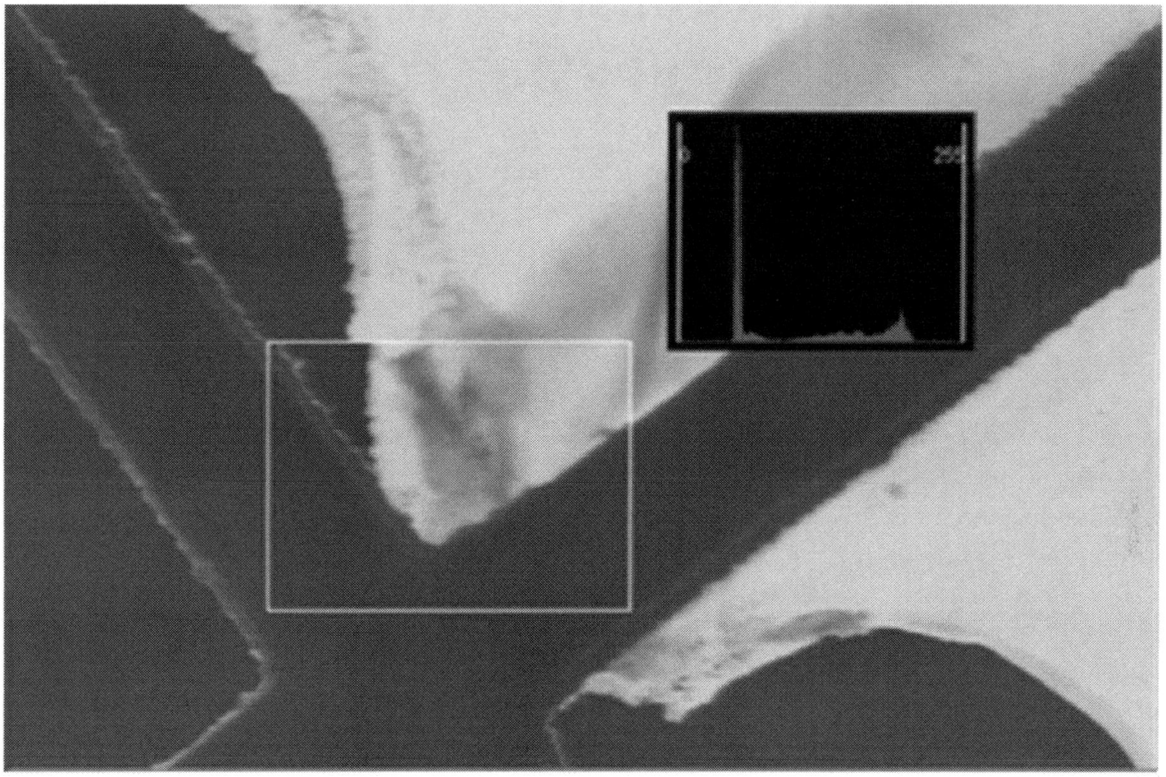

FIG. 3. 30kev HAADF STEM image of a Nafion sample showing the histogram obtained from the highlighted, rectangular, irradiated area. The image field of view is 70 mm in width.

See more of this Session: Poster Session: Nanoscale Science and Engineering
See more of this Group/Topical: Nanoscale Science and Engineering Forum

(693d) Polymer-Stabilized Graphene Dispersions At High Concentrations In Organic Solvents for Nanocomposite Production

Ahmed Wajid[1], Sriya Das[1], Fahmida Irin[1], Tanvir Ahmed[2], John Shelburne[1] and **Micah Green**[1], (1) Chemical Engineering, Texas Tech University, Lubbock, TX, (2)Mechanical Engineering, Texas Tech University, Lubbock, TX

Liquid phase exfoliation and dispersion of graphene, i.e. single layer graphite, is a critical challenge for bulk processing of graphene for use in advanced materials and devices. We demonstrate a simple and effective dispersion of pristine (unfunctionalized) graphene at high concentrations in a wide range of organic solvents by use of a stabilizing polymer (polyvinylpyrrolidone, PVP). These polymer-stabilized graphene dispersions are shown to be aggregation-resistant and readily redispersible even after freeze-drying. These dispersions are then used as a strong and conductive nano-filler for polymer nanocomposites. In particular, PVP/graphene nanocomposites may be produced via in situ polymerization of vinyl pyrrolidone loaded with dispersed graphene, resulting in excellent load transfer and improved mechanical properties. Also, graphene-based epoxy nanocomposites with enhanced mechanical and electrical properties at a very low loading have also been achieved. The electrical conductivity increases by 7 orders of magnitude and the tensile strength increases by 25 % in a 0.3 wt% graphene based epoxy nanocomposite compared to the baseline epoxy. Compared with the prior literature, we achieve higher concentrations and lower percolation threshold for our pristine graphene dispersions and composites, respectively.

See more of this Session: Graphene and Carbon Nanotubes: Characterization, Functionalization, and Dispersion
See more of this Group/Topical: Nanoscale Science and Engineering Forum

(662d) A Method to Form Molecular Nanorods Using Inorganic Nanoparticles As Nucleation Seeds

Li Li, Sunxi Wang and Guangzhao Mao, Chemical Engineering and Materials Science, Wayne State University, Detroit, MI

This paper describes a method to nucleate molecular crystalline nanorods from inorganic nanoparticles. The method is demonstrated using *n*-carboxylic acids that crystallize from organic solvents, such as ethanol and isopropanol, by solvent evaporation. Thiolate-protected inorganic nanoparticles including CdSe, CdS, and Au are used as nucleation seeds. *n*-Carboxylic acids form highly ordered 2-D crystalline layers on highly oriented pyrolytic graphite (HOPG). However, when crystallization occurs in the vicinity of the nanoparticles, the persistent HOPG-dominated nano-stripe pattern is disrupted, and nanorods attached to the nanoparticles become the dominant feature. In order to understand the underlying mechanism, the effects of film-forming conditions, carboxylic acid chain length, nanoparticle size, and chemical composition of the nanoparticle are examined. It is determined that high colloidal stability of the nanoparticles is a prerequisite for seed-mediated nucleation. A carboxylic chain length range amenable to the nanorod formation is identified, as is the relationship between the nanoparticle size and the number of nanorods per nanoparticle. This study contributes to solution-based bottom-up processes of nanocrystals that may be applicable to a variety of molecular crystalline compounds.

See more of this Session: Nanocrystal Science and Technologies
See more of this Group/Topical: Materials Engineering and Sciences Division

(420l) Flame-Made Silica-Coated Nanophosphors

Georgios A. Sotiriou, Melanie Schneider and Sotiris E. Pratsinis, Department of Mechanical and Process Engineering, Particle Technology Laboratory, ETH Zurich, Zurich, Switzerland

Nanophosphors are a promising new class of inorganic labels for bio-imaging applications, possessing a narrow emission bandwith, good photostability and low toxicity. Additionally, silica coating of nanoparticles enhances the dispersibility, bio-compatibility and the surface functionalization properties demonstrably. Combining these excellent properties in silica coated core-shell nanophosphors could lead to the replacement of disadvantageous existing bio-imaging probes such as organic dyes (photobleaching) and quantum dots (blinking, toxicity).

In this study, pure and silica coated $Y_2O_3:Tb^{3+}$ nanophosphors were synthesized by flame spray pyrolysis. The produced particles consisted of the monoclinic crystal phase and showed bright green phosphorescence [1]. By conducting an annealing study of the silica coated, as well as the uncoated samples, modifications in size and crystallinity could be observed. While the coated samples showed a general enhancement in crystallinity, the uncoated crystals expressed a phase transition from the monoclinic to the cubic crystal phase and an increase in crystal size. Analyzing the luminescent properties of these particles, a higher light output could be detected in case of the monoclinic $Y_2O_3:Tb^{3+}$. Consequently, in terms of Tb^{3+} phosphorescence in an Y_2O_3 host matrix, the monoclinic crystal phase can be considered superior to the cubic phase.

[1] G.A. Sotiriou, M. Schneider, S.E. Pratsinis, Color-Tunable Nanophosphors by Codoping Flame-Made Y_2O_3 with Tb and Eu, *J. Phys. Chem. C* 115, 1084 (2010).

See more of this Session: Poster Session: Nanoscale Science and Engineering
See more of this Group/Topical: Nanoscale Science and Engineering Forum

(105d) Spinodal Dewetting to Create Self-Assembled and Organized Nanoparticles: A High-Throughput Approach

William D. Michalak, James B. Miller and Andrew J. Gellman, Chemical Engineering, Carnegie Mellon University, Pittsburgh, PA

Metal nanoparticles on structured supports are used in many technologicalapplications such as biosensing, energy harvesting,and electronics. In every case, the functions and properties of themetallic nanostructures depend on their composition and structure (i.e. size,shape, and spatial distribution). Challenges to using metal nanoparticlesin these applications are the difficulties of optimizing the structure-propertyfunctionality over a large structural domain. In this work, a new methodis described to create a morphological gradient of particles on a substrate ina repeatable and controlled manner over a relatively large spatial domain usingPd on silicon nitride. The approach, suited for high-throughputfabrication and characterization, is based on inducing precursor thin films todewet from a substrate through spinodal dewetting. Spinodal dewettingallows the creation of particles that have well-defined structural propertiesby adjusting a few variables: initial film thickness, annealing temperature andannealing time. The morphologies of the particles were characterizedusing scanning tunneling and atomic force microscopies, and hydrodynamicstability and integral geometry analyses to confirm the dewettingmechanism. In addition, the hydrodynamic instability theory provides aconnection to the thermophysical properties of the system. The dewettingapproach is general to any metal/support system and provides an alternative,inexpensive, and robust means to rapidly create metal nanostructures. Itshows promise for large scale production of the metal structures, as well asunderstanding basic material properties.

See more of this Session: Nanofabrication and Nanoscale Processing
See more of this Group/Topical: Nanoscale Science and Engineering Forum

(746b) Plasmonic Biosensors with Biocompatible Nanosilver

Georgios A. Sotiriou[1], Takumi Sannomiya[2], Alexandra Teleki[1], Frank Krumeich[1], Janos Vrs[2] and Sotiris E. Pratsinis[1], (1)Department of Mechanical and Process Engineering, Particle Technology Laboratory, ETH Zurich, Zurich, Switzerland, (2)Department of Information Technology and Electrical Engineering, Laboratory of Biosensors and Biolelectronics, ETH Zurich

Noble metal nanoparticles and their unique plasmonic properties allow them to be used as agents for novel in-vivo bio-applications, such as targeted drug delivery and cancer cell therapy [1]. Nanosilver is the optimum such material because of its lowest losses of the plasmonic properties in the visible spectrum. The toxic nanosilver surface [2], however, can destroy the surrounding healthy tissue and thus, challenges its safe employment [3]. Therefore, "curing" nanosilver's toxicity would open up its use for such innovative biological applications. Here, nanosilver particles are hermetically coated with a thin SiO_2 layer using a scalable aerosol method [4]. Uncoated or partially coated nanosilver exhibits substantial toxicity against *E. coli* bacteria, while encapsulating it prevents this. The performance of these ready-to-use, biocompatible nanosilver particles as biosensors is demonstrated by measuring the adsorption of bovine serum albumin (BSA) in a model sensing experiment. The silica coating prevents nanosilver agglomeration and thus, apart from the limited toxicity, enhances the biosensitivity.

[1] Sotiriou, G. A., Hirt, A. M., Lozach, P. Y., Teleki, A., Krumeich, F. & Pratsinis, S. E. Hybrid, Silica-Coated, Janus-like Plasmonic-Magnetic Nanoparticles. *Chem. Mater.* **23**, 1985-1992 (2011).

[2] Sotiriou, G. A. & Pratsinis, S. E. Antibacterial activity of nanosilver ions and particles. *Environ. Sci. Technol.* **44**, 5649-5654 (2010).

[3] Sotiriou, G. A., Sannomiya, T., Teleki, A., Krumeich, F., Vörös, J. & Pratsinis, S. E. Non-Toxic Dry-Coated Nanosilver for Plasmonic Biosensors. *Adv. Funct. Mater.* **20**, 4250-4257 (2010).

See more of this Session: Nanotechnology and Nanobiotechnology for Sensors
See more of this Group/Topical: Nanoscale Science and Engineering Forum

(747c) Hybrid, Silica-Coated, Janus-Like Plasmonic-Magnetic Nanoparticles

Georgios A. Sotiriou[1], Ann M. Hirt[2], Pierre-Yves Lozach[3], Alexandra Teleki[1], Frank Krumeich[1] and Sotiris E. Pratsinis[1], (1)Department of Mechanical and Process Engineering, Particle Technology Laboratory, ETH Zurich, Zurich, Switzerland, (2)Department of Earth Sciences, ETH Zurich, Zurich, Switzerland, (3)Department of Biology, ETH Zurich, Zurich, Switzerland

Hybrid magnetic/plasmonic nanoparticles possess properties originating from each individual material. Such properties are beneficial for biological applications including bio-imaging, targeted drug delivery, *in vivo* diagnosis and therapy. Limitations regarding their stability and toxicity, however, challenge their safe use [1]. Here, the one-step flame synthesis of composite SiO_2-coated Ag/Fe_2O_3 nanoparticles is demonstrated. The hermetic SiO_2 coating [2] does not influence the morphology, the superparamagnetic properties of the iron oxide particles and the plasmonic optical properties of the silver particles. It does prevent, however, the release of toxic Ag^+ ions from the nanosilver surface [3] and reduces the interaction field among the iron oxide particles, resulting in a stable suspensions with no signs of agglomeration (flocculation) and sedimentation [4]. The feasibility of these multi-component nanoparticles with superior properties and performance is explored by their specific binding with live tagged cells, and the detection as well as magnetic manipulation of the latter is demonstrated. Therefore, the hybrid SiO_2-coated Ag/Fe_2O_3 nanoparticles do not exhibit the limiting physical properties of each individual component, while retain all the desired ones, facilitating their safe employment in such bio-applications.

[1] Sotiriou, G. A. & Pratsinis, S. E. Antibacterial activity of nanosilver ions and particles. *Environ. Sci. Technol.* **44**, 5649-5654 (2010).

[2] Sotiriou, G. A., Sannomiya, T., Teleki, A., Krumeich, F., Vörös, J. & Pratsinis, S. E. Non-Toxic Dry-Coated Nanosilver for Plasmonic Biosensors. *Adv. Funct. Mater.* **20**, 4250-4257 (2010).

[3] Sotiriou, G. A., Teleki, A., Camenzind, A., Krumeich, F., Meyer, A., Panke, S. & Pratsinis, S. E. Nanosilver on nanostructured silica: Antibacterial activity and Ag surface area. *Chem. Eng. J.* **in press**, DOI: 10.1016/j.cej.2011.1001.1099 (2011).

[4] Sotiriou, G. A., Hirt, A. M., Lozach, P. Y., Teleki, A., Krumeich, F. & Pratsinis, S. E. Hybrid, Silica-Coated, Janus-like Plasmonic-Magnetic Nanoparticles. *Chem. Mater.* **23**, 1985-1992 (2011).

See more of this Session: Nanotechnology for In Vivo and In Vitro Imaging
See more of this Group/Topical: Nanoscale Science and Engineering Forum

(353a) Award Submission: Hybrid, Silica-Coated, Janus-Like Plasmonic-Magnetic Nanoparticles

Georgios A. Sotiriou[1], Ann M. Hirt[2], Pierre-Yves Lozach[3], Alexandra Teleki[1], Frank Krumeich[1] and Sotiris E. Pratsinis[1], (1)Department of Mechanical and Process Engineering, Particle Technology Laboratory, ETH Zurich, Zurich, Switzerland, (2)Department of Earth Sciences, ETH Zurich, Zurich, Switzerland, (3)Department of Biology, ETH Zurich, Zurich, Switzerland

Hybrid magnetic/plasmonic nanoparticles possess properties originating from each individual material. Such properties are beneficial for biological applications including bio-imaging, targeted drug delivery, *in vivo* diagnosis and therapy. Limitations regarding their stability and toxicity, however, challenge their safe use [1,2]. Here, the one-step flame synthesis of composite SiO_2-coated Ag/Fe_2O_3 nanoparticles is demonstrated. The hermetic SiO_2 coating [3] does not influence the morphology, the superparamagnetic properties of the iron oxide particles and the plasmonic optical properties of the silver particles. It does prevent, however, the release of toxic Ag^+ ions from the nanosilver surface [4] and reduces the interaction field among the iron oxide particles, resulting in a stable suspensions with no signs of agglomeration (flocculation) and sedimentation [5]. The feasibility of these multi-component nanoparticles with superior properties and performance is explored by their specific binding with live tagged cells, and the detection as well as magnetic manipulation of the latter is demonstrated. Therefore, the hybrid SiO_2-coated Ag/Fe_2O_3 nanoparticles do not exhibit the limiting physical properties of each individual component, while retain all the desired ones, facilitating their safe employment in such bio-applications.

[1] Sotiriou, G. A. & Pratsinis, S. E. Antibacterial activity of nanosilver ions and particles. *Environ. Sci. Technol.* **44**, 5649-5654 (2010).

[2] Sotiriou, G. A. & Pratsinis, S. E. Engineering nanosilver as an antibacterial, biosensor and bioimaging material. *Curr. Opin. Chem. Eng.* **in press**, doi: 10.1016/j.coche.2011.1007.1001 (2011).

[3] Sotiriou, G. A., Sannomiya, T., Teleki, A., Krumeich, F., Vörös, J. & Pratsinis, S. E. Non-Toxic Dry-Coated Nanosilver for Plasmonic Biosensors. *Adv. Funct. Mater.* **20**, 4250-4257 (2010).

[4] Sotiriou, G. A., Teleki, A., Camenzind, A., Krumeich, F., Meyer, A., Panke, S. & Pratsinis, S. E. Nanosilver on nanostructured silica: Antibacterial activity and Ag surface area. *Chem. Eng. J.* **170**, 547-554 (2011).

[5] Sotiriou, G. A., Hirt, A. M., Lozach, P. Y., Teleki, A., Krumeich, F. & Pratsinis, S. E. Hybrid, Silica-Coated, Janus-like Plasmonic-Magnetic Nanoparticles. *Chem. Mater.* **23**, 1985-1992 (2011).

(369c) Removal of Nanoparticles In Semiconductor Manufacturing Effluents Using Porous Media Filtration

Jeff Rottman, Reyes Sierra and Farhang Shadman, Chemical and Environmental Engineering, University of Arizona, Tucson, AZ

The use of engineered nanoparticles in industrial manufacturing has rapidly increased due to their unique electronic, optical, thermal and photoactive properties. Nanoparticles are used in various fields including textiles, biomedical applications, food and agriculture and semiconductor manufacturing. Frequently used nanoparticles include titania (TiO_2), which is commonly used in pigments and as a photocatalyst, and silica (SiO_2), which is commonly used as an additive in chemical mechanical planarization (CMP) slurries in semiconductor manufacturing. Along with the benefits these nanoparticles provide, numerous concerns into the environmental, safety, and health effects of the released particles arise. Understanding the fundamental interactions of these nanoparticles with typical treatment technologies will allow for improved design of abatement technologies. Porous media filtration is a common technique used to remove particulates from water. The simplicity and ubiquity of this process makes it ideal for the removal of nanoparticles in waste streams. The principles of nanoparticle transport in porous media also relate to the environmental fate of these nanoparticles in soil and ground water.

Custom nanoparticles of silica have been developed which have a fluorescent core and controllable final size. These particles, due to their fluorescence, can be measured at significantly lower concentrations than in current studies. Additionally, the ability to control the final size of the particles provides a true comparison of size effects on nanoparticle retention. A test bed, equipped with real-time and on-line spectrometry, was developed and used to study the dynamics of capture and removal of both TiO_2 and these SiO_2 nanoparticles from wastewater. Studies were conducted comparing the effect of key solution, nanoparticle, and bed characteristics, including pH, ionic strength, the presence of surfactants, heavy metals, and dispersants, as well as particle and collector sizes. A process model is under development for analysis of data from this novel experimental method and for elucidating the fundamental parameters affecting nanoparticle retention and removal in such systems.

See more of this Session: Environmental Implications of Nanomaterials: Fate and Transport
See more of this Group/Topical: Topical C: Environmental Aspects, Applications, and Implications of Nanomaterials and Nanotechnology

(292b) Quantifying the Origin of Nanosilver Ions and Their Antibacterial Activity

Georgios A. Sotiriou[1], Andreas Meyer[2], Jesper Knijnenburg[1], Sven Panke[2] and Sotiris E. Pratsinis[1], (1) Department of Mechanical and Process Engineering, Particle Technology Laboratory, ETH Zurich, Zurich, Switzerland, (2)Department of Biosystems Science and Engineering, ETH Zurich

Nanosilver is one of the most prominent components in several nanotechnology products, primarily for its bactericidal and plasmonic properties [1,2]. Its use, however, is questioned as U.S. EPA has been petitioned to label it as pesticide. To better understand its antibacterial activity, nanosilver with closely controlled size immobilized on nanostructured silica is generated by flame aerosol technology [3]. The antibacterial activity of small (<10 nm) nanosilver particles is attributed mostly to released Ag^{+} ions from their surface upon contact with water [1]. The origin of these ions is explored systematically now by conditioning the surface of various nanosilver particles (4-9 nm) immobilized on nanostructured silica by wet-impregnation and dry flame-spray pyrolysis. Silica prevents nanosilver flocculation without the need of any surface functionalization (e.g. citrate) that would compromise the Ag^{+} ion release kinetics. Washing or reducing the Ag/SiO_2 nanoparticles removes the oxide layer from the nanosilver surface minimizing their ion leaching. The amount of leached Ag^{+} ions corresponds to the dissolution of one or two silver oxide monolayers. Finally, the antibacterial activity of small (< 10 nm) nanosilver particles that have their surface oxide layer removed or reduced is significantly lower than the freshly prepared ones by either wet or dry methods.

1. Sotiriou, G. A.; Pratsinis, S. E., Antibacterial activity of nanosilver ions and particles. *Environ. Sci. Technol.* **2010,** *44*, 5649.

2. Sotiriou, G. A.; Sannomiya, T.; Teleki, A.; Krumeich, F.; Vörös, J.; Pratsinis, S. E., Non-Toxic Dry-Coated Nanosilver for Plasmonic Biosensors. *Adv. Funct. Mater.* **2010,** *20*, 4250.

3. Sotiriou, G. A.; Teleki, A.; Camenzind, A.; Krumeich, F.; Meyer, A.; Panke, S.; Pratsinis, S. E. Nanosilver on nanostructured silica: Antibacterial activity and Ag surface area. *Chem. Eng. J.* **2011,** *170,* 547.

See more of this Session: Environmental Implications of Nanomaterials: Biological Interactions
See more of this Group/Topical: Topical C: Environmental Aspects, Applications, and Implications of Nanomaterials and Nanotechnology

(420m) Understanding the Stabilization of Liquid-Phase Exfoliated Graphene In Polar Solvents: Molecular Dynamics Simulations and Kinetic Theory of Colloid Aggregation

Chih-Jen Shih[1], Shangchao Lin[2], Michael Strano[3] and Daniel Blankschtein[1], (1)Department of Chemical Engineering, Massachusetts Institute of Technology, Cambridge, MA, (2)Mechanical Engineering, Massachusetts Institute of Technology, Cambridge, MA, (3)Chemical Engineering, Massachusetts Institute of Technology, Cambridge, MA

Understanding the solution phase dispersion of pristine, unfunctionalized graphene is important for the production of conducting inks and top-down approaches to electronics. It can also be used as higher-quality alternatives to chemical vapor deposition. In this talk, we discuss a theoretical framework which utilizes molecular dynamics simulation and the kinetic theory of colloid aggregation to inform the mechanism of stabilization of liquid-phase exfoliated graphene sheets in N-methylpyrrolidone (NMP), dimethylformamide (DMF), N,N'-dimethylsulphoxide (DMSO), γ-butyrolactone (GBL), and water (H_2O).

By calculating the potential of mean force between two solvated graphene sheets using molecular dynamics (MD) simulations, we found that the dominant barrier hindering the aggregation of graphene is the last layer of confined solvent molecules between the graphene sheets, resulting from the strong affinity of the solvent molecules for graphene. The origin of the energy barrier responsible for repelling the sheets is the steric repulsions between solvent molecules and graphene before the desorption of the confined single-layer solvent layer.

We have formulated a kinetic theory of colloid aggregation to model the aggregation of graphene sheets in the liquid phase in order to predict stability using the simulated potential of mean force. With only one adjustable parameter, the average collision area, which can be estimated from experimental data, our theory can describe the experimentally observed degradation of the single-layer graphene fraction in NMP. We use these results to rank potential solvents according to their ability to disperse pristine, unfunctionalized graphene and find that NMP ≈ DMSO > DMF > GBL > H_2O, consistent with the wide-spread use of the first three solvents for this purpose.

See more of this Session: Poster Session: Nanoscale Science and Engineering
See more of this Group/Topical: Nanoscale Science and Engineering Forum

(615d) Bi- and Tri- Layer Graphene Solutions

Chih-Jen Shih[1], Zhong Jin[1], Shangchao Lin[2], Geraldine LC Paulus[3], Nigel Forest Reuel[3], Qing Hua Wang[3], Daniel Blankschtein[3] and Michael Strano[1], (1)Chemical Engineering, Massachusetts Institute of Technology, Cambridge, MA, (2)Mechanical Engineering, Massachusetts Institute of Technology, Cambridge, MA, (3)Department of Chemical Engineering, Massachusetts Institute of Technology, Cambridge, MA

Many applications of pristine, AB-stacked bilayer and trilayer graphene require synthesis methods to control the number of stacked layers. Solution dispersion is promising for printable electronics and nanocomposites. However, no method is currently available for the predetermined exfoliation of AB-stacked or large area flakes such that conventional photolithography is possible. Here, we demonstrate that graphite intercalation with ionic, non-covalent intercalants to yield Stage-2 and Stage-3 compounds results in bilayer- and trilayer-enriched graphene dispersions, respectively. The intercalation and subsequent expansion of graphite allows mild sonication to yield stable dispersions, decreasing flake size reduction while maintaining AB stacking as confirmed by Raman spectroscopy. An on-chip micro-hydrodynamic separation method localizes large area graphene flakes (up to 50 μm^2). This constitutes the only viable route at this time for the mass-production of AB stacked bi- and tri- layer graphene. Unannealed bilayer flakes exhibit low resistivity (~1 kΩ) and mobilities which are higher (~400 $cm^2V^{-1}s^{-1}$) than those attained utilizing other solution methods.

See more of this Session: Synthesis of Graphene and Carbon Nanotubes: Kinetics, Mechanisms and Reactor Design
See more of this Group/Topical: Nanoscale Science and Engineering Forum

(321b) Characterization of Glass Transition Temperatures In Block Copolymer/Ionic Liquid Micelle Cores

Michelle M. Mok, Chemistry, University of Minnesota, Minneapolis, MN and Timothy P. Lodge, Department of Chemistry and Department of Chemical Engineering and Materials Science, University of Minnesota

Achieving equilibrium structures in block copolymer micelles is significantly more difficult than in small-molecule surfactant micelles, due to the much higher thermodynamic barriers and slower kinetics associated with the long-chain structures of their unimers. Polymer chains can also become trapped or frozen within micelles in cases when the core-forming block has a glass transition temperature (Tg) at the temperature ranges of interest. Measurements of micelle core glass transitions are challenging, given the nanoscale volumes and small volume fractions being probed. Such measurements would not only provide insight into block copolymer micellar equilibration, but also have the potential to yield contributions to the field of polymer systems confined at the nanoscale. The physical properties of such systems can be significantly different from those of bulk systems; this has been demonstrated in a number of cases in terms of changes to the Tg of polymer thin films, but few studies have been carried out in other geometries. Frozen micelles present an important opportunity for expanding the range of studies in low-dimensional structures

We discuss the development of a fluorescence-based technique for the investigation of Tgs in polystyrene (PS) micelles cores in ionic liquid solvents. Specifically, we demonstrate the ability to detect the transition from their fully glassy state to their fully rubbery state in micelles of poly(styrene-b-ethylene oxide) and poly(styrene-b-methylmethacrylate). The use of non-volatile ionic liquids as the solvent allowed for studies over extended temperature ranges while remaining in a stable, liquid state. The systems are explored as a function of copolymer block molecular weight and composition and micelle preparation protocol. We find that the PS core Tgs are shifted 20–45 °C below reported bulk PS homopolymer Tgs of equivalent molecular weight, and that they can also be significantly broader (up to 45 °C in breadth).

See more of this Session: Nanoscale Structure In Polymers II
See more of this Group/Topical: Materials Engineering and Sciences Division

(600c) Stability and Separation Performance of Highly c-Oriented AFI-Type Aluminophosphate Membranes

Jared A. Stoeger[1], Miguel Palomino[2], Charitomeni M. Veziri[3], Avelino Corma[2], Nick K. Kanellopoulos[3], Georgios N. Karanikolos[3] and Michael Tsapatsis[1], (1)Chemical Engineering and Materials Science, University of Minnesota, Minneapolis, MN, (2)Instituto de Tecnologia Quimica (CSIC-UPV), Universidad Politecnica de Valencia, Valencia, Spain, (3)Institute of Physical Chemistry, Demokritos National Research Center, Athens, Greece

Continuous,highly *c*-oriented aluminophosphate films(structure code AFI) were grown on porous α-Al_2O_3supports in order to investigate their potential for molecular separationapplications. Variations in metalsubstitution ($AlPO_4$-5, CoAPO-5, CoSAPO-5), growth method (forced airconvection oven, microwave), and calcination technique (conventional, rapidthermal processing) were chosen to examine the conditions by which membranequality was optimized. Followinghydrothermal growth in a forced air convection oven, $AlPO_4$-5 andCoAPO-5 membranes underwent densification to an AlPO-tridymitephase following heat treatment, resulting in reduced membrane functionalitywhich may be partially avoided through the use of rapid thermal processing. Onthe other hand, membranes prepared by an optimized method were found to bethermally stable with high membrane quality as inspected through pervaporationmeasurements consisting of a liquid hydrocarbon feed of n-heptaneand 1,3,5-triisopropylbenzene. We will discuss the new aspects thatallowed fabrication of oriented, continuous, high-quality aluminophosphatemembranes and we will provide useful insight into their potential forapplications such as molecular separations, catalysis, or host-guestassemblies.

See more of this Session: Nanostructured Thin Films
See more of this Group/Topical: Materials Engineering and Sciences Division

(125b) Smart Sustainable Manufacturing

Thomas F. Edgar, Chemical Engineering, The University of Texas at Austin, Austin, TX

Maximizing energy efficiency is receiving increased attention as a way to reduce the use of fossil fuels and the resulting production of greenhouse gases, thus providing an avenue to address possible future legislation of cap and trade or a carbon tax. Automation, process control, and optimization are critical technologies to operate plants in the most efficient way and to pursue the smart manufacturing paradigm. The increased use of renewable energy such as solar or wind power reduces carbon usage but adds a dynamic element to power production in process plants, which may involve interfacing with smart grids. Time of day pricing of power and use of demand response techniques to flatten load profiles will be important ingredients of smart grids. Increased usage of thermal and other energy storage systems will also give industrial energy users some additional degrees of freedom to deal with the dynamic power conditions.

See more of this Session: Sustainability Plenary
See more of this Group/Topical: Sustainable Engineering Forum

(704d) Peptide-Assisted Synthesis and Functionalization of Nanomaterials

Rajesh R. Naik, Materials and Manufacturing Directorate, Air Force Research Laboratory, Wright-Patterson-AFB, OH

Biological building blocks are invaluable in offering exceptional control, chemical specificity, and self-assembling properties, while nanomaterials exhibit unique optical, electronic and catalytic properties. Combination of both bio- and nano-materials can result in the creation of hybrid materials that exhibit multifunctional properties. One of the key approaches has been to exploit the ability of biomolecules to synthesize, assemble and/or functionalize nanostructures. The highly specific molecular recognition and self-assembly properties of biomolecules, along with their design flexibility, make them attractive as building blocks for fabricating and assembling nanostructures. Our research highlights the utility of biomolecules to control nucleation, growth and stabilization of nanoparticles, as well as in the assembly of nanomaterials. We have demonstrated the utility of biomolecules for synthesizing functionalization and assembly nanomaterials for catalytic, biosensing and other non-medical applications. In my talk I will highlight the use of peptide-enabled synthesis of nanomaterials for catalytic applications

See more of this Session: Nanoscale Science and Engineering In Biomolecular Catalysis I
See more of this Group/Topical: Nanoscale Science and Engineering Forum

(298b) Molecular Insights Into the Surface Morphology, Layering Structure, and Aggregation Kinetics of Surfactant-Stabilized Graphene Dispersions

Shangchao Lin[1], Chih-Jen Shih[2], Michael S. Strano[2] and Daniel Blankschtein[2], (1)Department of Chemical Engineering and Department of Mechanical Engineering, Massachusetts Institute of Technology, Cambridge, MA, (2)Department of Chemical Engineering, Massachusetts Institute of Technology, Cambridge, MA

The production of graphene with open bandgaps for the manufacturing of graphene-based electronic and optical devices requires synthesis methods to either control the number of layers in order to enrich AB-stacked bilayer or trilayer graphene, or to control the extent of functionalization of monolayer graphene. Solution-phase dispersion of graphene is promising for both methods to create printable electronics and nanocomposites. However, both methods face common challenges, including controlling the surface morphology, reducing the turbostratic layering, and enhancing the dispersion stability. To address these challenges at the molecular level, we successfully combined molecular simulations, theoretical modeling, and experimental measurements. First, we probed the surface structure and electrostatic potential of monolayer graphene dispersed in a sodium cholate (SC) surfactant aqueous solution, which exhibits 2D-sheets partially covered with a monolayer of negatively-charged cholate ions. Similar to the case of carbon nanotube functionalization, one may regulate the binding affinity of charged reactants for graphene functionalization by manipulating the surface morphology. Subsequently, we quantified the interactions between two graphene-surfactant assemblies by calculating the potential of mean force (PMF) between two surfactant-covered graphene sheets, which confirmed the existence of a metastable bilayer graphene structure due to the steric hindrance of the confined surfactant molecules. The traditional DLVO theory was found to be adequate to explain the long-range electrostatic repulsions between the ionic surfactant-covered graphene sheets, but was unable to account for the dominant, short-range steric hindrance imparted by the confined surfactant molecules. Interestingly, one faces a dilemma when using surfactants to disperse and stabilize graphene in aqueous solution: on the one hand, surfactants can stabilize graphene aqueous dispersions, but on the other hand, they prevent the formation of new AB-stacked bilayer and trilayer graphene resulting from the re-aggregation process. Finally, the lifetime and time-dependent distribution of various graphene layer types were predicted using a kinetic model of colloid aggregation, and each graphene layer type was further decomposed into subtypes, including the AB-stacked species and various turbostratic species. The kinetic model of colloid aggregation developed here can serve as a useful tool to evaluate the quality of graphene dispersions for subsequent substrate-transferring or functionalization processes.

See more of this Session: Graphene and Carbon Nanotubes: Absorption and Transport Processes

(693f) Advanced Separation of Empty and Water-Filled Nanotubes

Jeffrey A. Fagan[1], Vinayak Rastogi[2], Jeffrey R. Simpson[3] and Angela R Hight Walker[2], (1)Polymers Division, National Institute of Standards and Technology, Gaithersburg, MD, (2)National Institute of Standards and Technology, Gaithersburg, MD, (3)Towson University, Towson, MD

Water-filling of open-ended nanotubes has recently been shown to dramatically affect the properties of the nanotubes in solution (Fagan *et al. ACS NANO* **2011**). Here we will present strategies for the fine separation of empty and water-filled populations from mixed solutions and the effects of scaling up the processes to yield multiple milligrams/run of separated material. The separated populations are furthermore demonstrated to behave differently during subsequent purification on the basis of chirality and electronic type. Examples of the improved and altered properties of the empty and water-filled nanotubes will be given, and the resulting implications discussed.

See more of this Session: Graphene and Carbon Nanotubes: Characterization, Functionalization, and Dispersion
See more of this Group/Topical: Nanoscale Science and Engineering Forum

EFFECTS OF VARYING SURFACE FILM THICKNESS ON PARTICLE ADHESION IN SEMICONDUCTOR MATERIAL-BASED SYSTEMS

Katie M. Smith, Purdue University, West Lafayette, Indiana, USA
Jeffery W. Butterbaugh, FSI International, Chaska, Minnesota, USA
Stephen P. Beaudoin, Purdue University, West Lafayette, Indiana, USA

Abstract

The effects of surface film thickness on particle adhesion are examined for a number of common microelectronics device materials. These effects are investigated for a laboratory-scale model system, using atomic force microscopy, as well as an industrial-scale system, via particle removal efficiency experiments. Our results show that surface films below a critical thickness δ^* are effectively invisible, from an adhesion perspective, while those having thicknesses greater than δ^* exhibit bulk material behavior, essentially screening any influence from the underlying substrate. In systems with surface films of thicknesses (approximately) equal to δ^*, the net adhesion results from a combination of film and substrate contributions.

Introduction

The adhesion of contaminant particles to wafer and photomask surfaces is a problem of critical importance to the manufacture of microelectronics (1–10). As device dimensions continue to shrink and restrictions on material loss become more stringent, many are approaching operating regimes where the dimensions of the devices in one or more spatial directions are on the same length scale as van der Waals-London (vdW-L) forces, the primary forces that control particle adhesion. There is no well-accepted methodology for describing adhesion forces in systems with compositional variations at these length scales. As a result, the microelectronics industry does not have a theoretical framework to guide the development of new surface processing or cleaning technology. Research elucidating the fundamental origins and mechanisms of particle adhesion is essential, particularly in the context of wafer contamination prevention and removal.

Of particular interest here is the effect of varying surface film thickness on particle adhesion to layered substrates. The interaction phenomena arising between bodies in contact (e.g., a contaminant particle and a wafer surface) are governed predominantly by vdW-L dispersion forces (11–15). These forces have their origins in the electrostatic interactions between the dipoles within materials (14). A variety of approaches has been developed to describe the vdW-L forces between macroscopic bodies; however some of these appear in the literature without considering the appropriate regions of validity for the systems of interest. Furthermore, limited studies are available that focus on the adhesion of *inhomogeneous*, *layered* particles or substrates (16). In this work, we develop theory supporting the hypothesis that the thickness of the surface film material will affect the resultant interaction force between particle and substrate. We then examine a model system consisting of a pre-manufactured Si_3N_4 probe—representing a contaminant particle—and a Si substrate onto which GaN films had been deposited via atomic layer deposition (ALD). Finally, we present results from industrial-scale experiments involving the removal of model contaminant particles of SiO_2 from the surface of 300mm-diameter blanket Si wafers with surface films comprised of nano-scale TiN of varying thickness deposited over SiO_2. These results demonstrate the significant effect of the thickness of the nano-scale surface film on the resulting particle adhesion.

Theory

Model Development

In this work, we have used Vold's approach (17), with minor modifications in order to account for the non-additivity in multi-layered, highly polarized systems at small separation distances (i.e. <25 nm). Consider two multi-layered, semi-infinite slabs (named **L** and **R**) consisting of any number of different materials; a schematic of such a system appears in Figure 1.

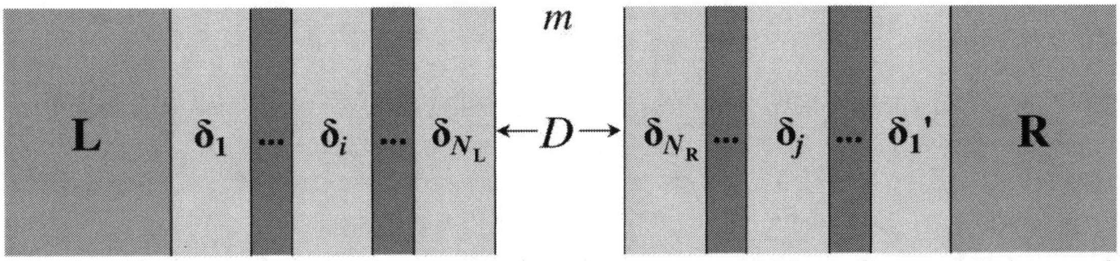

Figure 1. A schematic of two multi-layered parallel plates, separated by a distance D, interacting across a medium m.

The general expression for the vdW-L interaction energy between **L** and **R** is

$$E = -\sum_{i=1}^{N_L}\sum_{j=1}^{N_R} A(i,j)H_{ij} \qquad [1]$$

where N_L and N_R are the number of layers in the **L**- and **R**-side bodies; $A(i,j)$ is the 'effective' Hamaker constant for the interaction between any two layers i and j; and H_{ij} is a function that depends on the geometry of the layers as well as the distance between them. Derivations of H_{ij} for various ideal-geometry systems are available elsewhere (18). Modifying Equation [1] to correct for non-additivity and polarization effects, we arrive at

$$E_{\mathbf{L}\text{-}\delta_i\text{-}m\text{-}\delta_j\text{-}\mathbf{R}} = -\sum_{i=1}^{N_L}\sum_{j=1}^{N_R} A^{\text{eff}}(i,j)H_{ij} \qquad [2]$$

where $A^{\text{eff}}(i,j) = \alpha A(i,j)$ and α is a correction factor (19). For a system of layered materials, $A^{\text{eff}}(i,j)$ can be measured experimentally using instruments such as the surface forces apparatus (SFA) or atomic force microscope (AFM).

In this work, the idealized model system consists of a sphere {**1**} of radius R separated by a distance D across a medium {**0**} from a semi-infinite slab {**2**} having a surface film {**2'**} of thickness δ. See Figure 2 for a diagram of the idealized model system. The expression for the vdW-L interaction energy of this system is

$$E_{\{1\}\text{-}\{0\}\text{-}\{2'|2\}} = -\sum_{i=1}^{1}\sum_{j=1}^{2} A^{\text{eff}}(i,j)H_{ij} \qquad [3]$$

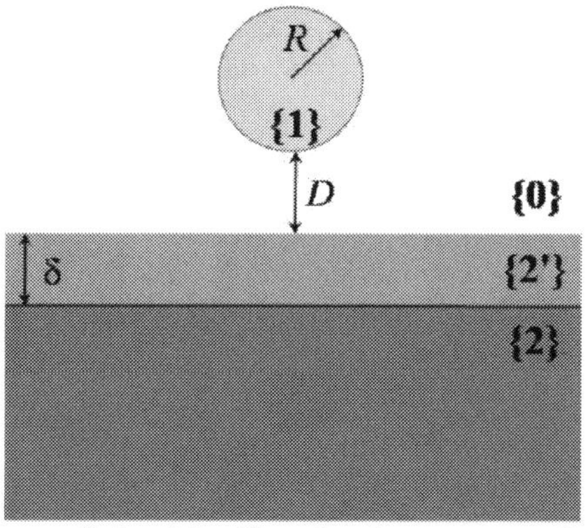

Figure 2. Diagram of the idealized sphere-coated flat plate system.

The vdW-L interaction force expression is obtained by differentiating Equation [3] with respect to separation distance D, which yields

$$F_{\{1\}\text{-}\{0\}\text{-}\{2'|2\}} = -\frac{\partial}{\partial D} E_{\{1\}\text{-}\{0\}\text{-}\{2'|2\}} = -\sum_{i=1}^{1}\sum_{j=1}^{2} A^{\text{eff}}(i,j)\frac{\partial}{\partial D}H_{ij} \qquad [4]$$

where $A^{\text{eff}}(i,j)$ is not affected by the differentiation as it is dependent only on the material properties of the multi-layer structure, which are taken to be unvarying. For the smooth sphere-flat plate system, it has been shown (18) that

$$\frac{\partial}{\partial D}H_{ij} = \frac{R}{6D_{ij}^2} \qquad [5]$$

where D_{ij} is the separation distance between the ith and j^{th} layers. Combining Equations [4] and [5], the final expression for the net vdW-L interaction force between a smooth sphere and a coated flat plate is given by

$$\frac{F_{\{1\}\text{-}\{0\}\text{-}\{2'|2\}}}{R} = \frac{1}{6}\left[A_{102'}^{\text{eff}}\left(\frac{1}{D^2} - \frac{1}{(D+\delta)^2}\right) + A_{102}^{\text{eff}}\left(\frac{1}{(D+\delta)^2}\right)\right] \qquad [6]$$

Critical Film Thickness

It is apparent from Equation [6] that in the limit of very small δ/D, the film will appear transparent to the particle such that as it approaches the substrate it will be interacting primarily with the underlying bulk material. However, as δ/D increases to the point where eventually $\delta >> D$, the dominant interaction transitions from that between the particle and the underlying substrate to that between the particle and the coating. This transition suggests the existence of three regimes of behavior: 1) the 'bulk substrate' region where the coating is so thin it is invisible to the approaching particle from an adhesion perspective; 2) the 'bulk coating' region where the coating is so thick that the adhesion is dominated by the properties of the coating; and 3) the 'transition region' where the adhesion is influenced by both the coating and the substrate. Coating thicknesses in the transition region are denoted by δ^*.

Consider a single spherical particle {**1**} approaching an isolated thin film {**2'**}, both of which are surrounded by a uniform medium {**0**}. The vdW-L interaction force between the particle and the film is obtained from Equation [6] by replacing material **2** with the medium **0** and recognizing that A_{100}^{eff} is equal to zero; this force expression is

$$\frac{F_{\{1\}\text{-}\{0\}\text{-}\{2'\}}}{R}=\frac{1}{6}A_{102'}^{\text{eff}}\left(\frac{1}{D^2}-\frac{1}{(D+\delta)^2}\right) \quad [7]$$

Taking the limit of Equation [7] as $\delta \rightarrow \infty$ yields the formula for the adhesion force between the spherical particle and, what is now, a planar slab of bulk film material (13):

$$\frac{F_{\{1\}\text{-}\{0\}\text{-}\{2'_{\text{bulk}}\}}}{R}=A_{102'}^{\text{eff}}\frac{1}{6D^2} \quad [8]$$

We define *critical film thickness* δ^* such that as $\delta \rightarrow \delta^*$, $F_{\{1\}\text{-}\{0\}\text{-}\{2'\}} \rightarrow F_{\{1\}\text{-}\{0\}\text{-}\{2'_{\text{bulk}}\}}$ i.e.

$$\frac{\Delta F}{F_{\{1\}\text{-}\{0\}\text{-}\{2'_{\text{bulk}}\}}}=\frac{F_{\{1\}\text{-}\{0\}\text{-}\{2'_{\text{bulk}}\}}-F_{\{1\}\text{-}\{0\}\text{-}\{2'\}}}{F_{\{1\}\text{-}\{0\}\text{-}\{2'_{\text{bulk}}\}}}\rightarrow \varepsilon \text{ as } \delta \rightarrow \delta^* \quad [9]$$

where ε is termed the *permissible fractional error*. Substituting the force expressions from Equations [7] and [8] into Equation [9] yields the following formula for δ^*:

$$\delta^*=\left(\varepsilon^{-1/2}-1\right)D \quad [10]$$

Experimental

Model System

The materials studied in these investigations include Si, SiO_2, Si_3N_4, and GaN. Sample substrates consisted of silicon coupons (~1 cm^2) having 5-, 11-, 12-, 16-, and 80nm thick GaN films, which were prepared via ALD, in the Anderson lab at the University of Florida (*Gainesville, FL*). Coupons of bare Si (with native oxide layer) and 'bulk' GaN (5μm on sapphire, prepared by University Wafer, *Boston, MA*) represented the limits of very thin coating/no coating (Si with native oxide), or very thick coating/'bulk' coating (5μm GaN). To simulate spherical contaminant particles, pre-manufactured AFM probed cantilevers, made of Si_3N_4 and having initial nominal radii of curvature (ROC) of ~15 nm, were employed (Veeco Probes, model OTR8-10, nominal spring constant 0.1 N/m). The exact ROC values of these probes were found using micrographs taken with a Hitachi S-4800 field emission scanning electron microscope (FE-SEM), in the Birck Nanotechnology Center at Purdue University.

Force measurements were made with a Veeco MultiMode® V Picoforce® scanning probe microscope (SPM) connected to a Veeco/Digital Instruments Nanoscope IV controller. All measurements were taken in a gaseous nitrogen ambient, maintained at a relative humidity of ~15%. A StaticMaster® ionizer (Amstat Industries, Inc.) was placed in the vicinity of the probe-substrate system to eliminate any electrostatic phenomena. For all samples, at least 1,000 approach-retract cycles were recorded in order to obtain statistically reliable average values for the adhesion force between probe and substrate.

Industrial-Scale System

This work was performed on six 300mm-diameter SiO_2-on-Si blanket wafers, onto which varying thicknesses (nominally: 5, 10, 15, and 20nm) of TiN had been deposited (prepared at SEMATECH, *Albany, NY*). Challenge particles were MSP dry-deposited SiO_2 particles having size distribution peaks at 55 and 110nm in diameter. After challenge particle deposition, the wafers were allowed to "age" for ~24 hours, and then analyzed using a KLA-Tencor SP1 to obtain pre-clean particle counts. Wafers were cleaned, at FSI International (*Chaska, MN*), using a less aggressive cryogenic aerosol particle removal process in order to clearly see the effects of TiN film thickness on particle removal efficiency (PRE).

Results

Model System: Force of Adhesion

The results from the AFM adhesion (pull-off) force experiments are shown in Figure 3. Note that the pull-off force values have been normalized with respect to the ROC of the Si_3N_4 probes.

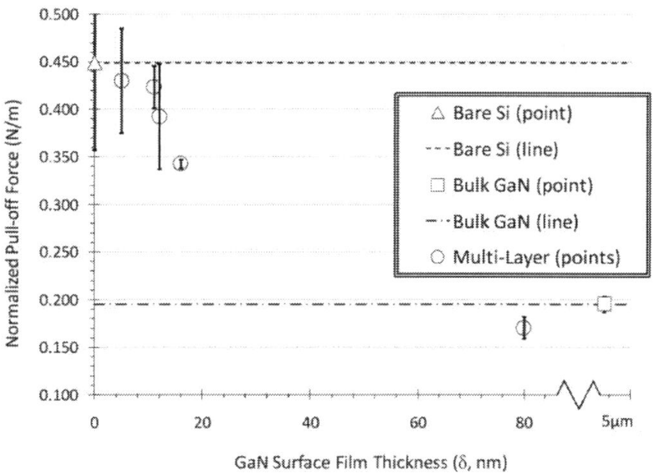

Figure 3. Variation in Si_3N_4 probe pull-off force with GaN surface film thickness.

We observe three regimes of behavior in Figure 3: Regime I, 'bulk substrate'; Regime II, 'transition'; and Regime III, 'bulk coating'. Regime I encompasses the bare Si data point (Δ) and the 5- and 11-nm thick GaN film data points. The pull-off force values for these two GaN film thicknesses vary from that of the bare Si by 4.2% (δ = 5nm) and 5.8% (δ = 11nm), both of which lie within the range of experimental error (~8%) for these measurements. Regime III includes the bulk GaN data point (□) and the 80-nm thick GaN film data point. At this GaN film thickness, the pull-off force value differs by 62% from the bare Si and by only 12% from the bulk GaN. Thus, an 80-nm thick GaN film on a Si substrate behaves much more like bulk GaN—since the contribution from the underlying Si is small—than bulk (bare) Si.

Regime II, the transition region, begins at δ = 12nm and extends beyond 20nm. For GaN film thicknesses of 12 and 16nm, the pull-off force values change by 12.5% and 23.6%, respectively, from the bare Si. This implies some combination of GaN film and underlying Si substrate contributions, both being significant to the determination of the net pull-off force. Regime II is also the region within which the greatest changes in pull-off force occur for the smallest changes in δ, which is seen clearly in Figure 3.

Industrial-Scale System: Particle Removal Efficiency

The PRE was measured when the cryogenic aerosol clean was applied. The results of the PRE experiments appear in Figure 4.

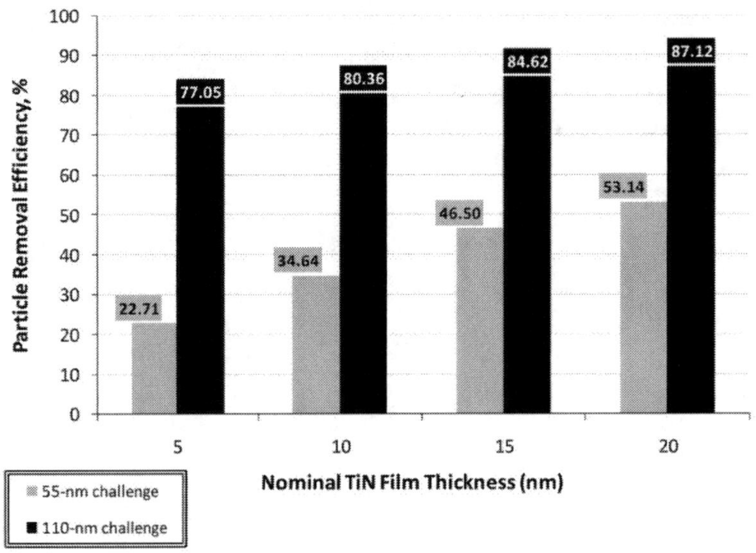

Figure 4. Variation in PRE with (nominal) TiN surface film thickness.

As seen in Figure 4, both 55- and 110-nm SiO_2 challenge PREs improve with TiN surface film thickness. Therefore, from a wafer cleaning perspective, a thin coating of TiN will result in enhanced wafer cleanliness (of SiO_2 particles) compared to that of the uncoated counterpart. The effect of particle size on PRE (for a cryogenic aerosol process) can also be seen in Figure 4: PRE increases as the contaminants become larger. This phenomenon is attributed to the fact that larger particles afford greater amounts of projected surface area for the transfer of momentum to occur from aerosolized ice crystals to contaminants thereby causing the latter's ejection from the wafer surface.

Figure 4 further elucidates the effect of particle size on the locations of the three regimes of behavior. For the 110nm challenge particles, a 15-nm change in TiN film thickness resulted in an improvement of the PRE just over 10%, while the same change improved the 55nm challenge PRE by more than 30%. We therefore conclude that nominal TiN film thicknesses of 5–20nm likely constitute some portion of the transition region (Regime II) for contaminants measuring 55nm in size, and for 110nm particles, comprise either of the bulk regions (Regime I or III, depending on the affinity of the contaminant particle material for TiN) instead.

Summary

The thickness of surface films on microelectronics devices has a significant impact on the magnitude of the forces with which particles will adhere to these surfaces, and therefore on the effectiveness of wafer cleaning processes to remove particulate contaminants. The range of influence imposed by the surface film thickness can be divided into three regions: at one extreme, the bulk substrate regime, in which the film is sufficiently thin to exhibit little to no contribution to the net force of adhesion between a particle and the surface; at the other extreme, the bulk coating regime, in which the film is so thick that any contribution from the underlying substrate to the adhesion force is small to effectively negligible; and a transition regime, lying between the two bulk regimes, in which the force of adhesion results from a combination of film and substrate contributions. The effects of surface film thickness are observed on the industrial scale when considering particle removal efficiency: depending on the affinity of the contaminant material for the film material and the size of the contaminant particles, the coating on a wafer can lead to considerable improvements in PRE.

Acknowledgments

The authors would like to thank Bill Taylor at SEMATECH for supplying the substrates for the PRE experiments, Murthy Peri and Greg Thomes, both of FSI International, for assistance with the PRE experiments, and Oh Hyun Kim, Joseph Revelli, and Tim Anderson, all of the University of Florida-Gainesville, for preparing the GaN thin film samples.

References

1. The International Technology Roadmap for Semiconductors: 2010, Yield Enhancement, Table ITWG14, pages 45–46.
2. J. W. Butterbaugh, S. Loper, G. Thomes, and D. Sheu, *MICRO*, **17**(6), 33 (1999).
3. J. W. Butterbaugh, S. Loper, and T. Wagener, in *Cleaning Technology in Semiconductor Device Manufacturing VIII/2003*, J. Ruzyllo, T. Hattori, R. L. Opila, and R. E. Novak, Editors, PV 2003-26, p. 116, The Electrochemical Society Proceedings Series, Pennington, NJ (2004).
4. T. J. Wagener and K. Kawaguchi, in *IEEE/SEMI Advanced Semiconductor Manufacturing Conference and Workshop/2004*, p. 467, IEEE ASMC Proceedings (2004).
5. T. J. Wagener, J. F. Weygand, and G. P. Thomes, *Solid State Phenomena*, **103–104**, 181 (2004).
6. J. M. Lauerhaas, J. F. Weygand, and G. P. Thomes, in *IEEE/SEMI Advanced Semiconductor Manufacturing Conference and Workshop/2005*, p. 11, IEEE ASMC Proceedings (2005).
7. J. M. Lauerhaas, C. Morote, and J. P. Plé, *Semicond. Int.*, **28**, 36 (2005).
8. J. W. Butterbaugh and C. K. Gould, *Semiconductor Manufacturing Magazine*, **7**(5), 33 (2006).
9. J. M. Lauerhaas, Y. S. Kok, A. Hamzah, and L. T. Tan, *ECS Trans.*, **11**(2), 33 (2007).
10. J. W. Butterbaugh, L. Larson, and R. Jammy, *Future Fab International*, **28**, 75 (2009).
11. F. London, *Zeit. Physik. Chem. B*, **11**, 222 (1931).
12. F. London, *Trans. Faraday Soc.*, **33**, 8b (1936).
13. H. C. Hamaker, *Physica*, **4**(10), 1058 (1937).
14. E. M. Lifshitz, *Soviet Physics JETP (English Trans.)*, **2**, 73 (1956).
15. I. E. Dzyaloshinskii, E. M. Lifshitz, and L. P. Pitaevskii, *Soviet Physics-Uspekhi (English Trans.)*, **4**(2), 153 (1961).
16. R. P. Jaiswal, *Adhesion Between Particles and Nano-structured Films (Doctoral Dissertation)*, p. 164, Purdue University Press, West Lafayette (2008).
17. M. J. Vold, *J. Coll. Sci.*, **16**, 1 (1961).
18. K. D. Danov, D. N. Petsev, N. D. Denkov, and R. Borwankar, *J. Chem. Phys.*, **99**, 7179 (1993).
19. R. P. Jaiswal, *Adhesion Between Particles and Nano-structured Films (Doctoral Dissertation)*, p. 164, Purdue University Press, West Lafayette (2008).

(704b) Engineered Quorum Sensing Systems for Protein Expression Regulation

Lianhong Sun, School of Life Sciences, University of Science and Technology of China, Hefei, China

A wildtype LuxI-LuxR quorum-sensing system enables autonomous activation of gene expression without the need of introducing inducing agents, a property particularly useful for industrial biotechnology and synthetic biology. However, the threshold cell density for activating gene expression is largely determined by the host genetic background and culture conditions. Such poor tunability exhibited by the wildtype system has significantly hampered its application in industrial microbiology. We have engineered the components of the LuxI-LuxR quorum-sensing system, and have used mutant components to construct highly tunable and auto-inducible systems for activation of gene expression at various cell densities. We further demonstrated the applications of these auto-inducible systems for the production of recombinant toxic proteins. By coupling with the *Cre-loxP* site-specific recombination system, tunable and auto-inducible deactivation of gene expression has also been achieved. Because this tunable and auto-inducible gene expression system enables choice of threshold cell densities for activating or deactivating gene expression, it is expected to be particularly useful for regulating gene expression in large-scale fermentations and synthetic biology.

See more of this Session: Nanoscale Science and Engineering In Biomolecular Catalysis I
See more of this Group/Topical: Nanoscale Science and Engineering Forum

(586a) Comparison of Photoreactor Designs for Oxidation of Dilute Aqueous Waste Contaminants

Amanda M. Grannas[1], **Dorothy W. Skaf**[2], Kevin C. Brodwater[2] and Montanna Herdemian[2], (1) Chemistry, Villanova University, Villanova, PA, (2)Chemical Engineering, Villanova University, Villanova, PA

The viability of photocatalytic reactions in advanced oxidation processes for removal of dilute aqueous contaminants was evaluated using a variety of reaction systems. Using acetaminophen as analyte, the oxidation reaction was studied using various operating conditions and reactor designs to better understand how the reaction conditions impact conversion and potentially, product selectivity. Catalyst concentration, reagent concentration, metal loading on the catalyst, oxygen purging, solution pH, solution temperature, addition of oxidant, batch versus continuous operation, light intensity and semiconductor stabilization were investigated. Literature results for reaction intermediates and reaction kinetics were used to interpret experimental results. A main focus was optimization of reaction conditions that can make photocatalytic oxidation of dilute aqueous contaminants more sustainable and economical. Catalyst stability, consumption of additional reagents, value and/or safety of the degradation products and energy consumption were considered for the various reaction options.

See more of this Session: Environmental Applications of Nanotechnology and Nanomaterials
See more of this Group/Topical: Topical C: Environmental Aspects, Applications, and Implications of Nanomaterials and Nanotechnology

Graphene-Based Molecular-Machine: Reversible and Robust Carrier Doping in Graphene *via* Mechanical Actuation of Tethered Azobenzene

*Phong Nguyen, Kabeer Jasuja, Nihar Mohanty, and Vikas Berry**

Department of Chemical Engineering, Kansas State University

1005 Durland Hall, Kansas State University, Manhattan, KS 66506

*vberry@ksu.edu

ABSTRACT

Molecular machines – molecules capable of mechanically responding (isomerizing) to external stimuli – are important components of molecular-actuation-based nanodevices. Now we demonstrate that ultrathin-graphene's sensitivity to interfaced-molecules' orientation, surface-vicinity, electronegativity, and density-of-states, can be employed to reversibly dope carriers in it *via* the mechanics of molecular machines tethered on its surface. Here, graphenic sheets are functionalized with electronegative and isomerizable azobenzene-molecules, which are optically transformed from their *trans* conformation (π–cloud-head ~ 1.1 nm away from graphene-surface) to *cis* conformation (π–cloud-head ~ 0.7 nm from graphene). This 0.42 nm displacement of the azo-head dopes 7.5 X 10^3 holes/μm^2 in the underlying graphene. We determine that ~20 azobenzene molecules produce 1 hole (hole-mobility of 301 $\mu m^2/V/s$). Due to their sub-nanoscale actuation, the molecular-machines are typically interfaced within nanoscale openings (nanopores or tunneling-junctions), which are challenging to fabricate. Here, we show that macroscale, sensitive and easy-to-fabricate graphene platform to achieve molecularly-actuated electrical-response. To compare azobenzene with directly doped graphene, we π–functionalized graphene with electronegative perylene tetracarboxylic acid (PCA), which led to ~3 fold increase in the hole density.

(420n) Environmentally Responsive Core-Shell Composite Nanoparticles - Synthesis, Characterization, and Applications

Sriya Sanyal[1], Huan Ma[1], Huang-Chiao Huang[2], Kaushal Rege[1] and Lenore L. Dai[1], (1)School for Engineering of Matter, Transport and Energy, Arizona State University, Tempe, AZ, (2)Chemical Engineering, Arizona State University, Tempe, AZ

Nanoparticles are receiving increasing attention since they have properties that are significantly different than those of bulk materials. Organic-inorganic composites are vital in biological, medical, and chemical applications. Among them, core-shell composite nanoparticles are a unique class of materials which are attractive for their potential applications as delivery vehicles (for drugs, dyes, cosmetics, ink etc.). In this presentation, we report the novel synthesis of environmentally responsive core-shell composite particles through one-step Pickering emulsion polymerization. The synthesized polystyrene/poly(N-isopropylacrylamide) (PNIPAAm)-silica core-shell nanoparticles are temperature sensitive, demonstrate effective drug release upon volume change, and show promising cytotoxicity results in prostate cancer therapy. In addition, we have incorporated 2-(N,N-dimethylamino) ethyl methacrylate into the Pickering emulsion polymerization and synthesized pH responsive core-shell composite nanoparticles which release rheological modifiers upon pH change.

See more of this Session: Poster Session: Nanoscale Science and Engineering
See more of this Group/Topical: Nanoscale Science and Engineering Forum

(266e) Development of Molecular and Mesoscopic Order In Mesostructured Zeolites

Robert J. Messinger[1], Kyungsu Na[2], Ryong Ryoo[2] and Bradley F. Chmelka[1], (1)Department of Chemical Engineering, University of California, Santa Barbara, CA, (2)Department of Chemistry, Korea Advanced Institute of Science and Technology, Daejeon, South Korea

Mesoporous zeolites represent a new and technologically important class of materials that exhibit improved transport, catalytic, and adsorption properties compared to conventional zeolites with sub-nanometer pore dimensions. During their syntheses, the transient development(s) of crystalline and mesoscopic order are closely coupled and difficult to control. Such processes are important to monitor and understand to optimize the compositions, structures, and properties of mesostructured zeolites. In particular, solid-state nuclear magnetic resonance (NMR) spectroscopy, in conjunction with small-angle X-ray scattering (SAXS), transmission electron microscopy (TEM), and adsorption measurements, establish new molecular-level insights on the local interactions and distributions of complicated organic structure-directing agents with respect to crystallizing inorganic frameworks. The analyses reveal the formation of intermediate layered frameworks, which subsequently transform into zeolitic structures with novel combinations of both nano- and mesoscale porosities. Such materials are shown to result from competition between simultaneous and coupled surfactant self-assembly and inorganic crystallization processes, the interplay between which governs the onset and development of framework structural order on different length scales. The results provide criteria for rational design strategies aimed at synthesizing hierarchically porous zeolites with different framework morphologies and improved transport, catalytic, and adsorption properties.

See more of this Session: Templated Assembly of Inorganic Nanomaterials I
See more of this Group/Topical: Nanoscale Science and Engineering Forum

(703c) Towards a Mechanistic Growth Model for Ionic Crystals

Preshit Dandekar, Department of Chemical Engineering, University of California Santa Barbara, Santa Barbara, CA and Michael F. Doherty, Department of Chemical Engineering, University of California, Santa Barbara, Santa Barbara, CA

The relative surface areas and shapes of the various faces of ionic crystals significantly affect their functionality in areas like catalysis, photovoltaics, zeolites, etc. The layered growth rates of the crystal faces determine the steady-state shape of the crystal. For crystals grown from solution, the growth rate depends on the rates of attachment and detachment of solute units on to and from the crystal surfaces. These rates are determined by the solid-solid as well as the solid-solvent interactions. A mechanistic understanding of these interactions is instrumental in engineering the shapes of ionic crystals to achieve better functionality. The Periodic Bond Chain (*PBC*) theory has been used here to study the solid-state interactions in ionic crystals by identifying directions of strong bonding. The *PBC*s are parallel to the edges of the growth spirals and their identification provides insight into the spiral growth mechanism of crystals. Bond chain networks that are stoichiometric and do not have any out-of-plane dipole moment have been identified for anatase (TiO_2) and calcite ($CaCO_3$) crystals using the crystal structure as input.

The electrostatic interactions in the *PBC*s were calculated by a simple columbic potential and the Madelung approach for a 1D ionic chain. The bond chain energies for all the *PBC*s in anatase have been calculated and the lattice energy has been accurately predicted. The kink and edge energies will be calculated for the spiral growth on these crystal faces to estimate the relative growth rates and eventually predict steady-state shapes of ionic crystals.

See more of this Session: Nanomanufacturing
See more of this Group/Topical: Materials Engineering and Sciences Division

(746f) Elucidation and Control of the Hybridization Chain Reaction

Victor A. Beck[1], Justin S. Bois[2], Robert M. Dirks[3] and Niles A. Pierce[3], (1)Bioengineering, California Institute of Technology, Pasadena, CA, (2)Chemistry and Biochemistry, UCLA, Los Angeles, CA, (3) Department of Bioengineering, California Institute of Technology, Pasadena, CA

The programmable chemistry of nucleic acid base pairing enables the design and construction of hybridization chain reactions (HCR) in which metastable DNA or RNA molecules undergo conditional polymerization in the presence of a target nucleic acid. Employed as programmable in situ amplifiers, HCR enables simultaneous mapping of multiple target mRNAs within intact vertebrate embryos. Employed as programmable mechanical transducers, HCR enables selective killing of cultured human cancer cells containing targeted cancer mutations. To elucidate and improve the properties of HCR systems, we use a combination of theory and experiments to probe and perturb HCR equilibrium behavior and self-assembly pathways. We find that HCR is appropriately modeled as a seeded nucleation-elongation polymerization and use this theoretical framework to re-engineer the equilibrium and kinetic properties of the mechanism.

See more of this Session: Nanotechnology and Nanobiotechnology for Sensors
See more of this Group/Topical: Nanoscale Science and Engineering Forum

(298a) The Outlook for Chemical and Biochemical Sensors Made of Graphitic Nanodevices

Yongki Choi[1], Patrick C. Sims[2], Issa S. Moody[3], Gregory A. Weiss[3] and **Philip G. Collins**[4], (1)Institute for Surface and Interface Science, University of California, Irvine, Irvine, CA, (2)Physics and Astronomy, University of California, Irvine, Irvine, CA, (3)Chemistry, University of California, Irvine, Irvine, CA, (4)Physics & Astronomy, University of California, Irvine, Irvine, CA

The enormous sensitivity of carbon nanotubes to their chemical environment prompted substantial excitement about their promise for commercial sensors and detectors. Subsequent research, using both nanotube and graphene electrodes, has generally demonstrated high sensitivity but poor stability and selectivity, as the electrodes are famously sensitive to most analytes including air, water, and noble gases. Much ongoing research has focused on improving selectivity and reducing unwanted responses, though the direct physiochemical mechanisms responsible for sensing remain poorly understood. In fact, the combination of possible interactions is daunting and remains under investigation on an analyte-by-analyte basis, as described in a recent review.[1] Part of the difficulty is that very small amounts of disorder, such as from atomic defects, chemical adducts, or contaminants, enormously enhance electron transfer between the environment and the otherwise inert graphitic surfaces. This enhancement is the underlying principle behind commercial carbon-based electrochemical sensors, after all, but it can be difficult to precisely control in a nanoelectrode of nanotubes or graphene. From this standpoint, nanographite electrodes may be unable to successfully compete with existing sensor products, even when defects are present that enhance their sensitivity.[2] An exception to this conclusion, and perhaps the main advantage of working with graphitic nanoelectrodes, appears to be their ability to monitor single molecule events. Because nanoelectrodes, and particularly one-dimensional single-walled nanotubes (SWNTs), can operate in the limit of having one single active site, there are opportunities for monitoring single molecule dynamics that are normally obscured in ensemble measurements. The direct interrogation of single molecule attachments, charge transfer events, or catalytic turnovers[4] provides truly new opportunities for electronic sensing and assaying of chemical and biochemical activity.

1. K.-J. Lee and J. Kong, "Chemical Sensing with SWNT-FETs," in Carbon Nanotube Electronics (A. Javey and J. Kong, Eds. New York: Springer, 2009).

2. V.R. Khalap, T. Sheps, A.A. Kane, P.G. Collins, Nano Lett. 10, 896 (2010).

4. B.R. Goldsmith, J.G. Coroneus, A.A. Kane, G.A. Weiss, P.G. Collins, Nano Lett. 8, 189 (2008).

See more of this Session: Graphene and Carbon Nanotubes: Absorption and Transport Processes
See more of this Group/Topical: Nanoscale Science and Engineering Forum

(744f) Toward the Retention of Enzyme Activity In High-Surface-Area Electrode Made of Redox Polymer Grafted Carbon Black

Takanori Tamaki, Tomoharu Sugiyama, Haruki Fujimoto, Hidenori Ohashi and Takeo Yamaguchi, Chemical Resources Laboratory, Tokyo Institute of Technology, Yokohama, Japan

Biofuel cells attracted attention as candidates for portable and implantable energy sources because a variety of fuels like glucose and ethanol can be used. One important issue to be addressed in enzyme-base biofuel cells is their low power density caused by their low current density. The rate-limiting step in mediated enzyme electrodes has been electron conduction via a redox polymer, which shuttles electrons from the enzyme to the electrode, due to the low apparent electron diffusion coefficient in the redox polymer (D_m).[1]

We have proposed a high-surface-area three-dimensional enzyme electrode made of redox-polymer-grafted carbon black with particle diameters of about 30 nm to obtain high current density.[2] In this electrode, the three-dimensional carbon electrode with high surface area and high electron conductivity plays the primary role in conducting electrons; thus, the electron conduction distance in the redox polymer was reduced, overcoming the rate-limiting step even when the redox polymer used has a low D_m. The electrode was used as a bioanode using glucose oxidase as the enzyme. The effectiveness of the high-surface-area electrode was verified by experiments and a mathematical model. The model calculation also suggests that an increase in the surface coverage of the enzyme allows an increase in the current density to the order of 10^2 mA/cm^2.[3] However, the low enzyme surface coverage, mainly due to the deactivation of enzymes upon adsorption on carbon black, limited the current density.[4]

In this work, we investigated two methods to retain the enzyme activity upon adsorption on carbon black. One was to utilize a hyperthermophilic laccase. In general, hyperthermophilc enzymes have rigid structures, and thus, have a potential to retain the activity during adsorption. The catalytic activity of laccases was measured after adsorption on carbon black. The activity of the hyperthermophilic enzyme did not change by adsorption, while the activity of the laccase with the optimal temperature around 60 C decreased to approximately 20% of that in solution after adsorption. In addition, for electrochemical measurements of laccases, a redox mediator for laccase was chemically immobilized to a polymer grafted on the carbon black. The other method for the retention of the enzyme activity was modification of carbon black to make its surface hydrophilic. Surface treatment of carbon black with a hydrophilic polymer reduced the deactivation of enzyme to some extent. Then, chemical modification of carbon surface was investigated to further retain the enzyme activity.

References [1] S. C. Barton, J. Gallaway et al., Chem. Rev., 2004, 104, 4867. [2] T. Tamaki and T.

Yamaguchi, Ind. Eng. Chem. Res., 2006, 45, 3050. etc. [3] T. Tamaki, T. Ito, T. Yamaguchi, Fuel Cells, 2009, 1, 37., [4] T. Tamaki, A. Hiraide, T. Ito, H. Ohashi, T. Yamaguchi, Ind. Eng. Chem. Res., 2010, 49, 36394.

See more of this Session: Nanoscale Science and Engineering In Biomolecular Catalysis II
See more of this Group/Topical: Nanoscale Science and Engineering Forum

(369a) Modeling Aggregation and Size Distribution of Nanoparticles In Aqueous Suspensions

Haoyang H. Liu, Chemical and Biomolecular Engineering, University of California, Los Angeles, Los Angeles, CA, Sirikarn Surawanvijit, Chemical&Biomolecular Engineering, University of California, Los Angeles, Los Angeles, CA, G. Orkoulas, Department of Chemical and Biomolecular Engineering, University of California, Los Angeles, Los Angeles, CA and Yoram Cohen, Chemical & Biomolecular Engineering, University of California, Los Angeles, Los Angeles, CA

The size distribution of the engineered nanoparticles (eNMs) and their aggregates in aqueous suspensions is crucial for predictions of the fate and transport of nanoparticles in the environment. Diffusion, sedimentation and interaction of nanoparticles with solid surfaces (including sedimentation), as well as living organisms are all affected by the size distribution of eNMs. In order to obtain the nanoparticle size distribution over the expected ranges of water chemistry and nanoparticle properties, predictive models are needed to assess the aggregation tendency of eNMs. In the present work, a constant-number kinetic Monte Carlo based model was developed to predict the aggregation rate and size distribution of nanoparticles. The modeling approach followed the "particles in a box" simulation method while allowing for particle sedimentation. The likelihood of nanoparticle aggregation was determined, based on the classical DLVO theory, via calculation of the aggregation probability of nanoparticle pairs. In developing the model, the reasonable minimum number of simulation particles and simulations instances were first evaluated to ensure convergence of the stationary statistics, demonstrating the ability to use parallel computing to dramatically decreasing simulation time. Model performance with respect to the predicted size distribution of eNMs was evaluated via comparison with experimental dynamic light scattering (DLS) measurements and with available literature data under varying conditions (e.g., pH, ionic strength, size distribution of the primary nanoparticles and nanoparticle surface charge). Model performance compared favorably with experimental DLS data demonstrating the potential of the present modeling approach for environmental analysis of the transport and fate of nanoparticles

See more of this Session: Environmental Implications of Nanomaterials: Fate and Transport
See more of this Group/Topical: Topical C: Environmental Aspects, Applications, and Implications of Nanomaterials and Nanotechnology

ENCAPSULATION OF MAGNETIC NANOPARTICLES WITHIN BIOFUNCTIONAL POLY (ETHYLENE GLYCOL) HYDROGEL FORMED VIA SURFACE INITIATED PHOTOPOLYMERIZATION

Caner Nazlı, T. Ipek Ergenc , H. Funda Yagci-Acar, Seda Kızılel, Koc University, Istanbul, Turkey

Introduction

Magnetic iron oxide nanoparticles (MIONPs) have been exploited widely in biotechnology and medicine as delivery systems for drugs, genes, biomolecules, stem cell tracking, and diabetes, hyperthermia and the detection of cancer [1] [2]. In addition to in vivo application, MIONPs are useful for various in vitro applications in medical diagnostics such as magnetic separation of cells, proteins, DNA/RNA, and other biomolecules [1]. Although MIONPs find extensive application areas in biomedicine and analysis methods, they have been hindered in clinical applications due to agglomeration in physiological condition and fast detection via immune cells [1, 3]. MIONPs without surface coating exhibit hydrophobic surface property with a large surface area to volume ratio [3-5]. So, these particles tend to agglomerate and form larger clusters [1, 6]. These structures are not appropriate for biomedical applications due to short circulation half-life [1] [3] and exhibition of ferromagnetic behavior without magnetic field [4]. Hence, MIONPs are not allowed to reach target tissue by some biological systems such as reticuloendothelial system, vascular endothelium or blood brain barrier [7]. Therefore, MIONPs should be covered with synthetic polymers in consideration of distribution in aqueous environment [8]. Here, we describe a new application of our previously developed technique [9] for coating of MIONPs within biofunctional PEG hydrogel using surface initiated photopolymerization in order to address difficulties involved in NPs biodistribution and cellular uptake. Figure 1 illustrates the process of the manufacturing biofunctional PEG hydrogel encapsulated MIONPs. This novel method allows for coating of MIONPs within biofunctional hydrogel layer to promote detection of tumors and accumulation of therapeutic agents over targeted tissues.

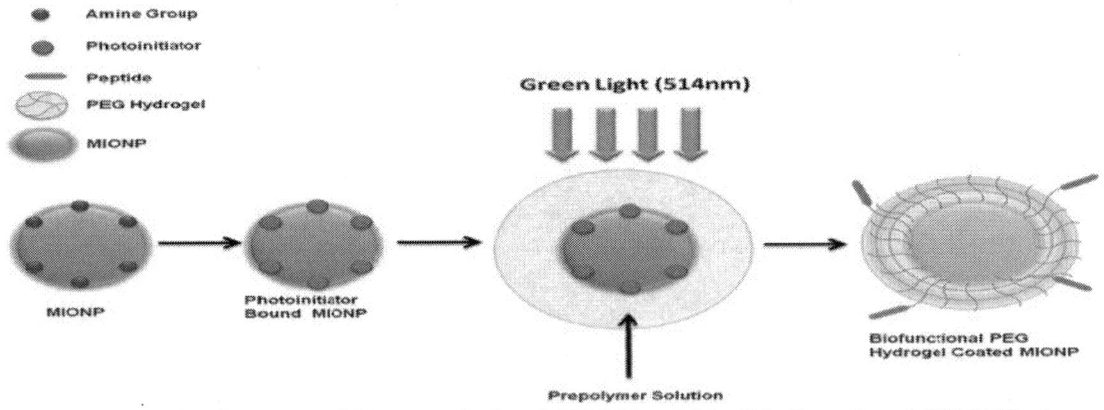

Figure 1. Schematic diagram of Encapsulating MIONP within Biofunctional PEG Hydrogel via

Materials and Methods

After covalent binding of EY to the surface of the 3-(aminopropyl)trimethoxysilane–tetramethoxysilane (APTMS) coated MIONPs, 0.5mg/ml modified MIONPs solution was mixed with the prepolymer solution (Final Concentrations: 5% PEGDA-575, 19mM VP, 113 mM TEA, 1.48*10-2 mM Acry-PEG(3400)-RGDS, 0.1mg/ml MIONP). The solution was adjusted to pH 8 using HCl. Next, the mixture was illuminated via green light (λ: 514 nm, P: 2.5 mW/cm2) using Argon ion laser (Coherent, Innova 70C) at varied time interval (20, 30, 60 sec.).

Encapsulated nanoparticles (ENPs) are characterized in terms of structure, size, stability and morphology applying Fourier transform infrared (FT-IR) spectroscopy, dynamic light scattering (DLS), zeta potential measurement and, atomic force microscopy (AFM) respectively. After obtaining desired ENPs, cytotoxicity is investigated through incubation of ENPs with HeLa cells via luminescent cell viability assay. Cellular uptakes of APTMS coated, PEG hydrogel coated and PEG hydrogel with peptide coated MIONPs are compared via Prussian blue staining method and inductively coupled plasma optical emission spectrometry (ICP-OES).

Results and Discussion

The binding of EY to APTMS coated MIONP was confirmed by FT-IR analysis. Figure 2 shows the FT-IR spectra of APTMS coated MIONP and EY-bound MIONP. The FT-IR spectra of APTMS coated MIONP exhibited two picks at 571 cm^{-1} and 590 cm^{-1} due to iron oxide skeleton [10-12]. The picks at 919, 1215, 1328, 1388, 1505, 1617 and 2931 cm-1 were assigned to C-H out-of-plane bending, aliphatic C–N stretching, aliphatic C-O stretching, Si-CH_2 scissoring, C-H bending, NH_2 scissoring & N-H bending and CH_2 stretching, respectively [10, 13-14]. After EY binding these picks were protected except the pick at 1505 cm^{-1}. The bands at 1456 cm^{-1} and 1541 cm^{-1} which corresponded aromatic C-C stretching and aromatic C=C stretching due to EY structure appeared after EY binding to MIONP. These new picks overlapped C-H bending pick at 1505 cm^{-1}. The picks at 3058 and 3352 cm^{-1} produced shoulder structure because of NH_2 stretching in the APTMS coated MIONP [14]. This shoulder structure disappeared mostly and one new pick at 3156 cm^{-1} which corresponded secondary amide N-H stretching appeared after the EY binding [14]. These two FT-IR spectra gave strong evidence about EY binding on MIONP covalently.

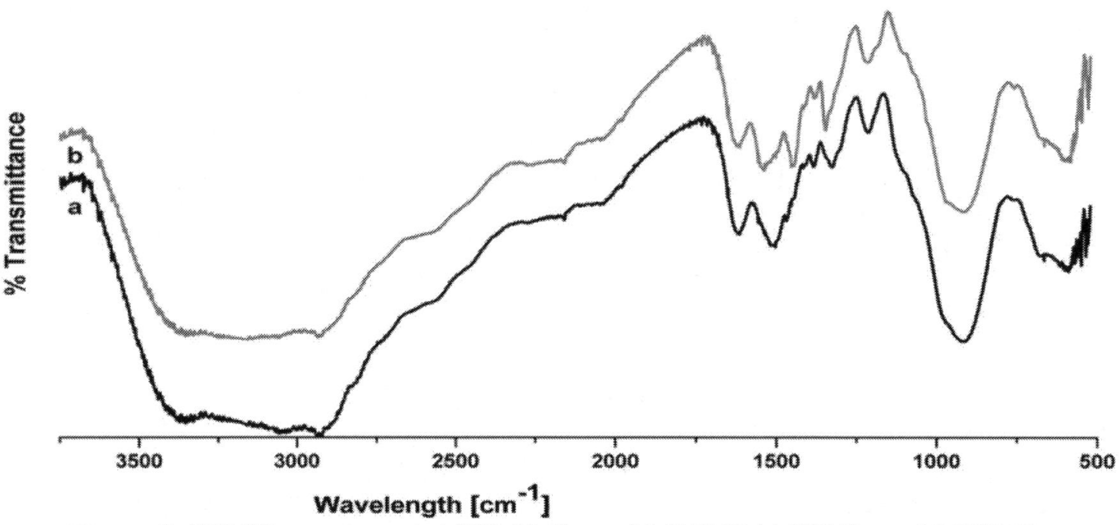

Figure 2. FT-IR spectra: a) APTMS Coated MIONP b) EY Bound MIONP

After EY binding on MIONPs, EY content of the product was determined using UV-Vis spectrophotometry measured at 537 nm (EY can absorb light at 537 nm [15]). After the binding reaction, the product contained 12 % EY by weight. This content corresponded that efficiency of EY binding on MIONPs was 57 %.

Pharmacokinetics and cellular uptake of NPs are primarily based on physicochemical features of them such as hydrodynamic size, surface charge and morphology [16]. In order to determine hydrodynamic diameter of different MIONPs, we applied DLS. Diameter of only APTMS coated NPs was 6.4 nm. After binding EY on NPs, diameter reached to 12.6 nm. When NPs were added into PPS, hydrodynamic size of NPs was 13.6 nm. The diameter of NPs became 23.3 nm after 20sec cross-linking reaction. Diameter increased to 58 nm and 122 nm, if reaction time was increased to 30 sec and 60 sec, respectively. When PEG-RGDS conjugate was added to the reaction, diameter of each sample was not changed significantly. Hydrodynamic size changes before and after each modification were compared at Figure 3.

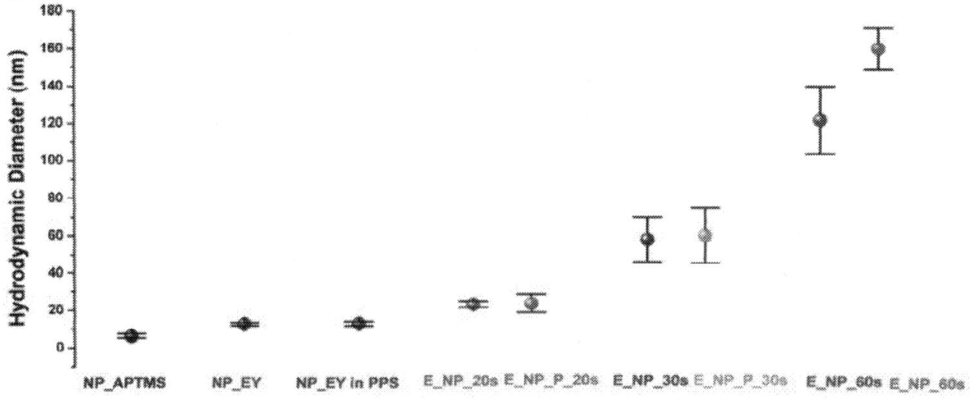

Figure 3. Change of hydrodynamic diameter before and after PEG Hydrogel coating obtained via DLS

Zeta potential measurement results show that changing surface charge of MIONPs during modifications (See Figure 3.7). After negatively charged EY was bound on the MIONPs, charge of NPs changed from positive charge (+22.3 mV) to neutral (+2.3 mV). After encapsulation with PEG hydrogel, coated NPs became negatively charged (about -30 mV) similar to PEGDA. Adding PEG-RGDS conjugate did not affect surface charge mostly. This zeta potential value is enough that encapsulated NPs could be stable in solution [17].

After AFM analysis, physically adsorbed RGDS functionalized PEG hydrogel encapsulated MIONPs on the silicon surface were seen at Figure 4. Height picture of PEG hydrogel coated MIONPs demonstrated that most of particles were encapsulated individually, while a few of them were encapsulated in cluster. When height picture and phase picture were compared, core-shell structure of nanoparticles could be seen. The particles had inner magnetite core and outer polymeric shell of PEG. So, this result gave important evidence that we encapsulated MIONPs within biofunctional PEG hydrogel successfully. Size and surface modification of NPs could result in dissimilar response of cells to NPs in terms of non-specific or targetting cellular uptake [18]. Our method allows controllable size adjustment by changing illumination time and entrapment varied biomolecules into PEG hydrogel coating over MIONPs. Therefore, size or surface functionalization can be adapted to the targeting cell type.

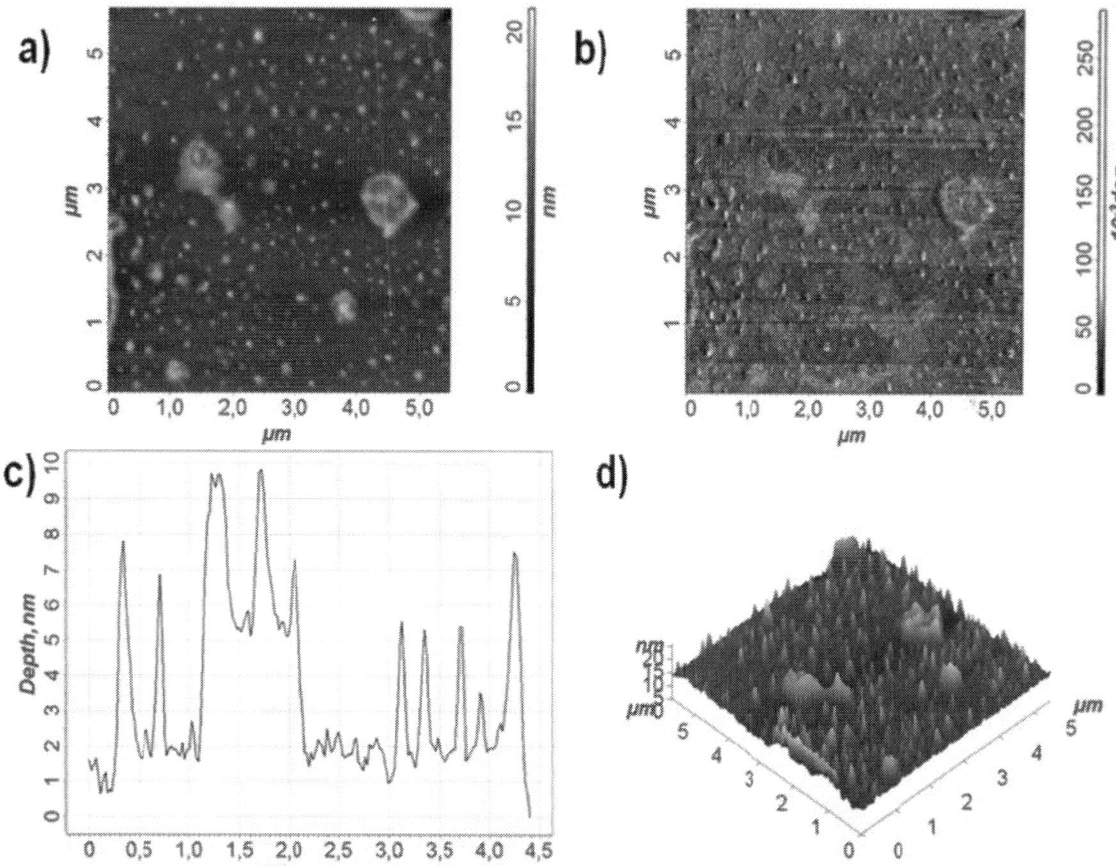

Figure 4. AFM Picture of Biofunctional PEG Hydrogel Coated MIONPs: a) Height Picture, b) Phase Picture, c) Cross Section Analysis, d) Three Dimensional Representation.

In order to determine the acute toxicity of APTMS coated MIONPs, EY-bound MIONPs, PEG hydrogel coated MIONPs and biofunctional PEG hydrogel coated MIONPs, the viability of HeLa cells incubated with all types of MIONPs at 0.01, 0.05 and 0.10 mg/ml concentrations was examined using Cell-Titer-GLO Luminescent Viability Assay after 24 and 48 h treatment (See Figure 5). When we compare them with control sample that not contains any NPs, they did not show toxic effect at low concentration (0.01mg/ml). At higher concentration samples (0.05 and 0.1 mg/ml), EY bound and 20 sec reacted PEG hydrogel coated MIONP samples had toxic effect on both 24hrs and 48hrs cultured HeLa cells. If samples were reacted enough during photopolymerization, PEG Hydrogel coating did not affect cell viability. As a result, cellular metabolic activity in cells that were treated 30 sec and 60 sec reacted PEG hydrogel coated MIONPs did not change much in comparison with control cells because no significant diminish in the viability of the treated HeLa cells existed at the examined nanoparticle concentration.

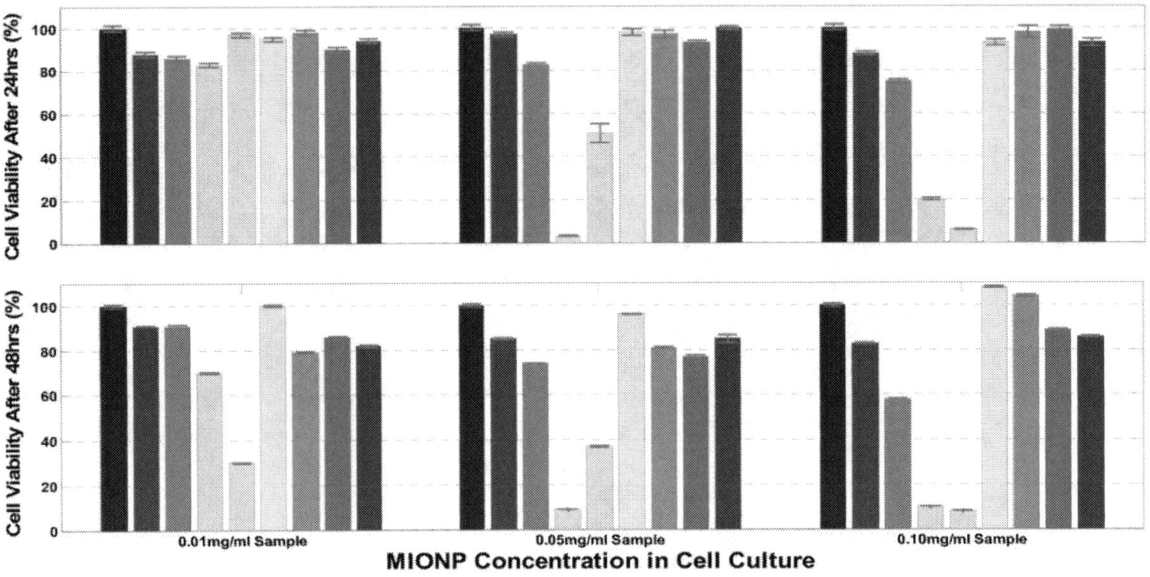

—MIONP_APTMS	—ENP_20	—ENP_P_30
—MIONP_EY	—ENP_P_20	—ENP_60
—MIONP_EY in PPS	—ENP_30	—ENP_P_60

Figure 5. Cytotoxicity profiles of varied MIONPs after 24 h and 48 h incubated with HELA cells.

The capability to accumulate effectual concentrations of the agents to tumor area is a significant step of cancer diagnosis and treatment [19]. Ligand-receptor approach has great potential to enhance the efficiency of therapeutic and contrast agents by maximizing the amount of the agents over the targeted site, and to diminish cytotoxicity or imaging background signal by decreasing systemic subjection [19]. In this study, to demonstrate the ability of biofunctional PEG hydrogel over MIONPs, intracellular uptake of various NPs was investigated ICP-OES measurement after 24 h treatment. ICP result showed that 60 sec photopolymerized RGDS functionalized PEG hydrogel coated MIONPs internalized much more into HeLa cells as compared both uncoated and only PEG hydrogel coated MIONPs (See Figure 6). Therefore, RGDS functionalization provides targeting cells specifically.

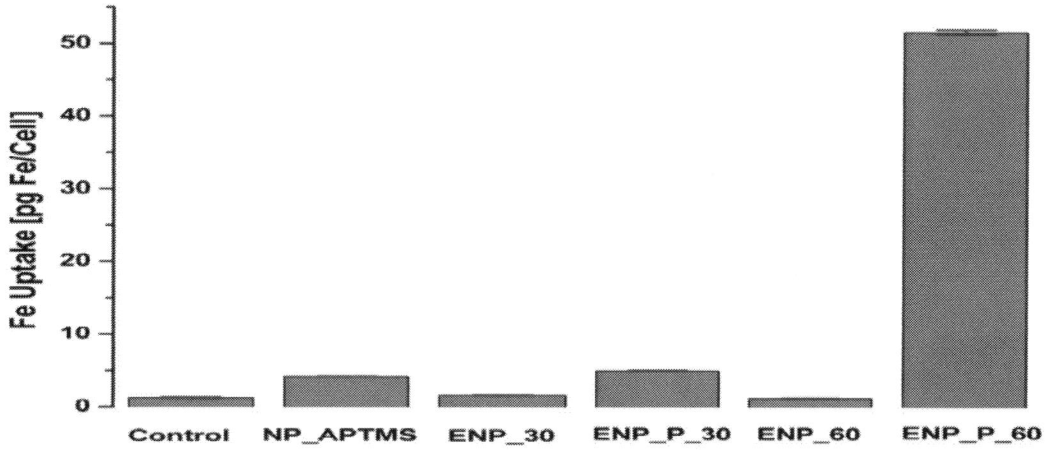

Figure 6. Cell Uptake of varied MIONPs after 24 h obtained via ICP

Conclusion

MIONPs hold enormous potential of creating new preferences for early cancer detection and targeted therapies. In order to overcome limitations of MIONP technologies, we have improved a novel strategy that is useful for encapsulation and functionalition of MINOPs. It is a significant candidate to develop desired MIONPs that demonstrate enhanced magnetic properties and biocompatible, biofunctional as well as multifunctional features. Encapsulated MIONPs within biofunctional PEG hydrogel could prevent agglomeration and increase blood circulation time which will be significant tool for diagnostic and therapeutic imaging technologies as well as targeted drug delivery area.

References

1. Cole, A. J., A. E. David, et al. (2011). "Polyethylene glycol modified, cross-linked starch-coated iron oxide nanoparticles for enhanced magnetic tumor targeting." Biomaterials 32(8): 2183-2193.
2. Gogo*t*si, I. U. G. (2006). Nanomaterials handbook. Boca Raton, CRC/Taylor & Francis.
3. Gupta, A. K. and M. Gupta (2005). "Synthesis and surface engineering of iron oxide nanoparticles for biomedical applications." Biomaterials 26(18): 3995-4021.
4. Gupta, A. K. and S. Wells (2004). "Surface-modified superparamagnetic nanoparticles for drug delivery: preparation, characterization, and cytotoxicity studies." IEEE Trans Nanobioscience 3(1): 66-73.
5. Hu, F., K. G. Neoh, et al. (2006). "Cellular response to magnetic nanoparticles "PEGylated" via surface-initiated atom transfer radical polymerization." Biomacromolecules 7(3): 809-816.

6. Jung Kwon Oh, J. M. P. (2010). "Iron Oxide-based Superparamagnetic Polymeric
7. Nanomaterials: Design, Preparation, and Biomedical
8. Application." Progress in Polymer Science.
9. Kizilel, S., V. H. Perez-Luna, et al. (2004). "Photopolymerization of poly(ethylene glycol) diacrylate on eosin-functionalized surfaces." Langmuir 20(20): 8652-8658.
10. Koo, Y. E., G. R. Reddy, et al. (2006). "Brain cancer diagnosis and therapy with nanoplatforms." Adv Drug Deliv Rev 58(14): 1556-1577.
11. Lee, H. Y., N. H. Lim, et al. (2006). "Preparation and magnetic resonance imaging effect of polyvinylpyrrolidone-coated iron oxide nanoparticles." J Biomed Mater Res B Appl Biomater 79(1): 142-150.
12. Ma, H. L., X. R. Qi, et al. (2007). "Preparation and characterization of superparamagnetic iron oxide nanoparticles stabilized by alginate." Int J Pharm 333(1-2): 177-186.
13. Mahmoudi, M., A. Simchi, et al. (2010). "Recent Advances in Surface Engineering of Superparamagnetic Iron Oxide Nanoparticles for Biomedical Applications." Journal of the Iranian Chemical Society 7: S1-S27.
14. Mikhaylova, M., D. K. Kim, et al. (2004). "BSA immobilization on amine-functionalized superparamagnetic iron oxide nanoparticles." Chemistry of Materials 16(12): 2344-2354.
15. Park, J. Y., P. Daksha, et al. (2008). "Highly water-dispersible PEG surface modified ultra small superparamagnetic iron oxide nanoparticles useful for target-specific biomedical applications." Nanotechnology 19(36): 365603.
16. Stuart, B. (2004). Infrared spectroscopy : fundamentals and applications. Chichester, West Sussex, England ; Hoboken, NJ, J. Wiley.
17. Sun, C., J. S. Lee, et al. (2008). "Magnetic nanoparticles in MR imaging and drug delivery." Adv Drug Deliv Rev 60(11): 1252-1265.
18. Tong, S., S. Hou, et al. (2010). "Coating Optimization of Superparamagnetic Iron Oxide Nanoparticles for High T(2) Relaxivity." Nano Lett.
19. Varo, G., L. S. Brown, et al. (1999). "Binding of calcium ions to bacteriorhodopsin." Biophys J 76(6): 3219-3226.
20. Veiseh, O., J. W. Gunn, et al. (2010). "Design and fabrication of magnetic nanoparticles for targeted drug delivery and imaging." Adv Drug Deliv Rev 62(3): 284-304.
21. Wu, X., Y. Tan, et al. (2010). "Toxic effects of iron oxide nanoparticles on human umbilical vein endothelial cells." Int J Nanomedicine 5: 385-399.

(704f) Nanoscale Enzyme Reactors and Enzyme Precipitate Coatings of Glucose Oxidase for Biosensor and Biofuel Cell Applications

Chulmin Jeon[1], Jae Kim[1], Jungbae Kim[1] and **Su Ha**[2], (1)Korea University, Seoul, South Korea, (2) Washington State University, Pullman, WA

In this study, we immobilize glucose oxidase (GOx) using two different enzyme immobilization techniques. First, nanoscale enzyme reactors (NERs) of GOx in conductive mesoporous carbons were prepared in a two-step process of enzyme adsorption and follow-up enzyme crosslinking. MSU-F-C, a mesoprous carbon, has a bottleneck pore structure with mesocellular pores of 26 nm connected with window mesopores of 17 nm. This structure enables the ship-in-a-bottle mechanism of NERs, which effectively prevents the crosslinked enzymes in mesocellular pores from leaching through the smaller window mesopores. Secondly, highly stable enzyme precipitate coatings (EPCs) on carbon nanotubes (CNTs) were prepared by precipitating GOx molecules in the presence of ammonium sulfate, then cross-linking the precipitated GOx aggregates on covalently attached enzyme molecules on the surface of CNTs. We used these NERs and EPCs of GOx as the enzyme electrode for both biosensor and biofuel cell applications. In this paper, their electrochemical performances and stability will be tested in terms of their sensitivity for detecting GOx and power density output for converting GOx into electrical power.

See more of this Session: Nanoscale Science and Engineering In Biomolecular Catalysis I
See more of this Group/Topical: Nanoscale Science and Engineering Forum

(599c) Modeling Protein-Surface Interactions for Biohybrid Solar Energy Utilization

Sndor . Kovcs, Energy, Environmental and Chemical Engineering, Washington University in St. Louis, St. Louis, MO and Cynthia S. Lo, Department of Energy, Environmental and Chemical Engineering, Washington University in St. Louis, St. Louis, MO

Photosynthesis is a robust and effective mechanism for harnessing solar energy in natural organisms. However, a closer look reveals photosynthetic systems possess inherently efficient energy capture and energy transfer mechanisms, which when coupled with engineered materials may provide mankind with a low-cost source of clean and renewable power. In this work, steered molecular dynamics (SMD) simulations are used to probe the structure and properties of the bionano interface in such a device. We compare the free energies of binding of native photosynthetic and engineered peptide sequences to silica and gold surfaces. These results can be extrapolated to predict how entire photosynthetic pigment protein complexes may be assembled and patterned in nanostructured biohybrid devices for solar energy utilization.

See more of this Session: Nanostructured Biomimetic and Biohybrid Materials and Devices
See more of this Group/Topical: Nanoscale Science and Engineering Forum

(693b) Role of Surfactant Molecular Structure On Carbon Nanotube Self Assembly: Insights From Atomistic Simulations

Naga Rajesh Tummala[1], Manaswee Suttipong[1], Boonyarach Kitiyanan[2] and **Alberto Striolo**[3], (1)School of Chemical, Biological and Materials Engineering, The University of Oklahoma, Norman, OK, (2)The Petroleum and Petrochemical College, Chulalongkorn University, Bangkok, Thailand, (3)School of Chemical Biological and Materials Engineering, The University of Oklahoma, Norman, OK

Stabilizing aqueous dispersions of carbon nanotubes mono-dispersed in diameter and chirality remains elusive. Surfactants have proven useful in deploying ultra-centrifugation techniques, but the molecular mechanism responsible for their effectiveness remains not fully understood. Based on a number of recent molecular simulation results, including those from our group, it appears that the morphology of the self-assembled surfactant aggregates on the carbon nanotubes strongly affects the effective potential of mean force between pairs of interacting carbon nanotubes. We will discuss here recent results obtained conducting equilibrium molecular dynamics simulations for systems composed of sodium dodecyl sulfate (SDS), sodium dodecyl benzene sulfonate (SDBS), flavin mononucleotide (FMN), and sodium bis(2-ethylhexyl) sulfosuccinate (AOT). The simulations are conducted at room conditions for different surface coverages on (6,6), (12,12), and (20,20) single walled carbon nanotubes. By comparing the different surfactant structures on the various nanotubes we will attempt to provide insights into the molecular mechanisms that could be used to stabilize aqueous dispersions of carbon nanotubes monodispersed in diameter.

See more of this Session: Graphene and Carbon Nanotubes: Characterization, Functionalization, and Dispersion
See more of this Group/Topical: Nanoscale Science and Engineering Forum

SYNTHESIS OF GOLD NANOWIRES WITH 2-D NETWORK STRUCTURE USING SODIUM CARBOXYMETHYL CELLULOSE BY SELF-REDUCTION METHOD

Chunrong Wang[*1], Huang Zhou[1], Yun Fang[2], Mingsheng Xu[1]*

[1]*School of Food Science and Engineering, Jiangxi Agricultural University, Nanchang, Jiangxi, P. R. China*

[2] *School of Chemical and Material Engineering, Jiangnan University, Wuxi, Jiangsu, P. R. China*

Abstract

A new green chemical method is reported to synthesize gold nanowires with 2-D network structure through reducing $HAuCl_4$ by sodium carboxymethyl cellulose (CMC). The concentration of CMC controlled the gold nanoparticles growth. TEM image displayed directly that the reduced products were gold nanowires. With a fixed concentration of $HAuCl_4$ of 0.2 mM, as the concentration of CMC was 0.05%, the reaction temperature was 80 °C, there were gold nanowires (with 10 nm wide) formed; as 0.5%, some multipodlike gold nanoparticles with equivalent diameter around 80 nm prepared. Selected electron diffraction suggested that the nanowires with 2-D network structrue were composed by single crystal nanowires. UV-Vis absorption spectrum of gold nanowires showed that there were anisotropic nanostures formed in the reacted solutions. The preferred noble metal nanowires with 2-D network may be obtained easily by this green chemical method.

Key Words : gold nanowires; self-reduction; sodium carboxymethyl cellulose; green chemistry

Instruction

Morphological control of nanoparitlces has attracted a great deal of attention because of its physical and chemical properties, and its applications in catalysis, chemical sensor, photoelectronic devices, and biomedicine [1-9]. There have been intensive efforts to synthesize ordered metallic nanowires using virous capping agents such as SDS [10], PVP) [11-12], blocok copolymer [13-14] and peptides [15-16]. However, synthesis of Au nanopwires is still the focus of this field. On the other hand, as the urgent need for environmental protection, green chemical methods, particularly the use of molecular self-reduction is being more and more popular. So far, to find simple control, multi-morphology of gold nanowires preparation method of green chemistry is still of great challenge.

As is well known, CMC is food additives and is environmentally friendly. Here we report a simple approach to synthesize gold nanowire network by CMC. Compared with ways about the organization of nanoparticles into two-dimensional or three-dimensional nanowires with a networked structure [10, 17-18], the present method is a one-step procedure and the morphology of the gold nanowire network can be controlled by carefully controlling the concentration of CMC. CMC molecules were chosen as capping and reducing agents. The study is meaningful to discover green chemistry method for synthesizing noble metal nanoparticles.

Experimental Parts

Chemicals

Hydrochloroauric acid trihydrate ($HAuCl_4 \cdot 3H_2O$, 99.9%), and CMC were purchased from China Medicine (Group) Shanghai Chemical Reagent Corporation. All reagents were used as received without further purification. The water used was deionized water.

Synthesis of Gold Nanoparticles and Characterization

In a typical experiment, the aqueous solution containing the reagents ($HAuCl_4$ and CMC) was placed in a 100 ml round-bottomed flask in a water bath maintained at a constant temperature (80 °C) for 8h without ambient light to avoid photoreactions of the gold complex. The products were separated by centrifugation (TGL-16G, Anting Scientific Instrument Factory), and washed with deionized water several times. Optical spectra were recorded using a SPECORD200 spectrophotometer (Analytik Jena AG, Germany). TEM images and electron diffraction patterns (ED) were recorded on JEM-2010 microscopy operated at 200 kV and H-600 microscopy operated at 100 kV.

Results and discussion

TEM images of gold nanowires with 2-D Network Structure

To investigate the influence of concentration of CMC on the gold nanowire network formation process, the concentration of CMC was varied from 0.05% to 0.7%. When the concentration of $HAuCl_4$ was 0.2mM, the initial light yellow solution turned light red, purple and finally dark blue upon the addition of CMC. Figure1 shows the typical gold nanowires with 2-D network structure prepared by reducing $HAuCl_4$ (0.2 mM) using CMC (0.05%) in aqueous solution. It was shown that the gold nanowires extended to about several square micrometers. Although the diameter of nanowires is not so uniform, most of them have diameters about 10 nm. Unlike nanostructures formed by the van de walls force of capping molecules on the nanoparticles surface, which are not so stable to survive environmental disturbance, the as-prepared gold nanowires could keep their morphology even under high-speed centrifugation (1,2000 rpm) and ultrasonic dispersion. The salient stability implies that the gold nanowire networks are formed not by the self-organization of the gold nanoparticles and nanowires via the hydrophilic force of the capping agents. The electron diffraction of gold nanowire (inset of the Figure 1) shows the diffraction rings, which are, correspond to (111), (200), (220), (311), and (420) of gold crystalline facets. The diffraction rings indicate that the gold nanowires may be composed of large amount of random orientated gold nanoparticles.

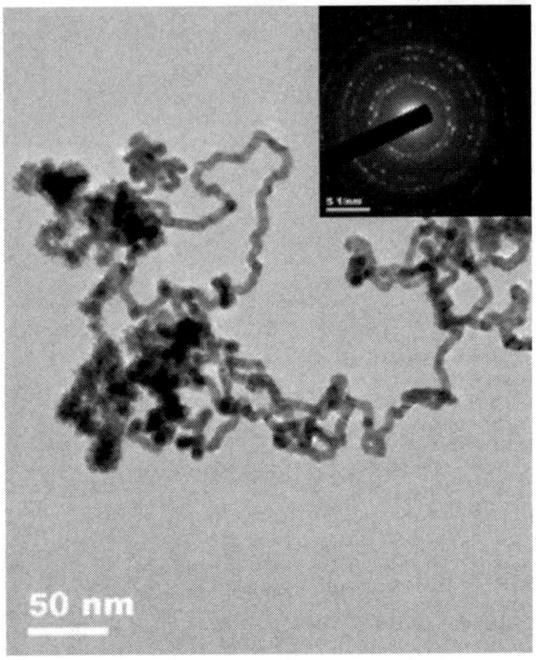

Figure1. TEM images of the gold nanoparticles formed in the $HAuCl_4$ (0.2 mM)-CMC (0.05%) solution and corresponding electron diffraction image.

Effect of CMC concentration on morphologies of gold nanoparticles

TEM images of gold nanoparticles synthesized in different concentration of CMC aqueous solutions, with a fixed concentration of $HAuCl_4$ (0.2 mM), are shown in Figure 2. Gold nanowire network had been synthesized in solution A concluding 0.1% CMC, which morphology is similar to that formed in solution added 0.05% CMC. In Figure 2a, the nanowires also extended to several square micrometers. Figure 2b shows that shorter gold nanowires (extended to about 150 nm) prepared in a 0.3% of CMC solution. It is obvious that the length of gold nanowire decreases as the concentration of CMC increases. When the CMC concentration in reacted solution C was 0.5%, some multipodlike gold nanoparticles with equivalent diameter around 80 nm were prepared (Figure 2c). Above results clearly indicates that high concentration of CMC is not helpful to synthesizing gold nanowires network. The capping agent reduced the gold ions in solution and controlled the growth orientation of gold nanoparticles. When the concentration of capping agent is high, the tiny gold nanoparticles were restricted to aggregate and form nanowire network, and multipodlike and quisi-spherical particles were prepared.

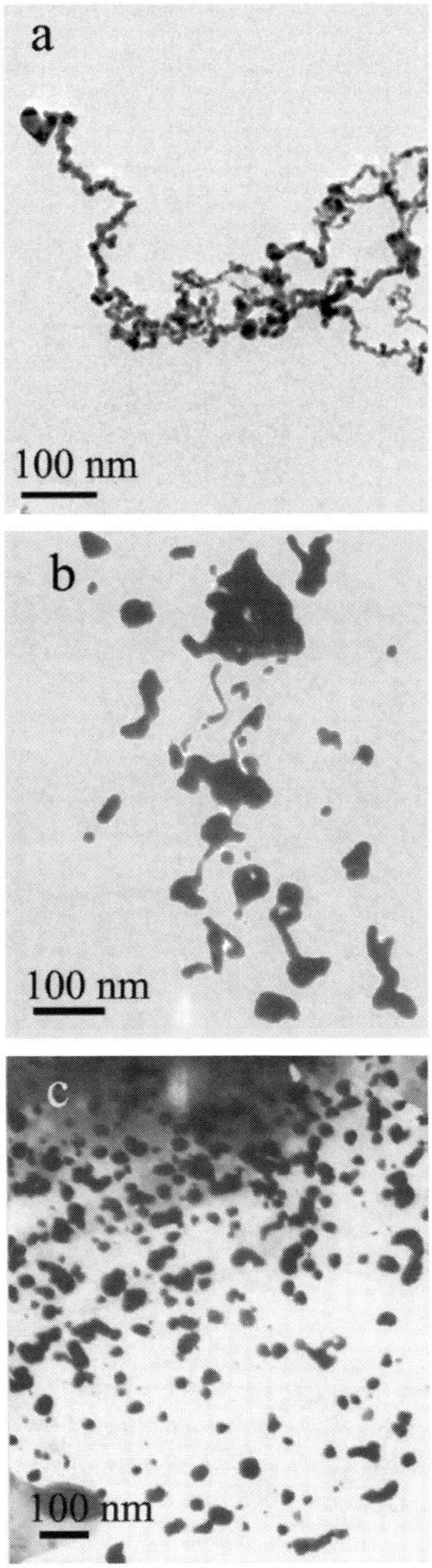

Figure2. TEM images of the gold nanoparticles formed in the $HAuCl_4$ (0.2 mM) solution with different concentration of CMC (a) 0.1%, (b) 0.3% and (c) 0.5%.

Optical Characterization

Gold nanoparticles exhibit strong plasmon resonance absorption that is dependent on the particle size and shape. Figure 3 shows the UV-vis absorption spectra of the gold nanostructures obtained from solutions containing different concentration of CMC with a fixed concentration of $HAuCl_4$ (0.2 mM). These spectra were taken for freshly prepared samples. Figure3 shows that the wavelength of the main absorption peak for all samples obtained in the CMC solutions are centered between 540 and 580 nm, which is typical of gold colloids. However, when the concentration of CMC below 0.1% , absorption bands above 700 nm are also stronger, indicted that gold nanowires formed[19].TEM images in Figure1 and Figure 2a approved this phenomenon. As the concentration of CMC increased, absorption bands above 700 nm decreased.

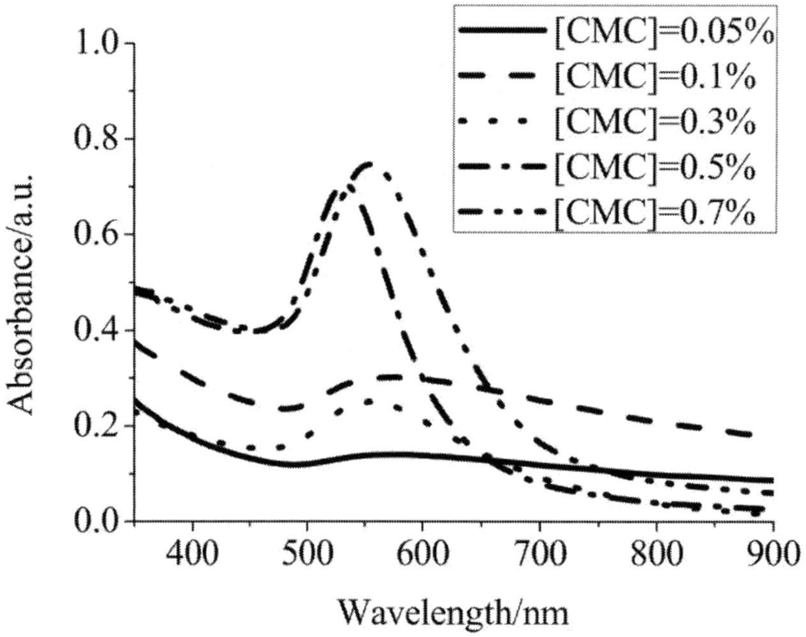

Figure 3. UV-vis spectra of gold hydrosols obtained with 0.2 mM of $HAuCl_4$ in different concentrations of CMC.

Conclusions

In this paper, Gold 2-D networked nanowires are prepared in CMC aqueous solution through a one-pot, facile and morphology-controlled synthesis route. Shape control was achieved by simply varying the concentration of CMC added to the reacted soltuion. Low concentration of CMC in solution is helpful to prepare the gold nanowire network, and high concentration of CMC solution restrict the aggregation of tiny gold paricles to synthesizing multipodlike particles. This method may be used to control other noble metal nanostructures.

Acknowledgement

We thank the National Science Foundation (31000820), the Jiangxi Provincial Office of Education Science Foundation (GJJ10109) and Jiangxi Agricultural University Doctor Start Foundation (2313) for funding this research.

References

1. Roucoux, A.; Schulz, J.; Patin, H.(2002), "Reduced Transition Metal Colloids: A Novel Family of Reusable Catalysts?", *Chemistry Reviews.*, 102, (10), pp3757-3778.
2. Tan, Y. N.; Su, X.; Liu, E. T.; Thomsen, J. S.(2010), "Gold-Nanoparticle-Based Assay for Instantaneous Detection of Nuclear Hormone Receptor-Response Elements Interactions", *Analytical Chemistry*, 82, (7), pp2759-2765.
3. Sierra-Sastre, Y.; Dayeh, S. A.; Picraux, S. T.; Batt, C. A.(2010), "Epitaxy of Ge Nanowires Grown from Biotemplated Au Nanoparticle Catalysts", *ACS Nano*, 4, (2), pp1209-1217.
4. Qu, S.; Li, H.; Peng, T.; Gao, Y.; Qiu, J.; Zhu, C.(2009), "Optical nonlinearities from transverse plasmon resonance in gold nano-rods", *Materials Letters*, 58, (9), pp1427-1430.
5. Patolsky, F.; Zheng, G.; Lieber, C. M.(2006), "Nanowire-Based Biosensors", *Analytical Chemistry*, 78, (13), pp4260-4269.
6. Huang, Y.; Fang, Y.; Sun, M.(2011), "Remote Excitation of Surface-Enhanced Raman Scattering on Single Au Nanowire with Quasi-Spherical Termini", *The Journal of Physical Chemistry C*, 115, (9), pp3558-3561.
7. Zhang, X.; Li, D.; Bourgeois, L.; Wang, H.; Webley, P. A.(2009), "Direct Electrodeposition of Porous Gold Nanowire Arrays for Biosensing Applications", *ChemPhysChem*, 10, (2), pp436-441.
8. Jung, M.; Noh, H.; Doh, Y. J.; Song, W.; Chong, Y.; Choi, M. S.; Yoo, Y.; Seo, K.; Kim, N.; Woo, B. C.; Kim, B.; Kim, J.(2011), "Superconducting Junction of a Single-Crystalline Au Nanowire for an Ideal Josephson Device", *ACS Nano*, 5, (3), pp2271-2276.
9. Ko, S. H.; Lee, D.; Kang, H. W.; Nam, K. H.; Yeo, J. Y.; Hong, S. J.; Grigoropoulos, C. P.; Sung, H. J.(2011), "Nanoforest of Hydrothermally Grown Hierarchical ZnO Nanowires for a High Efficiency Dye-Sensitized Solar Cell", *Nano Letters*, 11, (2), pp666-671.
10. Mafun, F.; Kohno, J.; Takeda, Y.; Kondow, T.(2003), "Formation of Gold Nanonetworks and Small Gold Nanoparticles by Irradiation of Intense Pulsed Laser onto Gold Nanoparticles", *The Journal of Physical Chemistry B*, 107, (46), pp12589-12596.
11. Sun, Y.; Yin, Y.; Mayers, B. T.; Herricks, T.; Xia, Y.(2002), "Uniform Silver Nanowires Synthesis by Reducing AgNO3 with Ethylene Glycol in the Presence of Seeds and Poly(Vinyl Pyrrolidone)", *Chemistry of Materials*, 14, (11), pp4736-4745.
12. Sun, Y.; Gates, B.; Mayers, B.; Xia, Y.(2002), "Crystalline Silver Nanowires by Soft Solution Processing", *Nano Letters*, 2, (2), pp165-168.
13. Jung, Y. S.; Lee, J. H.; Lee, J. Y.; Ross, C. A.(2010), "Fabrication of Diverse Metallic Nanowire Arrays Based on Block Copolymer Self-Assembly", *Nano Letters*, 10, (9), pp3722-3726.
14. Lu, W.; Shi, W.; Wang, W.; Jiang, L.(2008), "One-step synthesis of gold nanowire network capped by diacetylene molecules under ultrasonic irradiation", *Colloids and Surfaces A: Physicochemical and Engineering Aspects*, 317, (1-3), pp239-246.
15. Zhong, Z.; Luo, J.; Ang, T. P.; Highfield, J.; Lin, J.; Gedanken, A.(2004), "Controlled Organization of Au Colloids into Linear Assemblies", *The Journal of Physical Chemistry B*, 108, (47),

pp18119-18123.
16. Djalali, R.; Chen, Y.; Matsui, H.(2002), "Au Nanowire Fabrication from Sequenced Histidine-Rich Peptide", *Journal of the American Chemical Society*, 124, (46), pp13660-13661.
17. Ramanath, G.; D, A. G.; Maddanimath, T.; Ellis, A. V.; Ganesan, P. G.; Goswami, R.; Kumar, A.; Vijayamohanan, K.(2004), "Templateless Room-Temperature Assembly of Nanowire Networks from Nanoparticles", *Langmuir*, 20, (13), pp5583-5587.
18. Zhong, Z.; Luo, J.; Ang, T. P.; Highfield, J.; Lin, J.; Gedanken, A.(2004), "Controlled Organization of Au Colloids into Linear Assemblies", *The Journal of Physical Chemistry B*, 108, (47), pp18119-18123.
19. Pei, L.; Mori, K.; Adachi, M.(2004), "Formation Process of Two-Dimensional Networked Gold Nanowires by Citrate Reduction of $AuCl_4^-$ and the Shape Stabilization", *Langmuir*, 20, (18), pp7837-7843.

(246e) Janus Double Brush Copolymers

Yukun Li, Leela Christian-Tabak, Jiong Zou and **Chong Cheng**, Department of Chemical and Biological Engineering, SUNY-Buffalo, Buffalo, NY

Brush polymers have emerged as a class of nanostructured polymers. Representing a novel type of brush polymers, double brush copolymers (DBCs) have well-defined architectures with every grafting site carrying two different grafts. DBCs have been synthesized by tandem living polymerizations. With a variety of interesting nanostructures, DBCs have promising application potentials. For instance, they can serve as nano-surfactant for miniemulsions with remarkable stability.

See more of this Session: Nanoscale Structure In Polymers I
See more of this Group/Topical: Materials Engineering and Sciences Division

(121f) A Master-Equation Approach to Simulate Directed Self Assembly

R. Lakerveld, G. Stephanopoulos and P.I. Barton, Department of Chemical Engineering, Massachusetts Institute of Technology, Cambridge, MA

Robustfabrication of nanoscale structures is of great importance to enabletechnological breakthroughs in various fields such as molecular computing,diagnostics, and molecular factories [1]. Top-down methods such as e-beamlithography can be used to produce non-periodic nanoscale structures with aresolution in the range of 15 nm [2]. Bottom-up methods driven by self assemblycan be used to fabricate nanoscale structures with unmatched resolution at themolecular scale. Fabrication of nanoscale structures via self assembly has beendemonstrated particularly for periodic structures. A key challenge is also tofabricate robustly non-periodic nanoscale structure via directed self assembly.Top-down methods such as e-beam lithography can be used to fabricatenano-electrodes, which in turn can direct the self-assembly process by placingtime-varying controls in the form of electrostatic charges on a scaffold [3][4].Static charge can also be placed on a scaffold by using micro-contact printing[5]. Electrostatic forces are particularly relevant for directing self assemblydue to the tuneable direction and strength of these forces [6]. Furthermore,nanoparticles can be specifically functionalized to manipulate interactionsbetween nanoparticles. The often competing forces at the nanoscale have to beengineered such that the self assembly will be directed towards a desirednanostructure.

Soliset al. [7][8] addressed the influence of local point charges on directed selfassembly. The point charges interact with nanoparticles to direct their selfassembly into a desired configuration. First, Solis et al. [7] developed amethod to stabilize a desired configuration. Secondly, Solis et al. [8] developeda method to arrive at a desired configuration with maximum probability from anunknown initial structure. A sequence of pseudostatic problems approximates theoverall dynamic problem. Although such an approach is an important step forwardas it protects the system from favoring undesired configurations, it fails topredict the influence of kinetic traps during the course of a specificpseudostatic phase. Self-assembled systems are prone to kinetic traps [6],which can dynamically arrest the system in local minima of the potential energysurface.

Theaim of this contribution is to demonstrate the application of master equationsto simulate directed self assembly of nanoparticles including kinetic trapping.Master equations describe time-varying continuous-time Markov processes. Thestates represent the probabilities that the system will be in a certainconfiguration at a given time. Models based on master equations have beensuccessfully applied to simulate kinetic traps in related fields such as thefolding kinetics of proteins. A generic model framework will be presented basedon the Ising lattice model to describe self assembly. The number of states in thesemodels grows exponentially with the size of the system. Therefore, particularemphasis has been paid to the design of efficient computational algorithms. Thevariable-coefficient ODE solver DVODPK [9] with pre-conditioned iterativesolver GMRES is used to solve the resulting sparse and stiff system of ODEs.The complete algorithm runs in linear computational complexity, which makes thealgorithm favorable for large systems.

Theeffect of the various degrees of freedom to direct self assembly will be illustratedfor several case studies. First, the capability of the model to simulatekinetic trapping of a subset of state space will be illustrated. Second, theinfluence of temperature with respect to kinetic trapping will be illustrated.Third, the influence of the relative strengths between various inter-particleinteraction forces is illustrated. Fourth, several degrees of freedom will becombined to direct the self assembly of a system systematically towards adesired configuration by reducing the ergodicity of the system in a specificmanner. The strategy is to first direct the desired number of nanoparticles toa desired part of the domain. Subsequently, the ergodicity of the system isbroken by restricting transitions that involve the movement of particlesbetween the desired parts of the domain. Finally, the nanoparticles aredirected to their precise location. A strategy that utilizes ergodicitybreaking offers the prospect for approaching larger systems. The exponentialgrowth of the model inevitably requires coarse graining techniques to be ableto approach larger systems mathematically. The systematic decomposition ofsystem ergodicity allows for the application of such coarse graining techniquesin a natural way.

Themaster equations are solved explicitly for each state variable to yield thetime evolution of the entire probability distribution. Following this approach,the dynamic trajectories of parametric sensitivities can in principle becalculated, which are useful for dynamic optimization. An interesting directionfor future research is to apply rigorous dynamic optimization to reveal optimaland robust fabrication routes towards non-periodic nanoscale structures viaself assembly.

References

[1] Stephanopoulos,N., Solis, E.O.P., Stephanopoulos, G., Nanoscale Process Systems Engineering: Toward Molecular Factories, Synthetic Cells, and Adaptive Devices, *AIChE J.*,**2005**, 51, 1858-1869.

[2] Pires, D.,Hedrick, J.L., De Silva, A., Frommer, J., Gotsmann, B., Wolf, H., Despont, M.,Duerig, U. Knoll, A.W. Nanoscale Three-Dimensional Patterning of MolecularResists by Scanning Probes. *Science*, **2010**, 328, 732-735.

[3] Koh, S.J.Strategies for the Controlled Placement of Nanoscale Building Blocks. *NanoscaleRes. Lett.*, **2007**, 2, 519-545.

[4] Gates,B.D., Xu, Q., Stewart, M., Ryan, D., Wilson, C.G., Whitesides, G.M. NewApproaches to Nanofabrication: Molding, Printing, and Other Techniques. *Chem.Rev.*, **2005**, 105, 1171-1196.

[5] Jacobs,H.O., Whitesides, G.M. Submicrometer Patterning of Charge in Thin-FilmElectrets. *Science*, **2001**, 291, 1763-1766.

[6] Bishop,K.J.M., Wilmer, C.E., Soh, S., Grzybowski, B.A. Nanoscale Forces and Their Usesin Self-

Assembly. *Small*, **2009**, 5, 1600-1630.

[7] Solis,E.O.P., Barton, P.I., Stephanopoulos, G. Controlled Formation of Nanostructureswith Desired Geometries. 1. Robust Static Structures. *Ind.Eng.Chem.Res.*,**2010**, 49, 7728-7745

[8] Solis,E.O.P., Barton, P.I., Stephanopoulos, G. Controlled Formation of Nanostructureswith Desired Geometries. 2. Robust Dynamic Paths. *Ind.Eng.Chem.Res.*, **2010**,49, 7746-7757.

[9] Brown, P.N., Byrne, G. D, Hindmarsh, A. C. VODE, A Variable-Coefficient ODE Solver. *SIAM J. Sci. Stat. Comput.*, **1989**, 10, 1038-1051.

See more of this Session: Self and Directed Assembly at the Nanoscale
See more of this Group/Topical: Nanoscale Science and Engineering Forum

Organic Vesicles Captured by In-situ Reduction of Gold

Pingping Pang, Yun Fang, Hanting Chen

School of Chemical and material engineering, Jiangnan University, Wuxi, China

Abstract The partially hydrolyzed polyacrylamide (HPAM) and hydroxyethyl hydroxyethyl cellulose (HEC) can self-assemble into vesicles in pure aqueous solution at different pH. Because of low contrast of macromolecules, it seems difficult to get clear TEM images. Using the reducibility of HEC to $HAuCl_4$, the gold was in-situ reduced at the surface of self-assembled particles, so that we can get clear TEM images.

Keywords: nanotechnology, polymers, in-situ reduction, self-assembly, vesicle

Introduction

As a newly arisen science of nanotechnology, macromolecular self-assembly have been widely investigated because of the anticipate application in drug delivery[1], catalysis[2] and nanoreactors. In this field, block and graft copolymers have been deeply investigated.[3-5] Early researches have shown that amphiphilic copolymers can self-assemble into different morphologies in selective or common solvents such as spheres, rods, vesicles and compound micelles.[3,6-8] But these copolymers were synthesized difficultly and self-assembled in organic solvent which limited their application. Recent years, random copolymers have been given more attention gradually.[9-10] Moreover, researching macromolecular self-assembly in pure aqueous solution is more significative for the application in life science.[11-12]

Transmission electron microscope (TEM) is a main characterization method for the structure of polymeric micelles. However, it is often necessary to enhance contrast for polymeric micelles to get clear images. Using RuO_4 and phosphotungstic acid (PTA) as staining agents is a common method.[13-14] In this paper, we established a new contrast enhancing strategy by in-situ reduction of gold to increase the contrast of polymeric micelles. This strategy can provide the possibility of characterizing polymeric nanoparticles which smaller than 20 nm.

Experiments

$HAuCl_4$ and HCl were purchased from Sinopharm Chemical Reagent Co. Ltd. HEC (Mv=90000, DS=1.5, MS=2.5) were purchased from Sigma-Aldrich Co. USA. HPAM were synthesized by ourselves. Ultra pure water was used through out to prepare HPAM, HEC, HCl and $HAuCl_4$ aqueous solutions. In a typical preparation of micelles, aqueous solutions of 5 $mg{\cdot}mL^{-1}$ HPAM (2 mL) and 5 $mg{\cdot}mL^{-1}$ HEC (2 mL) were added into the reaction vessel. A certain amount of ultra pure water was added in the solution. Under magnetic stirring, 0.1 $mol{\cdot}L^{-1}$ HCl

(40 μL) were dropped per minute to adjust the solution to the desired pH. Finally, an aqueous solution of 1mmol·L^{-1} $HAuCl_4$ (1 mL) was added to the solution, and the total amount of the solution is 10 mL. The final concentration of HPAM and HEC were 1 mg·mL^{-1}.

The morphology and microstructure of the products were characterized using a transmission electron microscope (TEM, JEOL-JEM2100) with an accelerating voltage of 200 kV.

Results and Discussion

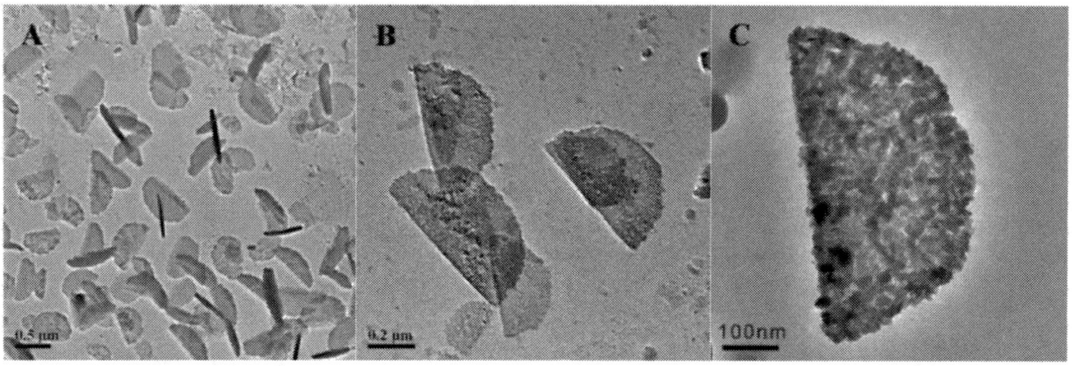

Fig.1 TEM images of HPAM-HEC large compound vesicles at pH=1.3 before (A, B) or after (C) in-situ reduction of gold

Figure 1 showed that HPAM-HEC self-assembled into 500 nm×300 nm×50 nm of half-moon micelles at pH=1.3. We added $HAuCl_4$ into the system. Using the reducibility of HEC, the gold atoms were in-situ reduced at the surface of self-assembled particles to increase the electron density of vesicles. Therefore, we got clear TEM image (Figure 1C) which morphology was the same to before in situ reduction of gold (Figure 1A,B).

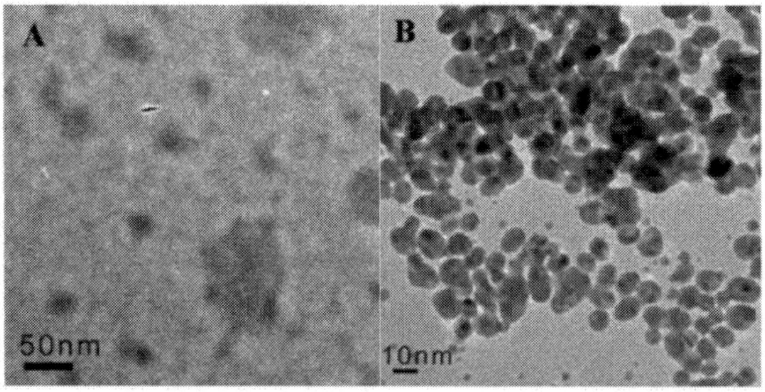

Fig.2 TEM images of HPAM-HEC spherical polymeric micelles at pH=0.9 before (A) or after (B) in-situ reduction of gold

When decreasing pH to a lower level, the large compound vesicles collapsed at pH=0.9 to disintegrate into 10 nm pseudo spherical polymeric micelles (Figure 2). Due to small and low contrast of the micelles, blurry images were obtained by TEM (Figure 2A). After in-situ reduction of gold, spherical polymeric micelles can be clearly observed (Figure 2B).

Conclusions

The partially hydrolyzed polyacrylamide (HPAM) and hydroxyethyl hydroxyethyl cellulose (HEC) would self-assemble into the half-moon miltivesicular vesicles of 500 nm ×300 nm ×50 nm at pH 1.3 and collapse to disintegrate into 10 nm pseudo spherical polymer micelles at pH 0.9. In-situ reduction of gold is efficient at observing the nanostructure of micelles with low contrast. This strategy can provide the possibility of characterizing polymeric nanoparticles which smaller than 20 nm.

References

[1] Weizhong Yuan, Xiaofei Li, Shuying Gu, Amin Cao, Jie Ren. Amphiphilic chitosan graft copolymer via combination of ROP, ATRP and click chemistry: Synthesis, self-assembly, thermosensitivity, fluorescence, and controlled drug release. *Polymer* 2011, 52, 658-666.

[2] Jong-UK Kim, Sang-Ho Cha, Kyusooon Shin, Jae Young Jho, Jong-Chan Lee. Preparation of Gold Nanowires and Nanosheets in Bulk Block Copolymer Phases under Mild Conditions. *Adv. Materials* 2004, 16, 459-464.

[3] Cristiano Giacomelli, Vanessa Schmidt, Karim Aissou, and Redouane Borsali. Block Copolymer Systems: From Single Chain to Self-Assembled Nanostructures. *Langmuir* 2010, 26, 15734-15744.

[4] Xiaohuan Zhang, Zhong Shen, Chun Feng, Dong Yang, Yaogong Li, Jianhua Hu, Guolin Lu, Xiaoyu Huang. PMHDO-g-PEG Double-Bond-Based Amphiphilic Graft Copolymer: Synthesis and Diverse Self-Assembled Nanostructures. *Macromolecules* 2009, 42, 4249-4256.

[5] Lin Ma, Ruigang Liu, Junjun Tan, Deqian Wang, Xin Jin, Hongliang Kang, Min Wu, Yong Huang. Self-Assembly and Dual-Stimuli Sensitivities of Hydroxypropylcellulose-graft-poly(N,N-dimethyl aminoethyl methacrylate) Copolymers in Aqueous Solution. *Langmuir* 2010, 26, 8697-8703.

[6] Lifeng Zhang, Adi Eisenberg. Multiple Morphologies of "Crew-cut" Aggregates of Polystyrene-b-Poly (Acrylic Acid) Block Copolymers. *Science* 1995, 268, 1728-1731.

[7] Prachur Bhargava, Yingfeng Tu, Joseph X. Zheng, Huiming Xiong, Roderic P. Quirk, Stephen Z. D. Cheng. Temperature-Induced Reversible Morphological Changes of Polystyrene-block-Poly(ethylene Oxide) Micelles in Solution. *J. Am. Chem. Soc.* 2007, 129, 1113-1121.

[8] Yapei Wang, Peng Han, Huaping Xu, Zhiqiang Wang, Xi Zhang, Alexander V. Kabanov. Photocontrolled Self-Assembly and Disassembly of Block Ionomer Complex Vesicles: A Facile Approach toward Supramolecular Polymer Nanocontainers. *Langmuir* 2010, 26, 709-715.

[9] Peide Guo, Wenyan Guan, Li Liang, Ping Yao. Self-assembly of pH-sensitive random copolymers: Poly(styrene-*co*-4-vinylpyridine). *Journal of Colloid and Interface Science* 2008, 323, 229-234.

[10] Xia Li, Hui Guo, Junliang Wang, Qi Wu, Xianfu Lin. Fabrication of novel hepatoma-targeting microdisks by hydrogen bond-assisted self-assembly of an azacitidine-conjugating amphiphilic random copolymer. *Acta Biomaterialia* 2010, 6, 511-518.

[11] Qiang Chen, Xiaoguang Liu, Kun Xu, Chunlei Song, Wende Zhang, Pixin Wang. Phase Behavior and Self-Assembly of Poly [N-vinylformamide-co-(acrylic acid)] Copolymers Under Highly Acidic Conditions. *Journal of Applied Polymer Science* 2008, 109, 2802-2807.

[12] Yun Fang, Zhongyu Lai, Pingping Pang, Ming Jiang. pH Responsive Self-Assembly of Low-Molecular-Weight HPAM/HEC in Aqueous Solution. *Acta Phys. -Chim. Sin.* 2011, 27, 1712-1718.

[13] John S. Trent, Jerry I. Scheinbeim, Peter R. Couchman. Ruthenium Tetraoxide Staining of Polymers for Electron Microscopy. *Macromolecules* 1983, 16, 589-598.

[14] T. M. Chou, P. Prayoonthong, A. Aitouchen, M. Libera. Nanoscale artifacts in RuO4-stained poly(styrene). *Polymer* 2002, 43, 2085-2088.

Partitioning of Fullerene between Water and Synthetic Membrane Materials: Effects of Temperature and Membrane Compositions

Yeonjeong Ha, Howard M. Liljestrand, Lynn E. Katz

Department of Civil, Architectural and Environmental Engineering

The University of Texas at Austin, 1 University Station C1786, Austin, Texas, 78712

1. Introduction

As the usage of engineered nanomaterials increases, concerns on the unknown risks of these nanoparticles towards humans and the environment have grown. Despite these growing concerns about nanomaterials in aquatic environments, many unknowns remain. Investigating bio-uptake and bio-accumulation of nanomaterials is of critical importance not only to determine bioavailability but also to evaluate fate and transport in the environment of these compounds.

For hydrophobic contaminants and ionized organic chemicals, many models have been developed to determine their bio-concentration and toxicity. For example, octanol-water partitioning (K_{ow}) has been successfully used to evaluate the bioavailability of chemicals. More recently, liposome-water partitioning (K_{lipw}) has been proposed as being more appropriate than octanol-water partitioning (K_{ow}) because synthetic membrane vesicles are better surrogates and K_{lipw} can reduce thermodynamic differences from actual biota. For engineered nanomaterials, K_{ow} values of C_{60} (Jafvert et al., 2008) and carbonnanotubes (Petersen et al., 2010) have been determined. However, few attempts have been made to measure K_{lipw} of nanoparticles to investigate bioavailability of engineered nanomaterials.

Carbon fullerene (C_{60}), which was discovered in 1985 (Kroto et al., 1985), is a major material of nanoscale science and engineering. The production and application of fullerene have steadily increased due to increasing commercial interest in its unique properties, such as cage like structure, strong hydrophobicity, and high reactivity. It is estimated that more than 1500 tons per year of carbon nanoparticles including fullerenes and nanotubes are produced (Ringwood et al., 2009). However, there is limited information concerning the impact of fullerene on the natural environment and human health.

Bioavailability of fullerene in aquatic environment can be investigated by measuring partition coefficient between liposome and water (K_{lipw}) similar to that for other chemicals. Many researchers (Yamamoto et al., 2004, Kwon et al., 2006) have determined K_{lipw} using an equilibrium dialysis technique in which water and liposome solutions are separated by a dialysis membrane through which the chemical can diffuse but the liposome cannot. However this approach is less useful for fullerene, because the dialysis membrane can prevent the diffusion of engineered nanoparticles after their aggregation in water to a particle size much bigger than the molecular scale of hydrophobic or ionizing organic chemicals.

In this study, solid supported lipid membrane (SSLM) was chosen as a model biological membranes. Nonporous silica beads were used as the solids, and these were coated with the lipids. K_{lipw} of fullerene was investigated with various types of lipid membrane materials and temperatures. In addition, ternary lipid (liquid ordered phase lipid, liquid disordered phase lipid and cholesterol) mixture membrane vesicles were prepared at two different temperatures (above and below the transition temperature) to investigate whether K_{lipw} of nanoparticles can be influenced by the phase height difference between liquid disordered and liquid ordered/cholesterol. Lipids used in this study are 1,2 dimyristoyl-sn-glycero-3-phosphochloline (DMPC), 1,2 dioleoyl-sn-glycero-3 phosphocholine (DOPC), 1,2 dierucoyl-sn-glycero-3-phosphochloline (DEruPC), DOPC/ sphinogosylphosphochloline (SM) /cholesterol, DMPC/SM/cholesterol, DEruPC/SM/cholesterol.

2. Materials and Methods

2.1 Chemicals

Fullerene (C_{60}, 99.5+ %) was purchased from SES Research (Houston, TX). Unsaturated lipids such as 1,2 dimyristoyl-sn-glycero-3-phosphochloline (DMPC, C14:1, C14:1), dioleoyl-sn-glycero-3 phosphocholine (DOPC, C18:1, C18:1), and 1,2 dierucoyl-sn-glycero-3-phosphochloline (DEruPC, C22:1, C22:1), and sphinogosylphosphochloline (SM) were obtained from Avanti Polar Lipids. Cholesterol was purchased from Aldrich Chemical Co., and non-functionalized silica microspheres (mean diameter of 5.2 μm, 100 % solids content) were obtained from Bang's Laboratories.

2.2 Aqueous C_{60} solution

Fullerene has negligible solubility in water due to its strong hydrophobicity. However, many studies have shown that fullerene can form highly stable, negatively charged colloidal aggregates (nC_{60}) when it is exposed to water. Stable aggregated forms of fullerene have more than 8-11 orders of magnitude greater solubility than its hydrophobic form (Ruoff et al., 1993). Therefore, it is evident that the transition from its extreme hydrophobic form to water stable forms have critical implications for fullerene transport, transformation and biological effects (Isaacson et al., 2009).

The water stable suspensions of fullerene can be formed by various preparation methods in the laboratory. In this study, a modified SON/nC_{60} method was applied to make the aqueous C_{60} solution. Briefly, the fullerenes nanoparticles were formed by first adding pure fullerene to toluene. The solution was then mixed with a large volume of the deionized water and the toluene was removed by evaporation under ultrasonic treatment. The fullerene suspension in water was filtered through a 0.8 μm membrane filter. The mean size of C_{60} particles in water is in the range between 50 and 100 nm as measured by Dynamic Light Scattering (Model Zetasizer nano ZS).

2.3 Preparation of solid supported lipid bilayer

The synthetic membrane vesicles of each lipid compositions can be prepared using the thin film hydration technique followed by the rapid extrusion process. The detailed procedure is described elsewhere (Escher et al., 2000, Kwon et al., 2006). Briefly, the organic solution of lipid was evaporated using a nitrogen purging and the thin film residue was dissolved in water. After storage and equilibration at 4°C for a day, the lipid suspension was extruded several times through a polycarbonate membrane to reduce polydiversity of vesicles.

Solid supported lipid bilayers were formed using methods described in Bayerl et al., (1990) and Baksh et al., (2004). Silica beads were washed three times with methanol. The lipid suspensions were incubated at 30 °C for 30 min, and the lipid dispersions were sonicated (Branson 2000) for 15 min. The resulting vesicle dispersions were incubated at 60 °C for 5 min and then poured onto the dry glass beads with 60 sec rigorous vortex mixing. After vortexing, the silica beads and lipid dispersions mixture were centrifuged for 10 min at 2,000 g and then supernatants were discarded. Excess vesicles not adsorbed at the solid surface were removed by redispersing in water, vortexing, and the supernatants were discarded. The mass of lipids which were adsorbed on the silica beads was determined by measuring the difference between initial lipid concentration and supernatant concentration. Each lipid concentration was measured by TOC Analyzer (Tekmar Dohrmann Apollo 9000).

For solid supported ternary lipid mixture membrane vesicles, weight ratio of unsaturated lipids: SM: cholesterol was 2:1:1. Two replicate samples were prepared. One sample was stored at below the transition temperature (T_m), and the other one was first stored at above T_m overnight and then placed at below T_m to create phase height difference between liquid disordered lipid and liquid ordered lipid/cholesterol.

2.4 Determination of K_{lipw}

The solid supported lipid bilayer and aqueous fullerene dispersion were placed into 1.8 mL amber vial with PTFE/silicon septa. The reactors were incubated for 80 hr using a custom-made tumbling device (4 °C, 11 °C, and 25 °C) or shaker (30 °C and 50 °C, at 180 rpm). For ternary lipid mixture, vials were incubated at 25 °C with a tumbling device. The preliminary experiment showed that equilibrium was obtained after 80 hrs using both tumbler and shaker. After 80 hrs, each vial was placed for 1 hr to settle down the solid supported lipid bilayer. Then supernatants contained only free fullerene without lipid vesicles, and concentration of fullerene in supernatants (C_w) were measured using a Waters 2690 high-performance liquid chromatography system equipped with a Waters 996 photodiode array detector (Milford, MA, USA) after liquid extraction using toluene and 0.1 M of $Mg(ClO_4)_2$. Double or Triplicate samples were prepared for each test. The partition coefficient (K_{lipw}) was calculated using

$$K_{lipw}(L/kg-liposome) = \frac{C_{lip}}{C_w m} = \frac{C_0 - C_w}{C_w m}$$

Where C_{lip} is the concentration of fullerene in the liposome side, which is same as the difference between the initial or control C_{60} concentration (C_0) and fullerene concentration in supernatants (C_w) and *m* is the concentration of liposome (kg-liposome/L).

3. Results and Discussion

Table 1 shows K_{lipw} values between water and DMPC ($K_{lipw,DMPC}$), and between water and DEruPC ($K_{lipw,DEruPC}$) at five different temperatures (4 °C, 11 °C, 25 °C, 30 °C, and 50 °C). The log K_{lipw} (L/kg) values range from 3.3 to 5.3. As far as we know, the K_{lipw} value of fullerene has not been reported. Some studies have concerned about Bioconcentration factor (BCF) of fullerene: Jafvert et al., (2008) estimated log BCF values of fullerene as 4.51, 5.37, 5.71 for cod, earthworms and salmon, respectively, and Tervonen et al., (2010) measured log BCF values of fullerene for *Daphnia magna* ranged from 3.30 and 3.88. Those reported values are generally consistent with the results of this study.

Table 1. Liposome – Water Partition Coefficients (K_{lipw}, L/kg-liposome) of fullerene obtained at five different temperatures using 1,2 dimyristoyl-sn-glycero-3-phosphochloline (DMPC) and 1,2 dierucoyl-sn-glycero-3-phosphochloline (DEruPC)*

	4 °C	11 °C	25 °C	30 °C	50 °C
$K_{lipw,DMPC}$	1.96×10^3	4.34×10^3	7.87×10^3	1.52×10^4	2.18×10^4
$K_{lipw,DEruPC}$	9.92×10^3	1.07×10^4	2.64×10^4	5.62×10^4	1.89×10^5

*Average values of duplicates analysis are used

Both $K_{lipw,DOPC}$ and $K_{lipw,DEruPC}$ values of fullerene increase with increasing temperatures (Figure 1). It has been reported that for most of hydrophobic chemicals such as EDCs (Kwon et al., 2007), K_{lipw} values with liquid crystalline phase liposomes decreased with increasing temperature. Therefore, partitioning processes of fullerene between water and liposome can be different from those of hydrophobic chemicals.

At all different temperatures, log $K_{lipw,DEruPC}$ values are higher than log $K_{lipw,DMPC}$ by more than 0.4 log units (Figure 1). Based on this result, it can be concluded that K_{lipw} values of fullerenes are larger for lipid bilayers composed of longer acyl chains. Wong-Ekkabut et al., (2008) indicated that when aggregated fullerenes are located close to the lipid head group, only small clusters form and rapidly penetrate into the bilayer and then disaggregate after entering the membrane. This result shows that, despite of large size of fullerene aggregates, which are from ten to a few hundreds of nanometers, they can penetrate through lipid bilayers. Therefore, the length of acyl chain of lipids can affect the K_{lipw} of fullerenes. However, more work is required to confirm this hypothesis.

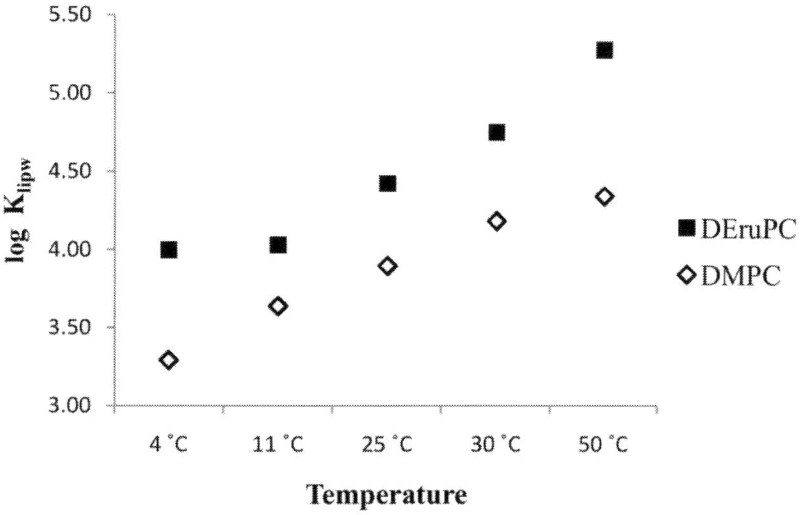

Figure 1. Effects of temperatures on lipid – water partitioning coefficients (K_{lipw}, L/kg-liposome) of fullerene using DMPC and DEruPC (average values of duplicates analysis are used).

4. References

1. Jafvert, C.T., Kulkarni, P.P. (2008), "Buckminsterfullerene's (C_{60}) octanol-water partition coefficient (Kow) and aqueous solubility," *Environ. Sci. Technol.* 42. pp. 5945-5950.
2. Petersen, E.J., Huang, Q., Weber, W.J.(2010), "Relevance of octanol-water distribution measurements to the potential ecological uptake of multi-walled carbon nanotubes," *Environ. Toxicol. Chem.* 29. pp. 1106-1112.
3. Kroto, H.W.,Heath, J.R., O'Brien, S.C., Curl, R.F., Smalley, R. E.(1985), "C60-Buckminsterfullerene," *Nature*. 318. pp. 162-163.
4. Ringwood, A.H., Levi-Polyachenko,N., Carroll, D.L. (2009), " Fullerene Exposures with Oysters: Embryonic, Adult, and Cellular Responses," *Environ. Sci. Technol.* 43. pp. 7136-7141.
5. Yamamoto, H., Liljestrand, H.M., Shimizu, Y. (2004), "Effects of dissolved organic matter surrogates on the partitioning of 17ß-Estradiol and p-Nonylphenol between synthetic membrane vesicles and water," *Environ. Sci. Technol.* 38(8). 2351-2358.
6. Kwon, JH., Liljestrand, H.M., Katz, L.E. (2006), "Partitioning of moderately hydrophobic endocrine discruptors between water and synthetic membrane vesicles," *Environ. Toxicol. Chem.* 25(8). pp.1984-1992.
7. Ruoff, R. S., Tse, D. S., Malhotra,R., Lorents, D.C. (1993), "Solubillity of C_{60} in a variety of solvents, " *J. Phys. Chem.* 97. pp. 3379-3383.
8. Isaacson, C.W., Kleber, M., Field, J.A. (2009), "Quantitative analysis of fullerene nanomaterials in environmental systems: A critical review," *Environ. Sci. Technol.* 43. pp. 6463-6474.

9. Escher, B.I., Schwarzenbach, R.P. (2000), "Evaluation of liposome – Water partitioning of organic acids and bases. 1. Development of a sorption model," *Environ. Sci. Technol.* 34(18). pp.3954-3961.
10. Bayerl, T.M., Bloom, M. (1990), "Physical properties of single phospholipid bilayers adsorbed to micro glass beads," *Biophys. J.* 58. pp. 357-362.
11. Baksh, M.M., Jaros, M., Groves, J.T. (2004), "Detection of molecular interactions at membrane surfaces through colloid phase transitions," *Nature.* 427. pp. 139-141.
12. Tervonen, K., Waissi, G., Petersen, E.J., akkanen, J., Kukkonen, J.V. (2010), "Analysis of fullerene-C60 and kinetic measurements for its accumulation and depuration in Daphnia magna," *Environ. Toxicol. Chem.* 29. pp. 1072-1078.
13. Kwon, JH., Liljestrand, H.M., Katz, L.E., Hiroshi, Y. (2007), "Partitioning thermodynamics of selected endocrine disruptors between water and synthetic membrane vesicles: Effect of membrane compositions", *Environ. Sci. Technol.* 41. pp. 4011-4018.
14. Wong-ekkabut, J., Baoukina, S., Triampo, W., Tang, I.M., Tieleman, D.P., Monticelli, L. (2008), "Computer simulation study of fullerene translocation through lipid membrane", *Nature Nanotechnology.* 3. pp. 363-368.

(543c) Synthesis and Characterization of Curcumin Based Poly (β-Amino Ester) Antioxidant Nanoparticles to Control Cellular Oxidative Stress

Prachi Gupta, Chemcial Engineering, University of Kentucky, Lexington, KY, Dr. Thomas Dziubla, Chemical and Materials Engineering, University of Kentucky, Lexington, KY and Dr. J. Zach Hilt, Chemical and Materials Engineering, The University of Kentucky, Lexington, KY

Curcumin, an anti-oxidant obtained from the turmeric plant, has recently been gaining interest in various biological fields as it has shown effectiveness as an anti-cancerous, anti-inflammatory agent and anti-oxidant for the suppression of oxidative stress. Oxidative stress is a pathophysiological condition defined by an increased production of reactive oxygen species (ROS, e.g.,DPPH radical, singlet oxygen ($^1O_2$), superoxide radicals O_2·–, peroxyl radical CCL_3O_2·), which react with and damage cell membranes, DNA, proteins, impairing normal cellular functioning. Antioxidants, like curcumin, are capable of intercepting ROS, thereby short-circuiting the self-propagating oxidative stress state. Despite this advantage, curcumin's poor water solubility and bioavailability makes its usage in medical applications more difficult. In order to overcome this pharmacological limitation, curcumin was conjugated with poly (β-amino-esters) (PBAE) to increase the bioavailability, and these polymers were further formulated into nanoparticles. These hydrophobic biodegradable polymeric nanoparticles are hypothesized to degrade hydrolytically within the endothelial cells and result in the release of curcumin, which in turn performs its anti-oxidant function. Degradation of nanoparticles with respect to time, cytotoxicity of nanoparticles towards the endothelial cells and their ability to prevent oxidative stress were studied in DCF florescence assays. Oxidative stress reduction effects were studied in two ways. First, degraded products of the hydrolyzed nanoparticles were introduced to the cells, and second, nanoparticles themselves were introduced to the cells. Curcumin loaded PBAE showed effective anti-oxidant activity through hydrolytic degradation of polymer releasing free curcumin, and varying the composition of hydrophilic and hydrophobic entities in the polymer was used as a tool to optimize its degradation time.

See more of this Session: Nanotechnology for Biotechnology and Pharmaceuticals II
See more of this Group/Topical: Nanoscale Science and Engineering Forum

(176a) Crystal Structure Engineering of Semiconductor Nanowires

Ildar Musin[1], Saujan Sivaram[2], Nae Chul Shin[2] and **Michael A. Filler**[3], (1)School of Chemical & Biomolecular Engineering, Georgia Institute of Technology, Atlanta, GA, (2)Chemical & Biomolecular Engineering, Georgia Institute of Technology, Atlanta, GA, (3)School of Chemial & Biomolecular Engineering, Georgia Institute of Technology, Atlanta, GA

Semiconductor nanowires provide a promising route to achieve next generation photonic, electronic, and energy conversion devices. Control of nanowire crystal structure (e.g. lattice, orientation, faceting) is critical to enable the appropriate function for a particular application. Unfortunately, this remains a challenging task with bottom-up syntheses. To this end, we demonstrate the ability to precisely engineer nanowire interface energetics via the addition of bifunctional precursors and, in doing so, rationally control Ge nanowire crystal growth orientation for the first time. More specifically, Ge nanowires are grown using the vapor-liquid-solid (VLS) technique with germane and a series of bifunctional alkylgermane species, including methylgermane, ethylgermane, and tert-butylgermane. Scanning electron microscopy (SEM) reveals dramatics change in growth direction as well as a reduction in tapering. In the case of methylgermane, transmission electron microscopy (TEM) confirms a transition from <111> to <110> oriented growth while maintaining single crystallinity throughout. Infrared and X-ray photoelectron spectroscopy (XPS) indicate that the nanowire surfaces are alkyl terminated. In addition to reducing the rate of oxidation, these surface species adjust the solid-vapor interface energy near the three-phase boundary, thereby influencing nucleation events and enabling changes in crystal orientation. The control of nuclei surface chemistry demonstrated by this work provides an important new handle for controlling nanowire properties for different applications. Furthermore the ability to effectively passivate sidewalls during growth is expected to enable more robust doping profiles by only permitting precursor incorporation through the catalyst tip.

See more of this Session: Nanostructured Materials for Electronic and Photonic Applications
See more of this Group/Topical: Materials Engineering and Sciences Division

(662e) Electroluminescence From Colloidal Nanocrystals (Quantum Dots) Via Field-Driven Ionization

Matthew J. Panzer[1], Vanessa Wood[2], Deniz Bozyigit[2], Yasuhiro Shirasaki[3], Ian Rousseau[3], Scott M. Geyer[4], Moungi G. Bawendi[4] and Vladimir Bulovic[3], (1)Chemical & Biological Engineering, Tufts University, Medford, MA, (2)Information Technology and Electrical Engineering, ETH Zurich, Zurich, Switzerland, (3)Electrical Engineering, Massachusetts Institute of Technology, Cambridge, MA, (4) Chemistry, Massachusetts Institute of Technology, Cambridge, MA

Colloidal semiconductor nanocrystals, or quantum dots (QDs), are a promising class of materials for the development of next-generation solid-state lighting technologies due to their facile solution-based syntheses, widely tunable and narrowband emission characteristics, and high photoluminescence efficiencies. However, obtaining efficient and sustained electroluminescence from QD thin films remains a significant challenge. A high degree of morphological and energetic disorder inherent to many nanosized lumophores, including QDs, places severe limitations on the charge injection into and transport rates through thin films of these materials. Here, we demonstrate that sustained electroluminescence from QD thin films can be achieved by the local generation and recombination of charges induced by the application of an alternating electric field, thereby eliminating the need for injection of charge carriers from the device electrodes and long-range carrier transport. We show electroluminescence from novel light-emitting structures containing thin films of QDs and other nanoscale emissive materials that do not support direct current excitation, and suggest a mechanism for the charge generation and electroluminescence that is consistent with our time-averaged and time-resolved experimental observations. The proposed field-driven ionization phenomenon may provide a new route toward the realization of efficient lighting and display platforms incorporating colloidal QDs.

See more of this Session: Nanocrystal Science and Technologies
See more of this Group/Topical: Materials Engineering and Sciences Division

(693c) Spontaneous Dissolution of Ultralong Carbon Nanotubes for Production of Neat CNT Fibers

A. Nicholas G. Parra-Vasquez[1], Natnael Behabtu[1], Micah J. Green[1], Cary L. Pint[2], Colin C. Young[1], Judith Schmidt[3], Ellina Kesselman[3], Anubha Goyal[4], Pulickel M. Ajayan[4], Yachin Cohen[3], Yeshayahu Talmon[3], Robert H. Hauge[2] and Matteo Pasquali[1], (1)Chemical and Biomolecular Engineering, Rice University, Houston, TX, (2)Richard E. Smalley Institute for Nanoscale Science and Technology, Rice University, Houston, TX, (3)Chemical Engineering, Technion-Israel Institute of Technology, Haifa 32000, Israel, (4)Department of Mechanical Engineering and Materials Science, Rice University, Houston, TX

We report that chlorosulfonic acid is a true solvent for a wide range of carbon nanotubes (CNTs), including single-walled (SWNTs), double-walled (DWNTs), multiwalled carbon nanotubes (MWNTs), and CNTs hundreds of micrometers long. The CNTs dissolve as individuals at low concentrations, as determined by cryo-TEM (cryogenic transmission electron microscopy), and form liquid-crystalline phases at high concentrations. The mechanism of dissolution is electrostatic stabilization through reversible protonation of the CNT side walls, as previously established for SWNTs. CNTs with highly defective side walls do not protonate sufficiently and, hence, do not dissolve. The dissolution and liquid-crystallinity of ultralong CNTs are critical advances in the liquid-phase processing of macroscopic CNT-based materials, such as fibers and films.

See more of this Session: Graphene and Carbon Nanotubes: Characterization, Functionalization, and Dispersion
See more of this Group/Topical: Nanoscale Science and Engineering Forum

(266f) Hydrothermal Synthesis of Zeolite with Three Dimensionally Ordered Mesoporous-Imprinted (3DOm-i) Structure

Huiyong Chen[1], James Wydra[1], Pyung-Soo Lee[1], Xueyi Zhang[2], **Wei Fan**[3] and Michael Tsapatsis[1], (1) Chemical Engineering and Materials Science, University of Minnesota, Minneapolis, MN, (2)Chemical Engineering & Materials Science, University of Minnesota, Minneapolis, MN, (3)Department of Chemical Engineering, University of Massachusetts Amherst, Amherst, MA

Due to the intrinsic ordered micropore structures and strong acid strength, zeolites are extensively used as heterogeneous acid catalysts in crude oil refineries and petrochemical processes. However, the micropore structures and high intrinsic activities frequently lead these materials to be subject to diffusion limitations that restricts reactant accessibility to the active sites on the interior surfaces of zeolites and inhibits the full utilization of zeolite catalysts. Nanofabrication of hierarchical zeolite catalysts with mesoporosity is a proved strategy for integrating shape selectivity provided by the intrinsic micropore structures and efficient mass transport facilitated by the mesopore structures. Considerable efforts have been devoted to the synthesis of zeolite catalysts with mesoporosity. Among the various approaches, the synthesis of hierarchical zeolites by the confined synthesis in hard templates holds exciting implications in term of creating ordered mesopore structures and controlling mesopore sizes. In principle, any desired mesopore structure can be achieved by manipulating the structure of the hard template. These advantages provide the unique opportunities to design the hierarchical pore structure and quantitatively investigate the effects of mesopore structure on the catalytic performance of zeolite catalysts.

Recently, we have demonstrated how a wide range of crystal morphologies of zeolite (MFI) can be realized through the confined growth within 3-dimensionally ordered mesoporous (3DOm) carbon by steam assisted crystallization (SAC) method[1]. Herein, we report a general synthesis route for the confined synthesis of zeolite within 3DOm carbon template through hydrothermal (HT) synthesis method. Different from the SAC approach, most zeolites have been successfully synthesized under hydrothermal (HT) conditions. The challenge for the zeolite growth inside the 3DOm carbon during the HT synthesis is to hinder the crystal growth outside the 3DOm carbon and allow the growth inside the 3DOm carbon. In the new approach, the formation of zeolite crystal outside 3DOm carbon is reduced by controlling the crystal growth conditions. Various zeolites including zeolite BEA, LTA, FAU and LTL with three dimensionally ordered mesoporous-imprinted (3DOm-i) structure have been synthesized by this approach. The approach exhibits versatile abilities for controlling the mesopore size (5 nm to 10 nm) and structure (ordered and disordered). The success in growth of zeolite in confined space by HT synthesis could provide a versatile alternative for the nanofabrication of hierarchical zeolites and other hierarchical materials.

(1).Fan, W.; M. A. Snyder; S. Kumar; P. S. Lee; W. C. Yoo; A. V. McCormick; R. L. Penn; A. Stein; M. Tsapatsis *Nat. Mater.* **2008**, *7*, 984-991.

See more of this Session: Templated Assembly of Inorganic Nanomaterials I
See more of this Group/Topical: Nanoscale Science and Engineering Forum

(665b) The Surface Hydrogen-Controlled Crystal Structure of Si Nanowires

Nae Chul Shin, Chemical & Biomolecular Engineering, Georgia Institute of Technology, Atlanta, GA and **Michael A. Filler**, School of Chemial & Biomolecular Engineering, Georgia Institute of Technology, Atlanta, GA

Semiconducting nanowires promise exciting advances in fields as diverse as photonics, optoelectronics, quantum computation, electrochemistry, and thermophysics. In the ideal situation, the combination of nanowire diameter, lattice structure (e.g. diamond cubic, wurtzite), crystal orientation (e.g. <111> vs. <110>), and sidewall faceting that yields the most robust device performance for a particular application would be known and could be rationally synthesized. Unfortunately, an inadequate understanding of nanowire process-structure and structure-property relationships prevents the accomplishment of this task at the present time. Although hydrogen is prevalent during the hydride-based vapor-liquid-solid (VLS) growth of semiconductor nanowires, its role is largely unknown. To this end, we systematically studied the effect of hydrogen during the growth of Si nanowires. We confirmed its influence on crystal growth orientation and also revealed a common chemical thread that connects multiple unresolved questions in the field for the first time. *In-situ* transmission infrared (IR) spectroscopy was used to identify the presence and bonding of hydrogen on Si nanowires as a function of growth conditions in the absence of any carrier gas. Si nanowires were grown via a two-step process: (1) brief nucleation followed by (2) elongation under different conditions. Vertically-oriented epitaxial Si nanowires with uniform densities, diameters, and lengths were obtained with this method. Real-time IR measurements reveal the evolution of surface Si-H stretching modes as a function of growth time and conditions. More specifically, the nanowire crystal orientation transitions from <111> to <112> upon reaching a critical hydrogen surface coverage, even for large diameter nanowires and a range of supersaturation conditions. The influence of hydrogen on interface energetics in the three-phase region is significant and opens new avenues to fabricate nanowire superstructures, which will also be presented. The extensive use of hydride chemistries for most group IV and III-V semiconductor nanowire syntheses suggests significant implications for a variety of systems.

See more of this Session: Nanowires I: Synthesis and Modeling
See more of this Group/Topical: Nanoscale Science and Engineering Forum

(172c) Electrical Characterization of Silicon Nanocrystal Films

Neema Rastgar[1], Dave Rowe[2], Lance M. Wheeler[2], Eray Aydil[1] and Uwe Kortshagen[2], (1)Chemical Engineering and Materials Science, University of Minnesota, Minneapolis, MN, (2)Mechanical Engineering, University of Minnesota, Minneapolis, MN

Electrical Characterization of Silicon Nanocrystal Films

Neema Rastgar, David Rowe, Lance Wheeler, Eray Aydil, Uwe Kortshagen

Thin films of semiconductor nanocrystals continue to receive attention as potential materials for making light-emitting diodes, photodiodes and solar cells. This approach to making optoelectronic devices may be promising because semiconductor nanocrystals are inexpensive to synthesize and their optoelectronic properties can be tuned by changing their size. However, devices based on thin films of nanocrystals typically show high electrical resistivity, and establishing control over electronic properties is difficult. The understanding of electronic transport in these nanocrystal films is in its infancy compared to bulk semiconductors. To improve this understanding and to learn how to manipulate charge carrier transport in semiconductor nanocrystal films, we study electronic transport in thin films of intrinsic and doped silicon nanocrystals.

Silicon nanocrystals with diameters ranging from 5-20 nm were synthesized through decomposition of silane in a radio-frequency plasma reactor. Thin films of these nanocrystals were deposited either through ballistic aerosol impaction onto substrates or through spin coating from colloidal dispersions of the nanocrystals. The former approach is in situ and the nanocrystals are deposited onto the substrate immediately after they leave the plasma. In the latter approach, the nanocrystals emerging from the plasma are collected, dispersed in a solvent and cast onto the substrate. In both cases, the nanocrystals are deposited between two 100 nm-thick thermally evaporated aluminum contacts to form thin films of randomly packed nanoparticles. Current-voltage characteristics of the nanocrystal films were measured as a function of doping and temperature between 100 and 300 K. Preliminary results show that the films exhibit space charge limited current above applied electric fields of 1000 V/cm, and Ohmic behavior at lower electric fields. The conductivity of annealed undoped films exhibits Arrhenius dependence on temperature, with an activation energy of 0.60 eV between room temperature and 225 K, indicative of conduction mediated by intrinsic carriers. Annealed boron-doped films, on the other hand, show moderate Arrhenius temperature dependence near room temperature, and weak temperature dependence below 225 K, characteristic of either dopant ionization or tunneling conduction.

Many films also demonstrate hysteresis in the current-voltage characteristics, ranging from insignificant to severe. The hysteresis is thought to arise from a parasitic capacitance in the film due to charging and appears to be most significant in films made of nanocrystals with ligands such as hexene. Resistance-capacitance (RC) time constants on the order of seconds describe the hysteresis. Conduction is sensitive to the surface conditions of the nanocrystals.

For example, the conductivity of nanocrystal films in vacuum increase by an order of magnitude when nitrogen, argon, and oxygen gases flow over the films and the chamber is purged continuously to reduce water partial pressure in the chamber. This change is reversible over a time scale of two hours and suggests that desorption of water from the film is the likely reason for improved conductivity.

This work was supported primarily by the MRSEC Program of the National Science Foundation under Award Number DMR-0819885.

See more of this Session: Nanoelectronic Materials
See more of this Group/Topical: Nanoscale Science and Engineering Forum

(609e) A Fundamental Investigation of Electron Transport and Recombination Characteristics of Titania Nanowire/Nanoparticle Hybrid Structures

Venkat Kalyan Vendra[1], Delaina A. Amos[1], Mahendra K. Sunkara[1] and Thad Druffel[2], (1)Chemical Engineering, University of Louisville, Louisville, KY, (2)Conn Center for Renewable Energy Research, Louisville

The electron transport and recombination of long titania nanowires and titania nanoparticle/nanowire composites are investigated. Understanding charge transport and recombination in 1D nanostructures of titania is critical for improving the performance of DSSC employing these 1D nanostructures as photoanodes in dye sensitized solar cells. 1D titania nanostructures are expected to have short transport times and high electron lifetimes. However, most of the 1D titania nanostructures have not shown these desirable properties. For instance, the electron diffusion coefficients and transport time constants in titania nanotubes were found to be very similar to that of titania nanoparticles.[1] But, the recombination time constants for titania nanotubes were 10 times higher than then nanoparticle based DSSCs. In contrast, DSSCs fabricated using titania nanowires showed both shorter transport time scales and higher electron lifetimes.[2] In a seminal paper, Richter and Schmuttenmaer showed that the lower the electron mobility in TiO2 nanotubes was due to the existence of exciton like trap states formed from Ti^{+3} donor states resulting from oxygen deficiencies, fluoride impurities and excess titanium introduced during synthesis.[3] Although thermal annealing is expected to remove the oxygen deficiencies, the structural integrity of the nanotubes would be compromised and hence a new synthesis technique is needed to prepare defect free TiO_2 nanotubes for enhanced DSSC performance.

Several approaches such as electrochemical anodization, solvothermal techniques, hydrothermal methods, thermal evaporation and template based approaches have been widely explored. However, most the methods produce short titania nanotubes (~ 2 μm in length) which is undesirable considering the low dye loading and slow electron transport in these architectures. In addition, all these synthesis techniques discussed above are very time consuming with the reaction time scales ranging from several hours to even days. To our knowledge there have not been any studies which have focused on the fast synthesis of single crystalline titania nanowires. Here we present a novel, scalable approach for making single crystalline titania nanowires arrays for dye sensitized solar cells.[4] Electron transport and recombination in these single crystalline nanowires and in single crystalline titania nanowires coated with titania nanoparticles are discussed.

[1] Zhu, K., Neale, N.R., Miedaner, A., Frank, A.J. Enhanced charge collection efficiencies and light scattering in dye sensitized solar cells using oriented TiO2 nanotubes. Nanoletters 2007, 7, 69-74.

[2] Tetreault, N., Horvath, E., Moehl, T., Brillet, J., Smajda, R., Bungener, S., Cai, N., Peng, W., Zakeeruddin, S.M., Forro, L., Magrez, A., Gratzel, M. High-Efficiency Solid-State Dye- Sensitized Solar Cells: Fast Charge Extraction through Self-Assembled 3D Fibrous Network of Crystalline TiO2 Nanowires. ACS Nano 2010, 4, 7644-7650.

[3] Richter, C., Schmuttenmaer, C.A., Exciton like trap states limit electron mobility in TiO2 nanotubes, Nature Nanotechnology 2010, 5, 769–772

[4] Kumar, V., Kim, J.H. Jasinski, J.B., Clark, E.L., Sunkara, M. Alkali metal assisted gas phase bulk production of titania nanowires. Journal of Crystal Growth and Design 2011, Article Accepted.

See more of this Session: Semiconductor Nanowires and Nanotubes for Next Generation Photovoltaics - From Fundamentals to Devices
See more of this Group/Topical: Topical 5: Nanomaterials for Energy Applications

(744b) Concentrically Electrospun Enzyme Fibers

Kenneth J. Balkus Jr. and Daniel N. Tran, Chemistry, University of Texas at Dallas, Richardson, TX

Cellulase and glucose oxidase fibers were successfully fabricated through core-shell electrospinning. Fibers provide a higher surface area when compared to protein immobilized on films, translating to higher active site accessibility and activity. Core-shell electrospining is used to allow the spinning of materials that do not readily electrospin. The outer shell material acts as a guiding sheath hat aids fiber formation of the inner core material. In our study, we have electrospun cellulase and glucose oxidase within the inner core of varying polymer sheaths. The fibers were then treated with glutaraldehyde vapor to cross-link the inner-core enzyme. The outer shell material was then washed away leaving only the cross-linked inner core enzyme. Fibers were still active after the cross-linking and washing procedure. These fibers have potential to be incorporated as reactive membranes for the production of biofuels (cellulase) or biofuel cells electrodes (glucose oxidase).

See more of this Session: Nanoscale Science and Engineering In Biomolecular Catalysis II
See more of this Group/Topical: Nanoscale Science and Engineering Forum

(703g) A Comparative Investigation of CO2 Removal In Two Different Hydrogen Redistribution Strategies for a Two-Stage Hydrogen-Permselective Membrane Methanol Reactor

M.S. Baktash[1], S. Mazinani[1], M. Zare[1], A. Najafi[1] and Mohammad Reza Rahimpour[2], (1)Department of Chemical engineering, Chemical and Petroleum Engineering School, Shiraz University, Shiraz, Iran, (2)School of Chemical and Petroleum Engineering, Shiraz University, Shiraz, Iran

One of the major problems facing mankind in 21st century is the global warming which is induced by the increasing concentration of carbon dioxide (CO_2) and other greenhouse gases in the atmosphere. One of the most promising processes for controlling the atmospheric CO_2 level is conversion of CO_2 to methanol by catalytic hydrogenation. In this paper, two different hydrogen redistribution strategies on CO_2 removal rate along a two-stage hydrogen-permselective membrane reactor have been compared. In the first strategy (co-current mode), fresh synthesis gas flows in the tube side of the second stage in same direction with reacting gas stream in the shell side, so that more hydrogen is provided in the first sections of the reactor. In the second strategy (counter-current mode), the gas streams are in the opposite direction, so that more hydrogen is provided in the last sections of the reactor. For this system, a dynamic plug flow one-dimensional model in the presence of catalyst deactivation was employed to compare both strategies in terms of activity, temperature, CO_2 removal and water production. It was recognized that the reactor in counter-current mode has great potential to higher conversion of CO_2 whereas in co-current mode, profile of temperature is more favorable and water production is lower. Enhancement of the CO_2 removal in counter-current mode causes a lower environmental impact. The lower water production rate in co-current mode reduces catalyst recrystallization.

See more of this Session: Nanomanufacturing
See more of this Group/Topical: Materials Engineering and Sciences Division

(600d) Synthesis of Zeolite Imidazolate Framework Films and Membranes On Metal Modified Supports

Miral N. Shah, Victor Varela-Guerrero and Hae-Kwon Jeong, Chemical Engineering, Texas A & M University, College Station, TX

Zeolitic imidazolate frameworks (ZIFs) are an exciting new class of hybrid organic-inorganic materials, known as metal-organic frameworks (MOFs), due to their exceptional thermal and chemical stability[1]. Their regular microporous structure makes them promising for a variety of gas separations[1-2].

To demonstrate this gas permeation performance in case of membranes, a general technique to fabricate a variety of ZIFs or MOFs membranes is required, thereby minimizing the time and resources required to develop a different method for each MOF[3]. We have previously reported a generally applicable thermal ligand modification technique which gives rise to continuous, well intergrown films and membranes in relatively short time (few hours as compared to days)[4]. Here we report a novel and general method for synthesizing MOF membranes. In this technique zinc nitrate solution is dropped on hot bare porous alumina support giving rise to metal modified supports (thermal metal modification). Zinc modified support promotes heterogeneous nucleation and growth and leads to continuous well intergrown films of ZIF-8. This technique offers certain advantages over the previously reported thermal ligand modification technique, wherein the necessity of ligand stability at high temperatures for thermal ligand modification is eliminated. Also it provides an environmentally greener route to membrane fabrication, due to reduced emissions from ligand decomposition at high temperature. Further, metal modification of support can be extended to the fabrication of any other ZIF membrane, or for any other MOF, in general. Here we investigate the gas permeation performance of a membrane grown on a zinc modified support as compared to a ligand modified support.

References:

1. Park, K. S.; Ni, Z.; Cote, A. P.; Choi, J. Y.; Huang, R. D.; Uribe-Romo, F. J.; Chae, H. K.; O'Keeffe, M.; Yaghi, O. M., Exceptional chemical and thermal stability of zeolitic imidazolate frameworks. *Proceedings of the National Academy of Sciences of the United States of America* **2006,** *103* (27), 10186-10191.

2. Liu, Y. Y.; Hu, E. P.; Khan, E. A.; Lai, Z. P., Synthesis and characterization of ZIF-69 membranes and separation for CO2/CO mixture. *Journal of Membrane Science* **2010,** *353* (1-2), 36-40.

3. Haldoupis, E.; Nair, S.; Sholl, D. S., Efficient Calculation of Diffusion Limitations in Metal Organic

Framework Materials: A Tool for Identifying Materials for Kinetic Separations. *Journal of the American Chemical Society* **2010,** *132* (21), 7528-7539.

4. McCarthy, M. C.; Varela-Guerrero, V.; Barnett, G. V.; Jeong, H. K., Synthesis of Zeolitic Imidazolate Framework Films and Membranes with Controlled Microstructures. *Langmuir* **2010,** *26* (18), 14636-14641.

See more of this Session: Nanostructured Thin Films
See more of this Group/Topical: Materials Engineering and Sciences Division

(172d) Electrostatic Coupling of Surface Charge to Bulk Defect Behavior In Metal Oxides

Prashun Gorai, Alice G. Hollister, Kristine Pangan-Okimoto and Edmund G. Seebauer, Department of Chemical and Biomolecular Engineering, University of Illinois, Urbana, IL

The technologically useful properties of semiconductor oxides such as titania and zinc oxide often depend on the concentration and diffusion of point defects. Near-surface effects are particularly important in nanoscale devices because most of the bulk is located in the vicinity of the surface. Past work in our laboratory with silicon and titania has shown that semiconductor surfaces serve as efficient pathways for generation and annihilation of point defects in the underlying bulk. Surfaces can, in addition, support electrically charged defects which create near-surface strong electric fields that can influence the local motion of charged defects resulting in the formation of space-charge layers. The electric field-driven accumulation or depletion of charged oxygen defects in such space-charge regions in metal oxides have direct implications on the performance of nanoscale devices such as gas sensors and memory resistors. Oxygen diffusion behavior was studied by exposing natural-abundance single-crystal rutile to isotopically labeled oxygen gas. The resulting profiles were measured by secondary ion mass spectrometry and subsequently modeled with continuum equations for the reaction, Fickian diffusion and electric field-driven diffusion of the key point defects. The degree of charge build-up at the surface can be quantified by an electric potential (V_i). The profiles calculated show a characteristic steep concentration gradient near the surface followed by a normal bulk diffusion profile. By identifying the charge-mediated field-driven diffusion mechanism as the controlling factor for the near-surface pile-up of oxygen, we demonstrate that this method allows the spatially resolved characterization of space charge regions near the surfaces of crystalline semiconductor metal oxides. The capability to predict oxygen pile-up in nanoscale metal oxide devices may be beneficial in device improvement via defect engineering of surfaces.

See more of this Session: Nanoelectronic Materials
See more of this Group/Topical: Nanoscale Science and Engineering Forum

(420o) The Rheological Responses of Binary Blended Perfluoropolyether Nano Film

Pil Seung Chung, Jungup Park and Myung S. Jhon, Chemical Engineering, Carnegie Mellon University, Pittsburgh, PA

During the intermittent contact between the head and the disk of the hard disk drive, the perfluoropolyether (PFPE) lubricant, which prevents tribological and mechanical damage causing from the direct head-disk contact, experiences the shear/elongation deformation and flow. Since the head media spacing, which represents the spacing between read/write head and the magnetic media in the hard disk drive (HDD), is decreasing less than 6.5 nm for Tbits/in^2 level HDDs [1], the total contact time also increases. Therefore, the molecular rheology of PFPE become critically important in designing effective lubricants that control the friction and wear. Recently, it was found that the application of nano blend PFPE film becomes one of the promising alternatives to enhance the performance and reliability [2]. In this paper, we examine for the first time the rheological responses of nano blend PFPE including dynamic moduli and complex viscosity, by monitoring the time-dependent strain-stress plots via nonequilibrium molecular dynamics (MD) simulations [3]. We observed that pure nonfunctional PFPEs exhibit liquid-like behavior, while "pseudoreptation-like" or "agglomeration" behavior was observed for the pure functional PFPEs, similar to our previous findings [4]. In addition endgroup couplings due to strong functional endgroup weaken as the temperature increases. For the binary blend PFPEs, we also observed "pseudoreptation-like" or agglomeration phenomena, which depend strongly on the blend ratio and imposed condition (*e.g.*, temperature and oscillatory frequency). We established finger print analysis between G¢-G^2 and nanoblend ratio. By constructing modified Cole-Cole plot [5] (*i.e.*, relationship between storage and loss moduli), we found that the nanostructural conformation depends on the blend ratio at low frequency, while weakly depends at high frequency. By analyzing the nano blend PFPE relaxation processes, the results provide us the optimization parameters of nano rheology of lubricant selection criteria to enhance the HDD performance.

References

1. J. Gui, "Tribology challenges for head-disk interface toward 1 Tb/in^2," *IEEE Trans. Magn.*, **39**, 716 (2003).
2. P.S. Chung, H. Park, and M.S. Jhon, "The static and dynamic responses of binary mixture perfluoropolyether lubricant films – Molecular structural effects," *IEEE Trans. Magn.*, **45**, 4644 (2009).
3. Q. Guo, P.S. Chung, H.G. Chen, and M.S. Jhon, "Molecular rheology of perfluoropolyether lubricant via nonequilibrium molecular dynamics simulation," *J. Appl. Phys.*, **99**, 08N105 (2006).
4. Q. Guo and M.S. Jhon, "The viscoelastic behavior of perfluoropolyether lubricants," *IEEE Trans. Magn.*, **42**, 2540 (2006).

5. M.S. Jhon, "Physicochemical properties of nanostructured perfluoropolyether films," *Adv. Chem. Phys.*, **129**, 1 (2004).

See more of this Session: Poster Session: Nanoscale Science and Engineering
See more of this Group/Topical: Nanoscale Science and Engineering Forum

(570f) Facile Method of Preparing Multifunctional Nanoparticles with Capabilities of Therapeutics, Targeting, and X-Ray and MRI Detectability

Ming Zhang and **Mustafa Akbulut**, Chemical Engineering, Texas A&M University, College Station, TX

A plethora of examples of nanoscale drug delivery systems with single function now exist in the literature. However, the major challenge facing the field is the development of nanoparticles that can incorporate multiple functions, such as therapeutics, targeting, imaging and stimulus responsiveness, into a single nanocarrier. The development of these types of multifunctional nanoparticles (MNPs) would further advance our capabilities for detection and treatment of diseases. However, methods for the efficient, cost-effective, scalable, and reproducible production of complex smart nanostructures remain elusive. Here, we descibe a facile method for the fabrication of multifunctional drug delivery systems with simultaneous CT and MRI detectability, therapeutic, and targeting capabilities.

See more of this Session: Bionanotechnology for Gene and Drug Delivery III
See more of this Group/Topical: Nanoscale Science and Engineering Forum

(570a) Interbilayer-Crosslinked Multilamellar Vesicles As Synthetic Vaccines for Potent Humoral and Cellular Immune Responses

James J. Moon[1], Heikyung Suh[1], Anjali Yadava[2] and Darrell J. Irvine[3], (1)Department of Biological Engineering and Materials Science and Engineering, Koch Institute for Integrative Cancer Research, Massachusetts Institute of Technology, Cambridge, MA, (2)Division of Malaria Vaccine Development, Walter Reed Army Institute of Research, Silver Spring, MD, (3)Departments of Bioengineering, Materials Science and Engineering, Koch Institute for Integrative Cancer Research, Ragon Institute of MGH, MIT, and Harvard, and the Howard Hughes Medical Institute, Massachusetts Institute of Technology, Cambridge, MA

Currently licensed vaccine adjuvants promote immunity by primarily eliciting humoral immune responses without stimulating cellular immunity. As concerted humoral and cellular immune responses are thought to be required for vaccines against cancer or intracellular pathogens such as HIV and malaria, there is a great interest in new vaccine technologies that can elicit both arms of the adaptive immunity. In this work, we describe development of interbilayer-crosslinked multilamellar vesicles (ICMVs) as a potent vaccine platform and present their efficacy in a malaria model. ICMVs were formed by fusing liposomes into multilamellar vesicles and subsequent crosslinking of adjacent lipid headgroups across lipid bilayers within multilamellar vesicles. These particles exhibited substantially enhanced protein antigen loading and extended drug release kinetics compared with traditional drug delivery vehicles (e.g. liposomes and polymeric particles). Crosslinking of lipid bilayers enhanced stability of particles in serum, allowing efficient delivery of cargo antigens to antigen-presenting cells. ICMVs encapsulating model antigen ovalbumin and FDA-approved adjuvant elicited potent antibody and CD8+ T cell responses *in vivo*, comparable to those for strong viral vector vaccines. With the goal of translating these results to an infectious disease model, we investigated the potency of ICMVs as a malaria vaccine. ICMVs carrying a malaria antigen and adjuvant elicited robust humoral immune responses with high-avidity antibody titers lasting more than a year, whereas soluble protein immunizations produced only transient, low-avidity antibody titers. Particle immunization also led to prominent germinal center formation in lymph nodes, accompanied with increased expansion of antigen-specific $CD4^+$ T cells and their differentiation into follicular helper T cells, partially explaining the long-lasting and high-avidity antibody titers elicited with ICMV immunization. These results suggest that these antigen/adjuvant-carrying ICMVs are potent particle vaccines that can elicit strong cellular and humoral immune responses, and highlight their potential as a vaccine platform against other infectious diseases.

See more of this Session: Bionanotechnology for Gene and Drug Delivery III
See more of this Group/Topical: Nanoscale Science and Engineering Forum

(600e) Control of Nano-Porosity In Plasma Deposited Low-k Diffusion Barrier and Inter-Layer Dielectrics for Nano-Electronic Applications

Sean King[1], Ebony Mays[1], Jeff Bielefeld[2], Ming Liu[3] and David Gidley[3], (1)Logic Technology Development, Intel Corporation, Hillsboro, OR, (2)Components Research, Intel Corporation, Hillsboro, OR, (3)Department of Physics, University of Michigan, Ann Arbor, MI

As the semiconductor industry strives to keep pace with Moore's Law, new materials with extreme properties are increasingly being introduced and tighter control of these material properties is being demanded. Low dielectric constant (i.e. low-k) materials are one specific example. Lower k materials are desired to replace SiO_2 (k=4.0) as the interlayer dielectric (ILD) and SiN_x:H (k = 7.0) as the Cu capping diffusion barrier layer in order to reduce resistance-capacitance (RC) delays in nano-electronic Cu interconnect structures. Typical methods for producing low-k materials consist of introducing controlled levels of nano-porosity via carbon doping during plasma enhanced chemical vapor deposition (PECVD) of SiO_2 and SiN_x:H matrix materials. While lowering k, the introduction of nano-porosity can seriously compromise the performance of these layers in their respective applications. In this presentation, we will demonstrate that critical thresholds in nano-porosity exist for the diffusion of water and solvents through low-k materials. Specifically, we utilize Fourier Transform Infra-Red (FTIR) spectroscopy, to show that the concentration and size of nano-pores formed in low-k a-SiC(N):H dielectric materials is controlled by the concentration of terminal Si-CH_3 bonding versus Si-C/N network bonding. We further combine moisture / solvent diffusivity measurement with x-ray reflectivity (XRR) and positron annihilation lifetime spectroscopy (PALS) to demonstrate that low-k a-SiC(N):H dielectrics become poor moisture diffusion barriers at mass densities < 2.0 g/cm^3 and when the pore size approaches that for the molecular diameter of water. Similarly, we show that low-k a-SiOC:H ILDs become easily penetrable by solvents and susceptible to downstream processing damage when the pore diameter approaches the size of the solvent and pores become interconnected. The implications of these critical nano-porosity thresholds on continued scaling of low-k diffusion barrier and ILD materials will be discussed as well as methods for overcoming these limitations.

See more of this Session: Nanostructured Thin Films
See more of this Group/Topical: Materials Engineering and Sciences Division

(439a) Hollow Gold Nanoshells for Gene and Drug Delivery

Joseph A. Zasadzinski[1], Natalie Forbes[2] and Gary Braun[2], (1)Department of Chemical Engineering and Materials Science, University of Minnesota, Minneapolis, MN, (2)Chemical Engineering, University of California, Santa Barbara, Santa Barbara, CA

The temporal and spatial control over the delivery of materials such as siRNA or drugs into cells remains a significant technical challenge. We demonstrate the pulsed near-infrared (NIR) laser dependent release of siRNA from coated 40 nm gold nanoshells. Tat-lipid coating mediates the cellular uptake of the nanomaterial at pM concentration, while spatiotemporal silencing of a reporter gene (green fluorescence protein) was studied using photomasking. The NIR laser induced release of siRNA from the nanoshells is found to be power and time dependent, through surface-linker bond cleavage, while the escape of the siRNA from endosomes occurs above a critical pulse energy attributed to local heating and cavitation. NIR laser controlled drug release from functional nanomaterials should facilitate more sophisticated developmental biology and therapeutic studies.

See more of this Session: Bionanotechnology for Gene and Drug Delivery I
See more of this Group/Topical: Nanoscale Science and Engineering Forum

(704a) Enzyme-Based Nanocomposites: Using Nature to Ward off Emerging Diseases

Jonathan Dordick, Chemical & Biological Engineering, Rensselaer Polytechnic Institute, Troy, NY

Nature is unparalleled in its structural and functional diversity. Living organisms make fantastic materials under myriad conditions with properties we cannot emulate today using conventional approaches. In many cases, nature has provided us with a blueprint to design and assemble both natural and synthetic building blocks to create a new generation of functional, organized, and responsive materials. We have taken cues from nature to design materials with unique structural and functional properties, along with new process technologies with the ability to produce a wide range of biomimetic structures. Specifically, we have focused on the generation of nanostructures that are functionalized with and in some cases constructed from biological molecules, complete with tailored selectivities and biocatalytic activities. For example, these nanostructures have been exploited in the generation of biocatalytically functional polymeric films, coatings, and paints that kill bacteria, prevent biofilm formation, and reduce fouling by bioorganic molecules. In this talk I will highlight our recent efforts to exploit the interface of biology with materials science, with a particular focus on enzyme-nanomaterial composites with a wide range of activities that endow surfaces with self-cleaning properties. In particular, surfaces have been generated with tailored activity against hospital-acquired infections (e.g., MRSA) and spores. Such activity provides a safe and potentially broadly applicable route to eliminating toxic compounds and pathogenic microorganisms from common surfaces.

See more of this Session: Nanoscale Science and Engineering In Biomolecular Catalysis I
See more of this Group/Topical: Nanoscale Science and Engineering Forum

(266a) Invited Lecture: Nanoporous Membranes for Energy Applications

Yushan Yan, Department of Chemical & Environmental Engineering, University of California, Riverside, Riverside, CA

Nanoporous membranes based on zeolite, polymer-zeolite nanocomposite, and pure polymer for energy applications will be discussed. Specific examples include zeolite membranes for purification of biofuels, polymer-zeolite nanocomposite membranes for sea water desalination, and pure polymer membranes for fuel cells and other electrochemical energy conversion devices such as electrolyzers, flow batteries, and solar hydrogen and solar fuels.

See more of this Session: Templated Assembly of Inorganic Nanomaterials I
See more of this Group/Topical: Nanoscale Science and Engineering Forum

KINETICALLY TRAPPED UNIFORM NANO-SIZE UNILAMELLAR VESICLES MADE OF THERMODYNAMICALLY STABLE MULTILAMELLAR VESICULAR PHOSPHOLIPID SOLUTIONS

Mu-Ping Nieh[1]*, Paul Dolinar*[2]*, Norbert Kučerka*[3,4]*, Steven R. Kline*[5]*, Lisa M. Debeer-Schmitt*[6]*, Kenneth C. Littrell*[6] *and John Katsaras*[3,6-8]

[1]University of Connecticut, Storrs, CT, USA
[2]University of Ottawa, Ottawa, ON, Canada
[3]Canadian Neutron Beam Centre, Chalk River, ON, Canada
[4]Comenius University, Bratislava, Slovakia,
[5]National Institute of Standards and Technology, Gaithersburg, Maryland, USA,
[6]Oak Ridge National Laboratory, Oak Ridge, TN, USA
[7]Brock University, St. Catharines, ON, Canada
[8]University of Guelph, Guelph, ON, Canada

Abstract

We report a study of kinetically trapped unilamellar vesicles (ULVs) of lipid mixtures composed of dimyristoyl phosphatidylcholine (DMPC), dihexanoyl phosphatidylcholine (DHPC) and (dimyristoyl phosphatidylglycerol) DMPG. Two membrane charge densities (namely, [DMPG]/[DMPC] = 0.01 and 0.001) and two solution salinities ([NaCl] = 0 and 0.2 M) are investigated. Small angle neutron scattering (SANS) is used to identify the morphology of aggregates. As high concentration samples are diluted at 50 °C, thermodynamically stable multi-lamellar vesicles (MLVs) are observed for both weakly charged and high salinity solution mixtures, implying that the electrostatic interactions between bilayers are insufficient to unbind the multilamellar structure. At low temperature (i.e. 10 °C) nanodiscs (also known as bicelles) or bilayered ribbons are observed and they slowly evolve into uniform-size nanoscopic unilamellar vesicles (ULVs) when incubated at 10 °C over a period of ~20 hours. The ULVs persist after being heated to 50 °C, where thermodynamically stable MLVs are observed. This result clearly indicates that these ULVs are kinetically trapped and that the mechanical properties (e.g. bending rigidity) of the nanodiscs at 10 °C favor the formation of nanoscopic ULVs over MLVs. From a practical point-of-view, this method of forming uniform-size ULVs may lend itself to their mass production, thus making them economically feasible for medical applications that depend on lipid-based systems for therapeutic and diagnostic purposes.

Introduction

Liposomes comprised of phospholipids have great potential as in-vivo delivery carriers due to their biocompatibility. However, the lamellarity of liposomes is an important parameter that dictates their loading capacity and in-vivo circulation half-life. Long-chain zwitterionic lipids generally form multilamellar vesicles (MLVs) in aqueous solutions. These MLVs have a lower payload capacity and a greater probability of being uptaken by the reticuloendothelial system - presumably because of their large size.[1-2] Thermodynamically, as the membrane undulation and Coulombic repulsive forces (e.g. in systems with sufficient amounts of charged lipids) begin to dominate over the van der Waals attraction force, the lamellar repeat spacing, d, of MLVs increases with increasing water content (i.e. decreased lipid concentration), until the lamellae unbind from each other resulting in the formation of unilamellar vesicles (ULVs). This thermodynamically driven transformation from MLVs to ULVs does not involve other intermediate morphologies. However, in the case of zwitterionic lipid bilayers, such morphological transformations are not observed as MLVs will only swell up to some limiting value of d, whereupon any further addition of water has no further effect [e.g. dimyristoyl (DMPC) and dipalmitoyl (DPPC) PC bilayers]. A different approach of forming ULVs from MLVs involves the addition of short-chain lipids or surfactants.[3-5] In the case of lipid mixtures containing long- and short-chain lipids, bilayered nanodiscs (also known as bicelles) are observed when the long-chain lipid is in the gel phase and the short-chain lipid is in the L_α phase.[6-7] The disc's planar bilayer is comprised of the long-chain lipid and the disc's highly-curved edge is occupied by the short-chain lipid to decrease the system's free energy.[8-11] At high temperatures (above the melting transition temperature of DMPC, T_M), perforated lamellae and MLVs are observed – presumably due to the increased miscibility between DMPC and DHPC.[12-15] In order to obtain uniform-size nanoscopic ULVs from a mixture of short- and long-chain phospholipids, a small amount of the long-chain charged lipid (~0.01 mole fraction of the total long-chain lipids) dimyristoyl phosphatidylglycerol (DMPG) is needed to induce sufficient Coulombic repulsion between bilayered membranes to unbind the MLVs.[13,16-18] The size and polydispersity of the resultant ULVs could vary extensively, and were highly dependent on their path of formation.[19] Here we report a study on how lipid aggregate morphology is affected by charge density (i.e. DMPG content) and salt concentration, as well as the system's kinetic pathway of formation.

Materials & Methods

Phospholipids (DMPC, DHPC and DMPG) were purchased from Avanti Polar Lipids and used as received. NaCl (≥ 99.5%) was purchased from Sigma-Aldrich Co. and dissolved in D_2O to produce a 0.2 M solution. Three series of samples with a constant molar ratio of long-chain to short-chain lipid, Q (i.e., ([DMPC]+[DMPG])/[DHPC]) = 4 and a total lipid concentration, C_{lp} = 10 % (mass fraction) were prepared in D_2O: (1) [DMPG]/[DMPC] (denoted as R) = 0.01 (control sample); (2) R = 0.001; and (3) R = 0.01, [NaCl] = 0.2 M. The required amount of lipids was first weighed according to the molar ratio, and thoroughly mixed in either D_2O or 0.2 M NaCl buffer until the solutions were transparent at low-T. The C_{lp} = 10% samples were then further diluted through a number of intermediate concentrations with either D_2O or a 0.2 M NaCl D_2O solution, always maintaining the same salinity as the initial 10% samples. Dilutions at both high (i.e. 50 °C) and low (i.e. 4 °C) temperatures were performed on various samples, and SANS measurements were taken at each condition within 1 hour after sample preparation. SANS data of low-T diluted samples were also collected after the temperature was raised to 50 °C. Moreover, in order to gain insight into the stability of the discoidal micelles, a SANS measurement was performed on a 0.2 % (R = 0.001, no salt) sample after 20 hours of incubation at 4 °C - to compare to those of non-incubated samples.

SANS experiments were conducted at either the NG3-SANS instrument located at the NIST (National Institute of Standards and Technology) Center for Neutron Research (NCNR, Gaithersburg, Maryland, USA), or the CG2-SANS instrument located at the High Flux Isotope Reactor (HFIR, ORNL, Oak Ridge, Tennessee, USA). At NG3-SANS, 6 Å wavelength (λ) neutrons with a $\Delta\lambda/\lambda$ = 15% were used in combination with sample-to-detector distances (SDD) of 13.17 m, 5 m and 1.33 m, yielding a range of scattering vectors, q [$=(4\pi/\lambda)\sin(\theta/2)$, where θ is the scattering angle], from 0.0035 $Å^{-1}$ to 0.4 $Å^{-1}$. At CG2-SANS, 4.75 Å ($\Delta\lambda/\lambda$ = 15%) neutrons and two SDDs (i.e. 4 m and 18.5 m) were used to cover a q range from 0.004 $Å^{-1}$ to 0.26 $Å^{-1}$. All 2-D raw data were corrected for sample and empty cell transmission, as well as background scattering, and then reduced to 1-D intensity plots as a function of q. SANS data were then fitted with the appropriate models (i.e. core-shell disc, cylinder with an elliptic cross-section, or spherical shell) that have been developed by NCNR using the software IGOR-Pro.[20]

Fitting curves were "smeared" using the appropriate instrumental resolution. However, a structure factor (inter-particle interaction) was not incorporated into the data analysis as it was not necessary in the case of the dilute samples – the data of non-dilute samples were not fitted.

Results & Discussion

High-T (50 °C) dilution

The 10 % control sample (R = 0.01) at 50 °C exhibits distinct first- and second-order quasi-Bragg peaks (at q_o = 0.019 and 0.038 Å^{-1}, respectively) indicative of long-range out-of-plane interactions with a lamellar repeat spacing, d, of 337 Å ($d = 2\pi/q_1$) (**Fig 1**). This lamellar spacing is similar to that of a DMPC/DHPC/DMPG mixture at the same C_{lp}[18], where perforated lamellae are observed. These results are also consistent with the aforementioned translucent appearance and a previous cryo-TEM report.[21] The Coulombic repulsion resulted from the charged DMPG, in the case of R = 0.01, is sufficient to unbind the DMPC MLVs. However the SANS patterns of the other two C_{lp} = 10% lipid mixtures (i.e. R=0.001 and R=0.01, [NaCl]=0.2 M) are drastically different from the control sample. For example, a sharp quasi-Bragg peak appears at q_o = 0.09 Å^{-1} and 0.085 Å^{-1} in the R=0.001 and R=0.01, 0.2 M [NaCl] samples, respectively, and correspond to d-spacings of 69.8 and 73.9 Å, respectively. Diluting both samples to 5% at 50 °C only leads to a marginal swelling of the lamellae (74.0 and 82.5 Å, respectively), implying either an insufficient Coulombic repulsion (in the case of R=0.001) or a strong screening of the charges (in the case of R=0.01, [NaCl] = 0.2 M), thus preventing the lamellae from freely swelling.

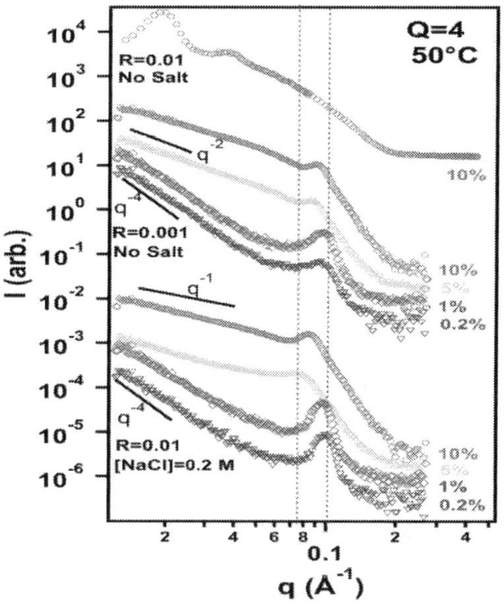

Figure 1 SANS data of 10% samples undergoing high-T dilution and measured at 50 °C. The control sample (R = 0.01, top circles) is characteristic of lamellae capable of swelling. The samples (R=0.001 no salt and R=0.01 with [NaCl] = 0.2 M) with C_{lp} = 10 % (circles), 5 % (top-pointed triangles), 1 % (diamonds) and 0.2 % (bottom-pointed triangles) show a non-

Although both systems exhibit a similar swelling behavior, their aggregate morphologies are most likely very different. For example, in the case of R=0.001 samples (for both C_{lp} = 5% and 10%), the low-q scattered intensity follows a q^{-2} dependence corresponding to a planar structure, while in the case of R= 0.01, [NaCl] = 0.2 M samples (for both C_{lp} = 5% and 10%) the scattered intensity decays as q^{-1}, indicative of elongated aggregates.[22] The planar structure is best described by a lamellar morphology, while the elongated structure is possibly made up of entangled, ribbon-like lamellae,[13] or tubular/elongated vesicles similar to those reported in octyl glucoside/egg phosphatidylcholine[23] or DMPC/geraniol[24] mixtures. Diluting the mixture further to 1% and 0.2% results in the lamellar repeat spacing peak shifting to higher q (~0.095 Å^{-1}), namely a d of 66 Å for both R=0.001 and R=0.001, [NaCl] = 0.2 M samples. The fact that this spacing is in good agreement with that of liquid crystalline (L_α) DMPC MLVs[25,26] implies that both lipid mixtures probably form MLVs that are comprised almost entirely of DMPC. Importantly, the Porod regime of these scattering curves exhibits a q^{-4} dependence, further proof of the presence of large MLVs. The resultant thermodynamically stable MLVs obtained from high-T dilution suggests that charge-reduced (R = 0.001) and salt-doped (R = 0.01, [NaCl] = 0.2 M) samples are in many ways similar to zwitterionic lipid systems (i.e. with no DMPG), where MLV is a thermodynamically equilibrium structure.

Low-T (10 °C) phases immediately after dilution

Nanodiscs normally form in amphiphilic mixtures composed of two molecules with distinct spontaneous curvatures. In this case, the long-chain lipids form the disc's planar region and the short-chain lipids coat the disc's rim.[21,27] Such nanodiscs composed of two lipids (e.g. long-chain and short-chain lipids) are usually found at low-T. This is presumably due to the immiscibility between the gel long-chain lipid and the liquid crystalline short-chain lipid. As such, discs are commonly thought to be thermodynamically stable at low temperatures.[19] However, to the best our knowledge a systematic study that has not yet been conducted on the stability of DMPC/DHPC discoidal micelles (at low-T) – except for a few kinetic studies on the micelle-to-vesicle transition in egg lecithin/bile salt mixtures, where the transition was triggered by dilution.[28]

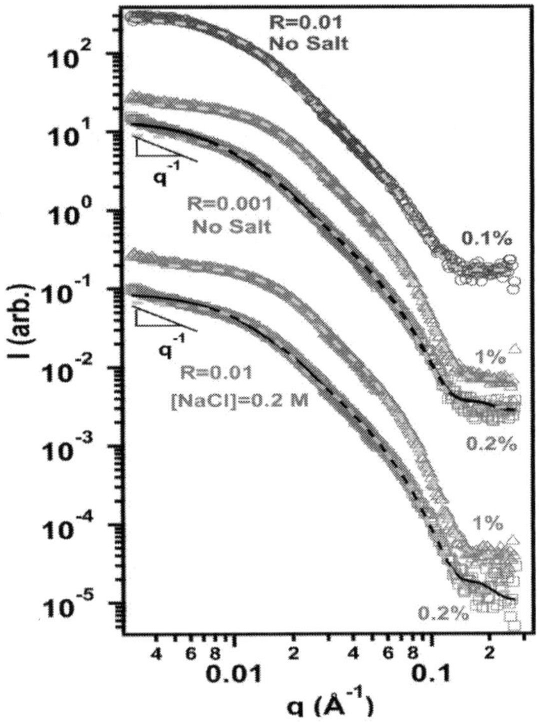

Figure 2 SANS data of samples diluted at 4 °C and measured as prepared. The 0.1 % R = 0.01 (circles), 1.0 % R = 0.001 (red triangles), and R = 0.01, [NaCl] = 0.2 M (green triangles) data are fitted reasonably well with the core-shell disc model (dashed lines). The 0.2 % of R = 0.001 (red squares) and R = 0.01, [NaCl] = 0.2 M (green squares) data are fitted using both the core-shell disc model (dashed lines) and the elongated ellipsoidal model (solid lines).

Low-T (10 °C) SANS data from all three samples (R=0.01, R=0.001, and R=0.01, [NaCl] = 0.2 M) at low lipid concentrations (≤ 1%) – used in order to minimize the possibility of interparticle interactions – are shown in Fig 2. It should be noted that all samples are diluted from the original 10% C_{lp} condition at low-T (4 °C) and studied at 10 °C *immediately after dilution*. The SANS data of the control sample (C_{lp} = 0.1% and R=0.01) are best fitted with the core-shell discoidal model, where the core and shell represent the hydrophobic acyl chains and hydrophilic headgroups, respectively. During the fitting process the shell and the core thicknesses are constrained between 5 - 15 Å, and 30 - 40 Å, respectively – realistic values for a PC lipid bilayer. The best fits (dashed line) to the data result in a shell thickness of 15 ± 1 Å, a core thickness of 34 ± 2 Å, and a disc radius of 186 ± 15 Å. In the case of 1% C_{lp} samples with a reduced charged density (R=0.001) and increased salinity (R=0.01, [NaCl]=0.2 M), an invariant shell thickness (15 Å) is obtained from the best fits to the SANS data, while the core thickness and disc radius are 32 ± 3 Å and 130 ± 12 Å, respectively for the R=0.001 sample, and 34 ± 3 Å and 123 ± 10 Å, respectively for the R=0.01, [NaCl] = 0.2 M sample. Compared to the R=0.01 and C_{lp} = 0.1% sample, the smaller disc radii in these 1% samples are presumably due to more DHPC being available to stabilize the disc's rim.

This disc model is, however, inadequate in fitting the lower C_{lp} (i.e. 0.2 %) data at low-*q* (i.e. < 0.006 $Å^{-1}$), where large deviations are observed for both weakly charged (R=0.001) and highly screened (R=0.01 and [NaCl] = 0.2 M) samples (Fig 2). The deviated low-*q* data reveal an onset of q^{-1} dependence, a possible sign of elongated aggregates. Although the range of q^{-1} dependence spans less than a decade (insufficient for conclusively identifying the existence of elongated objects), cryo-TEM studies have demonstrated the existence of elongated aggregates in a similar lipid mixture and at a similar temperature.[21] Moreover, the fact that these samples exhibit the same decay behavior as discoidal micelles at higher *q* suggests that the elongated micelles (presumably ribbon-like micelles) have a thickness similar to that of discoidal micelles. As such, the data from both samples are fitted using a cylindrical model with an elliptical cross section, which results in major and minor axes with values of 130 ± 15 Å and 26 ± 4 Å (for both samples), respectively. In this case, the constraints on the fitting parameters are released. From the best fits the length of these ribbons is determined to be about 610 Å for the charge-reduced sample, and about 500 Å for the salt-added sample. The fact that the best fit value for the minor axis corresponds to half of the disc's bilayer thickness [obtained from using the core-shell discoidal model, i.e. (15 Å + 34 Å)/2 = 24.5 Å], indicates that the model used is reasonable, and that the bilayer itself is not perturbed upon dilution.

Low-T Incubation

Fig 3 contains SANS data from a C_{lp} = 0.3%, R=0.001 sample at 10 °C, after more than 20 h of incubation at 4 °C. The SANS pattern contains oscillations completely different from that of a freshly prepared sample. The data from the incubated sample is well fit using a single spherical shell model, where the bilayered shell is assumed to have a uniform scattering length density and the enclosed core is assumed to be D_2O. The best fit result yields low polydispersity (p=25%) ULVs with an average inner radius, $\langle R_i \rangle$, of 164 ± 9 Å and a shell thickness of 37 ± 3 Å. This morphology is in sharp contrast to the above-mentioned discs and elongated ribbon structures that are formed immediately after preparation at 10 °C (Fig 2). Moreover, the ULV structure persists as the temperature is elevated to 50 °C, instead of transforming into thermodynamically stable MLVs (as mentioned in the "High-T dilution" section), indicating that the ULVs are kinetically trapped. Another interesting observation arising from these kinetically trapped ULVs is that the best fit to the high-T data results in a larger $\langle R_i \rangle$ (194 ± 15 Å) and a thinner shell (i.e. bilayer, 32 ± 3 Å). The thinner bilayer is expected as DMPC is presumably in the L_α phase ($T_M \sim 23$ °C for pure DMPC) at 50 °C. Although low-T ULVs have been previously observed in similar systems,[17,29] their mechanism of formation has not been understood. The observations made here pose two fundamental questions: (a) are all low-T discoidal micelles thermodynamically unstable even though they consistently form through temperature cycling? and (b) which physical parameter controls the rate of formation of these low-T ULVs?

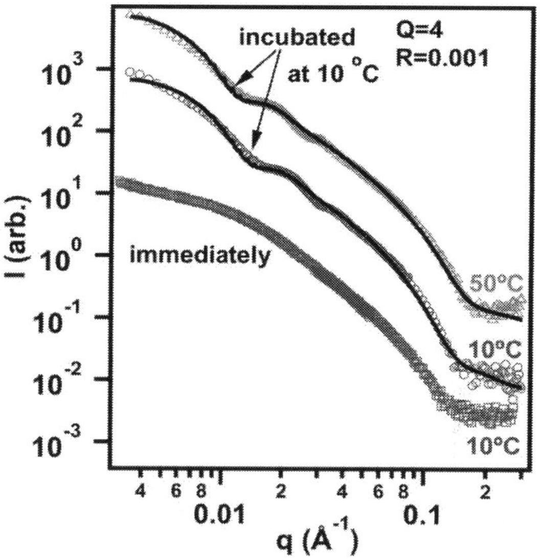

Figure 3 SANS data of R=0.001 (no salt) samples diluted from 10% to 0.2% (squares) and 0.3 % (circles) at 4 °C. Data for the 0.2 % sample were obtained as prepared, while data for the 0.3 % sample were taken after incubation at 4 °C for > 20 hours. The sample was then heated to 50 °C (triangles). Data for the incubated sample were fitted by the spherical shell model.

In the case of weakly charged lipid mixtures the answer to the first question is straight forward; i.e. discoidal micelles have been shown to be unstable as demonstrated by the formation of monodisperse ULVs at low-T. However, whether or not this applies to all such systems (i.e. including strongly charged systems) remains unknown, and a topic that requires further research. For example, a time-resolved study tracking the growth/transformation of discoidal micelles will be of great use in addressing this issue. Since the current study shows that charge density is one of the key parameters controlling the disc's folding process, monitoring the change in disc diameter as a function of charge density (at constant T) may reveal how the aggregate structure evolves with time. In the case of thermodynamically stable discoidal micelles their radius may grow initially after dilution, but should remain unaltered after equilibration.

With regard to which physical parameter controls the formation of these low-T ULVs, several models have been proposed to explain the spontaneous formation mechanism of ULVs, including the line tension associated with the disc's edge.[18,28,30]. These models invoke a growth stage that is triggered by insufficient amounts of the short-chain lipid (i.e. DHPC). This is because that as a result of increased temperature or decreased C_{lp}, DHPC's solubility in water and its miscibility with DMPC increases. The discs thus gradually increase in size through coalescence with their nearest neighboring discs in order to reduce the increased line tension that results from exposing hydrophobic chains to water. After a number of collisions with each other, the discs grow, up to a certain size, whereby they begin to self-fold into ULVs. The disc self folding process contains two essential components: (a) the continuous loss of DHPC that coats the disc's rim; and (b) the membrane being sufficiently flexible to fold. The loss of DHPC coupled with C_{lp} and temperature determines the effective collision frequency, F_{eff}, that results in the growth of the discs. It should be noted that the greater the F_{eff} the larger the discs. These larger discs are also more stable than their smaller counterparts because less "edge" lipid (i.e. DHPC) is required to stabilize them. Compared to the control sample (R=0.01), a higher collision frequency (not necessarily F_{eff}) is expected in either the weakly charged mixtures (R=0.001) or the highly charge-screened samples (R=0.01, [NaCl] = 0.2 M) - due to less inter-particle Coulombic repulsion. Hence, larger and more stable discs would be anticipated if there would be an insufficient amount of DHPC coating the disc's edge. The fact that the control sample (R = 0.01) yields more stable discs (in terms of longer time and lower C_{lp}) implies that the formation of low-T incubated ULVs is presumably dictated by a reduction in membrane rigidity. It has also been reported that lipid membranes become more flexible with decreased surface charge density.[31,32] For this reason, the bilayer of the control sample (R = 0.01) is expected to be more rigid than those of the other two mixtures (i.e. R=0.01, [NaCl] = 0.2 M and R=0.001), thus delaying, or possibly even prohibiting the transformation of discs into ULVs.

As a result of this analysis we propose the following kinetics scenario for low-T ULVs found in a weakly charged or high-salinity mixture. The original discoidal micelles (at low-T) in high-C_{lp} (10 %) samples transform into elongated ribbons after dilution to a lower C_{lp} at low-T. Because of the low solubility between gel-DMPC and L_α-DHPC, the discs/ribbons are initially metastable and slowly fuse into lamellae with other ribbons or discs in order to minimize rim tension. Due to an inherent lower membrane rigidity - compared to highly charged (R = 0.01) systems - the lamellae slowly self-fold into vesicles (even with DMPC in gel phase), achieving a lower free energy state (by elimination of the disc rim). Too low a membrane rigidity (e.g. zwitterioinc mixtures), however, may result in a fast self-folding process that gives rise to MLVs.

This dependence of membrane rigidity on resultant morphology can also explain previous observations of unstable discs at low-T (10 °C) and low C_{lp} ($\leq$ 1%) in zwitterionic DMPC/DHPC mixtures, where MLVs are observed immediately after dilution.[7] The bilayers of such a system (i.e. neutral) are presumably much more flexible resulting in an accelerated folding rate, and consequently the formation of MLVs. The results from this study also provide further insight into bicellar mixtures with Q = 4 and R = 0.03, where ULV size is found to be practically independent of C_{lp} and salt concentration after the samples had been incubated at 4 °C for over a period of one month.[33] In this case, ULVs might have already formed even prior to an increase in temperature (not confirmed) – a mechanism of ULV formation most likely similar to what is taking place in the current study. This also shows that ULVs are presumably stable at low-T for an extended period of time (e.g. months). However, the stability of high-T ULVs is not studied here, but based on a previous study we anticipated that they could be stable on the order of weeks.[16] It should also be noted that the thermodynamic unbinding of MLVs requires a lower bending rigidity of the membrane in order to enhance steric repulsion (i.e.

thermal undulation).[34] The difference between the two scenarios is mainly due to the formation mechanism of the ULVs, i.e. thermodynamically driven and kinetically trapped. The weakly charged system (R=0.001), though it forms ULVs, does not have sufficient Coulombic repulsion to unbind MLVs (Fig 1).

Conclusion

The present SANS study shows that MLVs obtained from high-T dilution are thermodynamically stable both upon the reduction of charge density and the addition of salt (NaCl). As the weakly charged and NaCl-doped mixtures are diluted at low-T (10 °C), nanodiscs and elongated ribbons are initially observed. However, after a prolonged period of incubation (> 20 hours), both morphologies transform into kinetically trapped ULVs, which remain unaltered even after heating to temperatures where MLVs are commonly observed. This is good evidence that both nanodiscs and ribbons are not thermodynamically stable morphologies, even though they reproducibly form when diluted at low-T from high C_{lp} samples. The formation mechanism of low polydispersity, kinetically trapped ULVs is fundamentally different from that of ULVs formed through the thermal unbinding of MLVs.[34] The present ULVs are most likely the result of increased membrane rigidity that retards the self-folding of the discs - a process unfavourable to MLV formation. However, in the case of higher charge density (R = 0.01) samples, discs seem to remain stable at low temperature for two reasons: (a) reduced effective collision frequency; and (b) a higher membrane rigidity. In other words, the mechanism that transforms discs to ULVs at low temperature is the result of a fine balance among membrane rigidity, C_{lp}, and charge density. This knowledge may prove to be useful when designing lipid-based delivery carriers for pharmaceuticals, where morphological changes are crucial in effectively delivering payloads.

Acknowledgments

The SANS work made use of facilities supported in part by the National Science Foundation under Agreement No. DMR-0944772. The mention of commercial products does not imply endorsement by NIST, nor does it imply that the materials or equipment identified are necessarily the best available for the purpose. The work at ORNL's High Flux Isotope Reactor was sponsored by the Scientific User Facilities Division, Office of Basic Energy Sciences, United States Department of Energy (U.S. DOE). ORNL is operated by UT–Battelle, LLC for the U.S. DOE under Contract No. DE-AC05-00OR22725. JK is partly supported through ORNL's Laboratory Directed Research and Development (LDRD) program. MPN would like to thank for the financial support from the UConn Faculty Large Grant and the Faculty Startup fund of the Institute of Materials Science (UConn).

References

1. Hwang, K. J., M. M. Padki, D. D. Chow, H. E. Essien, J. Y. Lai and P. L. Beaumier (1987), "Uptake of Small Liposomes by Non-Reticuloendothelial Tissues," *Biochim. Biophys. Acta, 901*, pp. 88-96.
2. Gabizon, A., H. Shmeeda and Y. Barenholz (2003), "Pharmacokinetics of Pegylated Liposomal Doxorubicin," *Clin. Pharmacokinet.*, *42*, pp. 419-436.
3. Gabriel, N. E. and M. E. Roberts (1984)," Spontaneous Formation of Stable Unilamellar Vesicles" *Biochemistry, 23,* pp. 4011-4015.

4. Gabriel, N. E. and M. E. Roberts (1986)," Interaction of Short-Chain Lecithin with Long-chain Phospholipids: Characterization of Vesicles That Form Spontaneously," *Biochemistry, 25,* pp. 2812-2821.
5. Hauser, H., N. Gains, H.-J. Eibl, M. Müller and E. Wehrli (1986) "Spontaneous Vesiculation of Aqueous Lipid Dispersions," *Biochemistry, 25,* pp. 2126-2134.
6. Nieh, M.-P., C. J. Glinka, S. Krueger, R. S. Prosser and J. Katsaras (2001),"SANS Study of the Structural Phases of Magnetically Alignable Lanthanide-Doped Phospholipid Mixtures," *Langmuir, 17,* pp. 2629-2638.
7. Nieh, M.-P., C. J. Glinka, S. Krueger, R. S. Prosser and J. Katsaras (2002), "SANS Study on the Effect of Lanthanide Ions and Charged Lipids on the Morphology of Phospholipid Mixtures," *Biophys. J., 82,* pp. 2487-2498.
8. Sanders II, C. R. and J. H. Prestegard (1991)," Orientation and Dynamics of β-Dodecyl Glucopyranoside in Phospholipid Bilayers by Oriented Sample NMR and Order Matrix Analysis," *J. Am. Chem. Soc., 113,* pp. 1987-1996.
9. Sanders II, C. R. and J. P. Schwonek (1992)," Characterization of Magnetically Orientable Bilayers in Mixtures of Dihexanoylphosphatidylcholine and Dimyristoylphosphatidylcholine by Solid-state NMR," *Biochemistry, 31,* pp. 8898-8905.
10. Struppe, J. and R. R. Vold (1998),"Dilute Bicellar Solutions for Structural NMR Work," *J. Magn. Reson., 135,* pp. 541-546.
11. Rowe, B. A. and S. L. Neal (2003)," Fluorescence Probe Study of Bicelle Structure as a Function of Temperature: Developing a Practical Bicelle Structure Model," *Langmuir,19,* pp. 2039-2048.
12. van Dam, L., G. Karlsson and K. Edwards (2004)," Direct Observation And Characterization of DMPC/DHPC Aggregates under Conditions Relevant for Biological Solution NMR," *Biochim. Biophys. Acta, 1664*, pp. 241-256.
13. Katsaras, J., T. A. Harroun, J. Pencer and M.-P. Nieh (2005)," 'Bicellar' lipid mixtures as used in biochemical and biophysical studies," *Naturwissenschaften, 92*, pp. 355–366.
14. Soong, R., M.-P. Nieh, E. Nicholson, J. Katsaras, and P. M. Macdonald (2010)," Bicellar Mixtures Containing Pluronic F68: Morphology and Lateral Diffusion from Combined SANS and PFG NMR Studies," *Langmuir, 26,* pp. 2630-2638.
15. Triba, M. N., D. E. Warschawski and P. F. Devaux (2005), "Reinvestigation by Phosphorus NMR of Lipid Distribution in Bicelles," *Biophys. J., 88,* pp. 1887-1901.
16. Nieh, M.-P., T. A. Harroun, V. A. Raghunathan, C. J. Glinka and J. Katsaras (2003)," Concentration-Independent Spontaneously Forming Biomimetic Vesicles," *Phys. Rev. Lett., 91,* 158105.
17. Nieh, M.-P., T. A. Harroun, V. A. Raghunathan, C. J. Glinka and J. Katsaras (2004)," Spontaneously Formed Monodisperse Biomimetic Unilamellar Vesicles: the Effect of Charge, Dilution, and Time," *Biophys. J., 86,* pp. 2615–2629.
18. Nieh, M.-P., N. Kučerka and J. Katsaras (2009)," Spontaneously Formed Unilamellar Vesicles," *Methods in Enzymol., 465*, pp. 3-20.
19. Nieh, M.-P., V. A. Raghunathan, S. R. Kline, T. A. Harroun, C.-Y. Huang, J. Pencer and J. Katsaras (2005),"Spontaneously Formed Unilamellar Vesicles with Path-Dependent Size Distribution," *Langmuir, 21,* pp. 6656-6661.
20. Kline, S. R. (2006), " Reduction and Analysis of SANS And USANS Data Using IGOR Pro," *J. Appl. Crystallogr., 39*, pp. 895–900.
21. van Dam, L., G. Karlsson and K. Edwards (2006)," Morphology of Magnetically Aligning DMPC/DHPC Aggregates – Perforated Sheets, Not Disks," *Langmuir, 22*, pp. 3280-3285.

22. Higgins, J. S.; Benoit, H. C. (1994), "Chapter 6 Form Factors," in *Polymers and Neutron Scattering*, Oxford University Press: New York, pp. 141-191.
23. Vinson, P. K., Y. Talmon and A. Walter (1989)," Vesicle-Micelle Transition of Hosphatidylcholine and Octylglucoside Elucidated by Cryo-Transmission Electron Microscopy," *Biophys. J.*, *56*, pp. 669-681.
24. Chiruvolu, S., H. E. Warriner, E. Naranjo, S. H. J. Idziak, J. O. Rädler, R. J. Plano, J. A. Zasadzinski and C. R. Safinya, C. R. (1994)," A Phase of Liposomes with Entangled Tubular Vesicles," *Science*, *266*, pp. 1222-1225.
25. Caffrey, M., J. Hogan and A. Mencke (1991), "Kinetics of the barotropic ripple ($P_{\beta'}$)/lamellar liquid crystal (L_{α}) phase transition in fully hydrated dimyristoylphosphatidylcholine (DMPC) monitored by time-resolved x-ray diffraction," *Biophys. J.*, *31*, pp. 8898-8905.
26. Kiselev, M. A., D. Lombardo, P. Leiseur, A. M. Kisselev, S. Borbely, T. N. Simonova and L. I. Barsukov (2008)," Membrane Self Assembly in Mixed DMPC/Nac Systems by SANS" *Chem. Phys.*, *345*, pp. 173-180.
27. Vinson, P. K., J. R. Bellare, H. T. Davis, W. G. Miller and L. E. Scriven, L. E. (1991),"Direct Imaging of Surfactant Micelles, Vesicles, Discs and Ripple Phase Structures by Cryo-Transmission Electron Microscopy," *J. Colloid & Interface Sci.*, *142*, pp. 74-91.
28. Leng, J., S. U. Egelhaaf and M. E. Cates (2003)," Kinetics of the Micelle-to-Vesicle Transition: Aqueous Lecithin-Bile Salt Mixtures," *Biophys. J.* 2003, 85, pp. 16214-1646.
29. Mahabir, S., W. K. Wan, J. Katsaras, M.-P. Nieh (2010)," Effects of Charge Density and Thermal History on the Morphologies of Spontaneously Formed Unilamellar Vesicles," *J. Phys. Chem. B*, *114*, pp. 5729-5735.
30. Fromherz, P. (1983),Lipid Vesicle Structure – Size Control by Edge-Active Agents," *Chem. Phys. Lett.*, *94*, 259-266.
31. Higgs, P. G. and J. F. Joanny (1990),"Enhanced Membrane Rigidity in Charged Lamellar Phases," *J. Phys. (France)*, *51*, pp. 2307-2320.
32. Claessens, M. M. A. E., B. F. van Oort, F. A. M. Leermakers, F. A. Hoekstra and M. A. Cohen Stuart (2004)," Charged Lipid Vesicles: Effects of Salts on Bending Rigidity, Stability, and Size," *Biophys. J.*, 87, pp. 3882-3892.
33. Yue, B., C. Y. Huang, M.-P. Nieh, C. J. Glinka and J. Katsaras (2005)," Highly Stable Phospholipid Unilamellar Vesicles from Spontaneous Vesiculation: a DLS and SANS Study," *J. Phys. Chem. B*, 109, pp. 609-616.
34. Pozo-Navas, B, V. A. Raghunathan, J. Katsaras, M. Rappolt, K. Lohner and G. Pabst (2003)," Discontinuous Unbinding Of Lipid Multibilayers," *Phys. Rev. Lett.*, *91*, 028101.

(248d) Control of Cell Migration Using a Mechanistically Tunable Fibrous Environment

Kevin Sheets, School of Biomedical Engineering and Sciences, Virginia Tech, Blacksburg, VA and **Amrinder S. Nain**, Mechanical Engineering, Virginia Tech, Blacksburg, VA

Cells attach to and interact with their immediate fibrous microenvironment known as the extracellular matrix (ECM), which provides region-specific biochemical and biophysical cues that drive cell behavior. Despite recent studies indicating the importance of controlling these factors which cause dramatic alterations in cell attachment, motility, and differentiation, most cell studies are still performed on substrates of flat tissue cultured glass, plastic, or gel. Here we present cell behavior the STEP (Spinneret-based Tunable Engineered Parameters) platform, which deposits high aspect ratio nano/micro fibers with tight control on diameter, spacing, porosity, and orientation. The resulting fibrous scaffolds, which mimic the mechanistic environment of ECM by providing biomechanical cues, can be tuned to elicit a desired cellular response. In this study, C2C12 mouse myoblast cells are seeded onto highly aligned fibrous STEP scaffolds and their migration speeds are shown to directly result from varying substrate stiffness in forms of fiber diameter and fiber length. Polystyrene fibers of three different diameters (200, 500, and 800 nm) were spun onto two substrate types (suspended single-layer (SS) and suspended double-layer (SD)) in addition to flat glass control. Using standard cell culture techniques, C2C12 mouse myoblasts were seeded onto the scaffolds and attached within 4-6 hours. After attaching, 2 mL DMEM media were added to keep cells viable over a 24 hour period. During this time, cells on fiber networks were time-lapse imaged by an incubating microscope. Since cells do not travel at a constant velocity, cell migration speeds were calculated by tracking the position of the cell's nucleus every hour and noting the maximum distance traveled per hour in a 16-hour window. Cells were observed to attach to and elongate along the fiber axis. On flat glass, cells were observed to travel at an average velocity of 30 μ m/hr. Cells constricted to SS fibers traveled at the highest speeds (55 – 80 μ m/hr) and onwe SD fibers were observed to probe each possible migration direction, thus, decreasing their overall migration velocity compared to SS substrates (40 – 50 μ m/hr). Similary, both fiber diameter and fiber length were observed to play a role in cell migration speed. In each case, the conditions which resulted in the lower fiber stiffness (smaller diameter, longer length) allowed cells to travel at the top end of the speed ranges previously listed. Greater fiber stiffness corresponds to an increased ability for the cell to form stable focal adhesions and a lower propensity to migrate at high speeds. STEP platform provides a unique tool to fine-tune the mechanical environment to which native cells attach and interact with so that scaffolds and drug delivery therapies may be improved upon in the future.

See more of this Session: Nanostructured Scaffolds for Tissue Engineering
See more of this Group/Topical: Nanoscale Science and Engineering Forum

(217c) Evaluation of the Toxicity of Nanomaterials Based On Knowledge Extraction From High Throughput Screening of Biological Toxicity Data

Rong Liu[1], Saji George[1], Bryan France[1], Robert Damoiseaux[1], Robert Rallo[2], Kenneth Bradley[1], Andre E. Nel[3] and Yoram Cohen[4], (1)California NanoSystems Institute, Center for the Environmental Implications of Nanotechnology (CEIN), Los Angeles, CA, (2)Enginyeria Informatica i Matematiques, Universitat Rovira Virgili, Tarragona, Spain, (3)Division of NanoMedicine, UCLA, Los Angeles, (4) Chemical & Biomolecular Engineering, University of California, Los Angeles, Los Angeles, CA

Nano-sized materials are increasingly utilized in many industrial products and processes, with over 1000 commercial products utilizing engineered nanomaterials (eNMs), primarily due to their unique nano-scale properties. Despite the many beneficial uses of nanomaterials, there is a growing public concern regarding the potential release of eNMs to the environment and the potential adverse impacts of exposures to eNMs. Although there are mounting studies on the toxicity of eNMs, understanding of the general principles governing eNMs toxicity potential and the long-term environmental health and safety associated with eNM containing products is in its infancy. In this regard, high throughput toxicity screening is critical for characterization of the potential hazard of eNMs, which is in turn indispensable information for subsequent risk assessment and development of environmental and health regulatory policies. In the present work an approach to knowledge extraction from high throughput screening (HTS) of nanoparticle toxicity is presented, based on RAW and BEAS-2B mammalian cell lines, focusing on identification of toxicity outcomes (i.e., "hit identification"), development of predictive nano-quantitative-structure-activity relations (nano-SARs), identification of pathway linkages and correlation of cell signaling pathways and cytotoxicity. Based on newly developed feature selection approach, a classification based nano-SAR was developed for cytotoxicity of metal and metal oxide nanoparticles. The approach was demonstrated for a library of metal and metal oxide nanoparticles to develop a classification based nano-SAR that can be used to identify decision boundaries with respect to hazard ranking of nanoparticles. In addition, self-organizing maps (SOM) analysis was performed to identify clusters corresponding to sub-lethal pro-inflammatory responses associated with ROS generation, lethal genotoxic responses, and cytotoxicity. Finally, complex network theory methods were applied to identify relationships between cell responses (pathway activation and cytotoxicity measures) as well as physicochemical properties of nanoparticles. For example, the analysis based on the above cell lines revealed a hierarchical activity-activity pattern where sub-lethal effects such as ROS generation and the intracellular Ca^{2+} flux are strongly related to lethal effects (e.g. cell membrane damage and cell death) and that eNM primary size, aggregation and its dissolution tendency were critical parameters affecting the observed toxicity behavior.

(121d) Exploiting Mixed Self-Assembled Monolayers for Design and Fabrication of Patchy Particles

Ines C. Pons-Siepermann, Chemical Engineering, University of Michigan, Ann Arbor, MI and Sharon C. Glotzer, Department of Chemical Engineering, University of Michigan, Ann Arbor, MI

Previous computational studies [1] explained the formation of patterns (stripes and patches) in binary mixtures of immiscible surfactants adsorbed on gold nanoparticles [2]. As an extension of those studies, we performed atomistic and mesoscale simulations of ternary and quaternary mixed self-assembled monolayers (SAMs) on nanoparticle surfaces Here we present predictions for new and unexpected patterns for patchy particles that could be synthesized through judicious choice of surfactant architecture, nanoparticle geometry, and SAM stoichiometry.

[1] C. Singh et al. Physical Review Letters 99, 226106 (2007)

[2] A.M. Jackson et al. Nature Materials, 3, 330-336 (2004)

See more of this Session: Self and Directed Assembly at the Nanoscale
See more of this Group/Topical: Nanoscale Science and Engineering Forum

(172e) Current-Driven Surface Morphological Stabilization of Coherently Strained Heteroepitaxial Thin Films

Georgios I. Sfyris, M. Rauf Gungor and Dimitrios Maroudas, Chemical Engineering, University of Massachusetts Amherst, Amhest, MA

The competition between elastic strain energy and surface energy is responsible for surface morphological instabilities in stressed elastic solids; these include the well-known Asaro-Tiller or Grinfeld (ATG) instability under conditions that promote surface diffusion and the Stanski-Krastanow (SK) morphological instability in the heteroepitaxial growth of thin films on solid substrates. In heteroepitaxial systems, the lattice mismatch between the thin film and the substrate materials induces a misfit strain, which is the source of elastic deformation. In recent studies, we have shown that applying on a stressed elastic conductor an external electric field of sufficient strength and proper direction can inhibit the ATG instability. Nevertheless, the role of surface electromigration in stabilizing the surface morphology of coherently strained epitaxial films remains largely unexplored.

In this presentation, we report the results of a morphological stability analysis for the driven morphological response of an epitaxial film surface. We have analyzed the surface morphological stability of a coherently strained thin film grown epitaxially on an elastic substrate and subjected simultaneously to an external electric field; both thick and thin (finite-thickness) elastic substrates have been examined. Due to its lattice mismatch with the substrate material, the film may undergo a SK instability, resulting in formation of islands on the film surface. We have developed a three-dimensional (3D) model for the surface morphological evolution of the thin film and conducted a linear stability analysis to examine the morphological stability of the epitaxial film's planar surface state. The analysis has shown that surface electromigration due to a properly applied and sufficiently strong electric field can inhibit SK-type instabilities, which can be used to control the onset of island formation on the film surface. We have also found that using a finite-thickness substrate can have the beneficial effect of reducing the critical strength of the electric field required to stabilize the planar surface morphology of the epitaxial film with respect to the field strength required in the case of an infinitely thick substrate.

We have determined the critical electric-field strength as a function of material properties and heteroepitaxial system parameters, as well as the electric field's optimal direction for the most efficient stabilization of the surface morphology. Furthermore, we have obtained detailed results for the effects of the finite-thickness substrate on the stabilization of the film's surface over a range of mechanical properties of the heteroepitaxial system's constituents. Perhaps more importantly, we have analyzed the surface morphological stabilization of heteroepitaxial films on substrates by combining the use of externally applied electric fields with well-known strain engineering techniques, such as the use of thin compliant substrates; such substrates provide some elastic accommodation of the lattice-mismatch strain in the epitaxial film due to the substrate's ability to relax parallel to its interface with the epitaxial film. It should be mentioned that employing substrate engineering techniques alone is not sufficient for stabilizing the planar morphology of the epitaxial films; for such stabilization, the simultaneous action of the externally applied electric field is required. Our analysis has demonstrated the benefits of the substrate compliance, as measured by the reduction in the critical electric-field strength required for stabilizing the epitaxial film surface; we have found this reduction in the critical field strength to be as high as two orders of magnitude with respect to that required for epitaxial films grown on conventional thick substrates. Our results generate experimentally testable hypotheses and motivate experimental measurements that can be compared directly with the theoretical predictions.

See more of this Session: Nanoelectronic Materials
See more of this Group/Topical: Nanoscale Science and Engineering Forum

TARGETED MRI AND OPTICAL MOLECULAR IMAGING USING GADOLINIUM LOADED SMALL UNILAMELLAR VESICLES

Umar Iqbal[1], Homam Albaghdadi[1], Mu-Ping Nieh[2], Ursula I. Tuor[3], Zoltan Mester[4], Danica Stanimirovic[1], John Katsaras[5,6], Abedelnasser Abulrob[1]

[1]Institute for Biological Science (IBS), National Research Council (NRC) Canada, Ottawa, ON, Canada, [2]University of Connecticut, Storrs, CT, USA [3]Institute for Biodiagnostics (IBD, West), NRC, Calgary, AB, Canada, [4]Institute for National Measurements Standards (INMS), NRC, Ottawa, ON, Canada, [5]Oak Ridge National Laboratory, Oak Ridge, TN, USA, [6]Canadian Neutron Beam Centre, NRC, Chalk River, Ontario, Canada

Abstract

Noninvasive investigation of cellular and molecular processes becomes possible through the novel techniques, one of which is molecular imaging, where enhanced sensitivity is a key component for clinic translation of the technique. In this presentation, spontaneously forming, small unilamellar vesicles (ULVs) (30 nm in diameter) were used as a platform to build a bi-modal [i.e., optical and Magnetic Resonance Imaging (MRI)] targeted contrast agent for the molecular imaging of brain tumors. Small ULVs were loaded with a gadolinium (Gd) chelated lipid (Gd-DPTA-BOA), functionalized with targeting antibodies (anti-EGFR monoclonal and anti-IGFBP7 single domain), and incorporated a near infrared dye (Cy5.5). The resultant ULVs were characterized *in vitro* using small angle neutron scattering (SANS), in phantom MRI and dynamic light scattering (DLS). Targeted (with antibodies) and nontargeted-Gd loaded sULVs labeled with Cy5.5 were assessed *in vivo* in a mice brain tumor model using both optical imaging and MRI. The results demonstrated that a spontaneously forming, nanosized ULV loaded with a high payload of Gd can selectively target and image, using MR and optical imaging, brain tumor vessels when functionalized with antibodies. The unique features of these targeting ULVs make them promising molecular MRI contrast agents.

Introduction

MRI is a non-invasive and powerful medical diagnostic technique that offers high-resolution anatomical information, and is frequently used for the detection of a variety of diseases (De Schepper and Bloem et al 2007). MRI creates images of the body using the principles of nuclear magnetic resonance. Images are usually generated using gadolinium (Gd-DTPA) as contrast agent, based on its free distribution in the body. While these images provide good anatomical information about the disease (e.g., tumor) localization and spread, to obtain information about molecular characteristics of the disease (e.g., expression of certain receptors that relate to disease prognosis, progression, etc.), a targeted contrast agent approach is required.[1]. Current gadolinium contrast agents (i.e. Gd-DPTA or Magnevist®) on the market require a high tissue concentration of 1-100 mM to obtain detectable image contrast.[2,3] To overcome these sensitivity issues, the use of liposomes that carry a high payload of gadolinium per liposome has been suggested as a viable strategy to increase imaging sensitivity.[4] A variety of targeted contrast agents consisting of gadolinium loaded liposomes decorated with targeting peptides or antibodies have been developed to date.[5,6] However, the clinical use of Gd loaded liposomes is hampered by slow clearance and toxicity *in vivo*.[7]

In industry, regular-sized liposomes are often produced by series of costly multistage filtration or extrusion to break up the multilamellar structure, which naturally forms in zwitterionic lipid mixtures.

Recently, we reported that a long- and short-chain phospholipid mixture (e.g., dimyristoyl phosphatidylcholine, DMPC, and dihexanoyl phosphatidylcholine, DHPC) doped with a long-chain charged lipid (dimyristoyl phosphatidylglycerol, DMPG) spontaneously form small unilamellar vesicles (sULV) with a low-polydispersity.[8-12] The size (diameter) of the sULV ranges from 10 to 40 nm and can be obtained in a controlled manner without multistage filtration.[8,9,13] They are also capable of encapsulating molecules.[14] This method has economic advantage over the traditional filtering or extrusion methods and producing a stable and monodisperse sULVs.

In this study, sULVs were used as a nano-sized platform for a novel MRI contrast agent by incorporating: PEGylated phospholipid, distearoyl phosphoethanolamine-N-[Amino (Polyethylene Glycol)2000] (DSPE-PEG2000-Amino) and gadolinium DTPA-bis-oleate (Gd-DTPA-BOA). Antibodies were attached to the distal end of the lipid- conjugated PEG molecules (5 mole%) on the surface of the sULV, whereas two molar ratios of Gd- DTPA-BOA to total lipids (0.2 and 0.4) were examined to evaluate the effect of Gd-DTPA-BOA on the sULV morphology. Small angle neutron scattering (SANS) was used to characterize the morphologies of the mixtures and in vitro phantom MR imaging was used to determine the Gd payload per nanoparticle. These novel Gd-sULV contrast agents were further linked with anti-epidermal growth factor receptor (EGFR) monoclonal antibodies (mAb) or anti-insulin-like-growth factor binding protein 7 (IGFBP7) single domain antibodies (sdAbs) and the near infrared dye, Cy5.5, to produce anti-EGFR mAb- or anti-IGFBP7 sdAb- targeted-Gd-sULVs-Cy5.5. Anti-EGFR mAb has been used previous to target EGFR-overexpressing tumor cells.[15] Anti-IGFBP7 sdAbs have been demonstrated previously to be effective at selectively binding to brain tumor vessels *in vivo*.[16] In this study, both anti-EGFR mAb- and anti-IGFBP7 sdAb-Gd-sULVs-Cy5.5 were assessed in an in vivo brain tumor model in nude mice using both optical and MR imaging.

MATERIALS & METHODS

Formulation of gadolinium loaded small unilamellar vesicles (Gd-sULV)

Dimyristoyl phosphatidylglycerol (DMPG), dihexanoyl phosphatidylcholine (DHPC) and distearoyl phosphoethanolamine-N-[Amino(Polyethylene Glycol)2000] (DSPE-PEG2000-Amino) were purchased from Avanti Polar Lipids (Alabaster, AL, USA). The Gd-DTPA-BOA was custom synthesized (IQSynthesis, Miramar, FL, USA). DMPG/DHPC/DSPE-PEG2000-Amine/Gd-DTPA-BOA at 31, 23.8, 5 and 40mol% respectively, were first dissolved in chloroform and dried by continuously flowing N_2 gas followed by evacuation with vacuum for 24 hours. The dried samples were then completely re-dispersed in water (D_2O for neutron scattering; H_2O for other tests) to form 10 % (by weight) solutions by temperature cycling and vortex between 4° and 50° C (4~5 cycles). The 10 wt. % solutions are liquid-like at low T (4°C) but gel-like at high T (50° C). All the samples were then diluted with cold D_2O at 4°C into a total lipid concentration of 2 %.

Dynamic light scattering (DLS) and small angle neutron scattering (SANS) measurements

Dynamic light scattering (DLS) on both Gd-sULV (non-targeting) and anti-IGFBP7 sdAb-Gd-sULV was performed using a zetasizer nano (Malvern Instruments Ltd). The DLS experiments were carried out at 23°C on samples dispersed (1:100 v/v) in milliQ water. Size distribution was determined by intensity-weighted data output.

For small angle neutron scattering (SANS), the lipid mixtures were dissolved in D_2O with a purity > 99.9 % (Chalk River Laboratories, ON, Canada) to enhance the neutron scattering contrast. All the SANS measurements were done at 50°C where the Gd-sULV is expected. SANS experiments were conducted at the 30m NG3 SANS instrument located at the NIST (National Institute of Standards and Technology) Center for Neutron Research (NCNR, Gaithersburg, MD). 6 Å wavelength (λ) of neutrons and three sample-to-detector distances (1m, 4m and 13m) were used, covering a scattering vector [$q = \frac{4\pi}{\lambda}\sin\left(\frac{\theta}{2}\right)$, θ being scattering angle] ranged from 0.003 $Å^{-1}$ to 0.3 $Å^{-1}$. The raw data were subtracted with the background (blocked beam), normalized with the monitored incident neutron flux and sample transmission, and finally subtracted with the equally treated empty cell scattering data. The reduced data then were circularly averaged with respect to the beam center and normalized to the absolute scale by the neutron flux and the sample thickness. The final scattering intensity, I is expressed as a function of q.

In vitro phantom MRI

The T_1 relaxivity properties of samples containing the Gd-sULV were compared to that of Gd-DTPA (Magnevist). Solutions of 50 μg/ml and 200 μg/ml Gd concentration were prepared of Gd-DTPA or Gd-sULVs prepared with either 20 mol% or 40 mol% Gd-DTPA-BOA. These solutions (270 μl) were aliquoted into tubes which were embedded in agarose within a container and scanned using a quadrature coil and a Bruker Biospec Avance II MRI system with a 9.4T magnet and Paravision 4 software (Bruker Biospin). T_1 maps of cross-sectional slices through the tubes were acquired using a RARE inversion-recovery sequence with variable repetition times. Three slices were acquired using a matrix of 128x128, field of view of 3cm^2, TE=10ms, flip angle of 180° and 10 different times of 135ms, 375ms, 630ms, 950ms, 1300ms, 1750ms, 2300ms, 3100ms, 4400ms, 10000 ms. T_1 for each sample was measured from the T_1 maps calculated using the Paravision 4 software (Bruker Biospin).

Bi-modal contrast agent (Gd-sULV-Cy5.5) and targeting ULVs

The Gd-sULVs were reacting with Cy5.5-NHS-ester (GE Healthcare; excitation 670nm, emission 690 nm) to form Gd-sULV-Cy5.5.. These Gd-sULV-Cy5.5 then were boiconjugated with anti-EGFR mAb or anti-IGFBP7 sdAb to have targeting effect. Detailed preparation method can be found in Ref 17.

Xenograft and intracranial models of U87MG.EGFRvIII glioblastoma in nude mice

The detailed description of xenograft and intracranial models of U87MG.EGFRvIII glioblastoma in nude mice can be also found in Ref 17.

In vivo near-infrared fluorescence imaging of antibody targeted Gd-sULVs in brain tumors.

Mice bearing 14 day-old U87MG.EGFRvIII xenograft brain tumors were injected with anti-EGFR mAb targeted-Gd-sULV-Cy5.5 or nontargeted Gd-sULV-Cy5.5 via the tail vein (0.2mls of 2% wt solution). In another set of animals, mice bearing 10 day old U87MG.EGFRvIII intracranial brain tumors were injected with anti-IGFBP7 sdAb targeted-Gd-sULV-Cy5.5 or nontargeted Gd-sULV-Cy5.5 via the tail vein (0.2mls of 2% wt solution). Animals were subjected to *in vivo* imaging studies using a small-animal time-domain eXplore Optix MX2 pre-clinical imager (Advanced Research Technologies) as described previously.[18-20] In all imaging experiments, a 670-nm pulsed laser diode was used for excitation and fluorescence emission collected at 700 nm. Each animal was imaged dorsally in the head region and whole body at various time points after i.v. injection. The data were recorded as temporal point-spread functions (TPSF) and the images were reconstructed as fluorescence concentration maps. Average fluorescence concentration data from ROIs placed around the tumor region or contralateral muscle (for xenograft tumor analysis only) were subsequently analyzed using the OptiView software package (Advanced Research Technologies). At the end of the experiment for mice bearing intracranial U87MG.EGFRvIII brain tumors, 40 μg of fluorescein labeled tomato lectin was injected i.v. for 10 mins before sacrifice to stain brain vessels. All animals were then perfused with heparanized saline and organs were excised and imaged *ex vivo*. *Ex vivo* organs were analyzed by placing an ROI around each organ and determining the total fluorescence concentration per gram tissue.

In vivo MRI of anti-IGFBP7 sdAb targeted and non-targeted Gd-sULVs

Tumor cells were implanted into CD-1 nude mice brains as described above. 7-8 days following intracranial implantation animals were scanned using standard T2 imaging (see below) to confirm successful tumor implantation. 10 days following injection, animals were anesthetized with isoflurane for contrast imaging using MRI. First, the femoral vein was isolated and a catheter was inserted into the vein for contrast administration. The mouse was then moved into a cradle for positioning in the centre of a 9.4T magnet equipped for MRI using a Bruker Biospec Avance II console. Animals were randomized to targeted or non-targeted contrast injection groups. In general, prior to contrast injection T2, T1 weighted and T1 map scans were acquired. For the T2 map, a spin-echo sequence was acquired having 16 echoes with 10 msec echo spacing, a 2x2 cm^2 field of view, a 128x128 data matrix, a repetition time of 5000 ms, for each of 10 slices 1 mm thick. T_1 maps were acquired using a single shot echo planar sequence with a 2x2 cm^2 field of view, a 128x128 data matrix, a repetition time of 8.5 s, an echo time of 38 msec and 22 inversion time points every 400 ms for a 1 mm thick slice through the tumor. T1 weighted images were acquired for 10 one mm thick slices with a RARE sequence using a 2x2cm^2 field of view, a 128x128 data matrix, a repetition time of 750 ms, an echo time of 7.56 ms and 7 averages. T2 and T1 maps were determined using local imaging software (Marevisi, NRC, Canada) and the Bruker Avance II software, respectively. After the pre-injection scans, either targeted or non-targeted contrast were injected intravenously (0.20 ml of 2% w/v Gd-sULV solution). In general, T1 maps were acquired repeatedly over the next 2 hours along with a T2 map and final T1 weighted scan. Effect of contrast was assessed using differences in intensity in the T1 weighted images by subtracting the pre-T1 weighted images from the final T1 weighted images. The effect of contrast on T1 values was also assessed by calculating T1 in tumor and contralateral unaffected brain prior to and following contrast injection.

RESULTS

Structural Characterization of Gd-sULV

SANS measurements on both 0.5 and 1.0 % samples incubated at either 50°C for 18 hours or 3 days were conducted to examine the stability of the sULVs. The SANS data (Fig. 1B) of both samples showed no smectic peaks associated with multi-lamellar vesicles, which otherwise would form in Gd-DTPA-BOA aqueous solutions. An ellipsoidal shell model (schematic shown in Fig. 1A) was applied to fit the SANS data. This model includes four structural parameters: a) the long core axis (a_{core}); b) the short core axis (b_{core}); c) shell thickness (t) and d) the polydispersity of b_{core}. All the best fits agree well

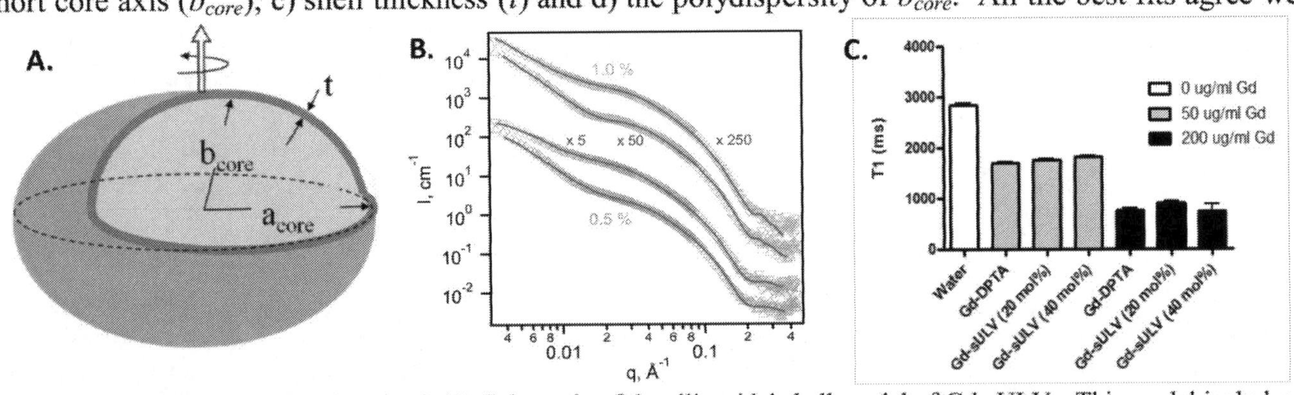

Figure 1. sULVs with high Gd payload. (A) Schematic of the ellipsoidal shell model of Gd-sULVs. This model includes four structural parameters: long core axis (a_{core}), short core axis (b_{core}), shell thickness (t), the polydispersity of b_{core}.(B) SANS data of a 40 mol% Gd-sULVs mixture with total lipid concentrations of 0.5% (green) and 1.0 % (orange) annealed at 50 °C for 18 h (inverted triangles) and 3 days (triangles). (C) The *in vitro* T1 relaxation of 20 mol% and 40 mol% Gd-sULV samples was compared to Gd-DPTA solutions at 0, 50 or 200 μg ml−1 Gd concentrations.

with the experimental data and the best fitted values of the parameters are listed in Table 1. Although the SANS data show significant differences in the low-q regime, the structural parameters obtained from the fits to the data do not differ dramatically, except for the value of a_{core} which is greater in the 1.0% sample. The best fitting results of 18 hrs and 3 days (0.5 wt.% and 1 wt.% samples) data show that they differ in contrast, the difference of neutron scattering length densities between core and shell as well as D_2O and core [(ρ_{core} - ρ_{shell}) and (ρ_{core} - ρ_{shell}) in Table 1]. Such difference in contrast between samples of two time period may result from the PEG chain rearranging with time at high temperature. However, the reason for the higher value of best fit result for scattering length density of the nanoparticle core, ρ_{core}, compared to that of D_2O, ρ_{D_2O}, is not understood so far.

Table 1. Best–fit data using ellipsoidal shell model for Gd-sULV samples at 50 °C for 18 hours and 3 days

	0.5 %		1.0 %	
	18 hours	3 days	18 hours	3 days
a_{core} (Å)	625	570	1070	1230
b_{core} (Å)	65	59	60	64
t (Å)	38	38	39	38
polydispersity	0.48	0.5	0.49	0.46
$(\rho_{core} - \rho_{shell})$, (Å^{-2})	6.6×10^{-6}	6.2×10^{-6}	6.7×10^{-6}	6.0×10^{-6}
$(\rho_{D2O} - \rho_{shell})$, (Å^{-2})	4.1×10^{-6}	4.6×10^{-6}	4.2×10^{-6}	4.7×10^{-6}

DLS of the nontargeted Gd-sULV revealed a hydrodynamic diameter of 30nm. Bioconjugation of sdAb to the Gd-sULV increased the average particle size to 41nm. These values seem to underestimate the dimension of the Gd-sULV, however lies in between $2(a_{core} + t)$ and $2(b_{core} + t)$ obtained from SANS data.

Effects of Gd-sULVs on in-vitro phantom T_1 relaxation

The T_1 signals of samples containing either Gd-DTPA or Gd-sULVs were similar when the concentration of total Gd within the sample was similar. The T_1 of samples containing 50ug/ml or 200ug/ml of Gd were similar irrespective of whether the Gd was free or loaded within vesicles (Fig. 1C). According to the observed T_1 effects, relaxivity of Gd-DTPA in water at 9.4T was 4.45±0.24 $mM^{-1}s^{-1}$ compared to 3.80±0.31 $mM^{-1}s^{-1}$ and 4.30±0.95 $mM^{-1}s^{-1}$ for the 20 mol% and 40 mol% solutions respectively, where these values were calculated according to the total Gd concentration in the samples. Because each vesicle contained a high payload of Gd, in order to obtain a similar T_1 effect for the 40mol% Gd-sULV, approximately 10-fold increase than 20 mol% Gd-sULV and 5540 fold increase than Gd-DTPA molecules were required. If relaxivity was considered proportional to the molar concentration of particles, the contrast solutions had a T_1 relaxivity of 2220±280 and 23800±5270 $mM^{-1}s^{-1}$ for the 20mol% and 40 mol% solutions, respectively.

Anti-EGFR-mAb targeted Gd-sULV-Cy5.5 in U87MG.EGFRvIII xenograft model in mice

The biodistribution of anti-EGFR mAb-Gd-sULVs-Cy5.5 and non-targeted Gd-sULVs-Cy5.5 after systemic injection was examined by in vivo optical imaging in mice bearing 14 day old xenograft GBM tumors. A peak in optical signal was observed approximately at 8 h post-injection for anti-EGFR mAb targeted Gd-sULVs (Figure 2A). The peak signal in the tumor appeared to be stable at 24 hours, at which point a statistically significant difference was demonstrated (Figure 2B). Anti-EGFR mAb targeted formulations achieved a higher tumor to muscle ratio of 2.7 compared to non-targeted formulations with a ratio of 1.6. This was determined by dividing the average fluorescence concentration in the tumor region to that of the contralateral muscle using a similarly sized ROI. Fluorescence from contralateral muscle tissue for both anti-EGFR mAb and non-targeted Gd-sULVs were similar.

Anti-IGFBP7 sdAb targeted Gd-sULV-Cy5.5 in an orthotopic U87MG brain tumor model in mice

In vivo biodistribution of the systemically-injected anti-IGFBP7 sdAb-Gd-ULV-Cy5.5 was compared with nontargeted-Gd-sULV-Cy5.5 in mice bearing 10-day old U87MG.EGFRvIII ogqrthotopic GBM tumors using *in vivo* optical imaging of the head (Fig. 3A). The anti-IGFBP7 sdAb-Gd-ULV-Cy5.5 targeted to brain tumor vessels with a peak signal at 2-4 hours (Fig. 3B). Nontargeted

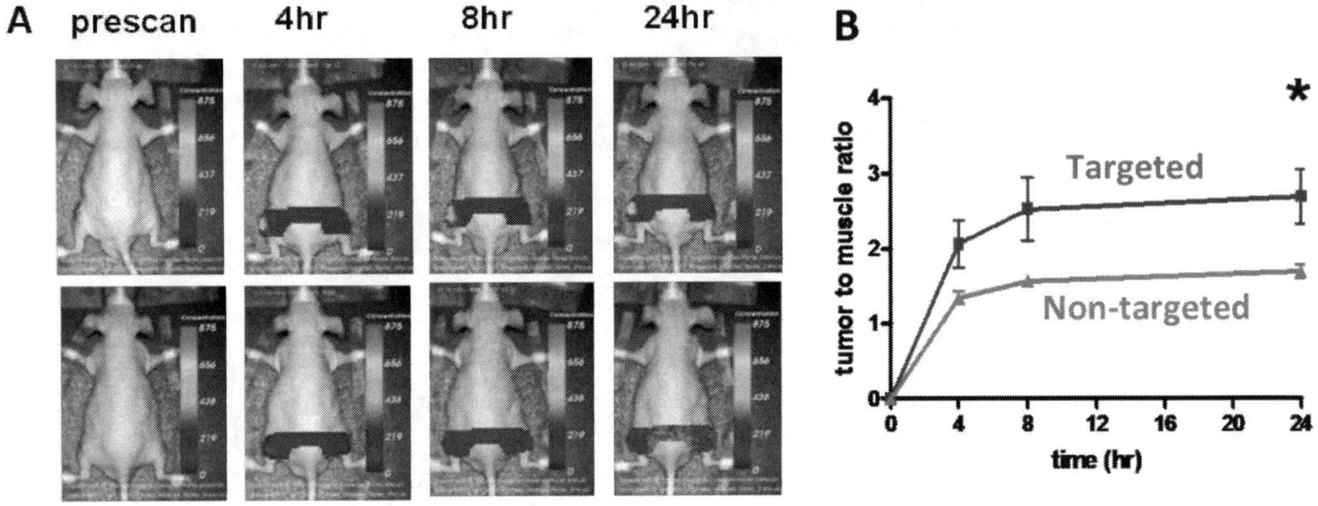

Figure 2. Enhanced accumulation and retention of anti-EGFR mAb targeted Gd-sULV-Cy5.5 in a xenograft tumor model. EGFR mAb targeted and non-targeted Gd-sULV-Cy5.5 were injected intravenously into nude mice bearing 14-day U87MG.EGFRvIII subcutaneous tumors in their left flank. (A) Mice were subjected to *in vivo* and *ex vivo* fluorescent imaging using the small-animal, time-domain eXplore Optix MX2 pre-clinical imager over 24 h. (B) Raw fluorescence *in vivo* image data from the tumor and contralateral flank region experiments were plotted as a function of time. Values correspond to mean ± SEM, n=5–6 per group. ∗ denotes a statistically significant difference $(p < 0.05)$.

Gd-sULVs-Cy5.5 demonstrated very little accumulation. Full body dorsal scan of the GBM bearing mice at 4hr demonstrated that tumor localization of anti-IGFBP7 sdAb targeted Gd-sULVs-Cy5.5 was a prominent signal in the body (Fig. 3C). At 24 hours the anti-IGFBP7 sdAb-Gd-sULV-Cy5.5 formulation was cleared to a similar level to that of the nontargeted ones in the tumor region of the mice (Fig. 3B).

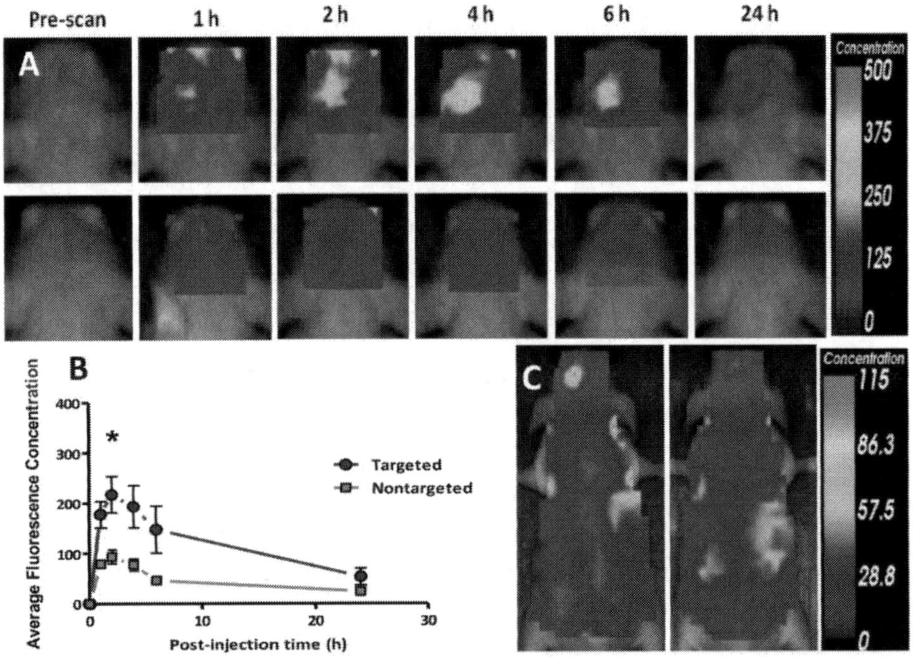

Figure 3. Anti-IGFBP7 sdAb targeted Gd-sULVs in an orthotopic brain tumor model. *In vivo* optical imaging of the biodistribution of nontargeted and anti-IGFBP7 sdAb targeted Gd-sULVs-Cy5.5 injected in mice bearing orthotopic glioblastoma tumors. (A) *In vivo* images of the head at various time points after intravenous injection of anti-IGFBP7 targeted (upper panels) and nontargeted (lower panels) Gd-sULVs-Cy5.5. (B) The average fluorescence concentration in the brain tumor region *in vivo* as a function of times after the injection. Data are expressed as mean ± SEM for n = 5 animals. ∗ indicates significant difference between the targeted Gd-sULV-Cy5.5 and nontargeted ones $(P < 0.01)$. (C) *In vivo* optical images of a whole animal body at 24 h after injection of targeted (left panel) and nontargeted (right panel) Gd-sULV-Cy5.5.

MRI of targeted Gd-sULV-Cy5.5 in the orthotopic U87MG.EGFRvIII brain tumor model in mice

Differences in intensity related to contrast accumulation or retention in the tissue were apparent in T1 weighted MRI images acquired at 9.4T. Images were acquired following injection of either anti-IGFBP7 sdAb-Gd-sULV (e.g. Fig. 4A, upper panel) or non-targeted

Gd-ULV (e.g. Fig. 4A, lower panel). In general, prior to contrast injection, the tumor was poorly visible in the T1 weighted scans. The distinction between brain and tumor became more evident after injection with the Gd-sULV labeled with anti-IGFBP7 but not the non-targeted Gd-sULV. A composite image obtained by subtracting the MRI scan obtained at 2hrs post injection from that prior to injection illustrates directly the contrast enhanced areas due to accumulation or retention of anti-IGFBP7 sdAb targeted-Gd-sULV (Fig 4A, right panels). Quantitative measurement of intensities in T1 weighted images for regions of tumor and comparable contralateral areas of normal brain (Fig 4B) confirm an increased contrast between brain and tumor at 2hr following injection of the anti-IGFBP7 sdAb labeled ULVs but not the non-targeted ones.

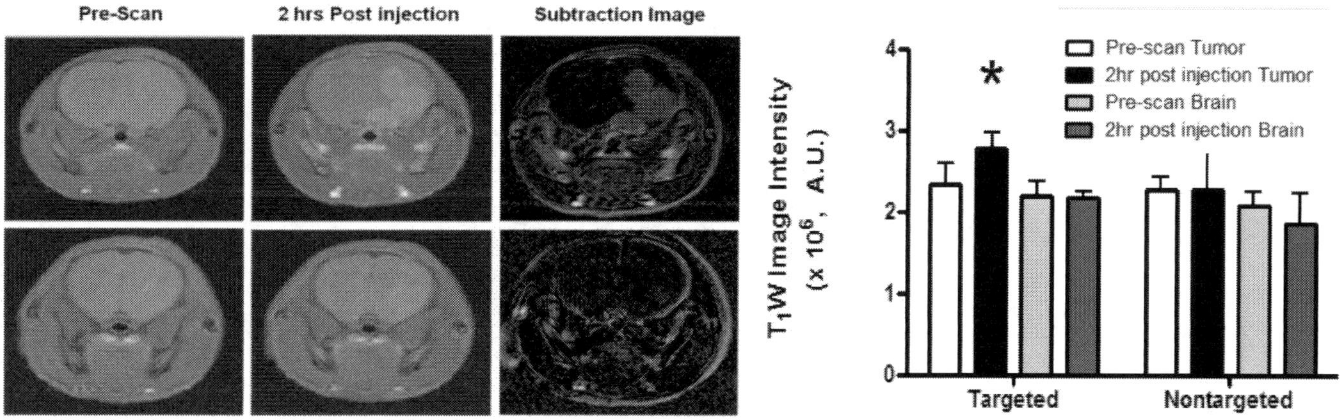

Figure 4. MRI of mice with brain tumors following injection of Gd loaded small unilamellar vesicles (Gd-sULVs). (A) T1 weighted (T1 w) images were acquired prior to and at 2 h following injection of either targeted anti-IGFBP7 sdAB labeled Gd-sULVs (upper panels) or non-targeted Gd-sULV's (lower panels). Subtraction images of the 2 h scan from pre-injection scan (A, right panels) show greater effects on T1 w intensities in tumor with targeted compared to non-targeted contrast injection. (B) Intensity measures in tumor and contralateral brain resulted in mean differences in intensity that differed depending on region and time following targeted contrast injection ($p < 0.03$, Two-way ANOVA; * $p < 0.001$, paired comparison of brain versus tumor at 2 h) whereas differences following non-targeted contrast injection were not significant ($p > 0.4$, Two-way ANOVA).

Conclusion

This study has shown that MRI sensitivity can be improved by increasing the targeting efficiency of Gd to the site of interest through the design of Gd loaded nano-sized sULVs conjugated with targeting molecules. These spontaneously forming sULVs have advantages over liposomes in their scalability, size and favorable pharmacokinetics. In this study, the IGFBP7 (tumor vessel specific), and to a lesser extent, EGFR (tumor cell specific) targets satisfy these requirements. Moreover, the bimodal nature of the Gd-sULV-Cy5.5 contrast agent affords the benefits of both optical and MR imaging. Optical imaging provides high sensitivity, while MRI provides high anatomical resolution. In this study, it was apparent that optical imaging gave rise to a larger increase in tumor signal compared to that achieved using the less sensitive MRI. However, MRI provided highly detailed anatomical information. Co-registration of optical and MRI signals would thus take advantage of the strengths of both these modalities. In this study, optical imaging allowed for rapid pre-clinical optimization of the Gd-ULVs pharmacokinetic and biodistribution parameters before assessment in the more expensive MR imaging. Also, the fluorescence of the Gd-sULV-Cy5.5 nanoparticle allows one to study the molecular localization of the contrast agent using fluorescent microscopy.

In conclusion, the targeted contrast agent, anti-IGFBP7 sdAb-Gd-sULV-Cy5.5 is an advanced paramagnetic lipid vesicles that is small in size leading to a more suitable pharmacokinetic profile for imaging, has high target avidity due to the presence of multiple small sdAbs on its surface, possesses bimodal optical-MRI capabilities and carries a high payload of Gd for improved MRI detection sensitivity. These features give promise to Gd-sULVs as a future MRI contrast agent in the clinic.

Acknowledgments

The authors acknowledge P Forsyth, X Lun and D Rushforth's contribution to producing the mice tumor model, and T Foniok and D Kirk for their assistance with the acquisition of MR images using these mice. We would also like to thank D Fatehi for her assistance with cell culture experiments and T Devesceri for his help with graphic and image processing. The authors acknowledge the support of NIST NCNR in providing the neutron research facilities used in this work. The SANS work made use of facilities supported in part by the National Science Foundation under Agreement No. DMR-0944772. MPN would like to thank for the financial support from the UConn Faculty Large Grant and the startup fund from the institute of materials science (IMS) at UConn.

References

1. De Schepper, A M and J. L. Bloem (2007),"Soft Tissue Tumors: Grading, Staging, And Tissue-Specific Diagnosis," *Top Magn. Reson. Imaging,* 18, pp. 431–433.
2. Glunde, K., A. P. Pathak and Z. M. Bhuiwalla (2007)," Molecular-Functional Imaging of Cancer: to Image and imaging," *Trends Mol. Med.,* 13, pp. 287–297.
3. Gupta, H. and R. Weissleder (1996),"Targeted Contrast Agents In MR Imaging," *Magn. Reson. Imaging Clin. N. Am.,* 4, pp. 171–184.
4. Glogard, C., G. Stensrud, R. Hovland, S. L. Fossheim and J. Klaveness (2002),"Liposomes as carriers of amphiphilic gadolinium chelates: the effect of membrane composition on incorporation efficacy and *in vitro* relaxivity," *Int. J. Pharm.,* 233, pp. 131–140.
5. Zhang, D., X. Y. Feng, T. D. Henning, L. Wen, W. Y. Lu, H. Pan, X. Wu and L. G. Zou (2009)," MR Imaging of Tumor Angiogenesis Using Sterically Stabilized Gd-DTPA Liposomes Targeted to CD105," *Eur. J. Radiol.,* 70, pp. 180–189.
6. Strijkers, G. J., E. Kluza, G. A. Van Tilborg, D. W. van der Schaft, A. W. Griffioen, W. J. Mulder and K. Nicolay (2010) "Paramagnetic And Fluorescent Liposomes For Target-Specific Imaging And Therapy Of Tumor Angiogenesis," *Angiogenesis,* 13, pp. 161–173.
7. Xu, R., Y. Wang, X. Wang, E. K. Jeong, D. L. Parker and Z. R. Lu (2007),"*In Vivo* Evaluation of a PAMAM-Cystamine-Gd-DO3A Conjugate as a Biodegradable Macromolecular MRI Contrast," Agent *Exp. Biol. Med.,* 232, pp. 1081–1089.
8. Nieh, M.-P., T. A. Harroun, V. A. Raghunathan, C. J. Glinka and J. Katsaras (2003), "Concentration Independent Spontaneously Forming Biomimetic Vesicles," *Phys. Rev. Lett.,* 91, 158105.
9. Nieh, M.-P., T. A. Harroun, V. A. Raghunathan, C. J. Glinka and J. Katsaras (2004), "Spontaneously Formed Monodisperse Biomimetic Unilamellar Vesicles: Effect Of Charge, Dilution and Time," *Biophys. J.,* 86, pp. 2615–2629.
10. Nieh, M.-P., V. A. Raghunathan, S. R. Kline, T. A. Harroun, C. Y. Huang, J. Pence and J. Katsaras (2005), "Spontaneously Formed Unilamellar Vesicles with Path Dependent Size Distribution," *Langmuir,* 21, pp. 6656–6661.
11. Nieh, M.-P., N. Kučerka and J. Katsaras (2009)," Spontaneously Formed Unilamellar Vesicles," *Methods in Enzymol., 465*, pp. 3-20.
12. Mahabir, S., W. K. Wan, J. Katsaras and M.-P. Nieh (2010),"The Effects of Charge Density and Thermal History on the Morphologies of Spontaneously Formed Unilamellar Vesicles," *J. Phys. Chem. B,* 114, pp. 5729–5735.

13. Yue, B., C. Y. Huang, M.-P. Nieh, C. J. Glinka and J. Katasaras (2005), "Highly Stable Phospholipid Unilamellar Vesicles from Spontaneous Vesiculation: a DLS And SANS Study," *J. Phys.Chem. B,* 109, pp. 609–616.
14. Nieh, M.-P., J. Katsaras and X. Qi (2008),"Controlled Release Mechanisms of Spontaneously Forming Unilamellar Vesicles," *Biochim. Biophys. Acta,* 1778, pp. 1467–1471.
15. Mamot, C., D. C. Drummond, C. O. Noble, V. Kallab, Z. Guo, K. Hong, D. B. Kirpotin and J. W. Park (2005), "Epidermal Growth Factor Receptor-Targeted Immunoliposomes Significantly Enhance the Efficacy of Multiple Anticancer Drugs *In Vivo,* " *Cancer Res.,*65, pp. 11631–11638.
16. Iqbal, U, U. Trojahn, H. Albaghdadi, J, Zhang, M. O'Connor-McCourt, D. Stanimirovic, B. Tomanek, G. Sutherland and A. Abulrob (2010),"Kinetic Analysis of Novel Mono- and Multivalent VHH-Fragments and Their Application for Molecular Imaging of Brain Tumors," *Br. J. Pharmacol.,* 160, pp. 1016–1028.
17. Iqbal, U, H. Albaghdadi, M.-P. Nieh, U. I. Tuor, Z. Mester, D. Stanimirovic, J. Katsaras and A. Abulrob (2011)," Small Unilamellar Vesicles: a Platform Technology for Molecular Imaging of Brain Tumors," *Nanotech.*, 22, 195102.
18. Abulrob, A., E. Brunette, J. Slinn, E. Baumann and D. Stanimirovic (2007), "*In Vivo* Time Domain Optical Imaging of Renal Ischemia-Reperfusion Injury: Discrimination Based on Fluorescence Lifetime," *Mol. Imaging,* 6, pp. 304–314.
19. Abulrob, A., E. Brunette, J. Slinn, E. Baumann and D. Stanimirovic (2008), "Dynamic Analysis of the Blood-Brain Barrier Disruption in Experimental Stroke Using Time Domain *In Vivo* Fluorescence Imaging" *Mol. Imaging,* 7, pp. 248–262.
20. Iqbal, U., H. Albaghdadi, Y. Luo, M. Arbabi, C. Desvaux, T. Veres, D. Stanimirovic and A. Abulrob (2010),"Molecular Imaging of Glioblastoma Multiforme Using Anti-Insulin-Like Growth Factor Binding Protein-7 Single Domain Antibodies," *Br. J. Cancer,* 103, pp. 1606–1616.

(439d) Tunable Non-Viral Gene Delivery Via Lbl Thin Films

Mary X. Wang[1], Raymond E. Samuel[1] and Paula Hammond[2], (1)Chemical Engineering, Massachusetts Institute of Technology, Cambridge, MA, (2)Chemical Engineering, MIT, Cambridge, MA

Multi-layered composite films fabricated of plasmid DNA and hydrolytically degradable polycation have considerable potential as a synthetic gene delivery system for localized non-viral transfection. Applications include various biomedical applications such as implant functionalization, specifically, as gene therapy in orthopedic implant-host bone integration. New bone formation can be promoted on the implant-bone interface if presented with appropriate growth factors. Bone formation and healing occurs through an extensive process of inflammation, reparation, formation, and remodeling. Development is controlled by various growth factors and can be promoted by transfection of wound sites with factors such as osteogenic BMPs. Since these factors play a role throughout the healing process, long-term gene delivery has great potential for promoting bone growth along different stages of reparation. We have developed polyelectrolyte multilayer (PEM) thin films whose initial and long-term release kinetics can be controlled via architecture modulation. These films consist of two distinct polymer sets. The first pair of polymers forms the baselayer of the film and is composed of cationic protamine sulfate and anionic sodium (4-polystyrene sulfonate) (SPS). Protamine sulfate is a natural polyamine that has been shown to support the adhesion, proliferation and differentiation of MC3T3-E1 pre-osteoblastic cells in vitro. These osteoconductive baselayers support the second polymer set fabricated from plasmid DNA and cationic Poly2, a poly-b-amino ester. Poly-b-amino esters are hydrolytically degradable and have been shown to complex with and condense DNA, resulting in improved receptor-mediated endocytosis. It was found that DNA release profiles can be changed by varying PEM film architecture (changing upper-layer number). Release profiles were found to be dual phase, with an initial burst release in the first week followed by linear release up through four weeks. Initial release kinetics within the first seven days were highly dependent on architecture. The ensuing linear release from 7 through 28 days was slow and relatively unchanged between film architectures. This has great potential for long-term gene therapy wherein specific architectures can be used for timed transfection, maximizing the effect of the therapy by allowing release to coincide with specific bone healing phases.

See more of this Session: Bionanotechnology for Gene and Drug Delivery I
See more of this Group/Topical: Nanoscale Science and Engineering Forum

(298f) Alignment Dynamics of Single-Walled Carbon Nanotubes In Pulsed Ultrahigh Magnetic Fields

A. Nicholas G. Parra-Vasquez[1], Jonah Shaver[2], Stefan Hansel[3], Oliver Portugall[3], Charles H. Mielke[4], Junichiro Kono[2], Robert H. Hauge[5] and Matteo Pasquali[1], (1)Chemical and Biomolecular Engineering, Rice University, Houston, TX, (2)Electrical and Computer Engineering, Rice University, Houston, TX, (3)Laboratoire National des Champs Magntiques Pulss, Toulouse 31400, France, (4)National High Magnetic Field Laboratory, Los Alamos, NM, (5)Richard E. Smalley Institute for Nanoscale Science and Technology, Rice University, Houston, TX

We have measured the dynamic alignment properties of single-walled carbon nanotube (SWNT) suspensions in pulsed high magnetic fields through linear dichroism spectroscopy. Millisecond-duration pulsed high magnetic fields up to 56 T as well as microsecond-duration pulsed ultrahigh magnetic fields up to 166 T were used. Because of their anisotropic magnetic properties, SWNTs align in an applied magnetic field, and because of their anisotropic optical properties, aligned SWNTs show linear dichroism. The characteristics of their overall alignment depend on several factors, including the viscosity and temperature of the suspending solvent, the degree of anisotropy of nanotube magnetic susceptibilities, the nanotube length distribution, the degree of nanotube bundling, and the strength and duration of the applied magnetic field. To explain our data, we have developed a theoretical model based on the Smoluchowski equation for rigid rods that accurately reproduces the salient features of the experimental data.

See more of this Session: Graphene and Carbon Nanotubes: Absorption and Transport Processes
See more of this Group/Topical: Nanoscale Science and Engineering Forum

(342f) Preparation of Anisotropic Silica Nanoparticles Via One-Dimensional Assembly of Presynthesized Spherical Seeds

Junzheng Wang, Ayae Sugawara-Narutaki, Atsushi Shimojima and Tatsuya Okubo, Department of Chemical System Engineering, The University of Tokyo, Tokyo, Japan

This presentation describes a facile solution process for the preparation of anisotropic silica nanoparticles (ASNPs).[1] ASNPs are prepared via one-dimensional assembly of spherical silica seeds (22 nm) in alcohol–water mixed media, followed by their in situ fixation and overgrowth with tetraethoxysilane (TEOS). Ethanol and $_{\mathrm{L}}$-arginine (Arg) are used to modify the dielectric constant and ionic strength of the reaction media, by which silica seed assembly is controlled through the adjustment of electrostatic interaction. Ethanol and Arg also serve as cosolvent and catalyst for hydrolysis and condensation of TEOS, respectively, which enables us to produce ASNPs in a simple one-pot process. In addition to ASNPs with wormlike structures, different kinds of NPs (bimodal spherical NPs, monodisperse spherical NPs, and spherical aggregates) have also been obtained by changing the concentrations of ethanol and Arg. The length, thickness, or both of ASNPs are controlled systematically by varying the concentrations of Arg, seed NPs, and TEOS. Other alcoholic cosolvents, such as methanol, 1-propanol, 2-propanol, and *t*-butanol, are also effective to give ASNPs when the dielectric constant of the alcohol–water mixed media is properly adjusted, showing the versatility of the present method.

References

(1) Wang, J. Z.; Sugawara, A.; Shimojima, A.; Okubo, T. *Langmuir*, **2010,** *26,* 18491–18498.

See more of this Session: Templated Assembly of Inorganic Nanomaterials II
See more of this Group/Topical: Nanoscale Science and Engineering Forum

(504a) Quantifying Interactions Between Drug Delivery Vehicles and Target Cells Using An Affinity-and Size-Tunable Model System

Bradley Harris, Maria Cekanova, Paul Dalhaimer

Drug delivery vehicles come in all shapes and sizes. We describe micron-length Cylindrically shaped vehicles that have high affinity for malignant B-lymphocytes Derived from mice. These long cylinders have unique flow properties in the vasculature of rodents, where they remain in the blood (and other areas) for up to One week. This presence in the blood, involving many passes through the spleen (and possibly lymph nodes), encouraged us to use the cylinders to help treat white blood cell cancers. We conjugated A20.1 peptide, which has nanomolar affinity for malignant B-lymphocytes, to the exteriors of the cylinders (A20.1---cylinders) using established poly(ethylene-oxide)-modifying chemistry. We loaded the A20.1-cylinders with the lipophilic membrane dye PKH26 and quantified the fluorescent signal of the cells after incubation with these cylinders. Malignant B-lymphocytes showed an A20.1 concentration-dependent increase in fluorescence. Splenocytes obtained from healthy mice have negligible fluorescent intensity when incubated with the PKH26-loaded A20.1-cylinders. Thus, our system is specific for target cells *in vitro*. Free A20.1 peptide can compete off the A20.1-cylinders, therefore the binding mechanism seems to be via the A20.1 receptor. Cylinders with lengths less than 2 microns or greater than 10 microns have the highest affinities for the cells. We will use this system to understand the interactions between drug delivery vehicles of varying lengths (sizes) and suspension cells.

(342c) Alignment and Orientation of ZnO Nanorod Assemblies

Stefan Schaefer, Martin Klaumuenzer, Michael Voigt and Wolfgang Peukert, Institute of Particle Technology, Friedrich-Alexander University of Erlangen Nuernberg, Erlangen, Germany

Nanoparticulate thin films are of high interest in various industries. In this contribution we focus on electronic applications, e.g. as semiconducting or conducting layer in field effect transistors (FETs), solar cells or displays[1, 2]. The properties of solution processed thin films are varying from highly porous to very densely packed nanoparticles according to their field of application. Despite the importance of these films, the mechanisms of film formation and assembly of nanoparticles is far from being understood. Guenther et al.[3, 4], Mahajeri et al. [5] and Reindl et al.[6] investigated dispersions and formation of thin films of spherical nanoparticles. Starting from these works we go one step forward to try to understand the basic concepts of tailoring nanoparticulate thin film properties which consist of anisotropic nanoparticles like nanorods. Therefore the control of the particle interactions, the drying behaviour of the dispersions and the wetting of the substrate is of high interest. Moreover, the structural evolution of the particulate network as a function of solid concentration and stability parameters is studied. ZnO nanorods produced by wet chemical synthesis with a mean diameter of about 90 nm in length and about 20 nm in width are dispersed in ethanolic and aqueous solutions. With steric stabilization mechanisms of short carbonic acids we produce stable dispersions against agglomeration and sedimentation which are investigated by dynamic light scattering and UV-vis spectroscopy. On patterned silica substrates with hydrophilic and hydrophobic areas droplets of these dispersions are deposited. The drying behaviour is investigated in ambient conditions and in controlled gas atmospheres. By monochromatic light interference we identify the thickness of the droplets in situ while drying occurs. The dried films are investigated by electron microscopy and profilometry. We show that the stability of the dispersion is strongly influencing self-assembly processes of the nanorods. With stabilized dispersions we produce oriented assemblies of nanorod structures in parallel and perpendicular to the substrate with lateral extension of several μm.

1. Marczak, R.; Werner, F.; Gnichwitz, J.-F.; Hirsch, A.; Guldi, D. M.; Peukert, W. Communication Via Electron and Energy Transfer between Zinc Oxide Nanoparticles and Organic Adsorbates. *The Journal of Physical Chemistry C* **2009,** *113* (11), 4669-4678.

2. Walther, S.; Schäfer, S.; Jank, M. P. M.; Thiem, H.; Peukert, W.; Frey, L.; Ryssel, H. Influence of Annealing Temperature and Measurement Ambient on Tfts Based on Gas Phase Synthesized Zno Nanoparticles. *Microelectronic Engineering 87* (11), 2312-2316.

3. Günther, L.; Peukert, W. Control of Coating Properties by Tailored Particle Interactions: Relation between Suspension Rheology and Film Structure. *Colloids and Surfaces A: Physicochemical and Engineering Aspects* **2003,** *225* (1-3), 49-61.

4. Günther, L.; Peukert, W. The Relevance of Interactions in Nanoparticle Systems – Application to Particulate Thin Films. *Particle & Particle Systems Characterization* **2002,** *19* (5), 312-320.

5. Mahajeri, M.; Voigt, M.; Klupp Taylor, R. N.; Reindl, A.; Peukert, W. Evaluation of the Film Formation and the Charge Transport Mechanism of Indium Tin Oxide Nanoparticle Films. *Thin Solid Films 518* (12), 3373-3381.

6. Reindl, A.; Mahajeri, M.; Hanft, J.; Peukert, W. The Influence of Dispersing and Stabilizing of Indium Tin Oxide Nanoparticles Upon the Characteristic Properties of Thin Films. *Thin Solid Films* **2009,** *517* (5), 1624-1629.

See more of this Session: Templated Assembly of Inorganic Nanomaterials II
See more of this Group/Topical: Nanoscale Science and Engineering Forum

(217b) A Novel Technique for In-Vivo Toxicological Characterization of Engineered Nanomaterials

Georgios A. Sotiriou[1], Edgar Diaz[2], Mark Long[2], John Godleski[2], Joseph Brain[2], Sotiris E. Pratsinis[1] and Philip Demokritou[2], (1)Department of Mechanical and Process Engineering, Particle Technology Laboratory, ETH Zurich, Zurich, Switzerland, (2)Department of Environmental Health, Harvard School of Public Health, Boston, MA

How do we balance the potential hazards from new and inadequately characterized engineered nanomaterials with the potential and power of nanotechnology? It has become apparent, that potential environmental health effects are critical factors in the likely success or failure of new nanotechnologies. The current nanotoxicology approach of evaluating the toxic properties of the large universe of existing and emerging ENMs at one material at the time, places a significant financial burden to an emerging industry and may slow down significantly the development of this emerging industry. Alternatively, a promising approach will be to examine families of engineered and rigorously characterized particles, and to study the role of such factors as particle size, composition, shape, charge and surface chemistry and their link to specific biological outcomes.

Here, a novel method is explored which is suitable for ENM *in-vivo* inhalation and *in-vitro* toxicological characterization studies. ENMs are produced continuously in the gas phase using industry relevant flame spray pyrolysis reactors, allowing their continuous transfer to inhalation chambers, without altering their state of agglomeration. Important properties of the generated aerosols (i.e. particle size, concentration, shape, state of agglomeration, surface chemistry) can be modified allowing for both *in-vitro* and *in-vivo* investigations of toxicity. The ability of the developed technique to generate a variety of industry relevant, property controlled exposure atmospheres for inhalation studies was systematically investigated. The suitability of the technique to characterize the toxicity of inhaled ENMs in intact animal models was also shown in an *in-vivo* study involving Sprague-Dawley rats, using freshly generated nano-iron oxide (Fe_2O_3) as a test aerosol. We demonstrated both pulmonary and systemic toxicity using a variety of toxicological assays under acute exposure conditions. The future use of this novel method will improve our understanding in terms of the health and safety of ENMs and help us to investigate further our central hypothesis that physical and chemical characteristics of ENMs determine their bioavailability, redistribution, and toxicity in the lungs.

See more of this Session: Environmental Health and Safety (EHS) Concerns of Nanomaterials
See more of this Group/Topical: Topical C: Environmental Aspects, Applications, and Implications of Nanomaterials and Nanotechnology

(172f) Formation of Thin Films of IZO and ITO Nanoparticles

Stefan Schaefer, Michael Voigt, Wolfgang Peukert and Mahdi Mahajeri, Institute of Particle Technology, Friedrich-Alexander University of Erlangen Nuernberg, Erlangen, Germany

Nanoparticulate thin films are of high interest in various industries. In this contribution we focus on electronic applications as conducting layers in solar cells or displays [1]. The properties of these solution processed films are varying from highly porous to very densely packed films according to their field of application. We present our work on transparent conductive oxide films of indium tin oxide nanoparticles (ITO) [2] which were prepared in the gas phase and indium zinc oxide nanoparticles (IZO) produced by wet chemical sol-gel synthesis. Both are dispersed in alcoholic solutions in order to deposit the particles by spin- and dip-coating on silica wafers and glass substrates. The stability of the dispersions is verified by either dynamic light scattering (DLS) or UV-vis spectroscopy. Analytical tools for investigation of morphology, surface and conductivity of the thin films were AFM, XRD, SEM, HRTEM, four-point probe measurements and impedance spectroscopy. The effects of the indium concentration in the IZO films on the morphology, the optical and electrical properties of the thin films were investigated. Moreover, the influence of the layer thickness and annealing temperature on the film properties are discussed. The ITO and IZO films deposited on glass substrates showed a high transparency in the visible range of at least 80% to 95%. With the ITO thin nanoparticulate films we achieved conductivities up to 120 Scm^{-1} simply by varying preparation conditions [3] and up to 196 Scm^{-1} with a treatment in forming gas [2], the latter being the best known value of ITO particles produced in the gas phase. IZO dispersions were stable over at least two months. With IZO nanoparticulate thin films we achieved conductivities up to 104 S/cm^{-1} by forming gas treatment of the films.

[1] R. Marczak, F. Werner, J.-F. Gnichwitz, A. Hirsch, D.M. Guldi, W. Peukert, Journal of Physical Chemistry C, 113 (2009) 4669-4678.

[2] M. Mahajeri, M. Voigt, R.N. Klupp Taylor, A. Reindl, W. Peukert, Thin Solid Films, 518 3373-3381.

[3] A. Reindl, M. Mahajeri, J. Hanft, W. Peukert, Thin Solid Films, 517 (2009) 1624-1629.

See more of this Session: Nanoelectronic Materials
See more of this Group/Topical: Nanoscale Science and Engineering Forum

(649d) A Fluorescent Single-Walled Carbon Nanotube °Chaperone Sensor- for Explosive and Pesticide Compounds

Daniel A. Heller[1], George W. Pratt[1], Jingqing Zhang[2], Nitish Nair[2], Adam J. Hansborough[1], Ardemis A. Boghossian[1], Nigel F. Reuel[1], Paul W. Barone[1] and Michael S. Strano[1], (1)Department of Chemical Engineering, Massachusetts Institute of Technology, Cambridge, MA, (2)Chemical Engineering, Massachusetts Institute of Technology, Cambridge, MA

Photoluminescent single walled carbon nanotubes (SWNT) transduce specific changes in peptide secondary structure, resulting in single-molecule detection of nitroaromatic compounds such as the explosive RDX and pesticides TFM and 2,4-dinitrophenol. Peptide-nanotube complexes report modulation of peptide conformation upon analyte binding via changes in SWNT photoluminescence wavelength. The fluorescence modulation is differentiable between analytes, resulting in compound identification via specific SWNT spectral fingerprint. A novel split-channel microscope constructed to image quantized spectral wavelength shifts in real-time, in response to nitroaromatic adsorption, results in single-molecule stochastic imaging of solvatochromic events. The indirect detection mechanism demonstrates that functionalization of the carbon nanotube surface can result in unique sites for molecular recognition, resolvable at the single molecule level.

See more of this Session: Graphene and Carbon Nanotubes: Applications
See more of this Group/Topical: Nanoscale Science and Engineering Forum

(246d) Photopolymerization In Lyotropic Liquid Crystal Templates for Improved Mechanical and Transport Properties

Bradley S. Forney and C. Allan Guymon, Chemical and Biochemical Engineering, University of Iowa, Iowa City, IA

A promising method of synthesizing polymers with useful property relationships utilizes self-assembling lyotropic liquid crystals (LLCs) as polymerization templates to direct polymer structure on the nanometer scale. Of particular importance is the potential to improve and tailor properties dependent on local order including mechanical strength and molecular transport by controlling polymer nanostructure. Herein, ordered nanostructures are generated in organic polymers through photopolymerization in LLC templates. Properties including solvent uptake, mechanical strength, solute release, and stimuli-response are significantly improved in polymer systems containing LLC nanostructure. These results could allow development of polymers with particular properties required for successful performance in a growing number of advanced biological and industrial applications.

See more of this Session: Nanoscale Structure In Polymers I
See more of this Group/Topical: Materials Engineering and Sciences Division

(586f) Improved Photocatalytic Activity Under Visible Light Irradiation of Nanosized-TiO2 Co-Doped with Vanadium and Nitrogen

Renuka Jaiswal[1], Rupali Dholam[1], Nainesh Patel[2], Antonio Miotello[2] and **Dushyant C. Kothari**[1], (1) Centre for Nanosciences & Nanotechnology, University of Mumbai,, Mumabi, India, (2)Physics, Universita degli Studi di Trento, Povo (Trento), Italy

Heterogeneous nanocrystalline titania (TiO_2) has been extensively investigated due to its notable photocatalytic capability for the industrial waste water treatment. However, an obvious problem hindering the application of TiO_2 is its poor response to visible light due to its wide bandgap. Herein, TiO_2 photo-catalyst is synthesized by co-doping vanadium and nitrogen by sol-gel method to sensitize TiO_2 in visible light. The concentrations of V and N were varied and optimized to enhance photo-catalytic activity. In order to increase the active surface area, the crystalline powder was ball milled to nanosize. The photo-catalysts were characterized by XRD, FT-IR, SEM, EDS, BET surface area, XPS, and UV–vis diffuse reflectance spectra. The co-doped TiO_2 nanocrystal has a narrower band gap than singly V-doped TiO_2 and singly N-doped TiO_2. The XPS detected vanadium (V^{4+} and V^{5+}) and nitrogen ions on the surface of photo-catalysts, but no diffractive peak of vanadium oxide or titanium nitride was observed in XRD spectra. High energy ball-milling was able to increase the surface area due to reduction in particle size below 50 nm as confirmed by SEM. The co-doped TiO_2 photo-catalyst showed significantly better photocatalytic activity for the degradation of organic dyes such as Rhodamine B and methylene blue under visible light irradiation ($\lambda > 420$ nm) as compared with V-doped TiO_2 and N-doped TiO_2. The incorporation of N and V in TiO_2 lattice induces the appearance of passivated shallow acceptor and donor levels near the valence and conduction band, respectively, causing narrowing of the band gap. In addition, these levels can also act as traps for photoexcited holes or electrons to reduce the recombination between photo-generated charges thus contributing to enhance the photocatalytic activity.

See more of this Session: Environmental Applications of Nanotechnology and Nanomaterials
See more of this Group/Topical: Topical C: Environmental Aspects, Applications, and Implications of Nanomaterials and Nanotechnology

(321f) Structures of Polyelectrolytes In Differently Charged Colloidal Solutions

Chongli Yuan, School of Chemical Engineering, Purdue University, West Lafayette, IN and Ian Smith, Chemical Engineerign, Purdue University, West Lafayette, IN

Polyelectrolytes are commonly used as a stabilizing agent in colloidal solutions, e.g., oil, lotion and etc. In addition to the electrostatic contribution, polyelectrolytes are capable of rearranging the nanoparticles into distinctive spatial organizations due to their unique conformational features, i.e., length, geometry and flexibility. This study explores the conformational changes of different types of model polyelectrolytes, e.g., flexible, semi-flexible and rigid polymers, when mixed with distinctively charged nanoparticles. The conformation and the dynamics of the model polymers will be characterized using fluorescence spectroscopy approach. This knowledge will help us to elucidate the interactions between polyelectrolytes and colloidal particles and lead to an unambiguous identification of an energetic term accounting for the effects of polyelectrolytes on colloidal suspensions.

See more of this Session: Nanoscale Structure In Polymers II
See more of this Group/Topical: Materials Engineering and Sciences Division

(599d) 3D Hybrid Nanospheres Via Assembly of CdTe Nanoparticles and Proteins

Sudhanshu Srivastava and Nicholas A. Kotov, Chemical Engineering, University of Michigan, Ann Arbor, MI

Architectures produced similar to the protein-protein or biomolecular interactions inspired from nature via synthetic processes and self-assembly of nanoparticles is a huge challenge. In this approach, the already known assemblies of positively charged CdTe NPs to create 2D nanosheets were employed for the insertion of native Cytochrome C (Cyt C) protein. Interestingly, positively charged CdTe nanoparticles (~4 nm) when assembled with Cyt C (~3nm size) protein under ambient conditions lead to form ~100 nm 3D spheres. Transmission electron microscopy (TEM) and scanning electron microscopy (SEM) were used to analyze the structure of the 3D spheres, which clearly indicated the formation of hybrid materials made from both the nanocomponents. The unique design supports the importance of protein conformation, which can dictate the morphology of the hybrid nanocomposites (CdTe nanoparticles/Cyt C) formed. The charge dipoles and other non-covalent interactions between Cyt C and nanoparticles were investigated and analyzed. The overall study enhances our understanding to integrate inorganic nanoparticles with proteins in a programmed manner to fabricate unique ordered hybrid nanocomposite materials.

See more of this Session: Nanostructured Biomimetic and Biohybrid Materials and Devices
See more of this Group/Topical: Nanoscale Science and Engineering Forum

(172g) Formation of Fullerene Superlattices by Interlayer Bonding In Twisted Bilayer Graphene

Andre R. Muniz and **Dimitrios Maroudas**, Chemical Engineering, University of Massachusetts Amherst, Amherst, MA

Graphene, a two-dimensional material consisting of one-atom-thick sheets of sp^2-hybridized carbon atoms arranged in a honeycomb lattice, possesses an outstanding and unique set of electronic and mechanical properties, which has potential to enable a broad range of technological applications. Chemical modification of graphene, such as hydrogenation, has been used to properly tailor graphene properties aiming at specific applications. Exposure of graphitic single- and multi-layered carbon nanomaterials to atomic hydrogen leads to formation of sp^3-hybridized C-H and C-C bonds, which alter the materials structure remarkably. In this presentation, based on a systematic computational analysis, we report a class of novel carbon nanostructures formed from rotationally misoriented graphene bilayers (twisted bilayer graphene), upon creation of interlayer covalent C-C bonds. These interlayer C-C bonds are formed due to chemisorption of H atoms onto the outer surfaces of the graphene layers for certain H coverages and hydrogenation patterns. The analysis is based on a combination of first-principles density functional theory (DFT) calculations and classical molecular-dynamics (MD) simulations. The DFT calculations are used for relaxing the atomic structure and determining the electronic band structure of the relaxed atomic configurations, while the MD simulations are based on a realistic reactive bond order potential and are used to relax nanostructures that require computational supercells much larger than those employed in the DFT calculations.

The carbon nanostructures that we have discovered consist of superlattices of caged fullerene-like configurations embedded within the graphene bilayer and are generated from pairs of graphene layers rotated by angles $q \sim 30°$ with respect to each other; in this limit, a zigzag- and an armchair-oriented pair of planes are aligned. We have relaxed the atomic structures and computed the corresponding electronic structures of these materials for both non-commensurate bilayers at $q = 30^o$ and commensurate bilayers at an angle q deviating slightly from the perfect value of 30°. For certain such structures, we show that the linear dispersion around the K point in the Brillouin zone (Dirac cones), characteristic of single-layer graphene and twisted bilayer graphene, is preserved in spite of the introduction of sp^3 bonds due to hydrogenation and interlayer C-C bonding. We also demonstrate that an entire class of similar caged structures of various sizes can be generated in an analogous manner.

Our findings have far reaching implications: 2-dimensional carbon nanostructures can be functionalized by formation of superstructures through embedding of 0-dimensional carbon structures to tailor their electronic properties without deviating from those of single-layer graphene. This embedding of fullerene-like configurations can be accomplished by chemical modification through hydrogenation to form certain patterns of chemisorbed H. We have found these structures to be less stable than their hydrogenated non-bonded counterparts at low pressure; under higher pressure, however, these structures are expected to become thermodynamically favorable. The structures proposed in this study may be related to the intermediate carbon phases observed in laser-induced shock wave loading and high-temperature shock compression experiments on graphitic samples, which were likely to exhibit a turbostratic stacking of planes.

See more of this Session: Nanoelectronic Materials
See more of this Group/Topical: Nanoscale Science and Engineering Forum

(266b) Invited Lecture: Bio-Inspired Design of Microporous Materials

Jeffrey D. Rimer, Chemical and Biomolecular Engineering, University of Houston, Houston, TX

Critical challenges in microporous materials design stem, in part, from an inability to *a priori* tune crystal habit (*i.e.* size, shape, and orientation) and a limited, often phenomenological, understanding of the relationships between crystal structure and function (*e.g.* catalytic activity). Given the prevalence of zeolites as industrial catalysts and their potential for emerging technologies, an approach to systematically engineer zeolite crystals with tunable properties can be economically viable. Moreover, fundamental studies that provide an improved molecular-level understanding of the underlying mechanism(s) of zeolite crystallization can be instrumental in the design and optimization of materials with superior properties for a variety of applications.

Tailoring anisotropic growth of zeolites is a ubiquitous design goal since suboptimal crystal morphology can often marginalize its utility in applications, such as catalysis and separations, by imposing internal mass transport limitations and/or limiting guest molecule access to the pores. Here we will discuss a method to selectively control the crystal habit and surface architecture of zeolites through a bio-inspired approach that mimics processes in biomineralization that mediate crystal growth and the self-assembly of hierarchically-structured materials [1]. We have employed this versatile approach to selectively tune pore surface area, internal diffusion pathlength, and surface topography of multiple zeolite framework types. The mechanism of this bio-inspired design will be discussed along with the results of colloidal and interfacial studies to characterize its effect on zeolite nucleation and growth, and the development of structure-function relationships to quantify the enhanced performance of zeolites in catalytic applications.

[1] Rimer et al., *Science* 330 (2010) 337-341.

See more of this Session: Templated Assembly of Inorganic Nanomaterials I
See more of this Group/Topical: Nanoscale Science and Engineering Forum

(586d) Visible-Light-Driven Photodegradation of Contaminants In Water Over Surface-Engineered BiOBr Semiconductor Micro/Nano-Structures

Zheng Jiang[1], Liang Kong[2], Tiancun Xiao[2] and Peter P. Edwards[2], (1)Jesus College, University of Oxford, Oxford, United Kingdom, (2)Department of Chemistry, Inorganic Chemistry Laboratory, University of Oxford, Oxford, United Kingdom

Visible-Light-Driven Photodegradation of Contaminants in Water over Surface-Engineered BiOBr Semiconductor Micro/Nano-Structures

Zheng Jiang[1,2], Liang Kong[1], Tiancun Xiao[1], Peter P. Edwards[1]

(1)Jesus College, University of Oxford, Oxford, OX1 3DW, U.K.; (2)Department of Chemistry, Inorganic Chemistry Laboratory, University of Oxford, Oxford, OX1 3QR, U.K.

Visible-Light-Driven Photocatalytic degradation of contaminants (PDC) over semiconductor photocatalysts represents an advanced, cost-efficient and promising water purification technology because it can utilize the large portion of energy in the free sun light[1,2]. However, the implementation of PDC technique is is still facing a great challenge over the effective and stable visible-light-responsive photocatalysts with desired structure and surface functionality. Extensive efforts have been made to enhance the adsorption of sunlight and depress the recombination of the photogenerated charges for maximum photocatalytic efficiency. Semiconductor BiOBr is an attractive photocatalyst with excellent Uv-Vis light illumination despite its adsorption of sunlight is limited by its relatively large bandgap[3,4]. This talk will focus on the strategies we developed to resolve the crucial issues over the BiOBr semiconductor by semiconductor surface-engineering. An efficient and controllable synthesis methodology has been developed to prepare BiOBr mico-/nano-structures with desired visible-light adsorption.

First, a rapid synthesis route was developed to prepare BiOBr nanoflowers with exposed highly reactive surfaces and it was further adopted to prepare heterojucntions composed of BiOBr and second semiconductors. The results reveal the surface-oriented BiOBr and its heterojunctions possess hierarchical architectures and ultrahigh activity under visible-light irradiation in photodegradation of several typical refractory model organic pollutants. The synergism effects of the enhanced surface absorption, enhanced light adsorption, mitigated light-scattering, and special hierarchical structures were unearthed through key characterizations. Second, visible-light-induced photodeposition was adopted to prepare the surface supported noble-metal/BiOBr photocatalysts and the effects of the NM on the BiOBr surface were explored. It was found that the surface NMs deteriorated the photocatalytic performance of the BiOBr photocatalysts, although the surface NMs significantly improve the stability of the BiOBr under long-term light exposure. The mechanism was proposed on basis of the surface engineering, and band structure of the multi-structures.

The correlation of the photocatalytic activity to its structure and surface will be discussed on basis of the detailed physicochemical characterization and theoretical analyses.

References:

1. Z Jiang, T Xiao,V Kuznetzov, P.P. Edwards, ***Phil. Trans. R. Soc. A***; 2010, 368; 3343-3364;

2. Z. Jiang, F. Yangb, N. Luo, B. T. T. Chu, D. Sun, H. Shia, T. Xiao, P. P. Edwards, ***Chem. Commun.***, 2008, 6372-6374

3. L. Kong, Z. Jiang, T. Xiao, L. Lu, M. O. Jones and P. P. Edwards, ***Chem. Commun.***, 2011, 47, 5512-5514

4. Z. Jiang, F. Yang, G. Yang, L. Kong, M. O. Jones, T. Xiao, P. P. Edwards, ***J PHOTOCHEM PHOTOBIOL A-CHEM***, 2010,212(1): 8-13.

See more of this Session: Environmental Applications of Nanotechnology and Nanomaterials
See more of this Group/Topical: Topical C: Environmental Aspects, Applications, and Implications of Nanomaterials and Nanotechnology

(298c) Determination of Co-Surfactant Coverage Density and Location for Density Gradient Ultracentrifugation Based Metallic-Semiconducting Separation of Carbon Nanotubes Using Analytical Ultracentrifugation

Vinayak Rastogi, National Institute of Standards and Technology, Gaithersburg, MD and Jeffrey A. Fagan, Polymers Division, National Institute of Standards and Technology, Gaithersburg, MD

We present an analytical ultracentrifugation (AUC) study to determine the density and location of an introduced co-surfactant, sodium-dodecylsulfate (SDS), mimicking the applied perturbation used in density gradient ultracentrifugation (DGU) based separation of metallic and semiconducting nanotubes. Previously isolated empty and water-filled carbon nanotubes[1] via a rate-zonal centrifugation were utilized to allow resolution of the co-surfactant effects. Analytical ultracentrifugation allows live spectroscopic observation of the sedimentation properties of nanotube and surfactant dispersions under the influence of high centrifugal field. Sedimentation velocity experiments were performed on the fractionated empty and water-filled nanotubes in sodium deoxycholate (Na-DOC) – SDS surfactant solutions in both H_2O and D_2O. The predictive ability of the AUC experiments were then tested with DGU experiments.

1. Fagan, J.A. et al. Separation of Empty and Water-Filled Single-Wall Carbon Nanotubes. *ACS Nano* (2011). doi:10.1021/nn200458t

See more of this Session: Graphene and Carbon Nanotubes: Absorption and Transport Processes
See more of this Group/Topical: Nanoscale Science and Engineering Forum

(609f) Vertically Aligned Silicon Radial p-n Junction Micropillar Array Solar Cells

Kane J. Miller, Chemical Engineering, University of Louisville, Louisville, KY, Kevin M. Walsh, Department of Electrical and Computer Engineering, University of Lousiville, Lousiville, KY and Xiao-an Fu, Chemical Engineering, University of Lousiville, Lousiville, KY

Solar cell applications using silicon have been extensively studied in the past. The theoretical limit of the efficiency of single-crystalline silicon solar cells is well known. Several methods for optimizing silicon solar cells and obtaining the maximum possible efficiency exist in the literature. Among these methods are changes in the geometry and configuration of the solar cells, such as introducing radial p-n junction solar cells.

Micro- and nanoscale radial p-n junction solar cells, in theory, result in increased solar cell efficiency compared to planar p-n junctions by reducing carrier collection length. The geometry of arrays of these radial p-n junctions also introduces internal reflection and light trapping effects, ultimately improving the photon absorption characteristics.

We have fabricated a first-pass vertically-aligned micropillar p-n junction solar cell array with a micropillar diameter of 2 μm, length of approximately 10 μm, and distance of 2 μm between adjacent pillars. These arrays were fabricated using conventional photolithography and deep reactive ion etching on a (100) n-type silicon wafer, followed by boron diffusion using a spin-on dopant. Initial measurements demonstrated an open-circuit voltage of 0.36 V, short-circuit current density of 24 mA/ cm^2, a fill factor of 0.6, and efficiency of 5.3%.

Nanopillar p-n junction solar cell arrays of varying pillar diameters and lengths, fabricated using the combination of electron beam lithography and deep reactive ion etching, will be discussed. The test results will be compared to planar silicon solar cells and to theoretical calculations. With various techniques, we expect to increase the efficiencies of our solar cells beyond 10%.

See more of this Session: Semiconductor Nanowires and Nanotubes for Next Generation Photovoltaics - From Fundamentals to Devices
See more of this Group/Topical: Topical 5: Nanomaterials for Energy Applications

(570e) Nanostructured Complexes Formed From DOPE and a Ferrocene-Containing Lipid Allow Redox-Based Control of Transfection In the Presence of Serum

John P. E. Muller[1], Burcu S. Aytar[1], Shinichi Hata[2], Hiro Takahashi[2], Yukishige Kondo[2], David M. Lynn[1] and Nicholas L. Abbott[1], (1)Department of Chemical and Biological Engineering, University of Wisconsin-Madison, Madison, WI, (2)Department of Industrial Chemistry, Tokyo University of Science, Tokyo, Japan

The ability to design nanostructured materials that deliver DNA to specific populations of cells at desired times, or to sub-populations of cells within a larger colony or tissue, would be broadly useful in a variety of contexts both *in vitro* (e.g., basic biological and biomedical research or tissue engineering) and *in vivo* (e.g., development of new gene-based therapies). Although a range of nanoscopic assemblies, including those formed from cationic lipids and polymers, have been reported to form complexes with DNA and enable the delivery of DNA to cells, the majority of these complexes are active from the time of their formation (and, thus, they do not readily permit control over the 'activation' or 'inactivation' in ways that provide methods for spatial and temporal control over delivery and/or internalization). In this presentation, we will report an advance in the design of nano-scale complexes of DNA and ferrocene-containing lipids that are stable in physiologically-relevant concentrations of serum and provide the basis of methods for redox-based control over the activation or inactivation of lipid/DNA assemblies in *extracellular* environments.

Past studies have established that the oxidation state of the ferrocene groups of the redox-active lipid bis (11-ferrocenylundecyl)dimethylammonium bromide (BFDMA) strongly influences the efficiency with which lipoplexes formed by BFDMA and DNA transfect cells. In this presentation, we will describe key findings that have emerged from an investigation of lipoplexes formed from mixtures of BFDMA, DNA, and dioleylphosphatidylethanolamine (DOPE, a zwitterionic lipid commonly used as a 'helper'-lipid for lipid-mediated DNA delivery). Whereas lipoplexes prepared from BFDMA and DNA exhibit low transfection efficiencies in the presence of serum, we have found that lipoplexes containing mixtures of BFDMA and DOPE expand the range of media compositions over which BFDMA can be used to control transfection efficiencies (e.g., from serum-free medium to media that contain physiologically-relevant concentrations of serum). Importantly, we also find the strong influence of the oxidation state of BFDMA on transfection efficiency observed for BFDMA/DNA lipoplexes in serum-free media to be maintained when this mixed lipid system is used in media containing up to 80% serum. Specifically, lipoplexes containing reduced BFDMA and DOPE promote high levels of cell transfection *in vitro* in serum-containing media, whereas lipoplexes formed using DOPE and oxidized BFDMA do not.

To provide fundamental insights into the influence of DOPE on the properties of lipoplexes formed from BFDMA and DOPE, we have also performed small angle X-ray scattering and small angle neutron scattering measurements. These measurements have revealed that incorporation of DOPE into lipoplexes containing reduced BFDMA leads to a change in the nanostructure of the lipoplex from multilamellar, L_{α}^{c} (BFDMA only), to inverse hexagonal, H_{II}^{c} (BFDMA and DOPE). Because lipoplexes with H_{II}^{c} nanostructures have been reported to fuse with cell membranes whereas L_{α}^{c} lipoplexes generally do not, we hypothesize that the change in nanostructure from L_{α}^{c} to H_{II}^{c} may underlie the high cell transfection efficiency of the BFDMA-DOPE/DNA lipoplexes that we have measured in the presence of serum. Overall, the results presented in this study suggest that lipoplexes formed from BFDMA and DOPE may offer the basis of approaches that permit active and external control of transfection of cells in the presence of serum via manipulation of the oxidation state of BFDMA.

See more of this Session: Bionanotechnology for Gene and Drug Delivery III
See more of this Group/Topical: Nanoscale Science and Engineering Forum

(532b) Life Cycle Energy Analysis and Midpoint Assessment of Multimegawatt Wind Turbines with Polymer Nanocomposite Blade Material

Laura A. Merugula[1], Bhavik R. Bakshi[1] and Vikas Khanna[2], (1)Lowrie Department of Chemical and Biomolecular Engineering, The Ohio State University, Columbus, OH, (2)Department of Civil and Environmental Engineering, University of Pittsburgh, Pittsburgh, PA

Life cycle analysis ofvapor-grown carbon nanofiber (VGCNF or CNF) manufacturing shows production ishighly energy-intensive with a cumulative energy demand that is influenced byfeedstock selection and dominated by processing energy needed to maintainreactor temperature for the catalytic pyrolysis of the hydrocarbon feed.However, at relatively low mass percentages, CNFs can reinforce polymercomposites yielding significant improvements in properties such as mode-1delamination resistance, vibrational damping, tensile strength, fatigueresistance and Young's modulus. A new method termed "pre-binding" has beendeveloped to reinforce polymer composites with CNFs at the interface of longfibers and resin. This method improves the manufacturability of large pieceswith CNF-reinforced polymer nanocomposites (PNCs) by overcoming processingissues inherent to previous methods of PNC fabrication. Materials scientistsare considering use of this emerging material for wind turbine blade productionto meet the demand for larger, more resilient utility-scale wind power plants.

Wind energy converters (WECs)on-line today are a relatively new technology with installations of multimegawattratings having been introduced to the modern market only in the last 15-20years. The largest 3-blade horizontal axis wind turbine (HAWT) rotor size inoperation is 126 meters in diameter with generator ratings from 5 to 7 MW whilethe growing market is dominated by new installations from 1 to 3 MW. Apparentlimits of fiberglass technology have been reached as rotor blades beyond about40 meters generally require some inclusion of carbon fiber to address stiffnessand edgewise fatigue, which become critical at this length. The less densecarbon fiber also leads to a positive deviation in trends of weight toincreasing size. Recent press releases communicate progress in R&D formeeting industrial-governmental collaborative goals for 20 MW plants possiblyby 2020 using available technologies with rotors expected to have diametersnear 200 meters.

In an even shorter timespan, numerous nanotechnology-enabled products have been introduced to theconsumer market with over 1,300 manufacturer-identifiedproducts populating The Project on Emerging Nanotechnologies consumer productinventory today. These products include automobile body parts with carbon nanotubes(CNTs) for enabling online electrostatic painting and a personal watercraftusing nanomaterials for reinforcement of the polymer composite for alightweight, stronger material.

While life cycle energyconsumption of CNF-reinforced PNCs for use in automobile body panels has beenevaluated, similar trends were not expected for life cycle energy in windturbines due to differences in operational phase energy consumption. Also,previous assessments of PNCs have not considered the method of pre-binding,which uses acetone as a vessel for dispersion and distribution of the CNFs ontoglass or carbon fiber matting.

Preliminary studies thatassumed replacement of polymer composite blade materials on a 2 MW and 5 MWplant with CNF pre-bound PNCs indicated a range of parameterization valuesunder which the energy return on investment (EROI) would not decrease. Thesestudies have been updated in a process-based life cycle assessment (LCA) toinclude more relevant process information and have been extended to a midpointanalysis using the CML and Traci methods to determine areas of focus foroptimization. An energy analysis was extended to a 5 MW WEC using cumulativeenergy demand and power generation figures from published LCA literature.

Impacts are reported perkilowatt-hour generation of electricity by the wind power plant and compared tothe base case, which is a 2 MW WEC in Middelgrund that is inventoried in the proprietaryEcoinvent database. In every feasible scenario, the EROI of wind power plants withthe PNC blade material remained competitive to thermoelectric power generation,and life cycle water consumption remained favorable. Where CNF production wasassumed to be based on benzene feedstock, energy analysis results were muchmore favorable than methane. Midpoint analysis, however, was assessed with methaneas the preferred hydrocarbon source for pyrolysis in nanofiber manufacture. Similartrends emerged under extension to the 5 MW system energy analysis.

Assuming a masscontribution of CNFs that ranges from 1 to 5 % by weight of each blade, theoverall mass of the nanomaterials is approximately 0.01-0.06% of the overallmass of the WEC system, yet there is direct correlation between the massfraction of the nanomaterials to nontrivial impacts per kWh compared to thebase case. Where studies have shown manufacturing of virgin nanomaterials tobe solvent-intensive, this study finds that the use phase of the nanomaterial,where CNFs are integrated into a product system, is highly solvent-intensive aswell. Current laboratory scale ratios of solvent spent to nanofibers dispersedare on the order of $5x10^2$ g solvent per g CNF. Maturation of thismaterial development must include a reduction of solvent use by optimization ofthe dispersion process and evaluation of solvent recovery systems for reusethat considers effects of nanomaterials in the operation. Photochemicaloxidation is a concern in terms of acetone emissions, the degree to which willbe dependent on recovery, reuse, and emissions controls. Also, the degree ofsolvent fresh feed required has an impact on the overall system's globalwarming potential and eutrophication potential due to the manufacturing phaseof the acetone.

Evaluating impacts of emergingtechnologies is complicated by the fast-pace of development and deployment,significant gaps in data and uncertainties about the future. Compilation oflife cycle data is unwieldy at best due to complex supply chains and regionalvariations in categories as energy supply and transportation. This is furthercomplicated by incomplete understanding of ecosystem service impacts, thoughwork is underway to quantify and compile ecosystem services for inclusion in LCA.System performance improvements will not occur with PNC integration inisolation of other technological changes within WEC design and operation, yet analysesas this are meant to give insight to areas of concern by identifying impactsand order of magnitude effects associated with this emerging material for theassumed use.

See more of this Session: Green and Renewable Energy Aspects of Nanotechnology
See more of this Group/Topical: Topical C: Environmental Aspects, Applications, and Implications of Nanomaterials and Nanotechnology

(462d) Effect of the Aggregation of TiO2 Nanoparticles On Their Fate and Transport In Natural Waters

Mara Marta Fidalgo[1], Marina B. Romanello[1], Liliana M. Bertini[1], Linna Du[2] and Jianhong Ren[2], (1) Department of Chemical Engineering, Instituto Tecnolgico de Buenos Aires, Buenos Aires, Argentina, (2)Department of Environmental Engineering, Texas A&M University Kingsville, Kingsville, TX

Nanotechnology is a growing area with many promising and revolutionary advances in a variety of fields, from medicine to sporting good design and personal care products. One of the most widely applied nanomaterial is nanoTiO_2. Although it is a well-known material at the macroscale, the environmental implications of the special properties exhibited by TiO_2 in its nanoform are not fully understood. Of particular interest are the aggregation processes in natural waters, the interaction with natural organic matter, and the effect of the presence of divalent ion such as calcium and magnesium. The different conditions of the water matrix that TiO_2 nanoparticles may encounter in surface and groundwater will influence their physico-chemical properties, and thus modify their surface charge, aggregation, and attachment characteristics. These properties are key to determine the fate and transport of nanomaterials in the environment, and therefore a complete understanding of the transformations that TiO_2 undergo in a natural setting is necessary to assess the risk posed to the environment by products containing this nanomaterial.

The size and surface charge of TiO_2 nanoparticles was investigated as a function of pH for a variety of relevant conditions, including different levels of ionic strength, concentration of natural organic matter and non-indiferent cations by Dynamic Light Scattering (DLS) using a Nanosizer Z90 (Malvern, UK) and by means of settling column experiments for those cases when DLS was not applicable. By size the authors refer to the measured equivalent hydrodynamic diameter of the aggregates formed by the nanoparticles. DLVO theory was applied to model the experimental conditions explored, and generally good agreement was observed when aggregate size is taken as effective particle size for the calculations.

Aggregate size and morphology were investigated by Scanning Electron Microscopy using a FESEM (JEOL JSM-6701F). To preserve the aggregates conditions as close as possible to their original form in suspension, a substrate was placed vertically in each suspension and the sample to be imaged was formed by aggregates that attached during the descent of the liquid level as the suspension evaporates. Special care was taken so that the change in concentration of the constituents in the solutions did not vary significantly during this process, taking into account the ratio of the relative change of volume to the initial volume of sample suspension and interrupting the process accordingly. Prior to imaging, the samples were gold coated with a sputter coater (Denton Vacuum Desk IV) to provide a conductive surface.

Sand column transport experiments were conducted for some characteristic conditions to investigate the mobility of nanoTiO_2 in porous media. The results were modeled applying filtration theory.

It was observed that aggregation is dependent on pH, shear (mixing conditions), presence of natural organic matter and divalent ions through their effect in the zeta potential of the particles although this parameter alone was insufficient to explain the different size of the aggregates. Attachment to the sand grains during flow in the porous media followed a similar trend.

This work provides critical insights on the physico-chemical properties of TiO_2 nanoparticles suspended in environmentally relevant aqueous matrixes. The information presented is key to understand the interaction with the natural environment: their fate and transport on groundwater and surface water, their bioavailability and their subsequent effect on the biota.

See more of this Session: Life Cycle Studies Associated with Nanomaterials and Nano-Enhanced Products
See more of this Group/Topical: Topical C: Environmental Aspects, Applications, and Implications of Nanomaterials and Nanotechnology

(504d) Conjugation of Anti-HER2 Monoclonal Antibody Onto PLGA-PEG Nanoparticles Using Click Chemistry

Joo-Youp Lee, Department of Chemical and Materials Engineering, University of Cincinnati, Cincinnati, OH and **Emily Smith**, Chemical Engineering, University of Cincinnati, Cincinnati, OH

Nanoparticles are known to passively target tumor tissue due to the Enhanced Permeability and Retention effect. Via this phenomenon the particles tend to naturally accumulate in the unhealthy tissue, it is why some cancer drugs have been formed into nano sized particles, allowing doctors to administer three times the dosage as compared to before. Actively targeted delivery continues to be a struggle. Antibodies, known as the “magic bullets”, have the most specific targeting abilities. They also are extremely difficult to work with as they tend to bind to the surface of particles in uncontrolled and unpredictable manners. Our intention is to create a biologically compatible polymer nanoparticle comprising of PLGA (poly lactic-co-glycolic acid) and PEG (poly ethylene glycol) in the 100-200 nm size range via emulsion method. PLGA is FDA approved for drug delivery purposes and PEG is widely used in a process called PEGylation to increase systemic circulation. An anti-HER2 monoclonal antibody will be selected for its ability to target an antigen overexpressed on the surface of most breast cancer cells and covalently attached to the nanoparticles via click chemistry. Click chemistry is a method that has been shown to be effective for this purpose with other types of nanoparticles and has shown improved efficacy in both ligand attachment and maintaining targeting activities as compared to other methods. The poly ethylene glycol will be modified to have an azide end group and the antibodies will have to be alkyne functionalized to facilitate the CuAAC, or Copper catalyzed azide-alkyne cycloaddition reaction. Final system targeting abilities will be tested via protein assay.

See more of this Session: Bionanotechnology for Gene and Drug Delivery II
See more of this Group/Topical: Nanoscale Science and Engineering Forum

FULLY ORGANIC ITO REPLACEMENT THROUGH ACID DOPING OF DOUBLE-WALLED CARBON NANOTUBE THIN FILM ASSEMBLIES

Yong Tae Park[a] and Jaime C. Grunlan[a,b]
[a]Department of Mechanical Engineering and [b]Department of Chemical Engineering
Texas A&M University, 3123 TAMU, College Station, TX, 77840 (USA)
E-mail: yong.t.park@gmail.com, jgrunlan@tamu.edu

Abstract

Transparent electrodes made from metal oxides, of which indium tin oxide (ITO) is most common, suffer from poor mechanical flexibility and electrochemical stability. Highly transparent and electrically conductive thin films based on double-walled carbon nanotubes (DWNTs) were assembled layer-by-layer (LbL) as a potential replacement for ITO. The alternate deposition of positively charged poly(diallyldimethyl ammonium chloride) [PDDA] and DWNTs, stabilized with negatively charged deoxycholate (DOC), exhibit linear film growth. A five bilayer (BL) assembly exhibits a sheet resistance of 309 Ω/sq, and visible light transmittance of 84%. In an effort to further reduce sheet resistance, these thin films were exposed to HNO_3 vapor. Sheet resistance of the 5 BL film was reduced to 104 Ω/sq (4200 S/cm conductivity, 22.9 nm thickness), with no change in transparency. Single-walled carbon nanotube assemblies show an even greater change when exposed to HNO_3 vapor, with a 5 BL assembly (%T > 85) decreasing from 1236 to 237 Ω/sq. The DWNT-based assemblies maintained their low sheet resistance after repeated bending and also showed excellent electrochemical stability relative to ITO. This work demonstrates the excellent optoelectronic performance, mechanical flexibility, and electrochemical stability of CNT-based assemblies, which are potentially useful as transparent electrodes for a variety of flexible electronics.

INTRODUCTION

Transparent electrodes continue to be studied because of increasing demand for optoelectronics, such as photodiodes for solar cells and electronic displays. Various materials have achieved a combination of conductivity and transparency to satisfy the needs of various optoelectronic applications.[1,2] Carbon nanotubes (CNTs) have been intensively studied because of their impressive electrical, thermal, mechanical, structural and chemical properties. Recently, CNT thin films have been investigated as electronic devices including sensors,[3] solar cells,[4] integrated circuits,[5] and transparent conductive thin films.[6-8] These CNT thin films show good optoelectronic performance, with values in the middle of the range of commercial ITO-coated PET, which typically has 50–200 Ω/sq sheet resistance and ~ 83% T at 550 nm wavelength. An LbL technique is examined here as a means of tailoring the optical transmittance and electrical conductivity of CNT-based thin films. This method produces thin films by harnessing electrostatic interactions through alternately exposing a substrate to positively and negatively-charged aqueous solutions at room temperature. In the present work, our goal is to accomplish CNT-based LbL thin films as one candidate of ITO replacements. The increase in electrical conductivity is believed to be a combination of acid anion doping and removal of insulating materials. This film is highly flexible, transparent, and electrochemically stable, making it a potential alternative to ITO. After 100 bending cycles, these films exhibit the same sheet resistance, but ITO-coated PET increases two orders of magnitude. Cyclic voltammetry was used to demonstrate the electrochemical stability of these transparent electrodes. This combination of optoelectronic performance, mechanical flexibility, and electrochemical stability make this a suitable candidate for a variety of applications requiring a transparent electrode.

Experimental section

Purified electric arc SWNTs (average 0.5–3 μm length and 1.4 nm diameter, $C \geq 90$ wt%), synthesized using Ni/Y catalysts, from Carbon Solutions, Inc. were used as a base of comparison for HiPCO DWNTs (+1 μm length and ~3.0 nm diameter, $C \geq 90$ wt%), purchased from Continental Carbon Nanotechnologies Inc., used in this study. PDDA and DOC (≥98%) were purchased from Sigma-Aldrich. Fuming sulfuric acid (H_2SO_4, 20% free SO_3 basis), fuming nitric acid (HNO_3, 99.5%), and fuming hydrochloric acid (HCl, 37%) were also used for acid treatment. All chemicals were used as received. PET (0.175 mm thick) and PS (0.125 mm thick) films were used as flexible substrates for CNT assemblies. A 100 Ω/sq ITO-coated PET sheet was purchased from Sigma-Aldrich and cut to size to measure the optical transmittance and the change in sheet resistance during bending cycles.

A cationic 0.25 wt% PDDA solution was prepared with 18.2 MΩ deionized water. 0.05 wt% CNTs were dissolved in deionized water containing 1 wt% DOC, followed by 10 W tip sonication for 20 min. Each cycle of LbL assembly consists of substrate immersion into an aqueous mixture, with rinsing and drying after each deposition, beginning with the cationic solution. Cycles were repeated to deposit the desired number of BLs. All films were stored in a dry box for a minimum of 12 hours prior to testing or acid treatment. A schematic of this LbL deposition process is shown in Scheme 1. For acid treatments, assembled CNT films were held in a saturated acid vapor environment. Three types of concentrated acids (fuming sulfuric, nitric, and hydrochloric) in petri dishes were maintained at 70 °C with a water bath. After a 2 to 30 min exposure to acid vapor, the CNT films were rinsed with deionized water and dried with filtered air.

Thickness of assemblies was determined by averaging the values of at least five different positions using a PHE-101 Ellipsometer equipped with a 632.8 nm laser. Absorbance and transmittance were measured between 300 and 850 nm with a USB2000 UV-vis spectrometer. All absorbance values in the text are for one-sided coating (*i.e.*, absorbance from slides coated on both sides was halved). Sheet resistance of nanotube assemblies on PET was measured with a Signatone Pro4 four-point probe system with 0.4 mm probe tip diameter and 1.0 mm tip spacing. The Fourier-transform infrared (FT-IR) spectra were obtained with an ALPHA FT-IR spectrometer. Scanning electron microscopy was performed with a Quanta 600 FE-SEM at an operating voltage of 10 kV. Energy dispersive X-ray (EDX) analysis was carried out using an Oxford EDX system attached to the FE-SEM. Thin film cross-sections were imaged with a JEOL 1200 EX TEM with an operating voltage of 100 kV. The specimens were sectioned down to ~90 nm and placed on 300 mesh nickel grids to dry prior to imaging. An Epsilon 851 electrochemical workstation was used for cyclic voltammetry. A CNT- or ITO-coated PET electrode, platinum wire, and an Ag/AgCl (3M KCl) electrode were used as the working, counter, and reference electrodes, respectively. These measurements were performed at ambient temperature (22±2 °C) in 0.1 M Na_2SO_4 solution that had been bubbled with N_2 gas for more than 20 min prior to measurement. Cyclic voltammetry was performed between two potential limits, −0.2 and 0.8 V, at a scan rate of 100 mV/s.

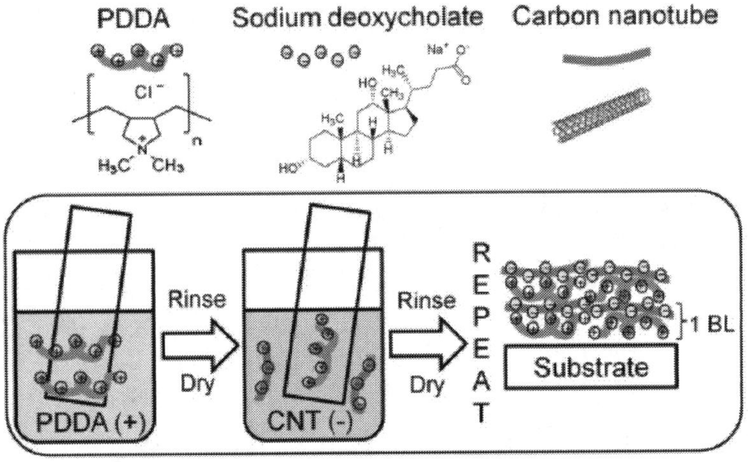

Scheme 1. Schematic of the LbL process. This recipe is represented by its own schematic: [PDDA/(CNT+DOC)]$_n$, where n is the number of cycles.

Results and discussion

Fig. 1 shows the linear growth of the following LbL assembly: cationic polyelectrolyte and anionic surfactant stabilized nanoparticles, with PDDA and DOC-stabilized DWNTs up to 10 BLs. Ellipsometric thickness of the CNT LbL thin film on a silicon wafer has the linear trend. The average thickness of one BL is 4.4 nm which is thicker than the DWNT thickness, 1.5−3 nm, suggesting that stabilizing surfactant and counterionic polymer molecules fully covered the DWNT network deposited at each BL step. Along with the thickness analysis, the graph of optical absorbance confirms that the CNT-based LbL assemblies of 10 BL grow in the same linear fashion as the graph of thickness does.

Also, Fig. 1 shows there is very little change in thickness and absorbance following acid vapor exposure. Previous studies indicate that swelling of CNT networks occurred with acid doping due to denser layers of acidic anions around the individual nanotubes.[9] In spite of this swelling effect, some CNT thin films have shown a significant decrease in thickness following acid doping.[6] The minimal reduction of thickness for the present LbL assemblies suggests a minimum amount of surfactant was associated with the DWNT during deposition. In some LbL systems, CNT assemblies with a polymeric stabilizer showed more than 30% thickness reduction after sulfuric acid treatment,[8] suggesting that these films contained excess polymer. No reduction of thickness for the LbL assembly indicates the minimum amount of surfactant associated with CNT during deposition.

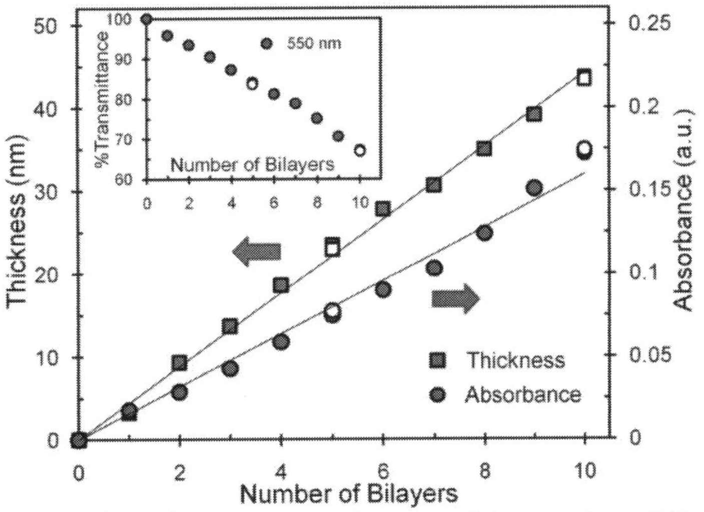

Fig. 1. Growth of PDDA/(DWNT+DOC) thin films, as a function of the number of BLs deposited. Transmittance was estimated from absorbance data. White ones with black maker lines are the values after 20 min nitric acid doping.

The morphology change in $[PDDA/(DWNT+DOC)]_6$ films before and after nitric acid treatment was analyzed using SEM, as shown in Fig. 2(A) and 2(B). DOC exfoliation of individual DWNTs in water results in a uniform distribution on the substrate.[9] Nanotubes, however, aggregate due to the high concentration in the deposited film, but they remain uniformly distributed, which creates a strong network that produces high electrical conductivity. Before treatment, the surface of the film was covered by a layer of insulators (polymer and surfactant) and perfectly exposed DWNTs could be hardly observed, while homogeneous DWNT networks were exposed due to removal of PDDA molecule after nitric acid treatment for 20 min. This removal of residual insulating materials enables direct contacts between conducting nanotubes. TEM cross-sectional micrographs (Fig. 2(C) and 2(D)) illustrate a highly inter-diffused nanostructure, in which the dark and light grey areas are DWNT-rich and PDDA-rich regions, respectively, and the darkest black dots are catalyst impurities from the HIPCO process. In addition to highlighting thin film structure, the TEM images verify the inner diameter of an individual DWNT (~3 nm) and the negligible change in the $[PDDA/(DWNT+DOC)]_{10}$ film thickness due to nitric acid treatment.

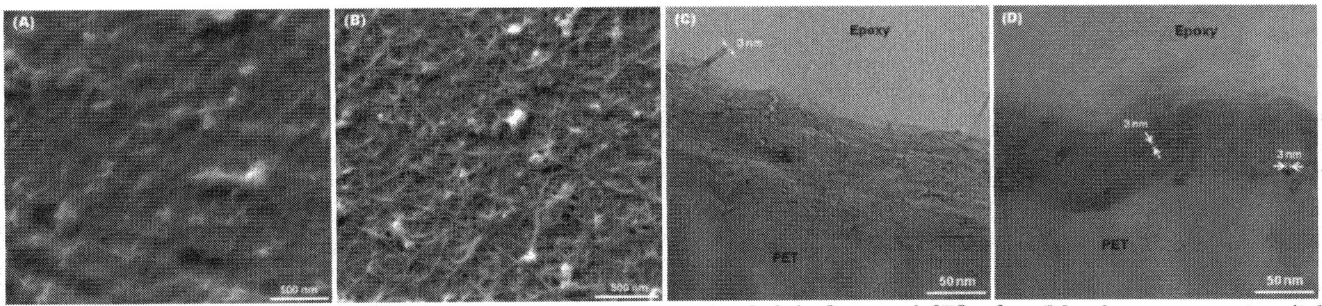

Fig. 2. SEM surface images of $[PDDA/(DWNT+DOC)]_6$ on PET **(A)** before and **(B)** after 20 min exposure to nitric acid vapor. TEM cross-sections of $[PDDA/(DWNT+DOC)]_{10}$ **(C)** before and **(D)** after 20 min treatment with nitric acid.

The sheet resistance and electrical conductivity of the [PDDA/(SWNT+DOC)]$_n$ and [PDDA/(SWNT+DOC)]$_n$ on PET substrates were measured as function of the optical transmittance according to the number of BLs deposited in Fig. 3. The sheet resistance of LbL films was shown to dramatically decrease with increasing number of BLs due to increased connectivity of CNT networks, along with gradual decrease in transmittance. A 10 BL DWNT thin film achieves sheet resistance of 127 Ω/sq with 67% T and 1180 Ω/sq with 94% T for just two BLs. The PDDA/(SWNT+DOC) system exhibits the same optoelectronic trend with higher sheet resistance and similar transmittance. A 5 BL DWNT-based film has the best optoelectronic performance of 309 Ω/sq with 84% T, but still higher sheet resistance than the requirement for ITO replacement.

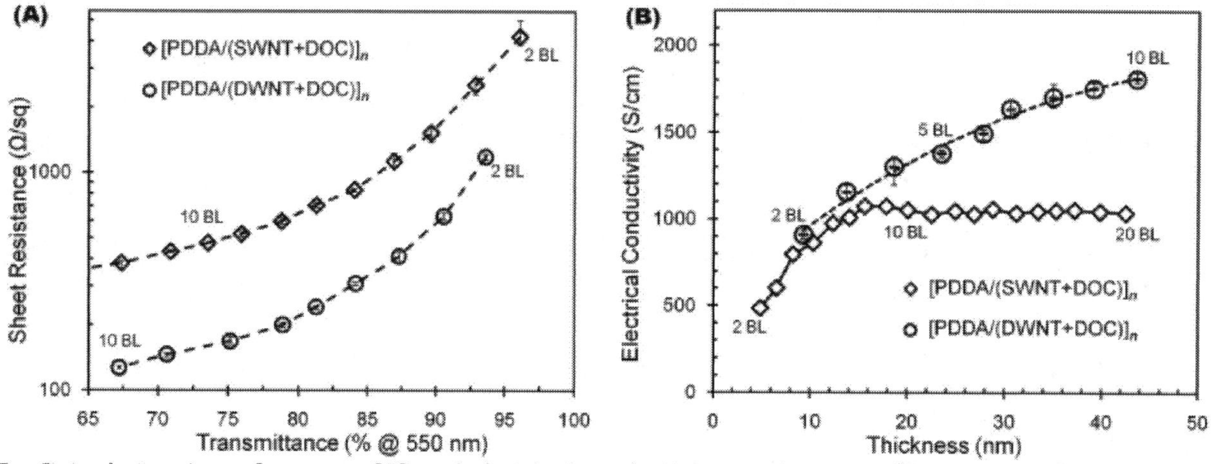

Fig. 3. Optoelectronic performance **(A)** and electrical conductivity vs. thickness **(B)** of both SWNT- and PDDA-base LbL thin films.

The sheet resistance of as-assembled films exhibit a higher value than the requirement for ITO replacement, due to the hindrance of the excessive polymer and surfactant to direct connections between CNTs. Strong acidic anions form a highly charge-transfer complex layer around individual CNTs and remove excessive polymer and surfactant.[6,9]

Most prior acid treatment studies have been performed with dip treatment (*i.e.*, immersing the films into a strong acid bath). Dip treating makes PET substrates brittle or even causes shrinkage. Moreover, dip treated CNT thin films readily de-adhere from substrates during rinsing. In the present study, films were exposed to acid vapor by heating the acid solution in a 70 °C water bath. Fig. 4 shows sheet resistance of DWNT assemblies after exposure to nitric, hydrochloric, or sulfuric acid vapor for increasing time periods. Nitric acid dramatically reduced the sheet resistance of as-assembled 5 BL PDDA/(DWNT + DOC) thin films, from 309 to 107 Ω/sq, after a 10 min exposure. Acid type plays a significant role in reducing sheet resistance. Nitric acid treatment provides the lowest resistance, while hydrochloric and sulfuric acids result in values that are nearly double that of HNO_3. This result suggests that PDDA and DOC molecules are more effectively removed by nitric acid vapor than the other two acids. The DWNT networks after each acid treatment were also characterized by SEM (not in this paper). 20 min nitric acid treatment exposes an extensive DWNT network, while more hazy DWNT networks are observed after 20 min hydrochloric and sulfuric acid treatments due to less removal of insulating molecules.

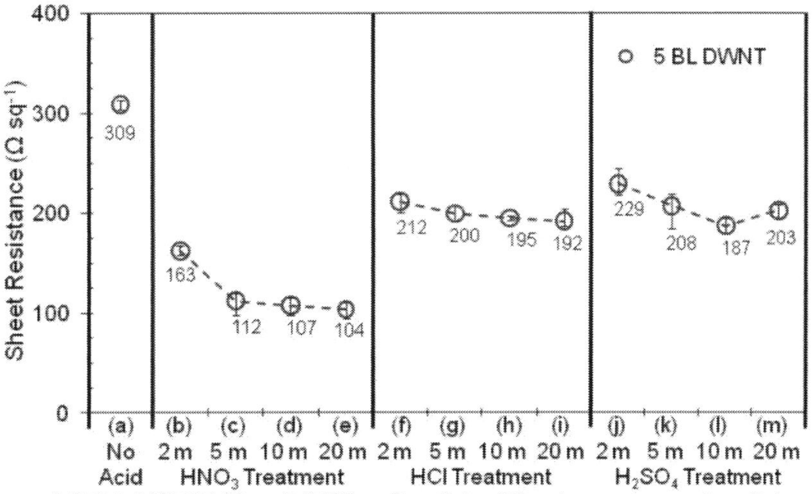

Fig. 4 Sheet resistance of [PDDA/(DWNT + DOC)]$_5$ after 2 to 20 min exposure to nitric acid, hydrochloric acid, and sulfuric acid. (a) is the sheet resistance of an as-assembled 5 BL DWNT thin film.

Fig. 5(A) shows the effect of nitric acid doping on CNT films according to the doping time for 2 to 20 min. The sheet resistance decreased significantly in the first few minutes when treated with 99.5% nitric acid. Upon the 2 min nitric acid doping to as-assembled CNT films, the sheet resistance of the 5 BL SWNT film exhibits an order of magnitude decrease, and that of the 5 BL DWNT film decreases moderately in half. However, further acid doping results in a very slight decrease, especially after 5 min. This may be because insulating materials on the surface of the film could be readily removed, while removal of DOC and PDDA inside the film took a longer time. The change in optoelectronic performances of several CNT-based LbL films due to acid doping is summarized in Fig. 5(B). [PDDA/(DWNT+DOC)]$_5$ films (107 Ω/sq and 84% T) had better optoelectronic performance than most of CNT-based films reported.[6-8]

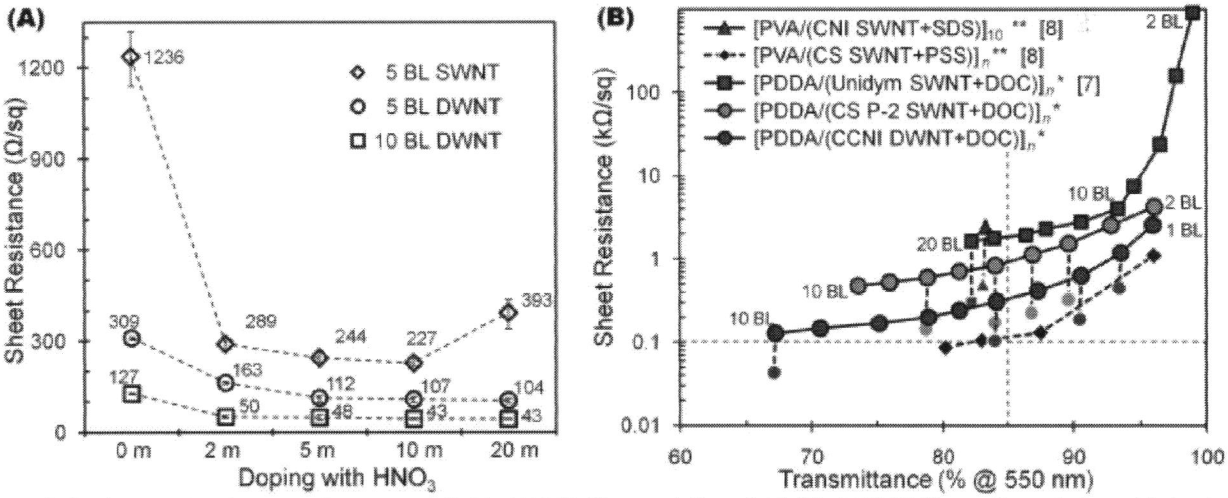

Fig. 5. (A) Change in sheet resistance of 5 BL SWNT film and 5 and 10 BL DWNT films after nitric acid doping for 2 to 20 min. **(B)** Optoelectronic performances of SWNT and DWNT LbL systems before and after nitric acid doping.

Recently, the bending performance of electrodes is very important to electronic applications. The bending property of CNT LbL films was compared to a commercial 100 Ω/sq ITO-coated PET sheet in Fig. 6(A). Their conductivity was maintained over 100 cycles of bending, while the conductivity of ITO coatings results permanent loss in conductivity. This outstanding feature of LbL films can be potentially used as a flexible electronics. Cyclic voltammograms (CV) of an HNO_3-treated 5 BL DWNT film and an ITO film (both on PET) were used to compare electrochemical stability, as shown in Fig. 6(B). The potential was cycled between -0.2 and 0.8 V vs. an Ag/AgCl reference electrode, with a scan rate of 100 mV/s, in 0.1 M Na_2SO_4 aqueous solution. ITO shows oxidation–reduction (redox) peaks at 0.1 and 0.3 V, respectively, indicating that ITO undergoes chemical changes under potentiodynamic stresses. A relatively large peak-to-peak separation of 0.2 V (between 0.1 and 0.3 V) is due to this transformation (irreversible change) that occurs in the ITO film during electrochemical cycling. The DWNT LbL films exhibit no redox peaks, suggesting they are electrochemically stable in this potential range.

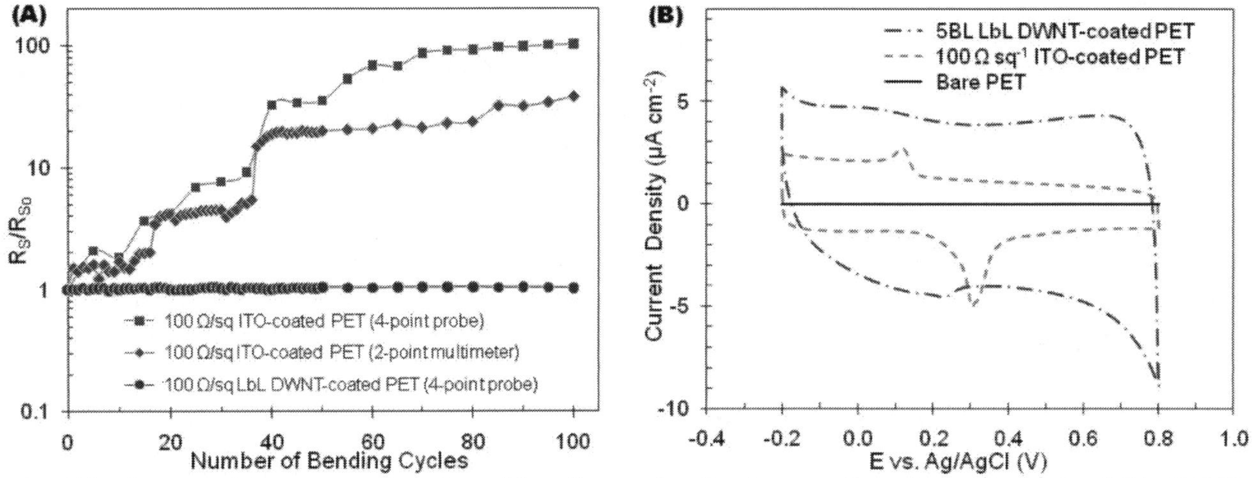

Fig. 6. (A) Sheet resistance as a function of bending cycles (to 1 cm radius of curvature) for 5 BL DWNT and 100 Ω/sq ITO-coated PET films. R_{So} indicates the resistance (~100 Ω/sq) prior to any bending. **(B)** Cyclic voltammograms of a $[PDDA/(DWNT + DOC)]_5$ thin film on PET and a commercial 100 Ω/sq ITO-coated PET.

CONCLUSIONS

Highly transparent and conductive thin films were prepared using LbL assembly with CNTs, DOC as a stabilizer and PDDA as polycations. This LbL adsorption of both species has linear increase independent on number of bilayers deposited. The self assembly of DOC-stabilized CNTs and PDDA was shown to realize the thinner and smoother films. Nitric acid doping was also treated for higher electrical conductivity. The optoelectronic performance of 5 BL DWNT LbL film here is much better than most other CNT thin films and capable of the ITO replacement. Additionally, the bending performance and electrochemical stability could suggest that these films potentially be used in a variety of electronic applications.

REFERENCES

1. Granqvist, C. G.; Hultaker, A. (2002) *Thin Solid Films*, 411, pp. 1-5.

2. Hu, L.; Hecht, D. S.; Gruuner, G. (2010) *Chem. Rev.*, 110, pp. 5790-5844.
3. Modi, A.; Koratkar, N.; Lass, E.; Wei, B. Q.; Ajayan, P. M. (2003) *Nature*, 424, pp. 171-174.
4. Lowman, G. M.; Hammond, P. T. (2005) *Small*, 1, pp. 1070-1073.
5. Friedman, R. S.; McAlpine, M. C.; Ricketts, D. S.; Ham, D.; Lieber, C. M. (2005) *Nature*, 434, pp. 1085.
6. Geng, H. Z.; Kim, K. K.; So, K. P.; Lee, Y. S.; Chang, Y.; Lee, Y. H. (2007) *J. Am. Chem. Soc.*, 129, pp. 7758-7759.
7. Park, Y. T.; Ham, A. Y.; Grunlan, J. C. (2011) *J. Mater. Chem.*, 21, pp. 363-368.
8. Shim, B. S.; Zhu, J.; Jan, E.; Critchley, K.; Kotov, N. A. (2010) *ACS Nano*, 4, pp. 3725-3734.
9. Ericson, L. M.; etc. (2004) *Science*, 305, pp. 1447-1450.
10. Wenseleers, W.; Vlasov, I. I.; Goovaerts, E.; Obraztsova, E. D.; Lobach, A. S.; Bouwen, A. (2004) *Adv. Funct. Mater.*, 14, pp. 1105-1112.

(570b) Rational Design of Pathogen Mimicking Amphiphilic Nanoparticle Adjuvants

Latrisha Petersen[1], Yashdeep Phanse[2], Amanda Ramer-Tait[2], Chang Sun Kong[3], Scott Broderick[3], Bret D. Ulery[1], Krishna Rajan[3], Bryan H. Bellaire[2], Michael J. Wannemuehler[2] and Balaji Narasimhan[1], (1)Chemical and Biological Engineering, Iowa State University, Ames, IA, (2)Veterinary Microbiology and Preventive Medicine, Iowa State University, Ames, IA, (3)Materials Science and Engineering, Iowa State University, Ames, IA

The natural infection of a pathogen stimulates a protective and enduring immune response. Thus, an ideal approach to vaccine design is to mimic the immune response associated with a natural infection while avoiding the undesirable side effects of the disease. Vaccines have proven to be the most effective method of preventing infectious diseases, yet many problems still exist including multi-dose administration, poor immunogenicity of vaccine antigens, and undesirable side effects of immunization with adjuvants which can all lead to poor patience compliance and poor vaccine efficacy. Thus there is a need for the development of new vaccine adjuvants and delivery vehicles capable of mimicking the properties of the natural infection while limiting the undesirable side effects and need for repeated administrations.

Polyanhydrides are a class of biomaterials that have demonstrated biocompatibility, controlled antigen release kinetics, antigen stabilization, and immune modulation, making them excellent vaccine delivery vehicles. Furthermore, they can be fabricated into nanoparticles capable of being inhaled or injected for ease of administration. This work is focused on polyanhydride copolymers based on sebacic acid (SA), 1,8-bis(*p*-carboxyphenoxy)-3,6-dioxaoctane (CPTEG) and 1,6-bis(*p*-carboxyphenoxy) hexane (CPH). While many chemistry-dependent trends associated with protein stabilization, protein release, phagocytosis, and immune modulation have been identified, little work to date has investigated the specific polymer characteristics responsible for these trends. By identifying these properties, novel adjuvants can be designed that more closely mimic infections with live pathogens. Therefore, this research is focused on the rational design of pathogen-mimicking adjuvants for vaccine delivery by utilizing flow cytometry and confocal microscopy in concert with materials informatics to identify key properties responsible for nanoparticle uptake and downstream activation of murine bone marrow derived dendritic cells (BMDCs). These internalization patterns were compared to that of live pathogens (*Yersinia pestis* and *Escherichia coli*) and the activation patterns compared to that of lipopolysaccharide (LPS, an endotoxin found on the surface of gram negative bacteria).

This investigation identified that the least hydrophilic (CPTEG-rich and the SA-rich) nanoparticles were readily internalized by BMDCs. Upon further assessment of the nanoparticle positive and negative BMDC populations, it was discovered that nanoparticle internalization was required for enhanced expression of CD40 and production of IL-12p40 and IL-6 but not for the expression of CD86 and MHC II. Interestingly, bacteria have been reported to have very similar internalization patters in which association is required for production of IL-12p40 and IL-6 and expression of CD40 but less dependent for expression of CD86 and MHC II. Additional flow cytometric analysis of the most adjuvanting nanoparticle chemistries (poly(SA) and 50:50 CPTEG:CPH) and the positive control (LPS) revealed stark similarities of double positive DC populations for cytokines and cell markers between the amphiphilic 50:50 CPTEG:CPH nanoparticles and LPS. Further pathogen mimicking patterns were observed with confocal microscopy in which internalization and persistence were compared between the bacteria, *E. coli* and *Y. pestis*, and the 50:50 CPTEG:CPH nanoparticles. Microscopic observations at both 2 and 48 h indicated that the nanoparticles were found inside late endosomes, consistent with phagosome–lysosome fusion events of antigen processing and presentation. When compared to the internalization of *Y. Pestis* and *E. coli*, the amphiphilic 50:50 CPTEG:CPH nanoparticles appeared to be similarly internalized and persist at 2 and 48 h, respectively, suggesting that the stimuli provided by these particles to activate DCs do not rapidly degrade, much like that provided by persisting bacteria.

To validate and assess these multiple pathogen mimicking observations, materials informatics analysis, including principal component analysis (PCA) and partial least squares (PLS), were employed. The PCA confirmed that 50:50 CPTEG:CPH nanoparticles were very similar in the overall activation, internalization, and persistence patterns to the pathogenic correlates. Furthermore, a PLS analysis was employed, which identified the polymer properties responsible for these pathogen mimicking patterns to be i) percentage of hydroxyl end groups, ii) hydrophobicity, iii) alkyl ethers, iv) aliphatic carbons, and v) backbone oxygen moieties. The approach described here will facilitate the rational design of single-dose, pathogen-mimicking vaccine adjuvants that could be tailored to provide immune modulation and protection against a wide range of emerging and re-emerging diseases.

See more of this Session: Bionanotechnology for Gene and Drug Delivery III
See more of this Group/Topical: Nanoscale Science and Engineering Forum

(395f) Controlling TiO_2 Nanoparticle Distribution within a Coating Film for Surface Mechanical Property Study

Rohan Uttarwar[1], Sunxi Wang[2], Guangzhao Mao[2] and Yinlun Huang[2], (1)Department of Chemical Engineering and Materials Science, Wayne State University, Detroit, MI, (2)Chemical Engineering and Materials Science, Wayne State University, Detroit, MI

Experimental studies have shown that nanocomposite coatings may show superior mechanical and barrier properties and new functionalities, such as self cleaning, self healing, super scratch resistance, super corrosion resistance. These smart functional coatings should have great application potentials in the automotive, semiconductor, defense, construction, aerospace, and other industries. In nanopaint design, a certain type of functional or non-functional nanoparticles needs to be dispersed in a coating matrix; a significantly increased surface area of these fillers at nanosize is the main reason behind improved mechanical and barrier properties of nanocoating films. It is known that layer-by-layer assembly of nanoparticles can be useful in incorporating self-cleaning phenomenon in coating thin films. [1],[2] The finding justifies the importance of nanoparticle size and its depth inside coating thin films to obtain superior performance properties. Hence, understanding and developing a correlation among them is critical to optimize nanocoating's performance properties.

In developing nanocoatings, the use of water-borne composites such as acrylics is increasing tremendously due to growing environmental concerns with solvent borne coatings. Some nanoparticles like titanium dioxide (TiO_2) can help in introducing smart functionality into coatings as they behave as oxidizing agents in the matrix. But formulating these nanocoatings with uniform dispersion without aggregation of nanoparticles is still a challenge.

This paper reports formation of nanocoating thin films of polymethyl methacrylate (PMMA) resin with dispersed TiO_2 nanoparticles. Thin films are prepared by sandwiching a nanoparticle film with known thickness and size of nanoparticles between two polymer nano-layers with varying thicknesses. The multilayer coat is characterized for its surface scratch resistance and hydrophilicity. A spin coating technique is employed to obtain all nanofilms of polymer and nanoparticles. A uniform dispersion of nanoparticles in selected solvent chloroform is achieved by modifying the surface of TiO_2.[3] The aggregation of particles on applied film is avoided by optimizing the TiO_2 solution concentration and spin coating parameters. The film thickness and surface smoothness are measured using X-ray diffraction studies of all polymer samples. A correlation between polymer solution concentration and nano-film thickness is then obtained. AFM images of TiO_2 films show uniform distribution of particles over the substrate. Various samples with nanoparticle film depth varying from substrate to top of the surface of the coating film are prepared by keeping the total coating film thickness constant at about 200nm. Surface mechanical properties are characterized by nanoindentation using round colloidal probe attached to AFM cantilever. The mechanical properties are found to be strongly affected by the depth of nanoparticles within the film and a correlation between them is obtained. The surface hydrophilicity is compared by measuring the change in contact angle of nanocoating films. The depth at which there is a maximum decrease in the contact angle due to TiO_2 incorporation is recorded. The experimental results are discussed in terms of betterment of nanocoating film performance properties by controlling the distribution of nanoparticles inside the film.

References

1. Wu Z, Lee D, Rubner MF, Cohen RE, Structural color in porous, superhydrophilic, and self-cleaning SiO_2/TiO_2 Bragg stacks, Small, 2008, 3(8), 1445-1451.
2. Zhang X, Sato O, Taguchi M, Einaga Y, Murakami T, Fujishima A, Self-Cleaning Particle Coating with Antireflection Properties, Chemistry of Materials, 2005, 17(3), 696-700.
3. Nakayama N, Hayashi T, Preparation of TiO_2 nanoparticles surface-modified by both carboxylic acid and amine: Dispersibility and stabilization in organic solvents, Colloids and Surfaces A, 2008, 317, 543-550.

See more of this Session: Nanoscale Structure In Polymers III
See more of this Group/Topical: Materials Engineering and Sciences Division

(744g) Novel High Surface-Area - Magnetoresponsive-Nano-Biocarriers for Efficient Saccharification of Biomass

Ankush A. Gokhale, Chemical Engineering & Materials Science, Michigan State University, East Lansing, MI and Ilsoon Lee, Chemical Engineering and Materials Science, Michigan State University, East Lansing, MI

Immobilization of enzymes on supports helps to improve the overall operational stability of enzymes by increasing their resilience towards changes in external factors like temperature and pH. Immobilization also facilitates easy separation, providing a facile route to recycle and reuse the enzyme, thus reducing the overall cost. The technique of alternate deposition of anionic and cationic polyelectrolytes has been used in the past, especially in the field of biosensors, to entrap metal nanoparticles and enzymes within the multilayer assembly. In our present work, we extend the application of the above technique to immobilization of cellulase on novel high surface area magnetic nanoparticle-modified-exfoliated graphite nanoplatelets. This work comprises of two phases. In the first phase, maghemite-magnetite decorated exfoliated graphite nanoplatelets (xGnP-MNP) were developed using polyelectrolyte based layer-by-layer assembly. In the next phase, we attach Accellerase-1000, a mixture of endoglucanase and β-glucosidase, on xGnP-MNP using a combination of weak polyacid brushes and spacer molecules. Weak polyacid brushes tend to show swelling behavior when the pH is increased beyond the pKa value. The degree of polyacid swelling can be easily tweaked by controlling the pH, thus providing us an effective handle to control the activity of immobilized enzymes. The effect of temperature variation on the activity of immobilized enzymes will also be shown. Facile separation of cellulase immobilized on maghemite-magnetite doped xGnP is possible by using small bar magnets. The activity of immobilized enzymes after repeated use would be shown. Cellulase activity is evaluated by using both soluble as well as insoluble substrates like 2% w/v carboxymethylcellulose (CMC) and avicel respectively.

See more of this Session: Nanoscale Science and Engineering In Biomolecular Catalysis II
See more of this Group/Topical: Nanoscale Science and Engineering Forum

(439b) Using Gold Nanoshells to Enhance Electroporation and Uptake Through the Cell Membrane

Alisha D. Peterson, Mark J. Jaroszeski and Vinay Gupta, Chemical and Biomedical Engineering, University of South Florida, Tampa, FL

Using Gold Nanoshells to Enhance Electroporation and Uptake Through the Cell Membrane

Alisha D. Peterson, Mark J. Jaroszeski, and Vinay K. Gupta

Department of Chemical & Biomedical Engineering, University of South Florida

Electroporation is a technique in which an external electric field is applied across a cell membrane. The applied field causes a disruption in the membrane creating pores, which ultimately allows for an increase in uptake of the surrounding media by the cell. The technique of electroporation has been studied for various applications including gene delivery and drug delivery. However, a continuing challenge is to enhance the delivery rates as well as the types of therapeutic agents that need to be delivered. In this context, gold nanoshells can form the basis of enhancing electroporation of the cell membrane. Gold nanoshells are nanoscale particles composed of a silica core and a thin gold shell. These particles are of great interest due to their unique and tunable optical properties, large surface area, and ability to heat locally. Nanoshells are being currently studied for applications in tumor imaging, drug delivery, and tumor ablation. Our research focuses on combining the conductive properties of gold nanoshells with the electroporation technique to enhance cellular uptake of therapeutic agents. The goal is to amplify the effects of the electric field on the cell membrane in order to increase permeability. Concurrently, the nanoshells can also serve as the carriers for delivery of the therapeutic agents by attach the biomolecules to the nanoshell surface. In this presentation we will present results from two phases of our research. First, the synthesis and characterization of gold nanoshells will be discussed. Secondly, the use of these nanoshells in electroporation of fibroblast cells (a model system) will be discussed. The influence of gold nanoshells on *in-vitro* cellular uptake of Sytox Green, an amino acid stain, will be shown. Results from the interplay of parameters such as voltage amplitude, number of electric pulses, pulse length, and ratio of particles to cells concentration will be presented.

See more of this Session: Bionanotechnology for Gene and Drug Delivery I
See more of this Group/Topical: Nanoscale Science and Engineering Forum

(600f) Zeolite Thin Films Prepared From Exfoliated MFI and MWW Nanosheets On Non-Porous Substrates

Christopher M. Lew, Kumar Varoon, Xueyi Zhang, Bahman Elyassi and Michael Tsapatsis, Chemical Engineering & Materials Science, University of Minnesota, Minneapolis, MN

We report on the synthesis of MFI- and MWW-type zeolite thin films prepared on non-porous substrates. Plate-like nanosheets were used as hierarchical building blocks for these films and were isolated from lamellar precursor materials by recently developed exfoliation methods. Film morphology was characterized by scanning electron microscopy (SEM), and X-ray diffraction (XRD) and transmission electron microscopy (TEM) were used to evaluate the zeolite structure. The use of lamellar nanosheets may create unique film properties and open new opportunities for thin film fabrication. These films are potentially useful in low-dielectric constant (low-*k*) applications for computer microprocessors.

See more of this Session: Nanostructured Thin Films
See more of this Group/Topical: Materials Engineering and Sciences Division

(755a) De Novo Design of Bioactive Protein-Resembling Nanospheres Via Dendrimer-Templated Peptide Amphiphile Assembly

Rachel S. Marullo, Chemical Engineering, University of California, Santa Barbara, Berkeley, CA, Brian Lin, Chemistry, University of California, Santa Barbara, Berkeley, CA and Matthew Tirrell, Bioengineering, UC Berkeley, Berkeley, CA

Peptide amphiphiles (PAs) are capable of self-assembly into micelles which can be used in the targeted delivery of peptide therapeutics and diagnostics. PA micelles exhibit a structural resemblance to proteins by having folded bioactive peptides displayed on the exterior of a hydrophobic core. However, PA self-assembly often results in high aspect ratio cylindrical micelles. Here we report a modular approach for controlling the self-assembly of biorelevant PAs into spherical, sub-100 nm structures by using dendrimers as supramolecular templates. The model PAs used in this investigation were: (1) di-C_{16}-bZip, a DNA binding PA sequence adapted from the alpha helical yeast transcription factor GCN4, which self-assembles into micron-long cylindrical micelles, and (2) di-C_{16}-NLS, a cell internalization PA, which self-assembles into polydisperse vesicles 100 to 400 nm in diameter.

The dendrimers were found to have a dramatic effect on the self-assembly of both model PAs, as the templating was successful for fabricating protein-resembling templated nanopsheres (PRTNs) averaging 50 nm in diameter. It was found that the PRTNs had a lower critical aggregation concentration than the PAs alone due to the presence of the hydrophobic dendrimer, which translated to an increase in PRTN stability in serum relative to the PA-only assemblies. Circular dichroism (CD) of di-C_{16}-bZip showed that the PAs within the PRTNs possessed enhanced alpha helical content compared to the unaggregated peptide. The bZip PRTNs were able to bind DNA and maintain their spherical geometry along the strands as observed with fluorescence assays and atomic force microscopy (AFM). Fluorescence resonance energy transfer (FRET) between fluorescently labeled di-C_{16}-bZip and di-C_{16}-NLS co-assembled onto dendrimers indicated that mixed 50 nm PRTNs possessing both DNA binding and cell internalizing functions were formed. Cell internalization studies of fluorescently labeled PRTNs showed distinct fluorescent spots, indicative of internalization via endocytosis. With these qualities, the PRTNs have potential in applications that necessitate the delivery of small multifunctional protein-like therapeutics and diagnostics.

See more of this Session: Self-Assembled Biomaterials
See more of this Group/Topical: Nanoscale Science and Engineering Forum

(420b) Fe-Ni Enzyme Inspired Sulfur-Tolerant Catalysts: A Combined Theoretical and Experimental Study

Debosruti Dutta, Department of Chemical and Biomedical Engineering, University of South Florida, Tampa, FL, Vanessa A. Castillo, University of South Florida, Tampa, FL, John N. Kuhn, Chemical and Biomedical Engineering, University of South Florida, Tampa, FL and Venkat R. Bhethanabotla, Department of Chemical Engineering, University of South Florida, Tampa, FL

A very promising technique to efficiently control sulfur poisoning of catalysts and to do hydrogenation is to mimic how nature does it. Excellent examples are naturally occurring iron- or nickel-containing metalloenzymes like Ni-Fe hydrogenase and Fe nitrogenase. These enzymes have sulfur atoms in their inherent chemical structure and yet, behave as extremely efficient catalysts. In the quest for biomimetic catalysts, scientists seek to replicate the coordination sphere that surrounds the metal centers, with the expectation that function will follow form. Hence, in this present work we have attempted to create Ni-Fe-S complexes of various compositions that have chemical structures similar to those of hydrogenases with the expectation that such structures will be efficient sulfur-tolerant hydrogenation catalysts. We have used density functional theory to predict the stability and viability of such chemical structures. We have also studied the role of CO and CN- groups in the hydrogenation reactions. Nickel/iron/sulfur [4Ni-4Fe-7S] clusters were synthesized *via* a macro-molecular templating technique. A host of characterization techniques like UV-vis absorption spectroscopy, EXAFS were used to establish the structure of the Ni-Fe-S clusters.

See more of this Session: Poster Session: Nanoscale Science and Engineering
See more of this Group/Topical: Nanoscale Science and Engineering Forum

(353c) Award Submission: Rational Design of Pathogen Mimicking Amphiphilic Nanoparticle Adjuvants

Latrisha Petersen[1], Yashdeep Phanse[2], Amanda Ramer-Tait[2], Chang Sun Kong[3], Scott Broderick[3], Bret D. Ulery[1], Rajan Krishna[3], Bryan H. Bellaire[2], Michael J. Wannemuehler[2] and Balaji Narasimhan[4], (1)Chemical and Biological Engineering, Iowa State University, Ames, IA, (2)Veterinary Microbiology and Preventive Medicine, Iowa State University, Ames, IA, (3)Materials Science and Engineering, Iowa State University, Ames, IA, (4)Department of Chemical and Biological Engineering, Iowa State University, Ames, IA

The natural infection of a pathogen stimulates a protective and enduring immune response. Thus, an ideal approach to vaccine design is to mimic the immune response associated with a natural infection while avoiding the undesirable side effects of the disease. Vaccines have proven to be the most effective method of preventing infectious diseases, yet many problems still exist including multi-dose administration, poor immunogenicity of vaccine antigens, and undesirable side effects of immunization with adjuvants which can all lead to poor patience compliance and poor vaccine efficacy. Thus there is a need for the development of new vaccine adjuvants and delivery vehicles capable of mimicking the properties of the natural infection while limiting the undesirable side effects and need for repeated administrations.

Polyanhydrides are a class of biomaterials that have demonstrated biocompatibility, controlled antigen release kinetics, antigen stabilization, and immune modulation, making them excellent vaccine delivery vehicles. Furthermore, they can be fabricated into nanoparticles capable of being inhaled or injected for ease of administration. This work is focused on polyanhydride copolymers based on sebacic acid (SA), 1,8-bis(*p*-carboxyphenoxy)-3,6-dioxaoctane (CPTEG) and 1,6-bis(*p*-carboxyphenoxy) hexane (CPH). While many chemistry-dependent trends associated with protein stabilization, protein release, phagocytosis, and immune modulation have been identified, little work to date has investigated the specific polymer characteristics responsible for these trends. By identifying these properties, novel adjuvants can be designed that more closely mimic infections with live pathogens. Therefore, this research is focused on the rational design of pathogen-mimicking adjuvants for vaccine delivery by utilizing flow cytometry and confocal microscopy in concert with materials informatics to identify key properties responsible for nanoparticle uptake and downstream activation of murine bone marrow derived dendritic cells (BMDCs). These internalization patterns were compared to that of live pathogens (*Yersinia pestis* and *Escherichia coli*) and the activation patterns compared to that of lipopolysaccharide (LPS, an endotoxin found on the surface of gram negative bacteria).

This investigation identified that the least hydrophilic (CPTEG-rich and the SA-rich) nanoparticles were readily internalized by BMDCs. Upon further assessment of the nanoparticle positive and negative

BMDC populations, it was discovered that nanoparticle internalization was required for enhanced expression of CD40 and production of IL-12p40 and IL-6 but not for the expression of CD86 and MHC II. Interestingly, bacteria have been reported to have very similar internalization patters in which association is required for production of IL-12p40 and IL-6 and expression of CD40 but less dependent for expression of CD86 and MHC II. Additional flow cytometric analysis of the most adjuvanting nanoparticle chemistries (poly(SA) and 50:50 CPTEG:CPH) and the positive control (LPS) revealed stark similarities of double positive DC populations for cytokines and cell markers between the amphiphilic 50:50 CPTEG:CPH nanoparticles and LPS. Further pathogen mimicking patterns were observed with confocal microscopy in which internalization and persistence were compared between the bacteria, *E. coli* and *Y. pestis*, and the 50:50 CPTEG:CPH nanoparticles. Microscopic observations at both 2 and 48 h indicated that the nanoparticles were found inside late endosomes, consistent with phagosome–lysosome fusion events of antigen processing and presentation. When compared to the internalization of *Y. Pestis* and *E. coli*, the amphiphilic 50:50 CPTEG:CPH nanoparticles appeared to be similarly internalized and persist at 2 and 48 h, respectively, suggesting that the stimuli provided by these particles to activate DCs do not rapidly degrade, much like that provided by persisting bacteria.

To validate and assess these multiple pathogen mimicking observations, materials informatics analysis, including principal component analysis (PCA) and partial least squares (PLS), were employed. The PCA confirmed that 50:50 CPTEG:CPH nanoparticles were very similar in the overall activation, internalization, and persistence patterns to the pathogenic correlates. Furthermore, a PLS analysis was employed, which identified the polymer properties responsible for these pathogen mimicking patterns to be i) percentage of hydroxyl end groups, ii) hydrophobicity, iii) alkyl ethers, iv) aliphatic carbons, and v) backbone oxygen moieties. The approach described here will facilitate the rational design of single-dose, pathogen-mimicking vaccine adjuvants that could be tailored to provide immune modulation and protection against a wide range of emerging and re-emerging diseases.

See more of this Session: Bionanotechnology Graduate Student Award Session
See more of this Group/Topical: Nanoscale Science and Engineering Forum

(374a) Encapsulation and Permeability Characteristics of Amorphous Hydrogenated Carbon Films Formed by Plasma Enhanced Chemical Vapor Deposition Technique

Anaram Shahravan and Themis Matsoukas, Department of Chemical Engineering, The Pennsylvania State University, University Park, PA

We have used plasma enhanced chemical vapor deposition technique to coat the materials and control the rate of their release. This powerful technique coat nanoparticles and nanopowders ranging from 15 nm to 1 micron and we can control the thickness (from 1nm to upwards of 100 nm) of the film by controlling the duration of the deposition time. We present results with two different core materials, silica nanopheres and KCl nanoparticles. By coating silica and etching the core in HF, we form hollow nanoshells whose dimensions depend on the size of the silica particle and the thickness of the shell. Permeation studies are conducted using KCl, whose release profile is measured easily by monitoring the conductivity of the solution. The thickness of the shell is one of the independent variables by which to control the rate of release. It has a strong effect on the rate of release, which increases from a six-hour release (shell thickness is 20 nm) to a long-term release over 30 days (shell thickness is 95 nm). The profile of release for any thickness follows a unique behavior: a fast release (35% of the final materials) during the first five minutes after the beginning of the dissolution, and a slower release till all of the core materials come out. In order to analyze this behavior a model based on an unsteady-state diffusion from sphere is developed. Very good agreement between experimental data and the model after half an hour from the beginning of the release is observed and based on that a diffusion coefficient for the shell is calculated. The diffusion coefficient of $(3.5\pm0.5)* 10^{-21}$ is obtained for four different samples with different shell thicknesses (20, 40, 75, 95 nm).

See more of this Session: Functional Nanoparticles and Nanocoatings on Particles I
See more of this Group/Topical: Particle Technology Forum

(420p) Multi-Particle Sintering Dynamics: From Fractal-Like Aggregates to Compact Structures

Max L. Eggersdorfer[1], Dirk Kadau[2], Hans J. Herrmann[2] and Sotiris E. Pratsinis[1], (1)Particle Technology Laboratory, Institute of Process Engineering, Department of Mechanical and Process Engineering, ETH Zurich, Zurich, Switzerland, (2)Department of Civil, Environmental and Geomatic Engineering, ETH Zurich, Zrich, Switzerland

Multi-Particle Sintering Dynamics: fromFractal-like Aggregates to Compact Structures

Multi-particle sintering is encountered inalmost all high temperature processes for material synthesis (titania, silica, nickel) and energy generation (e.g. flyash formation) resulting in aggregates of primary particles (hard- orsinter-bonded agglomerates). This mechanism of particle growth is investigatedquantitatively by mass and energy balances during viscous sintering ofamorphous aerosol materials (e.g. SiO_2, polymers) that typicallyhave a distribution of sizes and complex morphology[1]. Once coalescence or sintering startsbetween constituent primary particles, sinter necks are formed between themconverting the agglomerates to aggregates. During sintering, the latterprogressively densify until complete compact (e.g. fractaldimension $D_f = 3$) structures are formed atsufficiently long process times at high temperatures. In reality, however, itis rather seldom to have enough process time to complete particle coalescence.As a result, aggregates are formed with D_f in-betweenthose predicted by particle collision alone.

Here a new and rather simple model is introduced that describessintering of two differently sized particles and multi-particle aggregates of amorphous spherical particles (e.g.silica) by mass[2] (or volume) and energy balances[3]. This model is validated at limited cases of sintering between two (equallyor unequally sized) particles, and chains of particles. The evolution of morphology,surface area and radii of gyration of multi-particle aggregates are elucidatedfor various sizes and initial fractal dimension. For each of these structuresthat had been generated by diffusion limited (DLA), cluster-cluster (DLCA) andballistic particle-cluster agglomeration (BPCA) the surface area evolution ismonitored and found to scale differently than that of the radius of gyration(moment of inertia). Expressions are proposed for the evolution of fractaldimension and the surface area of aggregates undergoing viscous sintering. Theseexpressions are important in design of aerosol processes with populationbalance equations (PBE) and/or fluid dynamic simulations for material synthesisor minimization and even suppression of particle formation. Figure 1 shows thetemporal evolution of the effective fractal dimension of aDLCA agglomerate with 256 primary particles during viscous sintering. Atthe beginning of sintering (Fig. 1, at $t/t_0 = 0$ - 2) the highly ramifiedaggregate branches straighten when primary particles approach each other(reduction of center-to-center distance). This internal restructuringpractically unfolds the aggregate and D_f is reduced.However the branches continue shrinking while conserving mass. Further downstream $D_f$increased as aggregates compacted by sintering-coalescence consistent with in-situ measurements bysmall angle X-ray scattering (SAXS)[4].

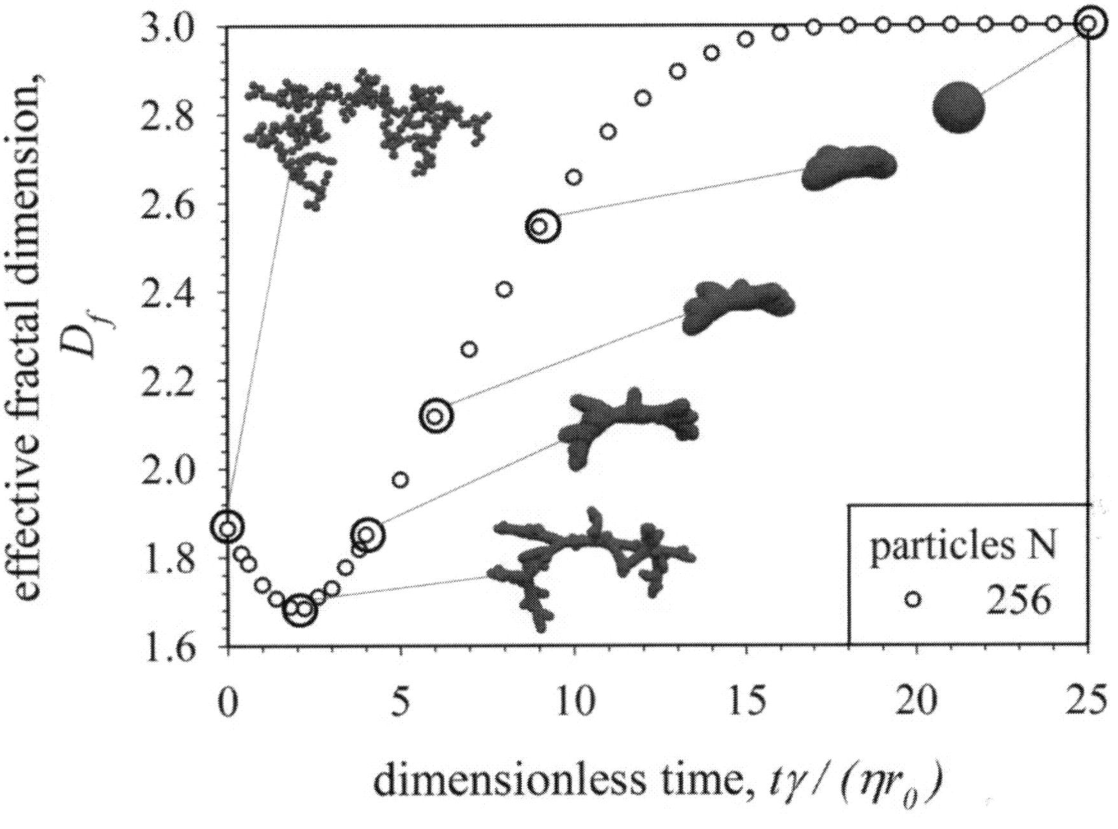

Figure 1: Evolution of the effectivefractal dimension D_f during viscous sintering of anagglomerate with 256 primary particles.

1. Eggersdorfer, M.L., Kadau, D.,Herrmann, H.J. and Pratsinis, S.E., Multi-particle sinteringdynamics: from fractal-like aggregates to compact structures, Langmuir in press(2011).

2. Kadushnikov,R.M., Skorokhod, V.V., Kamenin,I.G., Alievskii, V.M., Nurkanov,E.Y., Alievskii, D.M., Computer simulation ofspherical particle sintering, Powder Metall. Met. Ceram.40 (2001) 154-163.

3. Frenkel, J.,Viscous flow of crystalline bodies under the action of surface tension, J.Phys. (USSR) 9 (1945) 385-391.

4. Camenzind, A.,Schulz, H., Teleki, A., Beaucage,G., Narayanan, T., Pratsinis, S.E., Nanostructureevolution: from aggregated to spherical SiO2 particles made in diffusion flames,Eur. J. Inorg. Chem. (2008) 911-918.

(321e) Exploring Molecular Architecture Effects On the Microstructures of Block Copolymer Liquid Crystals

James A. Bergman and Jennifer M. O'Donnell, Chemical and Biological Engineering, Iowa State University, Ames, IA

Many amphiphilic block copolymers self-assemble in water to form periodic structures or liquid crystals. Liquid crystal microstructure greatly impacts the bulk material properties. For example, the viscosity of bicontinuous cubic microstructures is significantly greater than the viscosity of lamellar or hexagonal-close packed microstructures. In addition, bicontinuous cubic microstructures have superior transport properties relative to other microstructures. Therefore, certain microstructures are optimal for specific applications. Unfortunately, block copolymer chemistry and microstructure are coupled, and the optimal chemistry for a particular application often does not produce the optimal microstructure, so current applications require a compromise between chemistry and microstructure to optimize product performance. This work investigates the effect the copolyomer architecture at the hydrophilic – hydrophobic interface has on the self-assembled microstructure for a series of polymers with identical chemical composition.

Five polymer architectures, which differ only in the number of hydrophilic branches, have been synthesized by a combination of reversible addition-fragmentation chain transfer (RAFT) polymerization, and click chemistry. These methacrylic polymers contain the same number of hydrophobic butyl methacrylate (BMA) monomer units and the same number of hydrophilic hydroxyethyl methacrylate (HEMA) monomer units. The BMA block is grown as a straight polymer chain, and the HEMA block is grown as either a straight chain, or as two, three, four, or five branches. Branches are added via a 'graft from' scheme by clicking additional RAFT agents to the BMA backbone, then growing the HEMA block accordingly. The microstructure of each polymer architecture as a function of concentration in aqueous solution has been studied using small angle x-ray scattering (SAXS) and cryogenic transmission electron microscopy (cryo-TEM). The results indicate that altering the architecture of the copolymer without altering the chemical composition induces a change in the thermodynamically favorable microstructure.

See more of this Session: Nanoscale Structure In Polymers II
See more of this Group/Topical: Materials Engineering and Sciences Division

(599e) Synthesis of Metal Oxide Nanoparticles Through Facilitated Mineralization by Self-Assembled Biomimetic Molecules

Jinyoung Kwak, Inho Lee, Sungjun Ahn and **Sang-Yup Lee**, Department of Chemical & Biomolecular Engineering, Yonsei University, Seoul, South Korea

Metal oxide nanoparticles have been synthesized through hydrothermal and thermal decomposition methods to control their crystalline structure. Without the thermal treatment, amorphous structured minerals are conventionally obtained. Here we report the synthesis of semiconducting metal oxide nanoparticles with controlled crystalline structure in the self-assembled biomimetic bolaamphiphile molecules. The nanoparticle synthesis was achieved at room temperature through the catalytic function of the biomimetic molecule. Specifically, the nanoparticles were prepared by controlled minerlization of aqueous inorganic precursors which was mediated by the functional groups of biomimetic molecules. A biomimetic bolaamphiphile molecule of bis(N-α-amido-glycylglycine) heptane-1,7-dicarboxylate was synthesized and used to promote controlled mineralization of indium tin oxide (ITO) and zinc tin oxide (ZTO) nanoparticles. The prepared nanoparticles showed good electrical and optoelectrical properties which are comparable to those prepared by conventional methods. We also prepared new biomimetic molecule have surface-exposed mineralizing functionality and self-assembling property to make thin metal-biomimetic hybrid material. The outcomes indicate that designer biomimetic molecule is a promising molecule for the low-energy consuming fabrication method of functional nanomaterials.

See more of this Session: Nanostructured Biomimetic and Biohybrid Materials and Devices
See more of this Group/Topical: Nanoscale Science and Engineering Forum

(346c) Magnetic Quantum Dots Coupled with Magnetic Microarrays for Molecular Detection and Separation

Kalpesh D. Mahajan, William G. Lowrie Dept. of Chemical and Biomolecular engineering, The Ohio State University, Columbus, OH, Greg Vieira, Department of Physics, Ohio State University, Columbus, OH, Gang Ruan, Department of Chemical and Biomolecular Engineering, The Ohio State University, Columbus, OH, R. Sooryakumar, Department of Physics, The Ohio State University, Columbus, OH and Jessica O. Winter, Chemical and Biomolecular Engineering, The Ohio State University, Columbus, OH

Analysis of biomarker molecules is important not only in diagnostics, but also in monitoring disease progression, and response to therapy. Nanotechnology is revolutionalizing the field of biomarker detection with the development of ultrasensitive detection techniques. However, nanomaterials developed thus far have primarily been targeted towards molecular detection. An ideal chemical analysis scheme would also include separation of targeted molecules from the original location to permit further analysis and manipulation.

Here we describe a proof of principle study of the use of magnetic quantum dot nanoparticles for molecular detection and separation of biomarkers on the same microchip. A range of biomarkers: proteins, mutated DNAs, RNAs, circulating tumor cells can be detected with this assay. We have already demonstrated detection and separation of avidin from a 10^{-10} M solution; we have also successfully manipulated human leucocytes on the microchip. The assay is performed on a magnetic platform consisting of an array of magnetic discs or nanowires coupled to electromagnets that apply an external magnetic field. The target is detected by coalescence of magnetic and fluorescent nanoparticles to yield a composite nanoparticle with magnetic and fluorescent functionalities, whereas in the absence of the target these signals are not present. The response of a fluorescent probe to the external magnetic field constitutes a detection signal, and by controlling the direction of the external magnetic field the captured target is transported to the desired location on the array for further analysis. Detection and separation can be performed on the same platform. Further, because quantum dots are used as fluorescent probes, their size tunable and narrow emission spectra properties impart multiplexing capability. Ultimately this platform will be constructed in a lab on chip design incorporating nonmaterial-biomolecules mixing, detection, and magnetic separation. This technology could have tremendous impact on a number of fields including healthcare diagnostics, small-scale chemical synthesis, and molecular chromatography.

See more of this Session: Applications of Magnetic Nanoparticles
See more of this Group/Topical: Nanoscale Science and Engineering Forum

(420q) Oligodeoxynucleotide Conjugated Graft-Copolymer-Stabilized Gold Nanoparticle Scaffolds for In Situ PCR Diagnostic Assays

Jun Sung Kang, Issac S. Marks and T. Andrew Taton, Chemistry, University of Minnesota, Minneapolis, MN

Polymerase chain reaction (PCR) has been widely used in molecular biology for various applications, including gene sequencing, disease diagnostics, and gene functional analysis by amplification of target DNA sequences. Up to now, however, there have been no successful reports of gold nanoparticles (AuNP) as supports for PCR to detect target DNA. This has partly been because AuNP-DNA scaffolds are unstable to the thermal cycling and free thiols used in PCR, and partly because enzymes interact with AuNPs. We report that both of these problems are solved by using AuNPs stabilized by multiple thiolated poly(L-lysine)-*graft*-poly(ethylene glycol), PLL-*g*-[PEG:PSH]. The multi-thiolate graft copolymers were conjugated to oligodeoxynucleotide (ODN) primers via "click chemistry" and bound to AuNPs to yield AuNP-ODN scaffolds that successfully primed PCR reactions, as demonstrated by gel electrophoresis and fluorescent analysis. Moreover, the scaffolds were applied to rapid DNA diagnostics in a single PCR tube with magnetic particles through color change without any instrumental analysis. We anticipate that AuNP-ODN scaffolds might be integrated into microarray or sequencing technologies that currently rely on microscale surfaces and fluorescence signal analysis.

See more of this Session: Poster Session: Nanoscale Science and Engineering
See more of this Group/Topical: Nanoscale Science and Engineering Forum

(662c) Seeded Growth of Shape-Controlled Wurtzite CdSe Nanocrystals: Cubes, Hexagonal Platelets, and Bullets

Katherine P. Rice, Mark P. Stoykovich and Aaron E. Saunders, Chemical and Biological Engineering, University of Colorado, Boulder, CO

Previous investigations into the synthesis of wurtzite CdSe nanocrystals have given rise to well-developed methods for producing particles with anisotropic shapes such as rods, tetrapods and wires; however, the synthesis of other anisotropic shapes has proved challenging. Here, we demonstrate the use of a seeded-growth approach to produce colloidal, shape-controlled wurtzite CdSe nanoparticles with previously unobserved morphologies. The synthesis, which makes use of small CdSe nanocrystals as nucleation sites for subsequent growth, can be tuned to selectively yield colloidal CdSe cube- and hexagonal platelet-shaped nanocrystals, in addition to the previously observed rod- and bullet-shaped particles. We characterize the structure and discuss possible growth mechanisms for these new shapes, and demonstrate a quantitative analysis technique for shape classification based on Fourier descriptors obtained from transmission electron micrographs.

See more of this Session: Nanocrystal Science and Technologies
See more of this Group/Topical: Materials Engineering and Sciences Division

(693g) The Effects of Bingel Functionalization On CoMoCAT- SWNT and Their Dispersibility In Unsaturated Polyester Resin

Matthew J. Kayatin and Virginia A. Davis, Chemical Engineering, Auburn University, Auburn, AL

Sidewall esterification of single-walled carbon nanotubes (SWNT) can be achieved via the Bingel reaction. Stemming from the cyclopropanation of fullerenes, Bingel reaction conditions require sonication of nanotubes in 1,2-dichlorobenzene (o-DCB) followed by treatment with diethylbromomalonate in presence of the base 1,8-diazabicyclo[5.4.0]undec-7-ene. However, sonication of o-DCB generates a significant amount of sonopolymer. In contrast to previous studies with HiPco SWNT, we find that exposing CoMoCAT® SWNT (CG 200) to Bingel reaction conditions alters the content of carbonaceous species in the starting material. These effects must be carefully decoupled in order to study the reactivity of SWNT.

Sonopolymer generation and removal was studied using thermal gravimetric analysis (TGA) and attenuated total reflectance - Fourier transform infrared spectroscopy (ATR). The effect of sonopolymer removal on functional groups was monitored via resonant Raman spectroscopy (Raman), under low laser powers. The extent of reaction and inherent doping effect on the nanotubes is unknown for CoMoCAT® SWNT. We examine these effects with treatment time using Raman. In addition to ATR, coupled TGA-FTIR data was used to examine thermal stability of components and to provide confirmation of the surface functional groups present. Finally, factors controlling the high-shear dispersibility of the Bingel SWNT in to UPR were studied. Remarkably, SWNT sonicated in o-DCB alone show no re-dispersion in UPR while Bingel functionalized SWNT do disperse.

See more of this Session: Graphene and Carbon Nanotubes: Characterization, Functionalization, and Dispersion
See more of this Group/Topical: Nanoscale Science and Engineering Forum

(755c) CdTe/DNA Assembly to Create 2D Sheets

Sudhanshu Srivastava and Nicholas A. Kotov, Chemical Engineering, University of Michigan, Ann Arbor, MI

In this approach, the already known CdTe nanoparticle assemblies to create 1D nanowires were employed for the incorporation of linear DNA molecules. Two different lengths of DNA with different sequences were chosen as building blocks to self-assemble the CdTe nanoparticles. Results showed 2D sheet structures were formed in both the cases as compared to the known 1D nanowires. The idea proves that nanoparticle and DNA do interact and introduction of DNA dictates the final structure of the nanocomposite formed. On the contrary, the circular plasmid DNA when assembled with CdTe nanoparticles did not form any sheets, but lead to long wires. The temperature change and higher ionic strength lead were also investigated to dis-assemble the 2D sheets, which later formed 1D nanowires. The dynamic approach enhances our understanding to design CdTe/DNA hybrid materials with controlled architectures.

See more of this Session: Self-Assembled Biomaterials
See more of this Group/Topical: Nanoscale Science and Engineering Forum

(439g) Using Nanoelectroporation, Molecular Beacons and Single Cell PCR to Quantify the Relationship Between Sirna/MicroRNA Delivery and Gene Silencing In Cancer Cells

Pouyan E. Boukany, Nanoscale Science and Engineering Center for Affordable Nanoengineering of Polymeric Biomedical Devices, The Ohio State University, Columbus,, OH, Xinmei Wang, NSEC Center for Affordable Nanoengineering of Polymeric Biomedical Device (CANPBD), The Ohio State University, Columbus, OH, Bo Yu, Nanoscale Science and Engineering Center, Ohio State University, Columbus, OH, Yun Wu, Nanoscale Science and Engineering Center, The Ohio State University, Columbus, OH and Ly James Lee, The Ohio State University, Columbus, OH

Small RNAs such as siRNA and microRNA (miRNA) regulate intracellular pathway in cells and can also serve as a new molecular medicine in cancer therapy
by down- or up-regulation of targeted messenger RNA (mRNA) and proteins. However, our current understanding of the relationship between delivered siRNA/miRNA and
gene silencing in cancer cells is limited because of our inability to deliver the precise dosage and monitor how the delivered siRNA/miRNA affects the individual
cancer cells. The conventional nanoparticle and electroporation based gene delivery methods require a large number of cells and have poor dosage control.
Consequently, the obtained results are an average value from a large cell population and often with poor repeatability.

Recently, we developed a novel nanoelectroporation (NEP) method that can deliver biomolecules such as siRNA, miRNA and molecular beacons (MBs) into individual cells with controlled dosage and real-time monitoring of mRNA and miRNA expression. The precise delivery is achieved by the focused electric field through a nanochannel connecting two microscale channels with a single cell located at the outlet of the nanochannel in one microchannel and detecting/imaging agents located in the other microchannel. By adjusting the magnitude of electric field, pulse length and pulse number, dose of the delivery can be precisely controlled.

Using NEP, we were able to determine the critical dosage of siRNA (Mcl-1) and miR-29b to induce apoptosis of THP-1 (Human acute monocytic leukemia) cell line.The anti-apoptotic
protein Mcl-1 promotes the survival of lymphocytes and hematopoietic stem cells and is linked to drug resistance and poor treatment outcomes in many tumor types. Next, we delivered GAPDH (control) and Mcl-1 MBs into NEP treated cells to detect and monitor the Mcl-1 mRNA expression by confocal microscopy.Finally, we used single-Cell-qRT-PCR to establish the quantitative relationship between the copies of transfected miR-29b and the extent of MCl-1 mRNA down-regulation.

This work is the first to show that the quantitative relationship between siRNA/miRNA delivery and gene silencing/cell apoptosis in cancer cells can be established using a combination of nanoelectroporation, molecular beacons and single cell qRT-PCR.This analytical technique and resulting information are not only valuable for better understanding the cell biology but also useful for new drug development.

See more of this Session: Bionanotechnology for Gene and Drug Delivery I
See more of this Group/Topical: Nanoscale Science and Engineering Forum

RESONANT QUANTUM TUNNELING IN CARBON-NANOTUBE-BASED DEVICES

Meng-Mu Shih, University of Florida, Gainesville, FL

Abstract

This work discusses CNT-based electronic devices by using the resonant quantum tunneling in double-barrier quantum-well structures. This quantum model and numerical results demonstrate how the CNT diameter, quantum well width, barrier height, and barrier thickness affect the transmission probability through the double-barrier resonant tunneling structure. The energy of tunneling electrons needs to match with one of the resonant energies existed in the resonant tunneling quantum well in order to achieve sharp resonance and high transmission probability, which ideally approaches to unity. The number and the energy levels of resonant energy states can be affected and tuned by the CNT-based quantum structure. These results can be used to further explore applications such as in resonant tunneling diodes.

Model Description

Figure 1 shows the schematic structure for the resonant quantum tunneling [1]. This figure shows the cross-section view of a double-barrier quantum-well formed by semiconducting carbon nanotubes (CNTs). Each barrier can be formed by the junction of two crossed CNTs [2]. Key geometrical parameters in this model are CNT diameter (d), quantum well width (w), and barrier thickness (a). Physical parameters are energy (E), barrier height (Vo), and CNT effective mass (m_e) [3,4]. If the tunneling electron energy approaches tone of the resonant energies existed in the quantum well, the transmission probability or transmission probability (T2) can approach to unity. With proper design, this T2 can have a very sharp response to the tunneling electron energy.

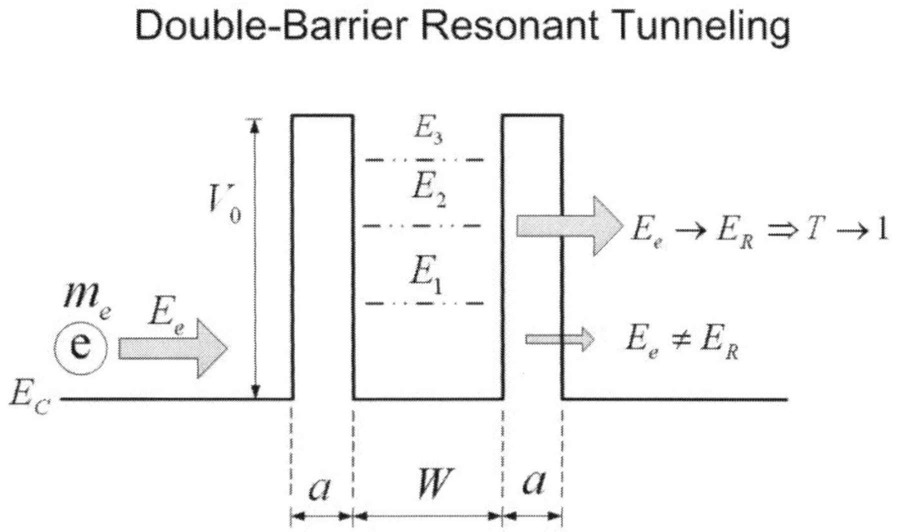

Figure 1 Schematic structure for double-barrier quantum-well resonant tunneling

Numerical Results and Discussion

The calculation results are based on the resonant model [1-5] described in Figure 1. Figure 2 shows the transmission coefficients versus the energy of tunneling electrons for CNT diameter d = 1 nm and 10 nm, when barrier height Vo = 0.3 eV, quantum well width w = 12 nm, barrier thickness a = 2.5 nm. Since the 10 nm CNT has smaller effective mass, the tunneling is easier and the overall tunneling probability is larger than the 1nm CNT. However, the 1 nm CNT has shaper response to the energy of tunneling electrons and will be used in the Figures 3-5 in order to demonstrate how the parameters affect the transmission probability. Figure 3 shows the transmission coefficients versus the energy of tunneling electrons for two quantum well width w = 12 nm and 6 nm, when CNT diameter d = 1 nm, barrier height Vo = 0.3 eV, and barrier thickness a = 2.5 nm. Smaller well width can have higher resonant energy in the well. Figure 4 shows the transmission coefficients versus the energy of tunneling electrons for two barrier height Vo = 0.3 eV and 0.15 eV, when CNT diameter d =1 nm, quantum well width w = 12 nm, and barrier thickness a = 2.5 nm. Smaller barrier height can reduce the number of resonant energy states. Figure 5 shows the transmission coefficients versus the energy of tunneling electrons for two barrier width a = 0.25 nm and 0.65 nm, when CNT diameter d = 1nm, quantum well width w = 12nm, and barrier height Vo = 0.3 eV. The larger barrier thickness will make the transmission probability curve has shaper peaks at the resonant energy levels. Since the thicker barrier will make the tunneling difficult, the electron energy needs to be more specific and closer to resonant energy in order to tunnel trough this structure.

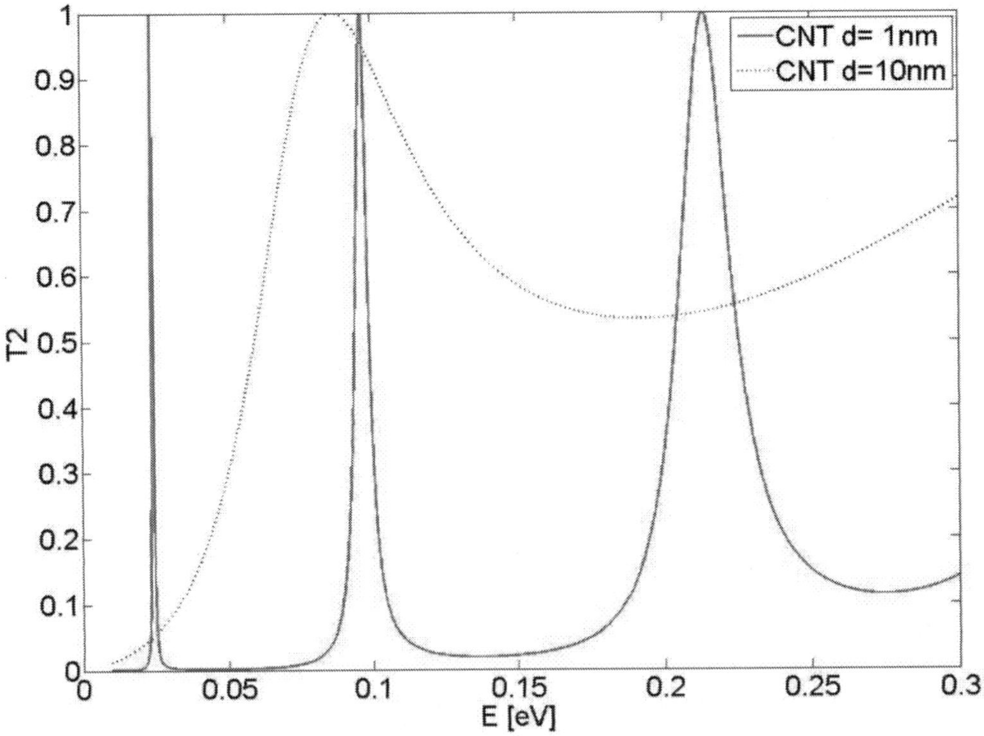

Figure 2 Transmission coefficients versus energy of tunneling electrons

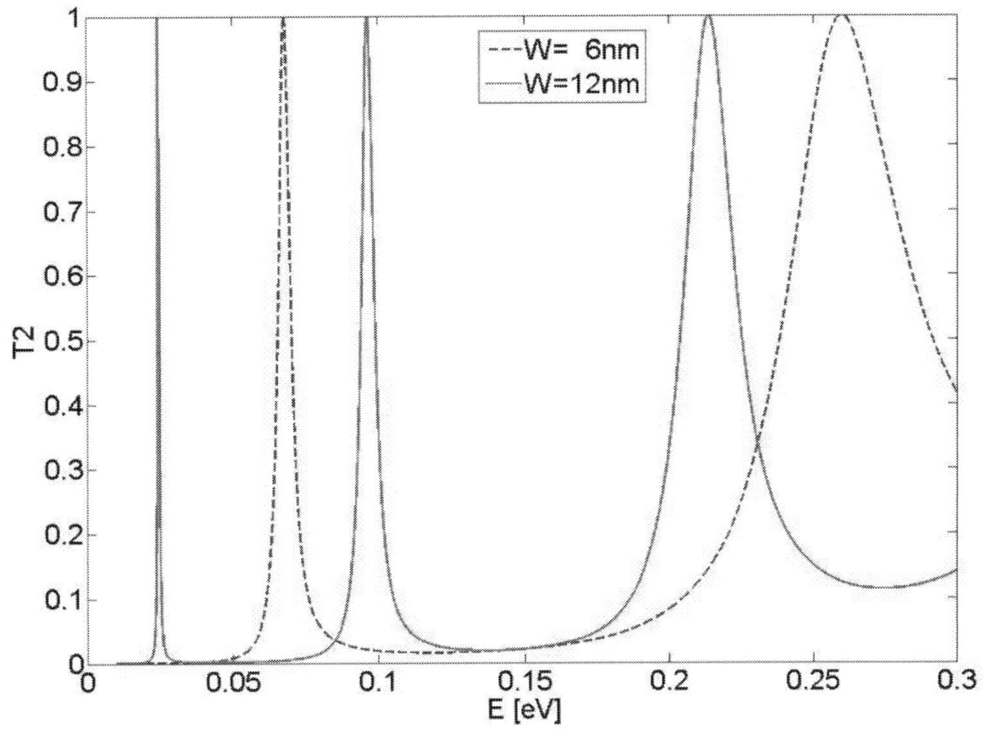

Figure 3 Transmission coefficients versus energy of tunneling electrons

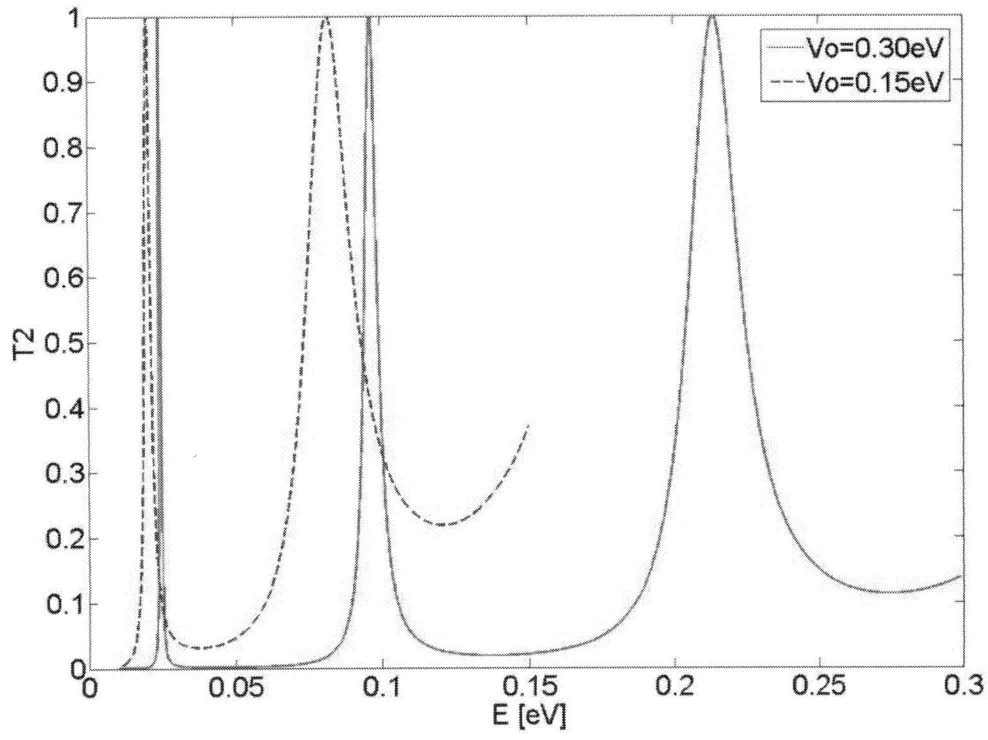

Figure 4 Transmission coefficients versus energy of tunneling electrons

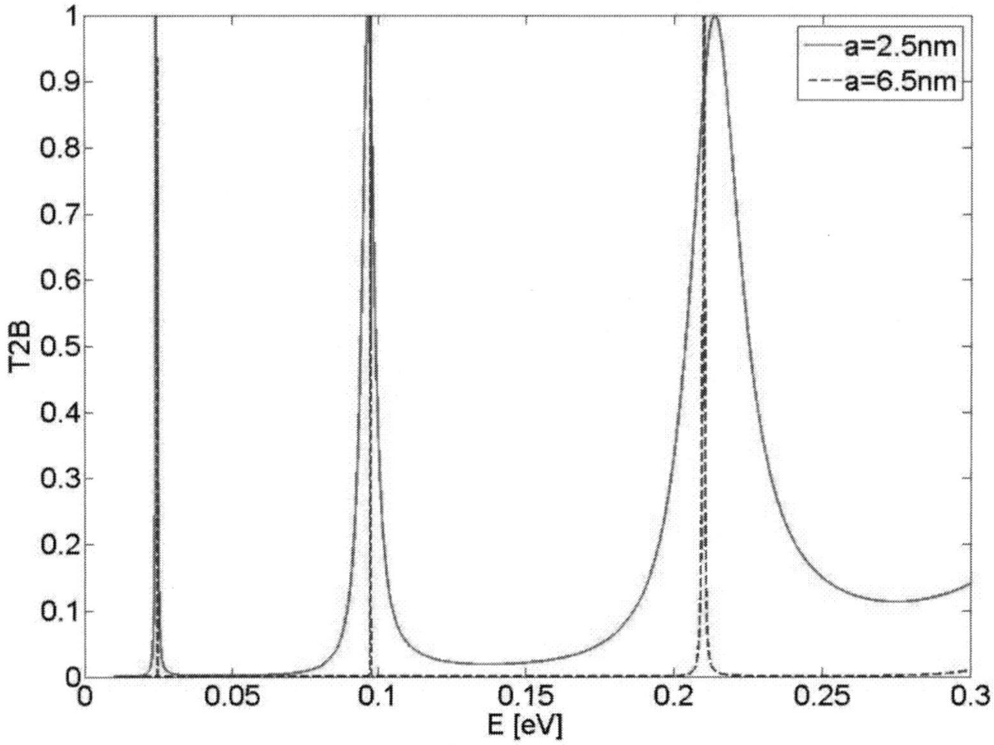

Figure 5 Transmission coefficients versus energy of tunneling electrons

References

1. Esaki, Leo (1973), "Long Journey to Tunneling," http://www.nobelprize.org/nobel_prizes/physics/laureates/1973/esaki-lecture.pdf
2. Biswas, Sujit et al. (2005), "Room-temperature Resonant Tunneling of Electrons in Carbon Nanotube Junction Quantum Wells," *Applied Physics Letters*, 86, 183101, pp. 01-03.
3. Pennington, Gary and Goldsman, Neil (2003), "Theory and Design of Field-Effect Carbon Nanotube Transistor," *International Conference on Simulation of Semiconductor Processes and Devices*, pp.167-170.
4. Marulanda, Jose M. and Srivastava, Ashok (2008), "Carrier Density and Effective Mass Calculations in Carbon Nanotubes," *Physica Status Solidi*, (b) Vol. 245, No. 11, pp. 2558–2562.
5. Singh, Jasprit (2004), "Modern Physics for Engineers," Weinheim, Germany, WILEY-VCH Verlag GmbH & KGaA.

(504c) Using Nanoelectroporation to Understand Intracellular Trafficking of Nanoparticles In Gene Delivery

Pouyan E. Boukany, Nanoscale Science and Engineering Center for Affordable Nanoengineering of Polymeric Biomedical Devices, The Ohio State University, Columbus,, OH, Yun Wu, Nanoscale Science and Engineering Center, The Ohio State University, Columbus, OH, Kam W. Leong, Biomedical Engineering, Duke University, Durham, NC and Ly James Lee, The Ohio State University, Columbus, OH

Most of bio-macromolecular agents and nanoparticles in conventional delivery methods are taken into cells through endocytosis process. Generally, bio-macromolecular agents and nanoparticles are targeted to cell receptors and internalized into endosome. Although the total amount of drugs/genes can be specified, the delivered dose dependent on the kinetics of cellular attachment, internalization and endosome escape, which is random due to biological heterogeneity.

To overcome this limitation, we recently developed a novel and yet simple nanoelectroporation (NEP) method for bio-macromolecular agents and nanoparticle delivery to individual cells with precise control of dosage. The delivery is achieved by the focused electric field through a nanochannel connecting two microscale channels with a single cell located at the outlet of the nanochannel in one microchannel and gene/drug/nanoparticles located in the other microchannel. The focused electric pulse porates the cell membrane to form a nanopore and also provides electrophoretic mobility of charged molecules/ nanoparticles to move them directry into the cytosol (i.e. by-passing the endocytothic pathway). By comparing the conventional deliverly method and the NEP delivery method, we can better understand cellular uptaking and intracellular trafficking of bio-macromolecular agents and nanoparticles in drug/ gene delivery because NEP delivery by-passed endocytosis and the consequent endosome escape.

In this work, QD605-amine or QD605-polycations and Cy5-oligonucelotides (Cy5-G3139) were chosen as the FRET pair for nanoparticle model.They were either encapsulated in lipoplex nanoparticles or form polyplex nanoparticles by the electrostatic interaction. The endosome of lung cancer cell (A549) was stained by transfection of plasmid EGFP-Rab7.The confocal microscopy was used to investigate the intracellular fate and unpacking of the nanoparticles in the lung cancer cells during/after NEP or conventional nanoparticle transfection. We demonstrated that NEP is able to deliver nanoparticles into cells without going through endocytosis and with minimum cell damage, which is not achieveabe in the conventional bulk electroporation delivery method. The in-situ FRET detection showed the time-dependent nanoparticle unpacking inside the cell. This technology allows us to quantify the effects of nanoparticle type and formulation on the kinetics of cellular uptaking, endocytosis, endosome escape, and nanoparticle unpacking in cancer cells.

QUANTUM TUNNELING IN CARBON-NANOTUBE-BASED FIELD EMISSION DISPLAY AND BAND-TO-BAND TUNNELING

Meng-Mu Shih, University of Florida, Gainesville, FL

Abstract

In today's competitive electronics market, consumer products are required to have SWaP features and considerations: size, weight, power and price. Although the traditional display technology has been replaced by the flat display technology, the CNT-based display products, due to several unique properties, are still considered to be the next-generation display technology to enable more creative applications in our daily life.

This work uses a quantum mechanical model to explain the working principle of such CNT-based field emission display (FED). This model is modified by considering parameters such as the CNT diameter, tunneling probability, and electrical field. In the CNT-based band-to-band tunneling model, results show how the CNT diameter and applied electrical field affect the quantum tunneling probability. Numerical results can be used to improve the design and performance of applications.

Field Emission Display Model

CNT-based field emission display was proposed by Samsung in 1999 [1]. Figure 1 shows the schematic structure for the quantum tunneling at the metal-vacuum interface [2, 3]. The modified CNT-based quantum tunneling model has parameters: CNT diameter (d), effective mass (m_e) [4, 5], barrier height (ϕ_w), tunneling width (w), and electrical field (E).

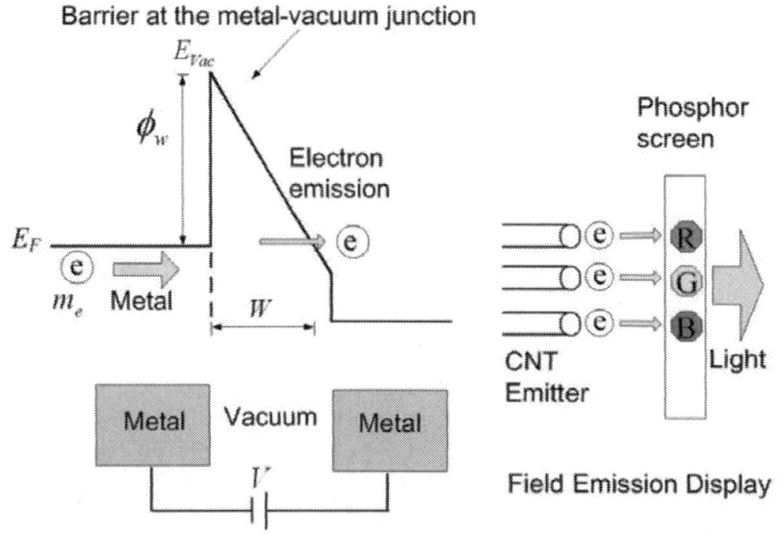

Figure 1 Schematic structure for quantum tunneling at the metal-vacuum interface

Band-to-Band Tunneling Model

CNT-based p-n junction with band-to-band tunneling in the transistor was proposed by IBM in 2004 [6]. Figure 2 shows the schematic structure for the band-to-band tunneling [3]. The modified CNT-based quantum tunneling model has parameters: CNT diameter (d), effective mass (m_e) [4, 5], CNT energy bandgap (Eg), tunneling width (w), and electrical field (E).

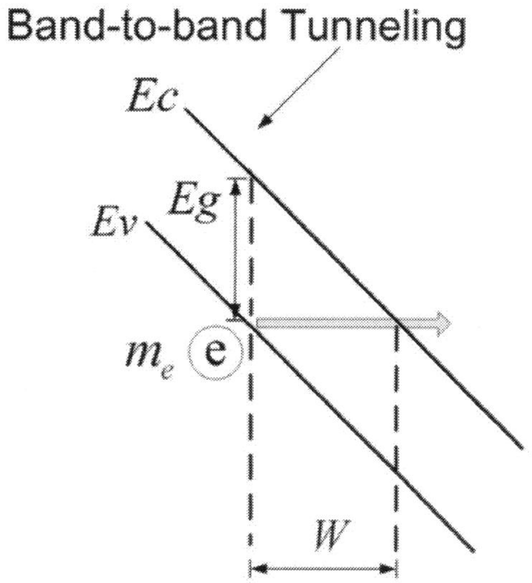

Figure 2 Schematic structure for quantum tunneling at the p-n junction

Numerical Results and Discussion

The calculation results for the field emission display model are based on the quantum tunneling model described in Figure 1. Figure 3 shows the electrical field (E) versus tunneling probability (T) for different materials. Since the metallic CNT with smaller diameter has larger work function as the tunneling barrier height, the tunneling is more difficult. The CNTs with diameters of 0.5 nm and 2 nm have the work function about 4.9 eV and 4.7 eV, respectively [7]. Tungsten has its work function of 4.55 eV and gold has its work function of 4.5 eV [3]. For a certain level of tunneling probability, the material with larger work function can have larger electrical field.

The calculation results for the band-to-band tunneling model are based on the quantum tunneling model described in Figure 2. Figure 4 shows the tunneling probability (T) versus CNT diameter (d) for electrical field E = 5×10^9 V/m and 10^9 V/m. Since the CNT with larger diameter has smaller energy bandgap as its tunneling barrier height and smaller effective mass, the tunneling is easier and the tunneling probability is higher than the CNT with smaller diameter. The larger electrical field can assist the tunneling and cause higher tunneling probability. Figure 5 shows the tunneling probability (T) versus electrical field (E) for CNT with diameter d = 6 nm and 2 nm. The CNT with larger diameter can reach tunneling probability of unity faster with smaller electrical field.

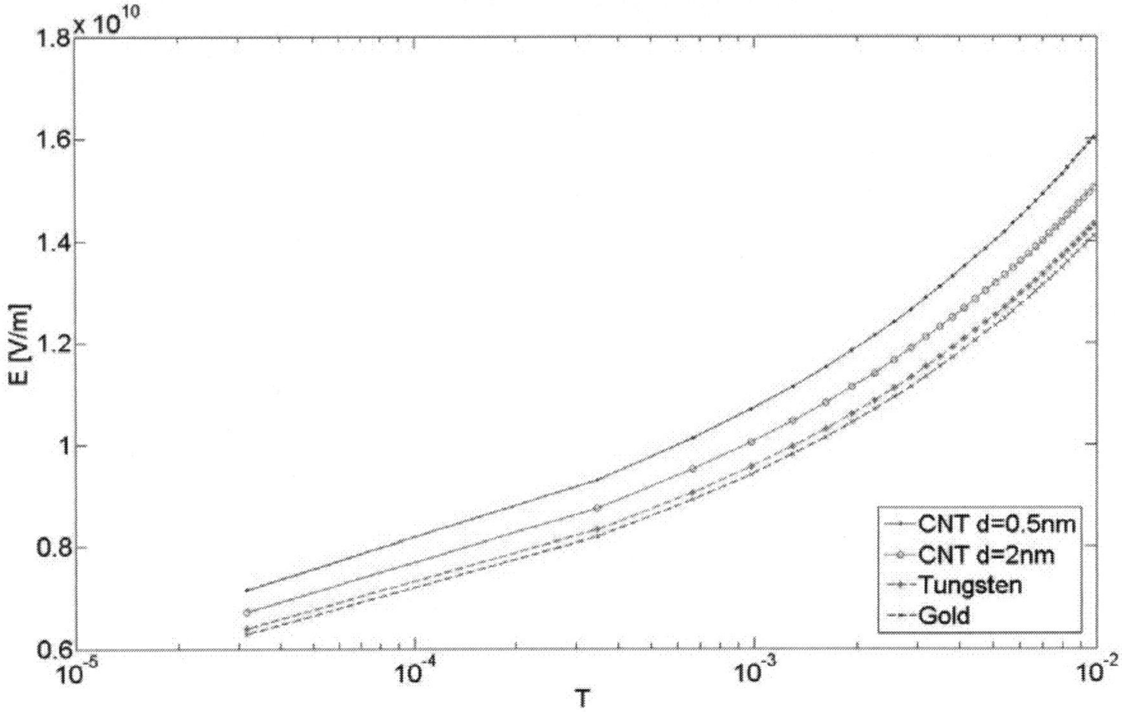

Figure 3 Electrical Field (E) versus tunneling probability (T) for different materials

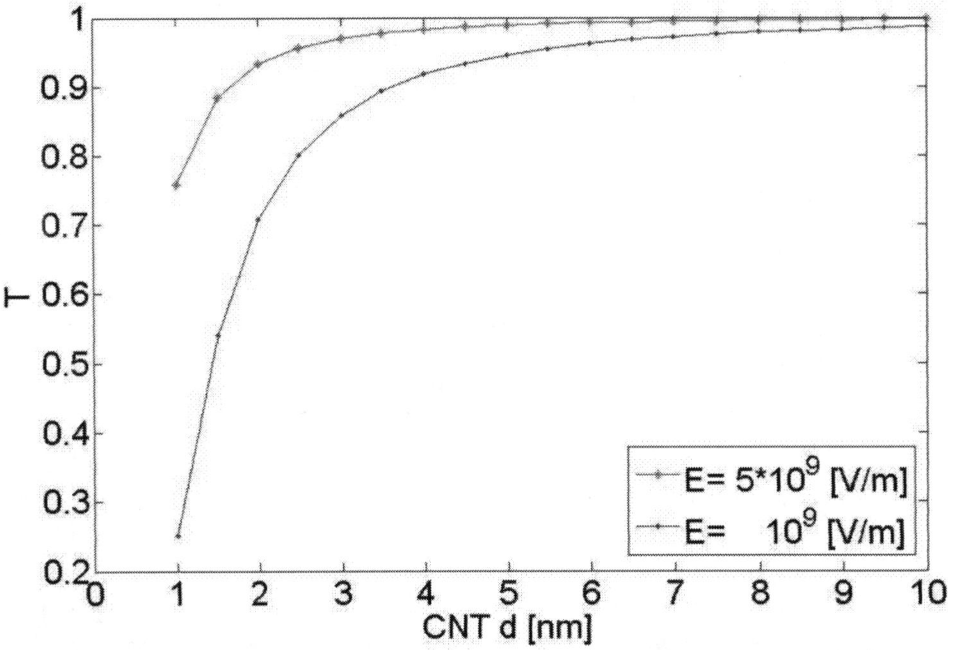

Figure 4 Tunneling probability (T) versus CNT diameter (d) for electrical field E = 5×10^9 V/m and 10^9 V/m

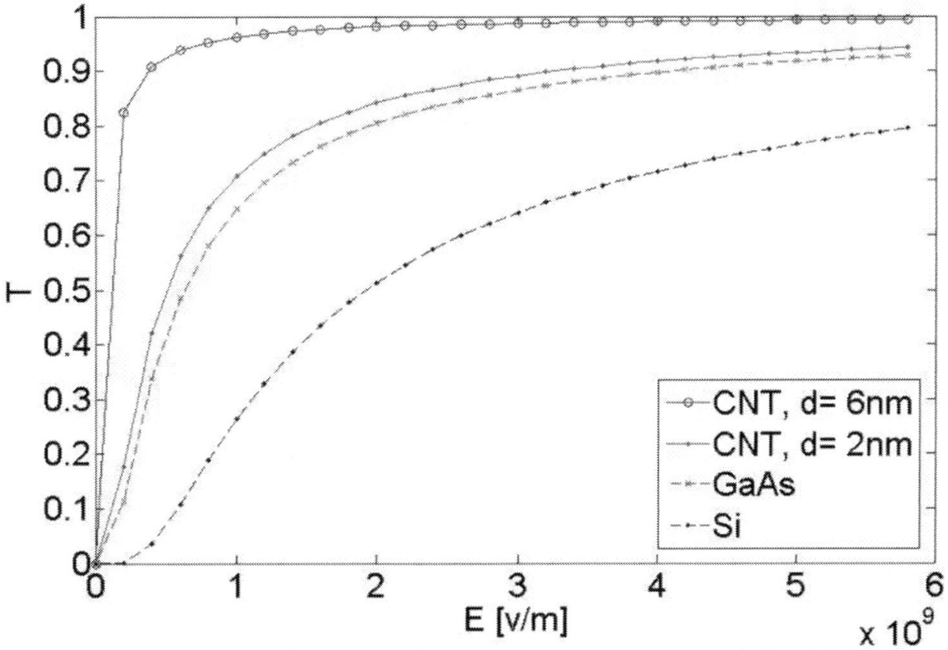

Figure 5 Tunneling probability (T) versus electrical field (E) for CNT with diameter d = 6 nm and 2 nm

References

1. Choi, W. B. et al. (1999), "Fully Sealed, High-brightness Carbon-nanotube Field-emission display," *Applied Physics Letters*, Vol. 75, No. 20, Nov., pp. 3129-3131.
2. Esaki, Leo (1973), "Long Journey to Tunneling," http://www.nobelprize.org/nobel_prizes/physics/laureates/1973/esaki-lecture.pdf
3. Singh, Jasprit (2004), "Modern Physics for Engineers," Weinheim, Germany, WILEY-VCH Verlag GmbH & KGaA.
4. Pennington, Gary and Goldsman, Neil (2003), "Theory and Design of Field-Effect Carbon Nanotube Transistor," in *International Conference on Simulation of Semiconductor Processes and Devices*, pp.167-170.
5. Marulanda, Jose M. and Srivastava, Ashok (2008), "Carrier Density and Effective Mass Calculations in Carbon Nanotubes," *Physica Status Solidi*, (b) Vol. 245, No. 11, pp. 2558–2562.
6. Appenzeller, J. et al. (2004), "Band-to-band tunneling in carbon nanotube field-effect transistors," *Physical Review Letters*, Vo1. 93, No. 19, pp. 196805-1-3.
7. Shan, Bin and Cho, Kyeongjae (2005), "First Principles Study of Work Functions of Single Wall Carbon Nanotubes," *Physical Review Letters*, Vol. 94, 236602.

(747b) Multimodal Bioimaging Agents Based On Luminescent Silicon Quantum Dots

Folarin Erogbogbo[1], Paras N. Prasad[1] and Mark T. Swihart[2], (1)Institute for Lasers, Photonics, and Biophotonics, University at Buffalo (SUNY), Buffalo, NY, (2)Chemical and Biological Engineering, University at Buffalo (SUNY), Buffalo, NY

Multimodal contrast agents are desirable for many biological imaging applications. Luminescent biocompatible silicon quantum dots (SiQDs) can be combined with MRI imaging agents, plasmonic nanoparticles, dyes, or therapeutics to create a new generation of such multifunctional probes. Here, we report the synthesis of different nanocomplexes incorporating SiQDs with magnetic, plasmonic, therapeutic and emission enhancing moieties. These multimodal nanoparobes are composed of PEGylated polymeric micelles, with hydrophobic SiQDs in their core, and another functional moiety co-encapsulated in the hydrophobic core or linked to the exterior. These moieties are characterized and applied to in vitro and in vivo applications. Demonstration of multimodal imaging with these newly developed probes opens the door for use of SiQDs in future diagnosis and therapy of cancer, atherosclerosis, and other diseases.

See more of this Session: Nanotechnology for In Vivo and In Vitro Imaging
See more of this Group/Topical: Nanoscale Science and Engineering Forum

(615b) Role of Alumina Type In the Growth of Single-Walled Carbon Nanotube Carpets From Alumina-Supported Fe Catalysts

Placidus B. Amama[1], Cary L. Pint[2], Seung Min Kim[3], Eric A. Stach[4], Robert H. Hauge[2] and Benji Maruyama[5], (1)Materials & Manufacturing Directorate, RXBT, Wright-Patterson Air Force Research Laboratory/UDRI, Dayton, OH, (2)Richard E. Smalley Institute for Nanoscale Science and Technology, Rice University, Houston, TX, (3)School of Materials Engineering & Birck Nanotechnology Center, Purdue University, West Lafayette, IN, (4)School of Materials Engineering and Birck Nanotechnology Center, Purdue University, West Lafayette, IN, (5)Materials & Manufacturing Directorate, RXBN, Wright Patterson Air Force Research Laboratory, Dayton, OH

The crucial role alumina plays as a catalyst support in the growth of single-walled carbon nanotube (SWNT) carpets is unsurprising as it is a well known structural modifier or catalyst promoter in hydrocarbon reforming and ammonia synthesis reactions. Although a lot of studies have been carried out hitherto to explain the role of alumina and the Fe-alumina interaction in the these reactions, an important area of SWNT carpet growth that has not yet been studied explicitly is how the type of alumina (based on deposition method) affects the activity and lifetime of the catalyst, SWNT quality and selectivity, and the growth termination process. Using a combination of microscopic and spectroscopic techniques, we have provided new insights into the Ostwald ripening behavior, activity, lifetime, and sub-surface diffusion of Fe catalysts supported on different alumina types.

See more of this Session: Synthesis of Graphene and Carbon Nanotubes: Kinetics, Mechanisms and Reactor Design
See more of this Group/Topical: Nanoscale Science and Engineering Forum

(703f) High Efficiency Rare Earth Doped Core-Shell Nanophosphors for Energy Applications

James Dorman[1], Gregory Kuzmanich[2], Abhijeet Joshi[3], Ju H. Choi[1] and Jane P. Chang[4], (1) Department of Chemical and Biomolecular Engineering, University of California at Los Angeles, Los Angeles, CA, (2)Chemisty and Biochemistry, University of California at Los Angeles, Los Angeles, CA, (3)Electrical Engineering, University of California at Los Angeles, Los Angeles, CA, (4)Department of Chemical and Biomolecular Engineering, Chemical Engineering Department, Cambridge, MA

The development of rare-earth (RE) doped phosphors allows for the conversion of photons to energies that are more usable for the desired application. Additionally, these RE phosphors have long lifetimes, on the order of ms, which offer potential in many energy conversion and energy transfer devices. Currently, RE phosphors are used in fiber optics amplifiers, modulated hybrid laser, optical interconnects and switches, optical displays, broad absorption solar cells and various other lighting applications. Energy transfer mechanisms of the excited RE states, such as defect quenching and sensitizer/emitter interactions, must be understood in order to achieve high efficiency energy conversion and propagation for future applications.

In order to synthesize high efficiency phosphors, trivalent RE ions are being doped into a core-shell metal oxide host lattice. The role of the core-shell structure reduces the effect of the surface quenching sites by increasing the distance between active ions and the surface hydroxyl groups. Secondly, the luminescent fingerprint can be further controlled with proper doping of the shell structure by either increasing the absorption spectrum or adding additional emission peaks. Primarily, this work focuses on the emission of visible photons through upconversion in Y_2O_3:Er^{3+}, Yb^{3+} nanophosphors, making them ideal components in broad absorption solar cells. By spatially controlling the position and concentration of the RE ions within the nanostructure, increased luminescence is observed due to energy transfer between the dopant ions within a critical interatomic distance. Passitvation of surface sites with increasing shell thickness was shown to increase luminescent lifetimes up to 53%, with a critical shell thickness of 8 nm, while lowering the theoretical lifetimes extracted from Judd-Ofelt parameters. The effect of the spatially controlled Yb ions was probed through the extraction of the upconversion photon requirement, showing a statistical decrease in photons from 2.16 to 1.43, or ~30 %. Finally, the effective energy transfer distance and energy transfer coefficients were studied by selectively exciting the Yb $^4F_{5/2}$-$^4F_{7/2}$ transition with a 904 nm laser diode while probing the Er^{3+} emission at 1540 nm as a Y_2O_3 spacer layer is added to the system. Measured results show the occurrence of energy transfer between the ions with a ~3 nm spacer layer, confirming the prediction of the Föester-Dexter theory. Additionally, these results are applied to a RE doped $LaPO_4$ system, producing white light from a single, high efficiency core-shell phosphor.

(374c) Atomic Layer Deposition As a Catalyst Synthesis Technique for Nickel Nanoparticles

Troy D. Gould[1], John L. Falconer[2], J. Will Medlin[3] and Alan W. Weimer[2], (1)Chemical and Biological Engineering, University of Colorado at Boulder, Boulder, CO, (2)Chemical & Biological Engineering, University of Colorado at Boulder, Boulder, CO, (3)Department of Chemical and Biological Engineering, College of Engineering and Applied Science, University of Colorado-Boulder, Colorado Center for Biorefining and Biofuels, Boulder, CO

Efforts to design new catalysts for renewable energy applications have revolved around increasing selectivity & activity by exploring advantages of catalysis at the nanoscale, mainly due to geometric & electronic effects. This work investigates the synthesis of Ni nanoparticle catalysts through the scalable gas phase synthesis route of Atomic Layer Deposition (ALD). Typically ALD deposits Ni as a NiO film, however; Ni nanoparticles were synthesized by using elevated reaction temperatures to produce particles with diameters <5nm. Smaller particles were created to have more catalytically active defects on the surface of the particles. The catalysts were characterized via Temperature Programed Reduction (TPR), chemisorption, Transmission Electron Microscopy (TEM), & Inductively Coupled Plasmon Mass Spectroscopy (ICP-MS) & compared to traditional incipient wetness Ni catalysts. The incipient wetness Ni catalysts were also used as a base comparison of catalytic activity for alkene hydrogenolysis reactions. We will also present the investigations of the controlled preparation of bimetallic catalysts using ALD.

See more of this Session: Functional Nanoparticles and Nanocoatings on Particles I
See more of this Group/Topical: Particle Technology Forum

(531a) Solution-Based Assembly of Large-Scale, Unidirectionally-Aligned Carbon Nanotubes for High-Performance Transistors

Guihua Yu[1], Melburne C. LeMieux[1], Benjamin C-K. Tee[2], Eric S.G. Shaqfeh[3] and Zhenan Bao[1], (1) Chemical Engineering, Stanford University, Stanford, CA, (2)Electrical Engineering, Stanford University, Stanford, CA, (3)Chemical Engineering and Mechanical Engineering, Stanford University, Stanford, CA

Central to many proposed electronic device-based applications of carbon nanotubes (CNTs) is the development of assembly methods that enable organization of these nanoscale building blocks over large areas with controlled orientation and density. In this talk, we will present a novel solution-based technique that combines spin assembly and molecular patterning for preparing large-scale, ultrathin CNT films with unidirectional alignment and controlled density. Guided by our simulation studies of particle adsorption to investigate both effects of the hydrodynamic force (spin rate) and surface interaction force (CNT-molecules), we have assembled well-ordered, unidirectionally-aligned CNTs over the wafer scale and directly integrated them as thin-film transistors (TFTs) that show promising device characteristics with large on/off ratio and high mobility. The scalable solution-based process for large-scale assembly of well-aligned CNT thin films can lead to many exciting applications ranging from large-scale, highly sensitive biological/chemical nanosensors to large-area flexible displays and solar cells.

See more of this Session: Graphene and Carbon Nanotube Based Devices
See more of this Group/Topical: Materials Engineering and Sciences Division

(504b) Polyplex Disassembly and DNA Release Observed by Real-Time AFM

Yi Zou, Lei Wan and Guangzhao Mao, Chemical Engineering and Materials Science, Wayne State University, Detroit, MI

This paper describes AFM studies of plasmid DNA release from individual polyplex nanoparticles. Three triggering mechanisms were used to cause polyplex disassembly: temperature gradient, redox reaction, and interpolyelectrolyte exchange reaction. First, AFM captured morphological change of the polyplex made by poly(ethylenimine)-graft-poly(N-isopropylacrylamide-co-N-dimethylacrylamide) (PEI-g-P(NIPA-co-DMA)) when temperature was moved across the lower critical solution temperature of the polycation. Second, in presence of glutathione or dithiothreitol, DNA release is triggered by a depolymerization of high-molecular-weight bioreducible polycations into low-molecular-weight oligocations via thiol and disulfide exchange reaction. The AFM images revealed a three-stage pathway beginning with a morphological change from metastable nanostructures into the more favorable toroid structure. Then toroids interact with each other by aggregation and fusion. Finally, DNA wormlike chains gradually unravel from the polyplex resulting in loose loops/tails that are held by a central compact core. Third, in DNA release induced by interpolyelectrolyte exchange, AFM images showed distinct intermediate polyplex structures including core-shell, toroid, nanoparticle decorated toroid, and loose chains held by a compact core structure. In all these studies, AFM is capable of capturing intermediate states of DNA release from polyplexes that point to some common features. For example, toroids and bundles tend to develop during early stages of polyplex disassembly. Loose DNA chains held by a central core tend to develop during late stages of polyplex disassembly. We were able to correlate some of the AFM results to *in vitro* observations.

See more of this Session: Bionanotechnology for Gene and Drug Delivery II
See more of this Group/Topical: Nanoscale Science and Engineering Forum

(176g) Palladium Nanowires Synthesized Via Templated Solid State Reduction for H2 Detection

Hector Mendez-Colberg and **Maria Martinez-Inesta**, Chemical Engineering Department, University of Puerto Rico, Mayaguez, PR

Pd nanowires have shown enhanced electrochemical detection of H_2 than the bulk due to higher surface to volume ratio. Here we show the results for ~5 nm Pd nanowires synthesized using zeolites as templates through a solid-state reduction method developed at our laboratory. After template removal, the Pd complex was characterized using TEM and hydrogen adsorption isotherms. Then it was deposited on a silicon based interdigitated electrode with platinum electrodes. The sensor showed the characteristic increase in resistance of bulk sensors, but with a mean response time of 10.12 seconds and sensitivity of 33% with exposure to 1% v/v H_2 that is orders of magnitude better than bulk sensors with recovery of the baseline. The hydrogen adsorption isotherm showed an α to β phase transition characteristic of nanoscale Pd.

See more of this Session: Nanostructured Materials for Electronic and Photonic Applications
See more of this Group/Topical: Materials Engineering and Sciences Division

(420r) Influence of Solvent Steric Effects On CO_2 Induced Nanoparticle Precipitation

Pranav S. Vengsarkar, Steven R. Saunders and Christopher B. Roberts, Department of Chemical Engineering, Auburn University, Auburn, AL

Nanoparticles have unique size-dependent chemical and physical properties that need to be harnessed for the proper application of these materials. However, the current methods used to produce monodisperse nanoparticles have several operational drawbacks such as the use of expensive equipment and large quantities of solvents or having low throughputs. A technique to size-selectively precipitate and fractionate organic phase gold and silver nanoparticle dispersions has been developed which utilizes the pressure tunable physico-chemical properties of CO_2 Gas-eXpanded Liquids (GXLs). This size-selective fractionation technique is based on the controlled reduction of the solvent strength through increases in the concentration of CO_2 (a known nonsolvent for aliphatic stabilizing ligands) via pressurization. These changes in solvent strength affect the subtle balance between the osmotic repulsive forces (due to the solvation of the stabilizing nanoparticle ligand tails) and the van der Waals forces of attraction between differently sized nanoparticles necessary to maintain a stable dispersion. Through modest changes in CO_2 pressure, increasingly smaller sized nanoparticles can be controllably precipitated from the dispersion.

The aim of this particular study is to investigate how the steric nature of a solvent affects the ligand-solvent interaction and thus the size dependent precipitation of nanoparticles using this GXL technique. Nanoparticles with a broad size distribution were synthesized by the arrested precipitation method using $HAuCl_4 \cdot 3H_2O$ as the precursor, TOAB as the phase transfer catalyst, $NaBH_4$ as the reducing agent and dodecanethiol as the stabilizing ligand. In order to investigate the steric effects of the solvent, these nanoparticles were dispersed in solvents with varying degrees of branching and then precipitated using the GXL technique. The pressures necessary to induce precipitation and the pressure range over which the nanoparticles precipitate in each of these solvents provide new physicochemical insights that help further our fundamental understanding of the phenomenon.

n-Hexane was chosen as the control solvent and the effect of several of its isomers on the precipitation and fractionation process will be discussed. The precipitation characteristics of dodecanethiol stabilized gold nanoparticles were quantified by measuring the intensity of the surface plasmon resonance band for the nanoparticle dispersion using UV-vis spectroscopy at various levels of CO_2 pressurization. The characterization of these nanoparticles was performed using transmission electron microscopy (TEM) to analyze their size distribution and hence judge the efficacy of the size-selective fractionation.

(369b) Stability of Engineered Nanoparticles In Aqueous Systems: Elucidating the Roles of Capping Agents and Natural Organic Matter

Jeffrey A. Nason, School of Chemical, Biological and Environmental Engineering, Oregon State University, Corvallis, OR

The rapid development of novel nanomaterials and their incorporation into consumer products has not been paralleled with an equal effort investigating the possible environmental implications. Increasingly, the properties of engineered nanoparticles (ENPs) are tailored for specific applications through the use of organic capping agents. Furthermore, when released into the environment it is probably that ENPs will interact with natural organic matter (NOM). The objective of this work is to develop an improved understanding of the roles of synthetic capping agents and NOM in controlling the aggregation of engineered nanoparticles (ENPs) in aquatic environmental systems. Using a suite of ligand-stabilized gold nanoparticles (AuNPs) and various NOM isolates, the specific aims of the project are to: (1) Correlate the physicochemical properties of capping agents with interactions between capped ENPs in aquatic systems; and (2) Correlate the physicochemical properties of capping agents and NOM with interactions between capped ENPs and NOM in aquatic systems. ENP stability with respect to aggregation is being assessed using the zeta potential and aggregation rates measured using time-resolved dynamic light scattering. Results indicate that NOM is adsorbed onto the AuNP surfaces, alterning surface charge and providing stability with respect to aggregation. Different classes of NOM stabilize AuNPs to different extents. The specific mechanisms of NOM-AuNP interactions are being probed using advanced surface and solution phases spectroscopic techniques (e.g., fluorescence, Raman, and EDX). By using a group of well-characterized ENPs that vary only with respect to their surface functionality, we aim to correlate environmental behavior with ENP properties in structure-activity type relationships. For example, results indicate that the stabilizing effects of NOM are related to the molecular weight of the specific NOM fractions. This work will provide an improved ability to predict environmental fate based on NP characteristics and yield the feedback necessary for the design of safer nanomaterials.

See more of this Session: Environmental Implications of Nanomaterials: Fate and Transport
See more of this Group/Topical: Topical C: Environmental Aspects, Applications, and Implications of Nanomaterials and Nanotechnology

(649c) Ionic Liquid Dispersed Nanocomposites

James A. Throckmorton, Department of Chemical and Biological Engineering, Drexel University, Philadelphia, PA and Giuseppe R. Palmese, Drexel University, Philadelphia, PA

Nanocomposites offer exciting ways to modify polymers and conventional composites by providing a moldable, flowable polymer-like composite with advantageous properties taken from embedded particles with desirable properties over small length scales. Increases in modulus, fracture toughness, and heat and electrical conductivity can be obtained with carefully selected particle types, geometries, loading fractions, processing methodologies, and polymer curing method.

Achieving effective macro-scale properties from these nanofillers requires good solvent chemistry and processing for nanoparticles dispersion, effective choice of nanoparticle material and geometry for the desired property improvements, and strong interfaces between the embedded particles and the polymer matrix. Our recent work has demonstrated that well-chosen ionic liquids can simultaneously disperse nanoparticles within an epoxy matrix and initiate epoxy cure reaction. This results in materials with potentially unique interphase properties resulting from interactions among the nanoparticle surface, ionic liquid, and epoxy resin.

This study explores the structure and properties of nanocomposites made with this unique processing method. Experimentally and conceptually, various methods of measuring dispersion for differing particle geometries, understanding the relationship between nano-scale dispersion and macroscopic properties, the effect of a percolated nanofiller network, and stress transfer through particles of different structures and geometries are explored.

See more of this Session: Graphene and Carbon Nanotubes: Applications
See more of this Group/Topical: Nanoscale Science and Engineering Forum

(662f) Manipulating Crystal Growth and Polymorphism by Confinement In Nanoscale Crystallization Chambers

Benjamin Hamilton, Artie McFerrin Department of Chemical Engineering, Texas A&M University, College Station, TX, Jeong-Myeong Ha, Department of Chemical Engineering, University of California, Berkeley, CA, Marc A. Hillmyer, Chemistry, University of Minnesota, Minneapolis, MN and Michael D. Ward, Molecular Design Institute, Department of Chemistry, New York University, New York, NY

The phase behaviors of crystalline solids embedded in nanoporous matrices have been studied for decades. Classic nucleation theory conjectures that phase stability is determined by the balance between a unfavorable surface free energy and stabilizing volume free energy. The size constraint imposed by nanometer-scale pores during crystallization results in large surface-to-volume ratios, which are reflected in crystal properties – for example, melting points and enthalpies of fusion – that differ drastically from their bulk-scale counterparts. Moreover, confinement within nanoscale pores can dramatically influence crystallization pathways and crystal polymorphism, particularly when the pore dimensions are comparable to the critical size of an emerging nucleus, at which the surface and volume free energies are in delicate balance and polymorph stability rankings may differ from bulk. Recent investigations have demonstrated that confined crystallization can be used to screen for and control polymorphism, which can impact food, pharmaceutical, explosive, and dye technological sectors, for which polymorph identity is critical for regulatory compliance and function.

We review our recent studies of the polymorphic and thermotropic properties of crystalline materials embedded in the nanometer-scale pores of porous glass powders and polymer monoliths, the latter generated from selective chemical etching of one component from shear-aligned diblock or triblock samples. The embedded nanocrystals exhibit an array of phase behaviors, including the selective formation of metastable amorphous and crystalline phases, thermodynamic stabilization of normally metastable phases, size-dependent polymorphism, formation of new polymorphs, and shifts of thermotropic relationships between polymorphs. Size confinement also permits the measurement of thermotropic properties that cannot be measured in bulk form using conventional methods. The well-aligned cylindrical pores of the polymer monoliths also allow determination and *manipulation* of nanocrystal orientation, wherein the constraints imposed by the pore walls result in a Darwinistic competition between nuclei that favors alignment of the fastest growth direction with the pore axis. Collectively, the examples described provide substantial insight into crystallization at a size scale that is difficult to realize by other means. Moreover, the behaviors resulting from nanoscopic confinement are remarkably consistent for a wide range of compounds, suggesting a reliable approach to studying the phase behaviors of compounds at the nanoscale. Newly emerging classes of porous materials promise expanded explorations of crystal growth under confinement and new routes to controlling crystallization outcomes.

Authors B.D. Hamilton and J.-M. Ha participated in this work while at the University of Minnesota Department of Chemical Engineering and Materials Science.

See more of this Session: Nanocrystal Science and Technologies
See more of this Group/Topical: Materials Engineering and Sciences Division

(342b) Molecular Simulation Studies of the Effect of Si/Al and Temperature On the Templated Synthesis of Pt Nanowires On VET-Type Zeolites

Javier Huertas-Miranda, Chemical Engineering, University of Puerto Rico - Mayaguez Campus, Mayaguez, PR and Maria Martinez-Inesta, Chemical Engineering Department, University of Puerto Rico, Mayaguez, PR

The synthesis of very thin metal nanowires is of special interest because of their unusual physical properties when compared with bulk materials. Among the synthesis methods for metal nanowires, the templated synthesis is a convenient and simple method that provides control over size and shape of these particles. The use of one-dimensional zeolites frameworks such as VET has been considered for these purposes. Here molecular simulations results are presented on the combined effect of Si/Al and temperature on the formation of stable one dimensional Pt structures inside the pores of VET zeolite framework. The results from grand canonical Monte Carlo simulations and geometry optimizations indicate that at high synthesis temperatures (623 K) and low Si/Al (3.25) quasicrystalline Pt nanowires are formed on the main pore channels of VET. These quasicrystalline structures were more thermally stable than configurations obtained at other simulation conditions.

See more of this Session: Templated Assembly of Inorganic Nanomaterials II
See more of this Group/Topical: Nanoscale Science and Engineering Forum

(346d) Analysis of Nanoparticle Transport In Magnetofection

Xiaozheng Xue and **Edward P. Furlani**, Dept. of Chemical and Biological Engineering, University at Buffalo (SUNY), Buffalo, NY

Magnetofection is a process in which a static magnetic field is used to direct and focus magnetic nanoparticles with surface-bound gene vectors towards target cells to enable transfection. The process enables rapid and highly efficient transfection and has been successfully demonstrated for all types of nucleic acids and across a broad range of cell lines. Magnetofection is well suited for multiwell culture plates wherein the magnetic force is produced by a separate rare-earth magnet positioned beneath each individual culture well. In this presentation we discuss a model for predicting particle transport in the magnetofection process. The model involves the solution of a drift-diffusion equation that governs the particle concentration in the culture well under the influence of a magnetic field. It accounts for the field-induced drift of the particles as well as fluidic drag and Brownian diffusion. The drift-diffusion equation is solved numerically using the finite volume method. We demonstrate the transport model via application to a conventional multiwell culture plate system and study particle transport and accumulation as a function of key system variables. We also study particle transport as the system is scaled from meso to microscale dimensions. The model provides insight into the physics of field-directed particle transport and enables rapid parametric analysis of particle accumulation on the cells, which is useful for optimizing novel magnetofection systems.

See more of this Session: Applications of Magnetic Nanoparticles
See more of this Group/Topical: Nanoscale Science and Engineering Forum

(217a) Assessment of Potential Exposure to Carbon Nanotubes In the Production of Polymer Nanocomposites

Drew J. Thompson[1], Lin Li[1], Jing Wang[2] and David Y.H. Pui[1], (1)Mechanical Engineering, University of Minnesota, Minneapolis, MN, (2)Institute of Environmental Engineering, ETH Zurich, Switzerland

Recent animal toxicology studies have shown that carbon nanotubes (CNTs) may pose a significant health risk to those exposed to this nanomaterial in the workplace. However, many questions regarding the toxicology of CNTs still remain unanswered. There has yet to be agreement on which dose metric would not only correlate with the negative health effects but also be able to be successfully implemented in field measurements. The roles of the inclusion of metal catalyst particles in and the morphology of CNTs and/or CNT structures in the toxicology of CNTs are still being investigated. While these matters are being researched further by toxicologists there remains a dearth of information on the actual exposure to CNTs. Even if CNTs are found to be hazardous, there will be no risk without exposure. To fully understand this potential risk effort must be taken to measure the occupational exposure to CNTs.

In this study the potential occupational exposure to CNTs in the production of polymer nanocomposites, where CNTs were used as a filler material, was assessed. The potential exposure to CNTs was quantified by analyzing respirable dust samples collected by personal sampling for elemental carbon. The mass concentration of elemental carbon, of which CNTs mainly consist, was determined by thermal-optical analysis of the sample. Potential exposure sources were identified and characterized by both real-time measurements and area sampling. The potential sources investigated were comprised of both production equipment and production related tasks, which included the handling of unbound CNTs. The real-time measurements were performed by instrumentation which provided number concentration, surface area concentration, respirable mass concentration, and size distributions for micron-sized particles. Area samples were analyzed by transmission electron microscopy to characterize the morphology and to calculate the number concentration of CNT structures. Other samples were taken for analysis by inductively coupled plasma mass spectrometry to measure the concentration of trace metals. The concentration of these metals was then used to estimate the CNT mass concentration based on known impurities. Thermal-optical analysis was performed on additional area samples to measure the elemental carbon mass concentration.

See more of this Session: Environmental Health and Safety (EHS) Concerns of Nanomaterials
See more of this Group/Topical: Topical C: Environmental Aspects, Applications, and Implications of Nanomaterials and Nanotechnology

(439f) Uniform In Vitro Gene Transfection Via High-Throughput Nano-Electroporation Chip

Xi Zhao[1], Lei Li[2], Yun Wu[2], Keliang Gao[2], Pouyan E. Boukany[2] and Ly James Lee[1], (1)William G. Lowrie Department of Chemical and Biomolecular Engineering, The Ohio State University, Columbus, OH, (2)Nanoscale Science and Engineering Center, The Ohio State University, Columbus, OH

A major disadvantage of most current drug/gene delivery systems is that dosage uniformity at cellular level cannot be achieved, with micro injection being the only exception. The standard deviation of gene copies transfected into individual cells through bulk electroporation, for example, can be as large as 250%. Therefore, the cellular responses to a single transfection will be different from each other and the overall result is an average, making it hard to draw precise and comprehensive conclusion.

We have developed a novel *in vitro* transfection method, nanoelectroporation (NEP), which provides controlled dosage delivery and thus can achieve uniform transfection at cellular-level. The NEP device has two arrays of micro channels facing each other, with their converging ends connected by nano tubes, which are created through etching gold-coated DNA wire away. Cells are placed at the end of micro channels and the unique dosage control characteristic is attributed to the close contact of cell membrane to nano channel, which condenses electric field and facilitates convection flow rather than diffusion. Being critical for NEP process, a precise cell placement is provided by optical trapping for small amount of cells, yet it becomes increasingly inefficient when cell population increases. Therefore, it is necessary to develop high-throughput NEP to transfect large amount of cells and extend the frontiers of NEP.

Our first design inherits the micro-nano-micro channel setup with greatly reduced cell side channel length allowing only one cell per micro channel, and replaces optical trapping with centrifugal force loading. By placing the chip at the periphery of a rotating disc, cells will spontaneously flow from a remote reservoir to the micro channel inlets and fill the channels with a solid contact. Excessive amount of cells will flow to a drain reservoir. After electroporation, de-loading can be achieved simply by turning the chip around for another centrifuge. Micro fluidic design on loading side and the operation conditions are optimized to ensure the best performance. For example, tethered antibody cell trapping spots can be deposited in front of micro channels to help capture one cell for each micro channel or even achieve selectivity on mixed cell sample. The overall design can be densely integrated so that one chip is capable of handling thousands of cells in several minutes.

Another new design replaces horizontal nanochannels with vertical nanonozzle array on a thin film produced by e-beam lithography or femtosecond laser fabrication. Such nanonozzle array provides similar electric field condensing function as aforementioned nanochannels. With this design, NEP can be performed on Z direction with cells sitting on top of the nanonozzle array and genes delivered from the bottom. The arrangement and distance between nozzles can be modified for optimal transfection efficiency. The nanonozzle array based NEP can transfect millions of cells at same time with controlled dosage delivery.

In this study, plasmid that encodes green fluorescence protein (pGFP) is used as the model gene and its delivery efficiency is characterized in mouse embryonic fibroblasts. With high throughput NEP, we are able to better understand the relationship between delivered dosage and transient/stable expression.

See more of this Session: Bionanotechnology for Gene and Drug Delivery I
See more of this Group/Topical: Nanoscale Science and Engineering Forum

(600g) Assembly and Scaffolding Strategies for Realizing Ultra-Thin Inorganic Films with Tunable Porosity

Shih-Chieh Kung, Zheng Tian and **Mark A. Snyder**, Department of Chemical Engineering, Lehigh University, Bethlehem, PA

Porous thin films scalable to tens of nanometers in thickness with tunable hierarchy in pore size and topology hold exciting technological implications spanning, among others, high-selectivity and high-flux membrane separations to rapid-response sensors selective for large and small target molecules. Owing to their inherent microporosity, inorganic materials such as zeolites, carbons, and some nanoparticle assemblies, hold specific promise in such applications where molecular selectivity is critical. Yet, control over porous film thickness has been a persistent challenge for films composed of such materials and supported on porous or non-porous substrates, leading in some cases to stress-induced crack formation and sluggish molecular flux. We address the challenge of controlling film thickness of porous inorganic films, and in some cases controlling underling porosity of these films, through a hierarchical templating strategy rooted in convective deposition of tunable nanoparticles and subsequent replication and materials scaffolding techniques. Specifically, we will address three classes of porous thin films scalable to just tens of nanometers in thickness: convectively deposited nanoparticulate colloidal crystalline films, carbon replica films characterized by either ordered mesoporosity or asymmetric mesopore-supported ultra-thin microporous layers, and (sacrificial) carbon-scaffolded ultra-thin zeolite films (e.g., silicalite-1, SAPO-34) composed of extensive high-aspect-ratio single-crystal domains. We will describe stages in film formation, sensitivity of film properties to synthesis conditions, transition from non-porous to highly-porous substrates as film supports, and the role of pore topology in thin film scaffolding. We will describe critical structure-property relations governing porous film performance with regards to achievable molecular selectivity and flux towards establishing proof-of-principle application of this synthesis approach to a broader class of porous inorganic materials.

See more of this Session: Nanostructured Thin Films
See more of this Group/Topical: Materials Engineering and Sciences Division

(499f) Lignin-Based Carbonaceous Nano-Fibrous Felts

Chuilin Lai[1], Lifeng Zhang[1], Hao Fong[1] and **Lew P. Christopher**[2], (1)Chemistry, South Dakota School of Mines & Technology, Rapid City, SD, (2)Center for Bioprocessing Research and Development and Department of Chemical and Biological Engineering, South Dakota School of Mines & Technology, Rapid City, SD

Lignin is a naturally existing biopolymer widely found in lignocellulosic plant biomass at an average of 20-30%. It is also a high-volume byproduct of wood pulping which exceeds 15 million tons per year in the US. Nonetheless, lignin has been utilized mainly as a low-grade fuel for production of heat and power to date. There is however a growing demand to identify applications with high economic value for lignin. Although production of carbon fibers from lignin has been reported in literature, the low mechanical properties obtained so far have prevented their commercial use in reinforcement and/or composites applications. Nano-fibrous felts consisting of porous carbon nanofibers derived from lignin, however, have potential applications that do not rely on mechanical properties but on large specific surface area. The use of low-cost lignin as the precursor to develop carbonaceous nano-fibrous felts is anticipated to have a great impact on these applications. The nanomaterials processing technique of electrospinning provides a straightforward approach for convenient preparation of fibers with diameters ranging from tens to hundreds of nanometers. In this research, innovative carbonaceous nano-fibrous felts with high specific surface area were developed by electrospinning aqueous mixtures of alkali lignin and poly(vinyl alcohol) with varied mass ratios followed by heat treatments for stabilization and carbonization. The developed carbonaceous nano-fibrous felts had a large pore volume and specific surface area and are therefore expected to find important applications as catalyst support, supercapacitor electrode, and hydrogen storage materials.

See more of this Session: Biobased Materials II: Lignin-Based Materials
See more of this Group/Topical: Forest and Plant Bioproducts Division

(420s) PEG-Based Functionalization of Iron Oxide Nanoparticles Using the ISOFURE Methodology

Robert J. Wydra, David S. Spencer and J. Zach Hilt, Chemical and Materials Engineering, University of Kentucky, Lexington, KY

PEG-basedFunctionalization of Iron Oxide Nanoparticles Using the ISOFURE Methodology

Robert J. Wydra,David S. Spencer, J. Zach Hilt

Department ofChemical and Materials Engineering, University of Kentucky

Magnetic nanoparticles, withtheir unique physical properties, are being studied for a wide range ofbiomedical applications such as imaging, targeted delivery, and thermal therapyof cancer. Surface modification plays an essential role in determining thesuccess of the particles in their application by improving the stability,preventing agglomeration, improving biocompatibility, and providing additionalfunctionality (e.g. targeting antibodies). Poly(ethylene glycol) (PEG) basedfunctionalization is common for biological applications as a mean to preventprotein adsorption and thus improve the circulation time and minimize hostresponse. Previous work has investigated using a solution based atom transferradical polymerization (ATRP) to successfully create PEG-iron oxide core-shellnanoparticles. In this study, a novel strategy to isolate, functionalize, andsubsequently release (ISOFURE) functionalized nanoparticles was harnessed.Initially iron oxide nanoparticles were synthesized using the facile co-precipitationtechnique. The bare particles were entrapped in a biodegradablepoly(β-amino ester) (PBAE) hydrogel. While isolated, the particles weremodified with poly(ethylene glycol) (MW=400) dimethacrylate (PEG400DMA) and other monomersusing ATRP. Following the reaction, the PBAE hydrogel network wasdegraded in water, and the functionalized nanoparticles were collected forcharacterization. The particles were characterized using Fourier transforminfrared spectroscopy, to verify surface functionalization; thermal gravimetricanalysis, to quantify mass percent of coating; dynamic light scattering, todetermine particle size; and UV-Vis spectroscopy, to determine particlestability in a variety of media.

See more of this Session: Poster Session: Nanoscale Science and Engineering
See more of this Group/Topical: Nanoscale Science and Engineering Forum

(176d) Multi-Scale Optical, Electrical, and Chemical Interrogation of Thiophene-Based Solar Cell Films

Chris Carach, Isaac Riisness and **Michael Gordon**, Chemical Engineering, University of California, Santa Barbara, Santa Barbara, CA

Understanding and controlling carrier transport in conjugated polymer films and composites is critical to the development and widespread application of plastic solar cells. Recent efforts have focused on "bulk heterojunction" (BHJ) structures where a conjugated polymer is mixed *at the nanoscale* with a fullerene acceptor to achieve large interfacial areas for exciton splitting. It has been seen that fabrication protocols drastically affect device efficiency and that charge transport is intimately tied to film morphology through local disorder, domain formation, and compositional heterogeneity. Since charge transport in these systems occurs over distances <100 nm, it is no surprise that *local* physicochemical properties have a tremendous impact on photovoltaic behavior. Unfortunately, our understanding of these issues is poor because conventional PV testing on devices tends to "average out" micro- and nanoscale heterogeneities. To this end, we combine device-level testing of BHJ films with confocal/near-field optical and electrical interrogation (conductive AFM and electrostatic force microscopy) to understand how nanoscale morphology, molecular alignment, chemical structure, and processing affect charge transport processes. In this work, we show that confocal Raman and photoluminescence (PL) imaging of polythiophene/fullerene blends, when combined with electrical AFM techniques, are highly descriptive means to study donor/acceptor mixing and nanoscale transport. For instance, spatially-correlated optical and electrical imaging are combined to differentiate between regions of phase separation, low donor concentration, and oxidative damage. The talk will also highlight how low temperature PL measurements and analysis of excitonic-vibronic coupling in polythiophene/blend films processed under different conditions (e.g. solvents, thermal annealing, etc.) can be used to probe local order and crystallinity. Preliminary results on near-field Raman investigations of P3HT-PCBM blends for localized chemical imaging at the nanoscale will be mentioned as well.

See more of this Session: Nanostructured Materials for Electronic and Photonic Applications
See more of this Group/Topical: Materials Engineering and Sciences Division

QUANTUM TUNNELING IN CARBON-NANOTUBE TRANSISTORS

Meng-Mu Shih, University of Florida, Gainesville, FL

Abstract

A modified model is used to discuss the quantum tunneling effects in carbon nanotube (CNT) transistors. In this work, CNTs with nano-meter diameters are studied. This modified prototype model can consider several CNT-related parameters such as CNT diameter, applied electrical field, effective mass, tunneling barrier height affected by metal contacts or strain, and tunneling width tuned by dopants. Numerical results will be discussed for design and performance issues of transistors. Some of the results show the possibilities of obtaining high currents more easily. However, some currents are not desirable or should be kept as low as possible. The on/off tunneling ratio can evaluate the transistor performance in circuit design. Some further results can be used to explain the working principles of CNT-transistor sensors.

Model Description

CNT–based transistors are proposed and modeled [1]. Figure 1 shows the schematic structure for the quantum tunneling at the metal-semiconductor interface [2, 3]. The modified CNT-based quantum tunneling model has parameters: CNT diameter (d), effective mass (m_e), barrier height (ϕ_b), tunneling width (w), and electrical field (E). The CNT effective mass can be calculated [4, 5]. The barrier height is assumed by mid-gap type of CNT material.

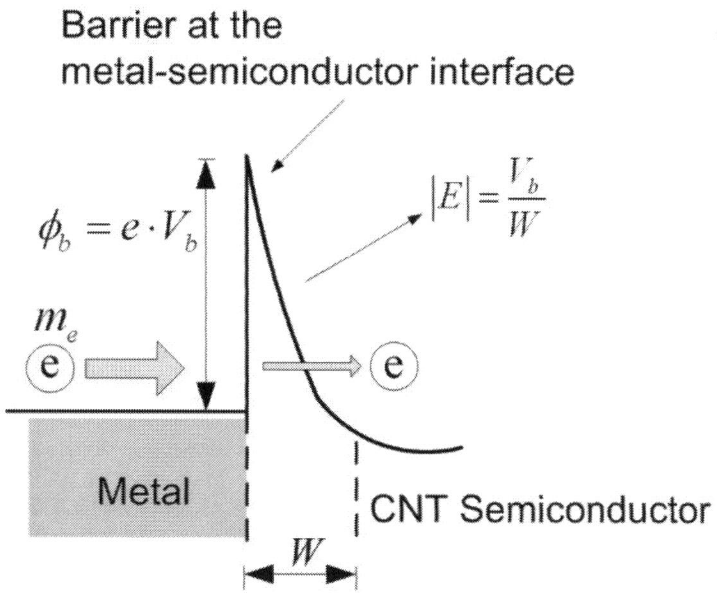

Figure 1 Schematic structure for quantum tunneling at the metal-semiconductor interface

Numerical Results and Discussion

The calculation results are based on the quantum tunneling model described in Figure 1. Figure 2 shows the tunneling probability (T) versus CNT diameter (d) for electrical field E = 5×10^9 V/m and 10^9 V/m. Since the CNT with larger diameter has smaller tunneling barrier height and smaller effective mass, the tunneling is easier and the tunneling probability is higher than the CNT with smaller diameter. The larger electrical field can assist the tunneling and cause higher tunneling probability. However, the CNT with larger diameter can have larger leakage tunneling. Figure 3 shows the tunneling probability (T) versus electrical field (E) for CNT with diameter d = 10 nm and 2 nm. The CNT with larger diameter can reach tunneling probability of unity faster with smaller electrical field. Figure 4 shows the tunneling probability (T) versus tunneling width (w) for CNT with diameter d = 1, 2, 4, 8 nm. This tunneling width can be affected by doping. The smaller tunneling width can make electrons easier to tunnel through and cause the tunneling probability to be higher. The CNT with larger diameter can make quantum tunneling much easier.

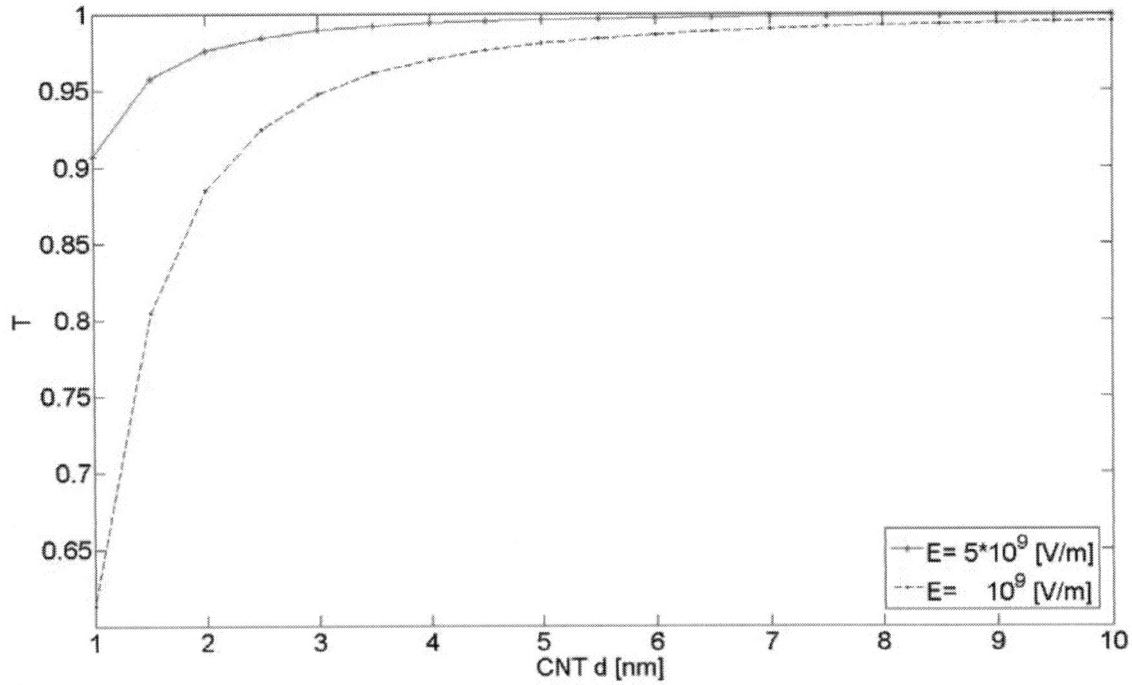

Figure 2 Tunneling probability (T) versus CNT diameter (d) for electrical field E = 5×10^9 V/m and 10^9 V/m

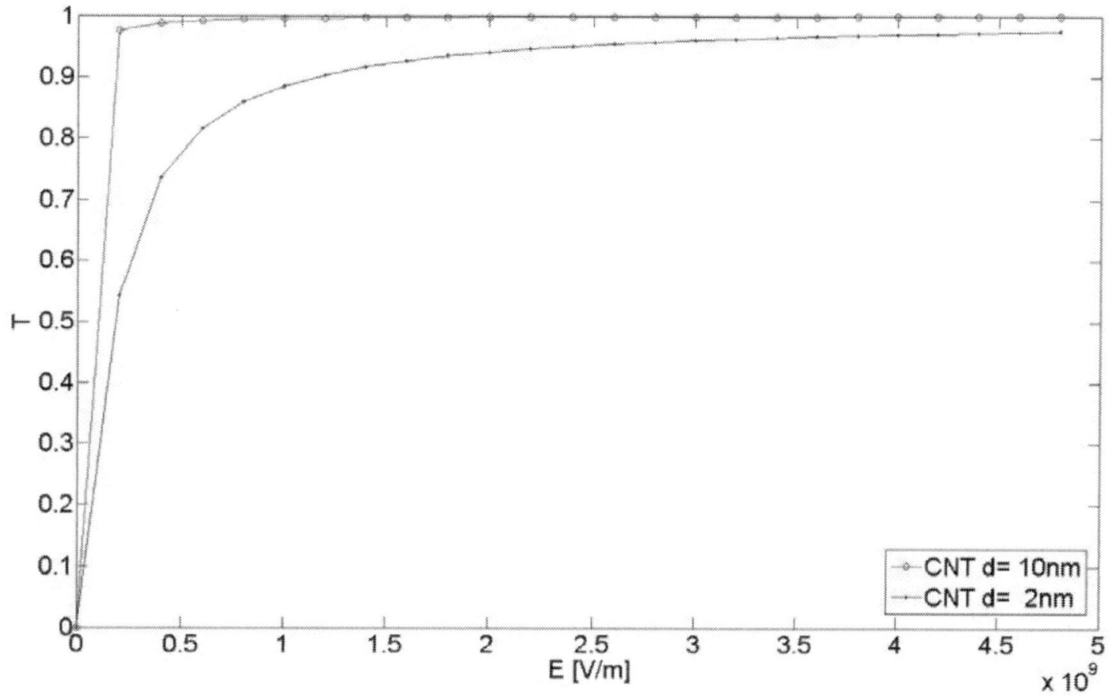

Figure 3 Tunneling probability (T) versus electrical field (E) for CNT with diameter d = 10 nm and 2 nm

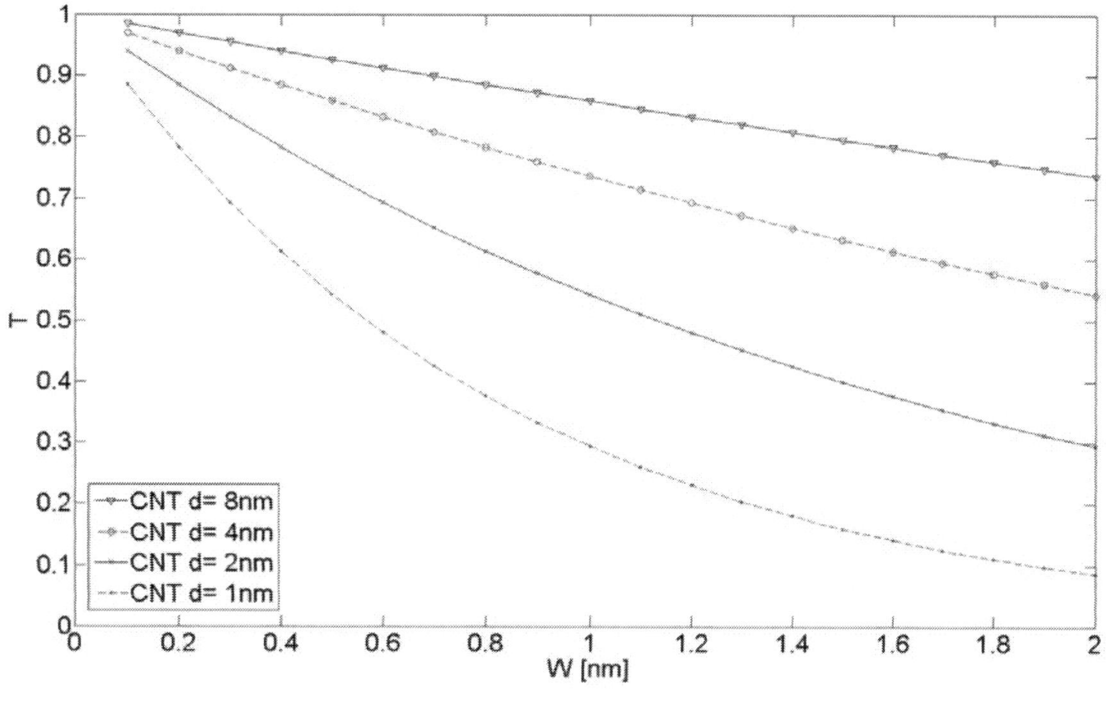

Figure 4 Tunneling probability (T) versus tunneling width (w) for CNT with diameter d = 1, 2, 4, 8 nm.

References

1. Guo, Jing et al. (2004), "A Numerical Study of Scaling Issues for Schottky-Barrier Carbon Nanotube Transistors," *IEEE Transaction on Electron Devices*, Vol. 51, No. 2, February, pp. 172-177.
2. Esaki, Leo (1973), "Long Journey to Tunneling," http://www.nobelprize.org/nobel_prizes/physics/laureates/1973/esaki-lecture.pdf
3. Singh, Jasprit (2004), "Modern Physics for Engineers," Weinheim, Germany, WILEY-VCH Verlag GmbH & KGaA.
4. Pennington, Gary and Goldsman, Neil (2003), "Theory and Design of Field-Effect Carbon Nanotube Transistor," in *International Conference on Simulation of Semiconductor Processes and Devices*, pp.167-170.
5. Marulanda, Jose M. and Srivastava, Ashok (2008), "Carrier Density and Effective Mass Calculations in Carbon Nanotubes," *Physica Status Solidi*, (b) Vol. 245, No. 11, pp. 2558–2562.

(420t) Sonolytic Synthesis of Single and Binary, Metal-Based Magnetic Nanostructured Materials

Gerard L. Moore[1], **Jayson P. Wicker**[1], Kenneth L. Roberts[2] and Dhananjay Kumar[3], (1)Chemical Engineering, North Carolina A&T State University, Greensboro, NC, (2)Chemical, Biological and Bio-Engineering, North Carolina A&T State University, Greensboro, NC, (3)Mechanical Engineering, North Carolina A&T State University, Greensboro, NC

Magnetic nanomaterials were synthesized using sonolytic techniques. Single metal- and binary-based nanostructured materials with various morphologies and magnetic properties were produced via sonochemical reactions with NH_3OH solutions and iron, magnesium and calcium chlorides. All magnetic nanomaterial products were characterized using x-ray diffraction (XRD), elemental analysis, scanning and transmission electron microscopies (SEM and TEM) and superconducting quantum interference device (SQUID) magnetometric analysis. Morphology and crystallite product sizes varied with the metal components, solution composition, and pH level. Crystallite sizes were observed to range from 5.92 to 72.13 nm. Various nanostructured products were observed to include nanoparticles, nanocrystallites, nanorods, microcubics and microdiscoids.

Magnetic data for sonolitically altered mixtures and alloyed materials were also measured in this work. Ultrasonic irradiation provided an effective pathway for the synthesis of the magnetic nanomaterials through redox reaction with metal chlorides and NH_3OH. The data revealed magnetic properties for all solid systems. $Mg(OH)_2$ and $CaCO_3$-based systems displayed an agreement with properties that coincided with superparamagnetic properties, where the plot of the magnetic field divided by the system temperature collapsed respectively. $Mg(OH)_2 + Fe(OH)_3$ system revealed an agreement with paramagnetic properties, meaning that the graph denoted that under applied field the magnetic moment was observed to be linear with a positive slope. As for, $Fe_3O_4 + CaCO_3$ mixtures, $MgFe_2O_3$ and Fe_3O_4 under applied magnetic field indicated a ferromagnetic hysteresis plot, meaning that the particles retained magnetic field after saturation. A coercivity study was performed to determine the intensity of an applied magnetic field to reduce the magnetization of the particles to zero for the ferromagnetic structures and this study showed a linear decline as temperature increased. It was also observed that $MgFe_2O_3$ required the highest magnetic field to reduce magnetization to zero.

See more of this Session: Poster Session: Nanoscale Science and Engineering
See more of this Group/Topical: Nanoscale Science and Engineering Forum

(385d) Synthesis and Characterization of PEG-Iron Oxide Core-Shell Nanoparticles for Dual Hyperthermia and Chemotherapy Treatment of Cancer

Robert J. Wydra[1], Anastasia M. Kruse[1], Younsoo Bae[2], Kimberly W. Anderson[1] and J. Zach Hilt[1], (1) Chemical and Materials Engineering, University of Kentucky, Lexington, KY, (2)Pharmaceutical Sciences, University of Kentucky, Lexington, KY

Synthesis andCharacterization of PEG-Iron Oxide Core-shell Nanoparticles for DualHyperthermia and Chemotherapy Treatment of Cancer

Robert J. Wydra[1],Anastasia M. Kruse[1], Younsoo Bae[2], Kimberly W. Anderson[1],J. Zach Hilt[1]

[1]Departmentof Chemical and Materials Engineering, University of Kentucky

[2]Departmentof Pharmaceutical Sciences, University of Kentucky

Hyperthermia,the heating of tissue in the 41-45C range, can induce cellular death on itsown or work in conjunction with chemotherapy for improved cancer therapy [1].In this study, core-shell nanocomposites were prepared with the intent ofco-delivery of a chemotherapeutic (Geldanamycin and17-N-Allylamino-17-demethoxygeldanamycin) and heat. The core-shellnanoparticles were prepared using atomic transfer radical polymerization (ATRP)to coat iron oxide (Fe_3O_4) nanoparticles with apoly(ethylene glycol) (PEG) based polymer shell. The iron oxide core allowsfor the remote heating of the particles in an alternating magnetic field(AMF). Nanoparticle mediated thermal therapy feasibility was demonstrated *invitro* with a thermoablation (55C) study on A549 lung carcinoma cells.Combinational therapy of chemotherapeutic and hyperthermia on A549 wasinvestigated to demonstrate an enhanced therapeutic effect.

FTIRmeasurements verified the PEG coating by observing peaks at 1715cm^{-1}and 1105cm^{-1}, which represent the carbonyl group (C=O) and ethergroup (C-O-C), respectively. TGA indicated different mass loss profiles andslight differences in overall mass loss between the core citrate coatedparticles and the polymer coated particles. The results from thermal therapydemonstration indicated that there is no toxic effect from the AMF coil orparticles in solution over the timeframe of the experiment. When the particleswere exposed to the AMF field, there was complete cellular death as a result ofthe heat generate by the nanoparticles.

ATRP wassuccessfully utilized to coat iron oxide nanoparticles with a PEG based polymershell. For the time frame of the thermal therapy experiments, there is aminimal toxic effect observed in A549 cells for the 10 mg/ml concentration ofPEG-coated Fe_3O_4 nanoparticles. Thermoablation of A549demonstrates the potential use of polymer coated particles for thermaltherapy. Future work involves the refinement of the dual therapy study todemonstrate a synergistic effect of a co-delivery of chemotherapeutics andhyperthermia from magnetic nanoparticles.

References:

[1] Issels RD.Eur J Cancer. 2008:44:2546-2554.

See more of this Session: Magnetic Nanoparticles In Biotechnology and Medicine
See more of this Group/Topical: Nanoscale Science and Engineering Forum

(420u) In Situ XPS Study of the Influence of Water Vapor On Catalytic Decomposition of Ethylene During Carbon Nanotube Growth

Placidus B. Amama, Materials & Manufacturing Directorate, RXBT, Wright-Patterson Air Force Research Laboratory/UDRI, Dayton, OH, Tyson C. Back, Materials and Manufacturing Directorate, Wright-Patterson Air Force Research Laboratory/UTC, Dayton, OH, Terry P. Murray, Materials Engineering, University of Dayton, Dayton, OH, Steven B. Fairchild, Materials & Manufacturing Directorate, Wright-Patterson Air Force Research Laboratory, Dayton, OH and Benji Maruyama, Materials & Manufacturing Directorate, RXBN, Wright Patterson Air Force Research Laboratory, Dayton, OH

Water-assisted chemical vapor deposition (CVD) has become one of the leading methods for producing vertically aligned carpets of single-walled carbon nanotube (SWNT) due to its high catalytic efficiency. The mechanism for the dissociative adsorption of the hydrocarbon on catalyst surfaces in the presence of water vapor is still not fully understood. In this work, we present critical evidence from in situ X-ray photoelectron spectroscopy (XPS) and Auger electron spectroscopy (AES) experiments using ethylene decomposition over Fe catalyst that further rationalizes the growth enhancement observed during water-assisted CVD growth of nanotubes.

See more of this Session: Poster Session: Nanoscale Science and Engineering
See more of this Group/Topical: Nanoscale Science and Engineering Forum

(420v) Characterization of Gas-Expanded Liquid-Deposited Gold Nanoparticle Films On Substrates of Varying Surface Energy

K. M. Hurst, **Jie Zhong**, Christopher B. Roberts and W. R. Ashurst, Department of Chemical Engineering, Auburn University, Auburn, AL

Alkanethiol-stabilized gold nanoparticles (AuNPs) were deposited via a gas-expanded liquid (GXL) technique utilizing CO_2-expanded hexane onto substrates of different surface energy. The different surface energies were achieved by coating silicon (100) substrates with various organic self-assembled monolayers (SAMs). Following the deposition of AuNP films, the films were characterized to determine the effect of substrate surface energy on nanoparticle film deposition and growth. Interestingly, the critical surface tension of a given substrate does not directly describe nanoparticle film morphology. However, the results in this study indicate a shift between layer-by-layer and island film growth based on the critical surface tension of the capping ligand. Additionally, the fraction of surface area covered by the AuNP film decreases as the oleophobic nature of the surfaces increases. Moreover, the influence of different capping ligands and the AuNP size and size distribution will also be discussed. On the basis of this information, the potential exists to engineer nanoparticle films with desired morphologies and characteristics.

See more of this Session: Poster Session: Nanoscale Science and Engineering
See more of this Group/Topical: Nanoscale Science and Engineering Forum

(609a) Exciton Engineering with Carbon Nanotubes and Graphene for Solar Energy Conversion: From Exciton Antennae to Nano-Heterojunctions

Michael Strano[1], Geraldine LC Paulus[2], Moon-Ho Ham[2], Chang Young Lee[2], Changsik Song[2], Kourosh Kalantar-Zadeh[3], Wonjoon Choi[4], Jae-Hee Han[1], Ryuichiro Maruyama[5], Esther S. Jeng[6], Daniel A. Heller[2], Woo-Jae Kim[1], Paul W. Barone[7] and Cristiano Fantini[8], (1)Chemical Engineering, Massachusetts Institute of Technology, Cambridge, MA, (2)Department of Chemical Engineering, Massachusetts Institute of Technology, Cambridge, MA, (3)School of Electrical and Computer Engineering, RMIT University, Melbourne, Australia, (4)Department of Mechanical Engineering, Massachusetts Institute of Technology, Cambridge, MA, (5)Materials Laboratories, Sony Corporation, Kanagawa, Japan, (6)Department of Chemistry, University of Illinois at Urbana-Champaign, Urbana, IL, (7)Chemical Engineereing, Massachusetts Institute of Technology, Cambridge, MA, (8)Departamento de Fisica, Universidade Federal de Minas Gerais, Belo Horizonte-MG, Brazil

Exciton engineering with carbon nanotubes and graphene for solar energy conversion: from exciton antennae to nano-heterojunctions

Michael S. Strano

Charles and Hilda Roddey Associate Professor of Chemical Engineering

66-566 Department of Chemical Engineering

77 Massachusetts Avenue

Cambridge, MA 02139-4307

Email: strano@MIT.EDU

phone: (617) 324-4323

fax: (617) 258-8224

http://web.mit.edu/stranogroup/

Our laboratory has been interested in how low dimensional materials, such as single walled carbon nanotubes or graphene, can augment and enhance solar conversion efficiencies and demonstrate new photonic concepts. We will discuss two recent examples of our work in this space.

In the first example, there is significant interest in combining carbon nanotubes with semiconducting polymers for photovoltaic applications because of potential advantages from smaller exciton transport lengths and enhanced charge separation. However, to date, bulk heterojunction (BHJ) devices have demonstrated relatively poor efficiencies, and little is understood about the polymer/nanotube junction. To investigate this interface, we fabricate a planar nano-heterojunction comprising well-isolated millimeter-long single-walled carbon nanotubes underneath a poly(3-hexylthiophene) (P3HT) layer (1). The resulting junctions display photovoltaic efficiencies per nanotube ranging from 3% to 3.82%, which exceed those of polymer/nanotube BHJs by a factor of 50-100. The increase is attributed to the absence of aggregate formation in this planar device geometry. It is shown that the polymer/nanotube interface itself is responsible for exciton dissociation. Typical open-circuit voltages are near 0.5 V with fill factors of 0.25-0.3, which are largely invariant with the number of nanotubes per device and P3HT thickness. A maximum efficiency is obtained for a 60 nm-thick P3HT layer, which is predicted by a Monte Carlo simulation that takes into account exciton generation, transport, recombination, and dissociation. We combine for the first time both optical T-matrix and kinetic Monte Carlo models to investigate the photocurrent generation in two state-of-the-art PHJ photovoltaics. The combined model takes into account the rates of exciton generation, transport, recombination and dissociation using literature values. By including the optical, electronic and structural properties of the different materials, we are able to predict the short-circuit current of recently reported P3HT/SWNT PHJ and also P3HT/PCBM PHJ solar cells from the literature. The experimental data for each of these devices show a maximum photocurrent output at a P3HT thickness of 60-65 nm, in contradiction to the expected value equal to the diffusion length of excitons in P3HT (8.5nm). The model demonstrates how a bulk exciton sink can explain this shifted maximum in the P3HT/SWNT case, whereas the maximum is mainly determined by PCBM interdiffusing in P3HT in the P3HT/PCBM case. This platform is promising for further understanding the potential role of polymer/nanotube interfaces for photovoltaic applications.

In the second example, there has been renewed interest in solar concentrators and optical antennas for improvements in photovoltaic energy harvesting and new optoelectronic devices. We dielectrophoretically assemble single-walled carbon nanotubes (SWNTs) of homogeneous composition into aligned filaments that can exchange excitation energy, concentrating it to the centre of core–shell structures with radial gradients in the optical bandgap (2). We find an unusually sharp, reversible decay in photoemission that occurs as such filaments are cycled from ambient temperature to only 357 K, attributed to the strongly temperature-dependent second-orderAuger process. Core–shell structures consisting of annular shells of mostly (6, 5) SWNTs (E_g = 1.21 eV) and cores with bandgaps smaller than those of the shell (E_g = 1.17 eV (7,5)–0.98 eV (8,7)) demonstrate the concentration concept: broadband absorption in the ultraviolet–near-infrared wavelength regime provides quasi-singular photoemission at the (8, 7) SWNTs. This approach demonstrates the potential of specifically designed collections of nanotubes to manipulate and concentrate excitons in unique ways.

1. Ham MH, Paulus GLC, Lee CY, Song C, Kalantar-zadeh K, Choi W, Han JH and **Strano MS**: Evidence for High-Efficiency Exciton Dissociation at Polymer/Single-Walled Carbon Nanotube Interfaces in Planar Nano-heterojunction Photovoltaics. ACS NANO, 4 (2010) 6251-6259

2. Han JH, Paulus GLC, Maruyama R, Jeng ES, Heller DA, Kim WJ, Barone PW, Lee CY, Choi JH, Ham MH, Song C, Fantini C, **Strano MS**: Exciton Antennas and Concentrators from Core-Shell and Corrugated Carbon Nanotube Filaments of Homogeneous Composition. NATURE MATERIALS, 9, 833 - 839 (2010).

See more of this Session: Semiconductor Nanowires and Nanotubes for Next Generation Photovoltaics From Fundamentals to Devices
See more of this Group/Topical: Topical 5: Nanomaterials for Energy Applications

(609d) Photoelectrochemical Complexes for Solar Energy Conversion That Dynamically, Reversibly and Autonomously Self-Assemble

Ardemis A. Boghossian[1], Jong Hyun Choi[2], Moon-Ho Ham[1], Esther S. Jeng[3], Rachel A. Graff[4], Daniel A. Heller[1], Alice C. Chang[3], Aidas Mattis[5], Timothy H. Bayburt[6], Yelena V. Grinkova[6], Adam Scott Zeiger[7], Krystyn J. Van Vliet[8], Erik K. Hobbie[9], Stephen G. Sligar[6], Colin A. Wraight[6] and Michael Strano[3], (1)Department of Chemical Engineering, Massachusetts Institute of Technology, Cambridge, MA, (2)Mechanical Engineering, Purdue University, West Lafayette, IN, (3)Chemical Engineering, Massachusetts Institute of Technology, Cambridge, MA, (4)Chemical and Biomolecular Engineering, University of Illinois at Urbana-Champaign, Urbana, IL, (5)Department of Biochemistry, University of Illinois at Urbana-Champaign, Urbana, IL, (6)Biochemistry, University of Illinois Urbana-Champaign, Urbana, IL, (7)Materials Science and Engineering, Massachusetts Institute of Technology, Cambridge, MA, (8)Materials Science and Engineering, MIT, Cambridge, MA, (9)Polymers Division, National Institute of Standards and Technology, Gaithersburg, MD

Photoelectrochemical complexes for solar energy conversion that dynamically, reversibly and autonomously self-assemble

Naturally occurring photosynthetic systems use elaborate pathways of self-repair to limit the impact of photo-damage. Here, we demonstrate a complex consisting of an aqueous solution containing photosynthetic reaction centers (RCs), membrane scaffold proteins (MSPs), phospholipids, and single-walled carbon nanotubes (SWCNTs) solubilized with the surfactant sodium cholate (SC) that mimics this process.(1) The components reversibly self-assemble into a highly ordered structure upon dialysis of the surfactant. The resulting assembled structure is photoelectrochemically active and consists of 4-nm-thick lipid bilayer disks (nanodisks, NDs) arranged parallel to the surface of the SWCNT with the RC housed within the bilayer such that its hole injecting site faces the nanotube surface. The structure can be assembled and disassembled autonomously with the addition or removal of surfactant. We model the kinetic and thermodynamic forces that drive the dynamics of this reversible self-assembly process.(2) The assembly is monitored using spectrofluorimetry during dialysis and subsequent surfactant addition and used to fit a kinetic model to determine the forward and reverse rate constants of ND and ND-SWCNT formation. The calculated ND and ND-SWCNT forward rate constants are 79 mM^{-1} s^{-1} and 5.4×10^2 mM^{-1} s^{-1}, respectively, and the reverse rate constants are negligible over the dialysis time scale. We find that the reaction is not diffusion-controlled since the ND-SWCNT reaction, which consists of entities with smaller diffusion coefficients, has a larger reaction rate constant. Using these rate parameters, we were able to develop a kinetic phase diagram for the formation of ND-SWCNT complexes, which indicates an optimal dialysis rate of approximately 8×10^{-4} s^{-1}.

The assembly is thermodynamically metastable and can only transition reversibly if the rate of surfactant removal exceeds a threshold value. Only in the assembled state do the complexes exhibit photoelectrochemical activity. We experimentally demonstrate a regeneration cycle that uses surfactant to switch between assembled and disassembled states, resulting in an increased photoconversion efficiency of more than 300% over 168 hours and an indefinite extension of the system lifetime. The system can disassemble and reassemble over an indefinite number of cycles. We also fit the model to these cyclic ND-SWCNT assembly and disassembly experiments and hence mimic the thermodynamic forces used in regeneration processes detailed previously. Such forces may form the basis of both synthetic and natural photoelectrochemical complexes capable of dynamic component replacement and repair.

1. Ham MH, Choi JH, Boghossian AA, Jeng ES, Graff R.A, Heller DA, Chang AC, Mattis A, Bayburt TH, Grinkova YV, Zeiger AS, Van Vliet KJ, Hobbie EK, Sligar SG, Wraight CA and **Strano MS**: Photoelectrochemical complexes for solar energy conversion that chemically and autonomously regenerate. NATURE CHEMISTRY, 2 (2010) 929-936.

2. Boghossian AA, Choi JH, Ham MH and **Strano MS**: Dynamical and Reversible Self-Assembly of Photoelectrochemical Complexes Based on Lipid Bilayer Disks, Photosynthetic Reaction Centers, and Single-Walled Carbon Nanotubes. Langmuir, 27 (2011) 1599-160

See more of this Session: Semiconductor Nanowires and Nanotubes for Next Generation Photovoltaics From Fundamentals to Devices
See more of this Group/Topical: Topical 5: Nanomaterials for Energy Applications

(543e) Monitoring of Dissolution of Poorly Water-Soluble Drugs During In Vitro Lipolysis by Electron Resonance Spectroscopy

Selena Di Maio[1], David E. Budil[2] and Rebecca L. Carrier[1], (1)Chemical Engineering, Northeastern University, Boston, MA, (2)Chemistry and Biological Chemistry, Northeastern University, Boston, MA

The recognized capability of ingested lipids to enhance the bioavailability of poorly water-soluble drugs is not well understood and is unpredictable. For this class of drug molecules, whose dissolution in water is likely the limiting-step of overall oral absorption, the influence of ingested lipids on oral absorption has been associated with complex, poorly characterized interactions between drugs and several colloidal nanostructures, taking place during lipid digestion. In particular, emulsion droplets, micelles and vesicles influence dissolution kinetics and are able to maintain a larger quantity of hydrophobic drug in solution, increasing the solubilization power of the gastrointestinal (GI) tract contents.

In order to quantitatively investigate how lipid digestion impacts dissolution of poorly water-soluble drugs via micelle/vesicle formation, an updated *in vitro* lipid digestion model has been designed, consisting of a bio-relevant medium, which reflects gastrointestinal contents in the fed state. Lipid digestion experiments were initiated by adding lipase/colipase enzymes and a lipid substrate – simple triglycerides (TG) or partially digested lipids (lipid mixture). The extent of lipid digestion was monitored indirectly by recording the volume of NaOH added during the experiment to titrate the fatty acids (FA) formed from the TG hydrolysis. Samples were collected at specific time intervals before and during the *in vitro* lipid digestion experiments and analyzed by dynamic light scattering (DLS), and electron paramagnetic resonance (EPR), to characterize respectively: size of colloidal particles, and dynamic drug partitioning between phases (aqueous, micellar, and oil). In lipolysis experiments coupled with EPR, the drug model (Tempol Benzoate) was also added to the bio-relevant medium and tracked between phases as the digestion proceeded. DLS analysis showed that the bio-relevant medium contained micelles with a hydrodynamic diameter less than 10 nm, which suddenly increased to 200 nm after the beginning of the lipid digestion, in agreement with previous *in vivo* studies concerning size of colloidal nanoparticles in human intestinal fluids. Before the addition of lipase, EPR spectra showed an increasing partitioning of the model drug Tempol Benzoate over time, initially added to the bio-relevant medium, into the oil and micellar phases over a time scale similar to that of lipid digestion. Interestingly, results showed that drug partitioning into the oil and the micellar phases further increased over time despite the parallel digestion, and therefore decreasing amount of the soybean oil. This might suggest a higher partitioning into the micellar phase as colloidal particles evolve from micelles to vesicles during the lipid digestion process. Multi-components EPR spectra analysis is being used to quantitatively resolve the amount of drug in each phase (aqueous, oil, micellar) over time. The kinetics of digestion, drug dissolution and partitioning, and, in future studies, drug absorption will be correlated in mathematical models enabling prediction of the overall oral absorption of drug molecules in the presence of lipids.

P. G. Welling, *Annu. Rev. Nutr.*, **1996**, *16*, pp. 383-415.

C. J. H. Porter, N.L. Trevaskis, W.N. Charman, *Nature Rev.*, **2007**, *6*, pp. 231-248.

O. Hernell, J.E. Staggers, M.C. Carey, *Biochemistry*, **1990**, *29*, 2041–2056.

See more of this Session: Nanotechnology for Biotechnology and Pharmaceuticals II
See more of this Group/Topical: Nanoscale Science and Engineering Forum

(499c) Effect of Lignin Obtained From Ionic Liquid Pretreated Poplar Wood On the Thermomechanical Properties of Polypropylene

Srikanth Pilla, Wisconsin Institute for Discovery, University of Wisconsin-Madison, Milwaukee, WI, Anantharam P. Dadi, SuGanit Systems, Toledo, OH, Balakrishna Maddi, Chemical Engineering, University of Toledo, Toledo, OH, Craig Clemons, USDA-Forest Products Laboratory and Joseph Lawrence, Bioengineering, University of Toledo, Toledo, OH

PP-Lignin Composites

See more of this Session: Biobased Materials II: Lignin-Based Materials
See more of this Group/Topical: Forest and Plant Bioproducts Division

(105e) Lead-Free Nanosolders for Nanowires Assembly and Packaging

Fan Gao, **Xiaopeng Li**, Yang Shu and **Zhiyong Gu**, Chemical Engineering, University of Massachusetts Lowell, Lowell, MA

The demand for high-performance interconnects in advanced packaging pushes the dramatic scaling down of the electronics packaging feature size from microscale to nanoscale. In the last several years, the approach of nanosoldering has shown promise in the construction of functional and interconnected nanostructures from nano-building blocks by utilizing nanosolder reflow processes. In this research, the potential of nanosolder materials and their applications in nanowires assembly are investigated. Nanosolders have been directly fabricated onto nanowires with different functional segments. By following a reflow process, the reliable nanosolder joints are formed between nanowires. The melting behavior and diffusion of nanosolders between nanowires have been studied for nanocomponent joining and interconnection. The interaction of nanosolders with metallic surfaces of copper nanowires and substrate are studied. The effect of industrially relevant fluxes has been studied on nanosolder reflow processes. These studies provide necessary first-hand knowledge in developing the nanosoldering techniques and their potential applications in nanomaterials assembly and nanodevice fabrication.

See more of this Session: Nanofabrication and Nanoscale Processing
See more of this Group/Topical: Nanoscale Science and Engineering Forum

(420w) Char Layer Mechanical Properties for Thermoplastic Polyurethane Elastomer Nanocomposites

Preejith Ambuken[1], Holly A. Stretz[2], Joseph H. Koo[3] and Jason C. Lee[3], (1)Chemical Engineering, Tennessee Technological University, Cookeville, TN, (2)Department of Chemical Engineering, Tennessee Technological University, Cookeville, TN, (3)Dept. of Mechanical Engineering, University of Texas - Austin, Austin, TX

This work describes the strength of thermoplastic polyurethane (TPU) nanocomposite chars at high temperature using Dynamic Mechanical Analysis. Different compositions of PU nanocomposites were prepared using a Montmorillonite organo-clay, carbon nano fiber and multiwall carbon nanotubes. The materials were compounded by twin screw extrusion. The rheological and mechanical properties of different PU nanocomposite chars at temperatures up to 600°C have been correlated with their fire retardance properties, including UL-94 and dynamic TGA.

See more of this Session: Poster Session: Nanoscale Science and Engineering
See more of this Group/Topical: Nanoscale Science and Engineering Forum

(532c) Advanced Oxidation Processes with Carbon Nanotubes: Surface-Promoted Formation of Hydroxyl Radical During Ozonation

Rebekah L. Oulton[1], Michael J. Nalbandian[1], Howard A. Fairbrother[2], Kevin Wepasnick[2] and David M. Cwiertny[1], (1)Chemical & Environmental Engineering, University of California, Riverside, Riverside, CA, (2)Department of Chemistry, Johns Hopkins University, Baltimore, MD

Advanced Oxidation Processes (AOPs), which utilize hydroxyl radical (•OH) to chemically degrade contaminants, are becoming increasingly utilized in water and wastewater treatment for disinfection and removal of organic micropollutants via chemical oxidation reactions. Catalytic ozonation represents an alternative AOP in which a solid substrate such as activated carbon or common metal oxides are used to promote •OH formation from ozone. This project explores the efficacy of using multi-walled carbon nanotubes (MWCNTs) as new materials for promoting •OH production during ozonation. Using para-chlorobenzoic acid (*p*-CBA) as a hydroxyl radical probe, aqueous suspensions of functionalized (i.e., surface oxidized) and nonfunctionalized MWCNTs increased •OH concentrations in bulk solution by an order of magnitude or more over levels observed with ozone alone and resulted in •OH levels comparable to more traditional ozone-based AOPs (e.g., hydrogen peroxide with ozone). Production of •OH was greatest on oxidized MWCNTs, and evidence suggests this enhanced •OH production cannot be explained solely by the increased dispersivity of oxidized MWCNTs in aqueous suspensions. Chemical routes for MWCNT functionalization were used to systematically vary the surface oxygen content of MWCNTs, and the ability of these functionalized materials to generate •OH was subsequently determined in batch ozonation experiments, again using *p*-CBA as a radical probe. Complementary surface analysis with X-ray photoelectron spectroscopy (XPS) combined with chemical derivatization was used to correlate •OH production not only to total MWCNT surface oxygen content, but also to the concentration of specific functional groups (hydroxyl, carbonyl and carboxylic acid groups) resulting from oxidation. Results to date show that •OH production increases with MWCNT surface oxidation, and that carboxyl surface groups appear largely responsible for this increase in •OH production.

See more of this Session: Green and Renewable Energy Aspects of Nanotechnology
See more of this Group/Topical: Topical C: Environmental Aspects, Applications, and Implications of Nanomaterials and Nanotechnology

(292f) Single Wall Carbon Nanotubes Enter Cells by Endocytosis and Not Membrane Penetration

Peter N. Yaron[1], Brian D. Holt[2], Phillip A. Short[2], Mathias Lsche[3], Mohammad F. Islam[4] and Kris N. Dahl[5], (1)Department of Chemical Engineering, Carnegie Mellon University, Pittsburgh, PA, (2) Biomedical Engineering, Carnegie Mellon University, Pittsburgh, PA, (3)Physics, Carnegie Mellon University, Pittsburgh, PA, (4)Material Science, Carnegie Mellon University, Pittsburgh, PA, (5) Chemical Engineering, Carnegie Mellon University, Pittsburgh, PA

Carbon nanotubes are increasingly being tested for use in cellular applications. Determining the mode of entry is essential to control and regulate specific interactions with cells, to understand toxicological effects of nanotubes, and to develop nanotube-based cellular technologies. We investigated cellular uptake of Pluronic copolymer–stabilized, purified ~145 nm long single wall carbon nanotubes (SWCNTs) through a series of complementary cellular, cell-mimetic, and *in vitro* model membrane experiments. SWCNTs localized within fluorescently labeled endosomes as measured by fluorescence lifetime imaging microscopy. Confocal Raman spectroscopy showed a dramatic reduction in SWCNT uptake into cells at 4 °C compared with 37 °C. Electrochemical impedance spectroscopy and Langmuir monolayer film balance measurements showed that Pluronic-stabilized SWCNTs associated with membranes but did not possess sufficient insertion energy to penetrate through the membrane. SWCNTs associated with vesicles made from plasma membranes but did not rupture the vesicles. These measurements, combined, demonstrated that Pluronic-stabilized SWCNTs did not penetrate the membrane via non-specific physical interactions but entered cells via endocytosis, and association of SWCNTs to membrane likely increased uptake. These results suggest that specific targeting of SWCNTs for use in therapeutics should focus on endocytosis as the preferential method for the cellular uptake.

See more of this Session: Environmental Implications of Nanomaterials: Biological Interactions
See more of this Group/Topical: Topical C: Environmental Aspects, Applications, and Implications of Nanomaterials and Nanotechnology

(504f) Site-Specific Modification of AAV2 Vector by Using the Genetically Encoded Aldehyde Tag

Yarong Liu[1], Kye Il Joo[2], Yun Fang[1], Chi-Lin Lee[1] and Pin Wang[1], (1)Chemical Engineering, University of Southern California, Los Angeles, CA, (2)Chemical Engineering and Materials Science, University of Southern California, Los Angeles, CA

Adeno-associated virus (AAV) vectors have attracted considerable interest because of its great premise as a vector for human gene therapy. However, their applicability is limited due to the restricted range of cells that they can efficiently transduce. Although alternative AAV serotypes might be available for enhancing gene delivery to certain cells/tissues, this does not provide selective tropism of virus for specific target cells or tissues. Therefore, it is desirable to redirect AAV vector tropism that specifically transduces selected target cells in vivo. Currently, the majority of targeting efforts rely upon genetic modification of AAV capsid proteins and chemical conjugation utilizing the reactivity of functional groups in order to introduce targeting ligands onto AAV. However, genetic insertion of targeting peptide motifs into AAV capsid usually results in low production yield, dramatic reduction of vector titer or significant drop of DNA packing ability. Moreover, the lysine modification, which is wildly used for chemical conjugation of targeting ligands onto the viral surface, has been known to be often problematic because lysine residues are likely involved in conformational change of virus and/or host-virus interactions that are necessary for infection.

To overcome these limitations, we have developed a new method for site-specific modification of AAV2 in order to generate a targeted AAV2 vector. This approach involves the insertion of a genetically encoded aldehyde tag (13-residues: LCTPSRAALLTGR) onto AAV2 capsids, which can be metabolically modified to generate an aldehyde group as a unique chemical handle for further site-specific introduction of targeting ligands without causing a significant loss of viral titer. We demonstrated that this aldehyde tag enables covalent attachment of hydrazide-functionalized molecules including antibodies and peptides to AAV2 vector in a site-specific manner. It showed that antibody conjugation to AAV2 could significantly enhance viral transduction in both permissive and non-permissive cell lines expressing alternative cell-surface receptors. In addition, RGD-peptide conjugated AAV2 exhibited tumor targeting in vitro. These results demonstrated a general and efficient means for targeting AAV vectors.

See more of this Session: Bionanotechnology for Gene and Drug Delivery II
See more of this Group/Topical: Nanoscale Science and Engineering Forum

(176b) Solvothermal Route to the Synthesis of Iron Sulfide Nanomaterials

Leize Zhu[1], Jessica Tanumihardja[2] and Qiuming Yu[2], (1)Dept. of Chemical Engineering, University of Washington, Seattle, WA, (2)Chemical Engineering, University of Washington, Seattle, WA

Recently, nanomaterials have employed as the new building blocks to construct solar energy harvesting devices. Of particular interest are the size dependent properties such as size quantization effects in semiconductor nanomaterials. Many semiconductor nanomaterials have been developed for solar cells application [1]. However, the use of toxic elements such as cadmium or lead remains a limitation as the environmental issues need to be addressed and these elements are not abundant [2]. Herein, we proposed a cheap, non-toxic and abundant semiconductor nanomaterials-pyrite nanowires, which has an indirect energy transition at 0.95 eV, a direct transition at 1.03 eV and an integrated absorption coefficient of 3.3×10^5 cm^{-1} for the energy spectrum of wavelength values (λ) between 300 and 750 nm [3]. With a high absorption coefficient from visible to near-infrared region, material consumption is significantly reduced in semiconductors for thin layer solar cells compared to conventional silicon-based solar cells. Pyrite has previously been prepared using several high temperature approaches including sulfurization of iron films, sulfurization of iron oxide films, reactive sputtering, and spray pyrolysis, yet at elevated temperatures, segregation of iron and sulfur species is unavoidable, which could change the stoichiometry and material phase of the deposited film. Thus, solvothermal process is a prefered method used for producing pyrite nanocrystals [4]. Despite the recent advancement of solvothermal routes to the synthesis of pyrite FeS_2 nanocrystals, to obtain pure pyrite phase and controllable shape and size of pyrite nanocrystals still remain challenges. In this work, we have synthesized iron sulfur nanowires, microflowers and microplates by using $FeCl_3$•$6H_2O$, $FeCl_2$ • $4H_2O$, $Fe(NO_3)_3$ • $9H_2O$ and $FeSO_4$ • $7H_2O$ as the iron source and thiourea, sulfur and $Na_2S_2O_3$ as the sulfur source. The solvent used is limited to ethylenediamine and ethylenediamine+ethyleneglycol mixture. Scanning electron microscope (SEM) was used to determine the morphology of the materials and energy dispersive x-ray analysis (EDAX) was used to determine the chemical composition of these materials. By tuning the concentration, we can obtain the length of wires ranging from 100-500 nanometers to 1-10 micrometers with 100-200 nm in diameter. X-ray photoelectron spectroscopy (XPS) results indicate that these materials are iron monosulfide along with some oxidized products. These wire-like materials are not well-crystallized which can be seen from high-resolution transmission electron microscopy (HRTEM) images. We found temperature is also an important factor in controlling the morphology of the iron sulfur materials. Microplates can be obtained at high temperature (210 • C). Only wires can be obtained at 180 • C . We also found that the counter-ion and valence state of the iron source play an important role in determining the morphology and phase of the final products. By changing the solvent from ethylenediamine to water, significant change in the phase of the product was observed (Pyrite along with macarsite), which is verified by X-Ray Diffraction (XRD). Further characterization of the magnetic and electronic property of these iron sulfur materials is carrying out.

References

[1] Kamat, P.V. J. Phys. Chem. C 2008, 112, 18737–18753.

[2] Alivisatos, A.P. et al. Chem. Mater. 2009, 21, 2568–2570.

[3] Chen, C.W. et al. Nanotechnology, 2009, 20, 405207-405212.

[4] Chaudhuri, S. et al. Chem. Phys. Lett. 2004, 398, 22–26.

See more of this Session: Nanostructured Materials for Electronic and Photonic Applications
See more of this Group/Topical: Materials Engineering and Sciences Division

Studying the Adsorption Mechanism of Hydrogen on the Carbon Based Adsorbents for Storage Purposes

Ali Qajar [1], Maryam Peer [1], Ramakrishnan Rajagopalan [2], Henry C. Foley [1,2]
[1] Department of Chemical Engineering, PennState University, State College, PA 16802
[2] Materials Research Institute, PennState University, State College, PA 16802

Abstract:

Hydrogen adsorption on carbon based adsorbents was studied via synthesizing modified version of carbons by incorporating Boron and other heteroatoms. Presence of heteroatoms changed force field energy in the pores and the interaction energy distribution between H_2 molecules and the adsorbent surface. At room temperature carbon adsorbents are believed to be able to adsorb H_2 with 9 KJ/mol. Theoretically, incorporation of boron can increased with value to 12 KJ/mol. This changes is attributed to the electon deficiency of boron. XRD, XPS and TEM techniques were utilized to characterize the synthesized materials. XPS confirmed incorporation of ~ 17 (atomic)% boron in carbon structure. A sample with 700 m^2/g surface area and ~12 KJ/mol was synthesized. Besides, presence of turbostratic regions was verified by XRD patterns and TEM images.

(586c) Synthesis and Characterization of Magnetic Nanoparticles for Enhanced Gas-Liquid Mass Transfer

Alexander P. Mathews, Department of Civil Engineering, Kansas State University, Manhattan, KS, Dambar B. Hamal, Department of Chemistry, Kansas State University, Manhattan, KS, Paul Owings, Civil Engineering, Kansas State University, Manhattan, KS and Ken J. Klabunde, Chemistry, Kansas State University, Manhattan, KY

Abstract

Magnetite particles weresynthesized by the coprecipitation of iron (II) and iron (III) chlorides in thepresence of a surfactant and ammine base. These particles had relatively lowinternal surface areas, but high external surface due to the small particlediameter. Silica coated magnetite particles were synthesized by the hydrolysisof tetraorthosilicate (TEOS) and the coprecipitation of iron chlorides in thepresence of different surfactants and an ammine base. Commercial magnetitenanoparticles were also coated with silica using TEOS and a base in thepresence or absence of a dispersant. All samples were characterized using XRD,SEM, TEM, and nitrogen adsorption to obtain diffraction patterns, particlemorphology data, and BET surface areas. Adsorption isotherm studies wereconducted for the adsorption of trichloroethylene from aqueous samples on tolow surface area magnetite nanoparticles and high surface area silica coatedmagnetite nanoparticles. Studies were also conducted on the effect of nanoparticleson the effect of high and low surface area nanoparticles on gas-liquid masstransfer in aqueous systems. This paper will present data on the synthesis andcharacterization of these nanoparticles and their potential to enhance masstransfer in gas-liquid systems.

See more of this Session: Environmental Applications of Nanotechnology and Nanomaterials
See more of this Group/Topical: Topical C: Environmental Aspects, Applications, and Implications of Nanomaterials and Nanotechnology

(704c) Development of Methods to Engineer Microbial Biocatalytic Coating Microstructure for Optimal Reactivity

Michael Flickinger, Chemical and Biomolecular Engineering, North Carolina state University, Raleigh, NC and Jessica Jenkins, Department of Chemical and Biomolecular Engineering, North Carolina State University, Raleigh, NC

Optimization of the reactivity of microbial biocatalytic coatings depends on microstructure (nanoporosity, adhesion, cell alignment or packing), physical intensification (cell density/ surface area), reduction of diffusion path (thinness, pore structure) and genetic manipulation to minimize loss of reactivity resulting from cellular stresses associated with coating, drying and rehydration. Up-regulation of gene expression of coating-entrapped microbes can further increase reactivity and allow coatings to adapt or "self-tune" to changing conditions. In order to optimize, model and predict the reactivity of multi-layer and multi-organism biocatalytic coatings, we study the microstructure of uniform monolayers of polymer particle + cell blends deposited onto flexible polyester by convective sedimentation assembly (CSA) and dielectrophoresis (DEP). CSA orders cells on surfaces by evaporation of the meniscus; coating microstructure is effected by evaporation rate, sedimentation, and particle or cell properties (size, density, buoyancy, charge) and convective transport. DEP aligns microbes in an electric field as charged polarizable particles. CSA and DEP methods used for ordering polymer particle are modified to screen polymer emulsion + cell blends to generate permanent cell adhesion to the substrate and preserve reactivity after drying and rehydration. Model systems investigated are: H_2 production by *Rhodopseudomonas palustris*, CO_2 adsorption and O_2 evolution by *Chlamydomonas reinhardtii* or *Synechococcus sps.* Understanding the relationships between coating microstructure and reactivity will lead to engineering highly reactive nanoporous biocatalytic coatings which incorporate both H_2 and O_2 producing microbial coatings (as fuel cells), or CO_x adsorbing coatings for recycling carbon emissions to liquid fuels or chemical intermediates.

See more of this Session: Nanoscale Science and Engineering In Biomolecular Catalysis I
See more of this Group/Topical: Nanoscale Science and Engineering Forum

(504e) Paclitaxel-Conjugated Virus Nanoparticles for Targeted Breast Cancer Treatment

Fang Wei[1], Kellie I. McConnell[1], Vamseedhar Rayaprolu[2], Brian Bothner[2], Tse-Kuan Yu[3] and Junghae Suh[1], (1)Bioengineering, Rice University, Houston, TX, (2)Department of Chemistry and Biochemistry, Montana State University, Bozeman, MT, (3)Radiation Oncology, The University of Texas M. D. Anderson Cancer Center, Houston, TX

Acritical difficulty in designing nanotherapeutics is in developing highlytargeted platforms that can reach target sites with high efficiency. Purelyrational improvements to nanoparticle design can be extremely difficult.Additionally, due to the high degree of heterogeneity in breast tumors within apatient as well as between patients, there is currently no one biomarkeravailable that specifically identifies all breast cancer cells. To overcomethese obstacles, we are developing a novel drug delivery system based onadeno-associated virus (AAV) nanoparticles to deliver chemotherapeutic agentsspecifically to breast cells and limit the exposure of drugs to normalnon-breast tissues. AAV is a 25 nm virus nanoparticle that is currently inclinical trials for a variety of gene therapy applications. The AAV capsid is asupramolecular assembly of 60 protein subunits, lending itself well tomultivalent conjugation of drug molecules. We have used directed evolution tocreate virus nanoparticles that can specifically bind breast tissue.Paclitaxel (taxol), chemically modified with N-hydroxysuccinimide, wascovalently attached to the virus capsid surface. Free taxol was successfullyremoved from the AAV-taxol conjugation. AAV-taxol was assayed for degree ofconjugation with UV-vis, Electrospray Ionization Mass Spectrometry (ESI-MS) andimmunoblotting; proper capsid assembly with Enzyme-Linked Immunosorbent Assay (ELISA) and *Transmission Electron Microscopy (*TEM);genomic packaging with Quantitative Polymerase Chain Reaction (Q-PCR); andcytotoxicity with cell viability assays. This work will offer superiortargeting of breast tissues and provide greater drug payload. It has thepotential of significantly decreasing the morbidity associated with systemicchemotherapy.

See more of this Session: Bionanotechnology for Gene and Drug Delivery II
See more of this Group/Topical: Nanoscale Science and Engineering Forum

(703b) Self-Assembled, Nanostructured Carbon for Energy Storage and Water Treatment

Richard T. Mayes[1], James O. Kiggans[2], Costas Tsouris[3], Sheng Dai[1] and **David W. DePaoli**[3], (1) Chemical Sciences Division, Oak Ridge National Laboratory, Oak Ridge, TN, (2)Materials Science and Technology Division, Oak Ridge National Laboratory, Oak Ridge, TN, (3)Energy and Transportation Science Division, Oak Ridge National Laboratory, Oak Ridge, TN

This talk will present progress in efforts aimed at translating a unique approach for synthesis of self-assembled nanostructured carbon [1] into industrially viable technologies for two important applications: electrochemical double-layer capacitors for energy storage, and capacitive deionization systems for water treatment. Self-assembly through templating is one way to tailor carbon-based materials, enabling the formation of a uniform porous network in the mesopore size regime via manipulation and control of macroscopic process variables. These tailorable, nanostructured materials are synthesized by self assembly in solution, followed by heat treatment. Further activation results in a bimodal pore structure where micropores are present with the mesopores. The larger mesopores allow for faster electrolyte transport. Characteristics of templated carbons having relatively high surface areas and hierarchical pore structures will be presented, as well as measurements of their specific capacitance under conditions relevant to applications.

A focus of the presentation is on recent efforts to convert the mesoporous carbon into electrode materials that may be used practically in capacitive deionization devices. Previous work had indicated that ORNL's mesoporous carbon (OMC) had attractive electrical performance; bench scale testing of the material showed significantly better performance in terms of capacity than commercial aerogel material [2]. However, a previous approach to generate electrodes suffered from variability and mechanical problems. More recent advances have generated materials that utilize the mesoporous carbon in a form more suitable for CDI electrodes in industrial application. Results of prototype testing will be presented.

1. C. Liang and S. Dai, *J. Am. Chem. Soc.*, **2006**, *128* (16), pp 5316–5317. **DOI:** 10.1021/ja060242k
2. Mayes, R.T., C. Tsouris, J.O. Kiggans, S.M. Mahurin, D.W. DePaoli, and S. Dai, *J. Mater. Chem.,* 2010, **20**(39), 8674-8678; **DOI:** 10.1039/C0JM01911A.

ACKNOWLEDGMENT: This research was supported by the Industrial Technologies Program of the U. S. DOE Office of Energy Efficiency and Renewable Energy (EERE), under Contract DE-AC05-0096OR22725 with Oak Ridge National Laboratory, managed by UT-Battelle, LLC.

See more of this Session: Nanomanufacturing
See more of this Group/Topical: Materials Engineering and Sciences Division

(420y) The Investigation of Polymer Coated Magnetic Nanocomposites for the Immobilization of Carbonic Anhydrase for CO2 Capture

Joo Seob Lee, Chemical & Petroleum Engineering, University of wyoming, Laramie, WY, Kevin John Schilling Jr., Chemical and Petroleum Engineering, University of Wyoming, Laramie, WY and Patrick A. Johnson, Chemical & Petroleum Engineering, University of Wyoming, Laramie, WY

Polymer coated iron oxide (Fe_3O_4) such as poly vinyl alcohol (PVA) coated Fe_3O_4 nanocomposites and β-Cyclodextrin (β-CD) coated Fe_3O_4 spherical aggregates have been investigated. We have studied each material for the immobilization of carbonic anhydrase (CA) for CO_2 capture applications. CA-PVA-Fe_3O_4 was prepared by enzyme encapsulation, in which CA was embedded into PVA-Fe_3O_4 matrix. Fe_3O_4, PVA-Fe_3O_4, and CA-PVA-Fe_3O_4 were characterized with FT-IR, TEM, SEM, DSC, DLS, and zeta-potential, which confirmed the surface interactions, enzyme immobilization, colloidal stability, and enzyme activity. β-CD-Fe_3O_4 was fabricated with two surfactants, β-CD and polyethylene glycol nonylphenyl ether (NP^{30}). The products were studied by SEM, TEM, DLS and zeta-potential measurement.

See more of this Session: Poster Session: Nanoscale Science and Engineering
See more of this Group/Topical: Nanoscale Science and Engineering Forum

(121a) Coarse-Grained Molecular Dynamics Simulations of Self-Assembled Structures of Cylindrical Micelles and Charged Nanoparticles

Abhinanden Sambasivam, **Ashish Sangwai** and Radhakrishna Sureshkumar, Department of Biomedical and Chemical Engineering, Syracuse University, Syracuse, NY

Wormlike micelles in solutions are 'living polymers' which exhibit reversible self-assembly and rich rheological behavior due to shear-induced conformational and phase changes. In addition, entangled micelles are excellent templates for creating stable and 3-D quasi-ordered active nanostructures suitable for a variety of practical applications ranging from plasmonics to medicine. In this work, equilibrium and non-equilibrium coarse-grained molecular dynamics (CG MD) simulations are employed to decipher the fundamental mechanisms of self-assembly between nanoparticles and cylindrical micelles and how such hybrid structures are perturbed by flow shear. MARTINI force fields are employed to describe micelle structures as done in our recent work [1]. Specifically, we investigate the interaction between cylindrical micelles of cationic cetyltrimethylammonium chloride (CTAC) surfactant and negatively charged nanoparticles. Surfactant molecules were observed to form self-assembled structures such as vesicles or surfactant coronas around the nanoparticles at zeta potentials of -40 mV or larger in magnitude. Potential of mean force (PMF) between the nanoparticle and cylindrical micelles evaluated using umbrella sampling suggests that the end caps of cylindrical micelles are more susceptible to opening up and forming junctions with nanoparticles compared to relatively stiff central domains. This is due to the fact that higher curvature and less ordering of surfactants at the end caps result in higher free energy, which can be reduced by the formation of a micelle-particle bridge. Effects of governing parameters such as the zeta potential of the nanoparticle, added salt concentration, and flow shear on the association and phase behavior of micelle-nanoparticle structures will be also discussed.

Acknowledgement: We acknowledge National Science Foundation grant CBET-1049454 and the New York Center for Computational Sciences at Stony Brook University/Brookhaven National Laboratory for support of this research.

1. A.V. Sangwai and R. Sureshkumar, Coarse-Grained Molecular Dynamics Simulations of the Sphere to Rod Transition in Surfactant Micelles, *Langmuir*, Article ASAP, **http://pubs.acs.org/doi/full/10.1021/la2006315.**

See more of this Session: Self and Directed Assembly at the Nanoscale
See more of this Group/Topical: Nanoscale Science and Engineering Forum

Study of the Photodeposition of Noble Metals on BiOCl for the Photocatalytic Decomposition of Rhodamine B

Liang Kong, Zheng Jiang, Henry H.-C. Lai, Rebecca Nicholls, Tiancun Xiao, Peter P. Edwards**

Department of Chemistry, Inorganic Chemistry Laboratory, University of Oxford, U.K.

Introduction

Extensive efforts have been made to discover robust semiconductor photocatalysts for efficient utilization of the most sustainable solar energy in photocatalytic water splitting, CO_2 reduction and environment cleanup[1-4]. The p-type BiOCl has been recently emerged as an active photocatalysts only under the illustration of UV light which accounts for approximately 5% energy of sunlight spectrum[5]. This photocatalyst is still facing the similar major challenges encountered by most photocatalysts: catalyst stability and energy loss associated with the recombination of photogenerated charge carriers (electrons and holes). A variety of strategies, such as the formation of heterojunctions[6], impurity doping and surface metallisation, have been attempted to extend the harvesting of sunlight into visible light region[7]. Surface metallisation using noble metals is one of the most effective methodologies for trapping photogenerated electrons due to the high conductivity, excellent electron-trap capability and additional plasmonic effect of the nano-sized noble metals. To date, the semiconductor surface metallisation mainly concentrated on the n-type semiconductors[8] rather than p-type semiconductor photocatalysts. Most recently, Ag/BiOBr and Pt/BiOI have demonstrated enhanced photocatalytic performance[9,10], whereas, the effects of noble metals deposition on the photocatalytic performance BiOCl has not been explored.

In this work, we deposited a small amount of various noble metals on BiOCl *via* a simple photo-deposition method and found that the noble metals have different influences on the structure, properties and photocatalytic performance of BiOCl. The interaction between noble metal and BiOCl has been discussed to interpret the effects of surface-deposited noble metal.

Results

Crystallinity and morphology of the metal/BiOCl (here after NM/BOC)

The XRD patterns (Fig. 1A) of the NM/BOC samples are well consistent with the tetragonal BiOCl phase (JCPDS no. 01-082-0485[11]) without obvious change of the characteristic diffraction peaks, suggesting the surface deposition of noble metals does not affect the crystal structure of BiOCl. There is no diffraction peaks of noble metals observed, which can be attributed to the low loading (0.5 wt%) and the high dispersion of noble metals on the surface of BiOCl. Although the lattice parameters of the BiOCl are not affected significantly, as shown in Table 1, the deposited noble metal species drastically decrease the crystallite size in following order of BiOCl > Rh/BOC > Pt/BOC > Pd/BOC > Ag/BOC.

Table 1. Crystal parameters, band-gap and photodegradation rate constant of NM/BOC

Sample ID	Crystallite size (Å)	Lattice Parameters		k_{RhB} (min^{-1})
		a (Å)	*c* (Å)	
BiOCl	451	3.892	7.376	0.247
0.5 wt% Pt-BiOCl	355	3.889	7.361	0.169
0.5 wt% Pd-BiOCl	325	3.887	7.365	0.194
0.5 wt% Rh-BiOCl	384	3.893	7.383	0.140
0.5 wt% Ag-BiOCl	292	3.888	7.387	0.253

As shown in Fig.1(B), Ag/BiOCl sample remains the flake-like morphology as same as the pure BiOCl (SEM not shown here). The average diameter of the flakes is approximately 200 nm. The TEM image (Fig. 1C)of the Ag/BiOCl sample further evidences its flake-like morphology and the nano-sized metal Ag particles in diameter approximately 5nm were highly dispersed on the surface of the BiOCl flakes in agreement with the XRD results. Other NM/BiOCl samples have the similar flake morphologies.

Fig. 2 (A)XRD patterns of as-syntheisised undoped BiOCl and samples with 0.5 wt% photodeposited Pt, Pd, Ag and Rh, (B)SEM and (C)TEM image of 0.5 wt% Ag/BiOCl sample

XPS analysis

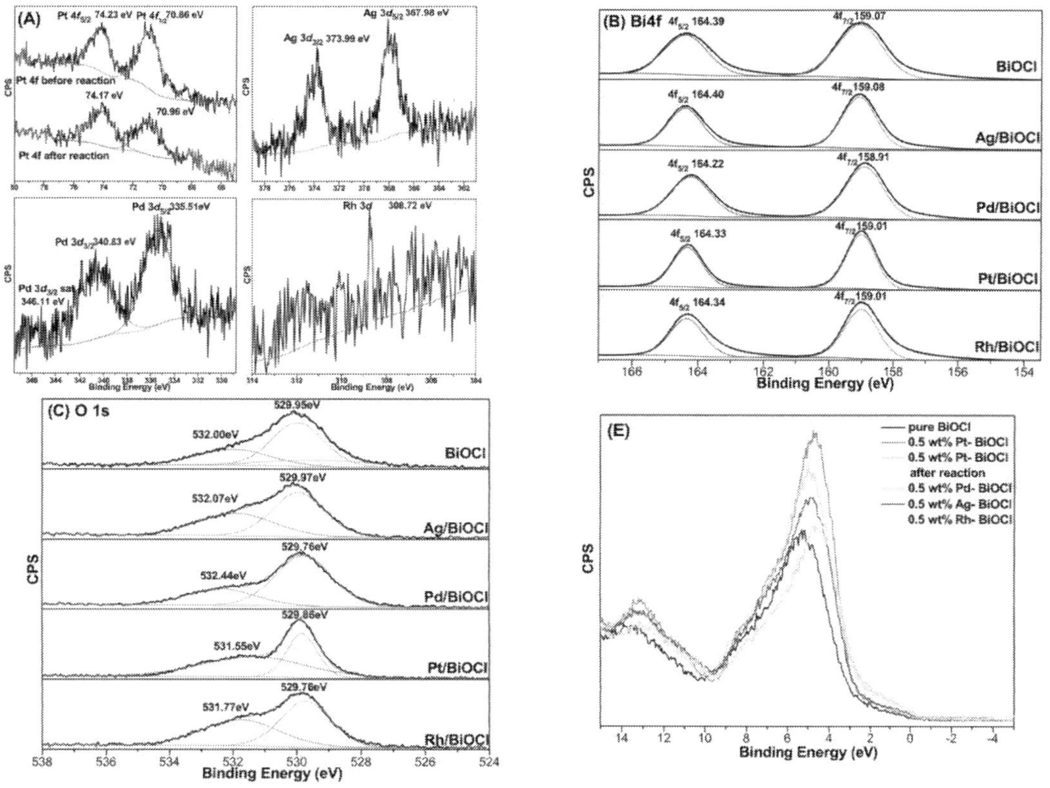

Fig.2 XPS spectra of (A) Bi 4*f*, (B) O 1*s*, (C) Pt 4*f* (before and after photocatalytic reaction), Ag 3*d*, Pd 3*d* and Rh 3*d*, and (E) VBM of all the NM/BiOCl samples

The characteristic binding energies (B.E.) of Bi, Cl, O and noble metals can be well identified from the XPS results of the corresponding samples. The XPS-derived surface atomic ratios of noble metal, Cl, O and Bi elements the samples are listed in Table 2. These results confirm the desired compositions of the BiOCl and NM/BOC samples. The noble metals in the study are all in metallic states as shown in Fig 2(A) without any signals associated with noble metal cations observed, suggesting the photoreduction is a successful strategy in deposition of noble metals on the BiOCl photocatalyst. Fig 2(B) shows the Bi $4f_{5/2}$ and Bi $4f_{7/2}$ XPS core lines with B.E located around 164.3 eV and 159.0 eV, revealing the Bi in BiOCl is Bi^{3+} [12]. The Bi 4*f* core lines shifted towards lower energies for the NM/BOC samples except for Ag/BOC whose Bi 4*f* core lines increased about 0.2 eV. The most significant red-shifts of Bi 4*f* B.E. are observed on the Pd/BOC sample. The detailed B.E. changes may be found in Table 3. There is no obvious changes can be observed from Cl $2p_{1/2}$ (199.4 eV) and Cl $2p_{3/2}$ (197.8 eV) peaks of Cl^- in the BiOCl and NM/BOC, suggesting the supported noble metal can not affect the inter-layered Cl^- [13,14]. There are two different oxygen species presented on the surface of the series samples (Fig. 2C). The O 1*s* B.E. around 532.0 eV is assigned to the Bi-O bond in BiOCl, while the O 1*s* peak at 529.9 eV due to the surface hydroxyl groups of the sample[15,16]. Moreover, the O 1s of Bi-O shift towards higher energy against the trend of Bi 4f on the NM/BOC samples except for that of Ag/BOC, which further confirms that surface metallisation affect the chemical environment of the composite. The valence band XPS of the samples are compared in Fig.2D, in which the shifts of VB edges (VBMs) towards lower energy on the NM/BOC than VB of BOC are clearly observed and the shifts significantly depend on the noble metal specials.

Table 2. Surface elemental composition (%) of the NM-BiOCl samples

Atomic Ratio	Bi	O	Cl	NM
BiOCl	39.2	38.2	22.6	-
Rh-BiOCl	36.3	52.4	10.9	0.49
Pd-BiOCl	37.4	36.1	25.9	0.56
Pt-BiOCl	38.3	34.2	27.3	0.18
Pt-BiOCl after reaction	33.4	43.8	22.7	0.19
Ag-BiOCl	36.8	41.9	20.9	0.46

Table 3. XPS characteristic peak positions of the NM-BiOCl samples

Sample ID	Peak position (eV)				
	Bi $4f_{5/2}$	Bi $4f_{7/2}$	Cl $2p_{1/2}$	Cl $2p_{3/2}$	O 1*s*
BiOCl	164.4	159	199.5	197.8	530.1
Pt-BiOCl	164.2	158.9	199.3	197.7	529.8
Pt-BiOCl after reaction	164.3	159.0	199.4	197.8	529.9
Pd-BiOCl	164.2	158.9	199.3	197.7	529.7
Rh-BiOCl	164.3	159.0	199.3	197.7	529.8
Ag-BiOCl	164.4	159.1	199.6	197.9	530.1

Uv-Vis Spectra

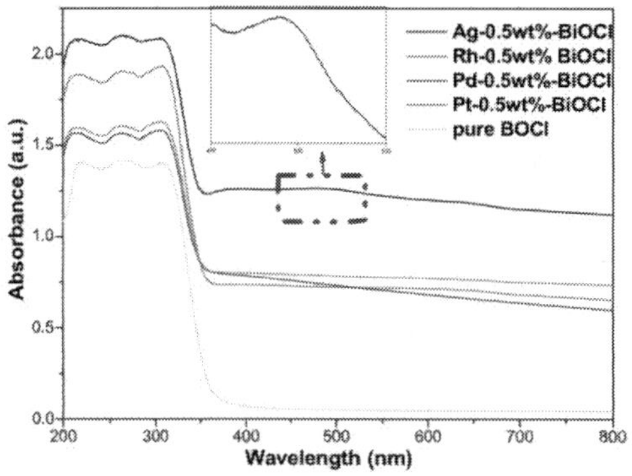

Fig. 3 UV-vis diffuse reflectance spectra of NM- BOC photocatalysts

Fig. 3 shows the UV/vis diffuse reflectance spectra of BiOCl and NM/BOC samples. T BiOCl can only absorb UV light with wavelength below 360 nm, while all the NM/BOC samples can harvest light through the whole UV-Vis region, suggesting even small amount of noble metals can greatly improve the light absorption in comparison to pure BiOCl. The enhanced absorption of NM/BiOCl samples both in UV and Visible regions may be attributed to the close field surface plasmonic effect (or the scattering effect)[9]. The Ag/BOC sample exhibits the strongest plasmonic effect among the NM/BOC samples. The indirect and direct band-gap energy of NM/BOC can thus be estimated from the data plot of $(\alpha \cdot hv)^{1/2}$ and $(\alpha \cdot hv)^2$ *versus* photo energy (hv) [5] and listed in Table 2, from which the introducing of noble metals decrease both the indirect band-gap and direct band-gap.

Photocatalytic activity

Photocatalytic decomposition of RhB, a chemically stable organic dye with a characteristic absorption peak at 554 nm, was chosen as a probe reaction to clarify the effects of deposited noble metals on the BiOCl. As shown in Fig. 4(A) and (B), all the samples can completely decompose Rh.B within 30 minutes UV light irradiation under the same reaction conditions. The photocatalytic activity highly dependes on the type of noble metal, with the activity changes in the order of Ag/BOC > pure BiOCl > Pd/BOC > Pt/BOC > Rh/BOC. The results suggest that the modification of noble metals cannot enhance the photocatalytic activity of the BiOCl samples even though they have enhanced absorption of irradiation.

Although most of the NM/BOC are less active than BOC except for Ag/BOC, the NM/BiOCl samples showed higher stability after UV-light irradiation. For example, the fresh and used 0.5 wt% Pt/BiOCl samples shows the very similar XRD patterns (Fig.4C) except for the stronger Bragg diffraction intentisty for the sample post photocatalysis. The enhanced intensity may be attributed to the growth of the crystallite during the photocatalytic reaction. The XPS spectra of fresh and used Pt/BiOCl samples were also compared, as shown in Table 2, indicating there is no obvious difference between the samples before and after the photocatalytic reaction. Both the XRD pattern and XPS spectra of the samples after reaction confirmed that the metal-modified samples are considerably stable under UV light irradiation.

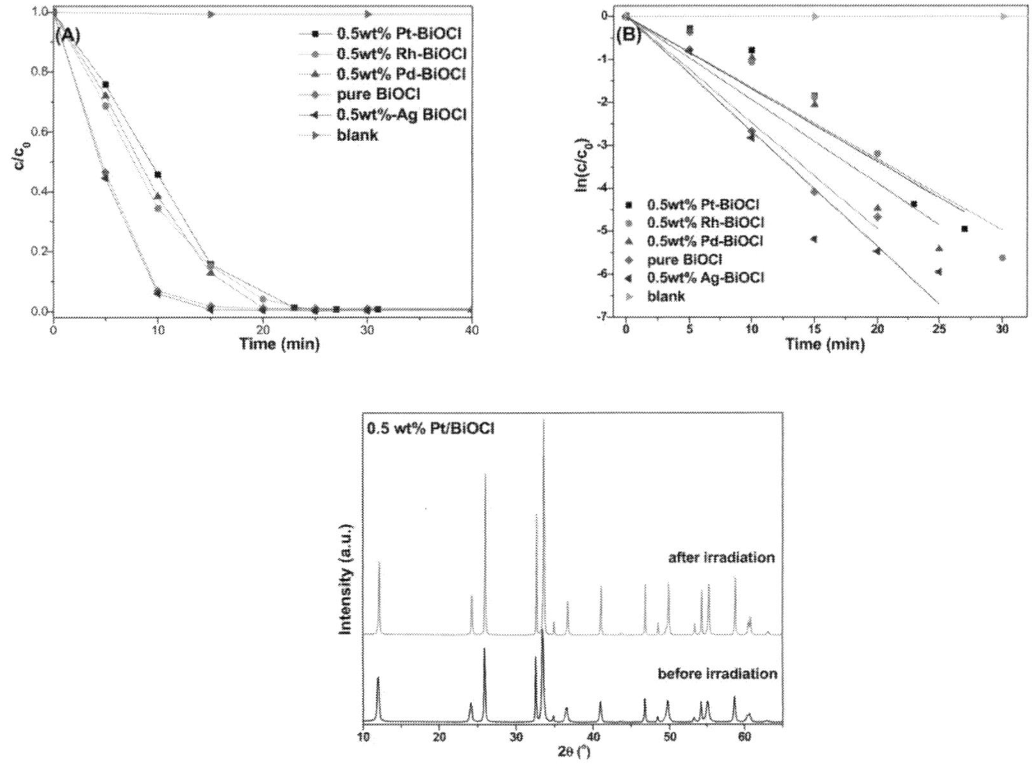

Fig. 4 (A) Temporal course of the degradation of RhB in the presence of different noble metal- BiOCl; (B) Kinetic linear curves of RhB phpotocatalytic degrdation over noble metal- BiOCl and (C) XRD patterns of 0.5 wt% Pt- BiOCl sample before and after UV-light irradiation

Working mechanism

Understanding the reason of the superior photocatalytic activity and the working mechanism of the Nm/BiOCl composite catalysts is essential for fundamental investigation of semiconductor photocatalysts. The formation of Schottky barrier between metals and semiconductor could enhance the photocatalytic activity due to their different Fermi levels, characterised by the work function of the metals and the band structure of the semiconductors[8]. Photoexcited electrons in the conduction band of the semiconductor will raise its Fermi level to more negative values[17], usually leading to enhanced photo-activity of most photocatlysts, such as TiO_2, while our experimental results differ from previous reports that the modification of noble metals decrease the activity of photocatalysts. According to the different work functions of the noble metals, we obtain the height of the noble metals in the reducing order of Pd > Pt > Ag > Rh. This trend is in accordance with the photocatalytic activity consequence of NM/BiOCl samples, i.e., the noble metal with higher Fermi level shows better photocatlaytic activity. However, the work functions of the NM are more positive than the CB of BiOCl, which leads to photogenerated electrons from the VB easily flow into noble metal rather than the CB of BOC and decreased photocatalytic activity of the NM/BOC due to the less opportunity to generate reactive O_2^{-}•. At the same time, the chemical potential of the photogenerated holes are less oxidative due to the shift of VB towards to more negative positions, which means the decreased oxidative capability of VB holes to oxidize the dye molecules. The overall results are decreased activity of the NM/BOC in comparison with pure BiOCl.

Conclusion

Noble metal-modified BiOCl photocatalysts have been synthesised *via* a two-step precipitation-photodeposition method. The metallic state modification on the surface of BiOCl significantly enhances the absorbance in the visible-light region, *i.e.* decreases the band-gap to about 2.7 eV, however, the modification cannot effectively enhance their photocatalytic activity with respect to the pure BiOCl in photodecomposion of RhB. The decreased photocatalytic activity of the NM/BOC can be well interpret by the work function difference between the NMs and BiOCl.

Acknowledgement

This work was financially supported by the principal's major fund at Jesus College, University of Oxford, Oxford Challenge Seed Fund, UK Photocatalysis Network, UKSHEC (EPSRC). Dr Jiang appreciates John Houghton Fellowship at Jesus College (Oxford) and the grants of Royal Society for international conference (TG092414) and international collaboration (TG101750).

Reference

(1)Ohtani, B. *Journal of Photochemistry and Photobiology C: Photochemistry Reviews* **2010**, *11*, 157-178.
(2)Jiang, Z.; Yang, F.; Yang, G.; Kong, L.; Jones, M. O.; Xiao, T.; Edwards, P. P. *Journal of Photochemistry and Photobiology A: Chemistry* **2010**, *212*, 8-13.
(3)Jiang, Z.; Al-Shahrani, F.; Lin, T.-W.; Cui, Y.; Xiao, T. In *Studies in Surface Science and Catalysis*; Dongyuan Zhao, S. Q. Y. T., Chengzhong, Y., Eds.; Elsevier, 2007; Vol. Volume 165; pp 355-359.
(4)Jiang, Z.; Xiao, T.; Kuznetsov, V. L.; Edwards, P. P. *Philosophical Transactions of the Royal Society A: Mathematical, Physical and Engineering Sciences* **2010**, *368*, 3343-3364.
(5)Zhang, K.-L.; Liu, C.-M.; Huang, F.-Q.; Zheng, C.; Wang, W.-D. *Applied Catalysis B: Environmental* **2006**, *68*, 125-129.
(6)Shamaila, S.; Sajjad, A. K. L.; Chen, F.; Zhang, J. *Journal of Colloid and Interface Science* **2011**, *356*, 465-472.
(7)Kong, L.; Jiang, Z.; Xiao, T.; Lu, L.; Jones, M. O.; Edwards, P. P. *Chemical Communications* **2011**, *47*, 5512-5514.
(8)Ohtani, B.; Kakimoto, M.; Nishimoto, S.; Kagiya, T. *Journal of Photochemistry and Photobiology A: Chemistry* **1993**, *70*, 265-272.
(9)Yu, C.; Fan, C.; Meng, X.; Yang, K.; Cao, F.; Li, X. *Reaction Kinetics, Mechanisms and Catalysis* **2011**, *103*, 141-151.
(10)Yu, C.; Yu, J. C.; Fan, C.; Wen, H.; Hu, S. *Materials Science and Engineering: B* **2010**, *166*, 213-219.
(11)Keramidas, K. G.; Voutsas, G. P.; Rentzeperis, P. I. *Zeitschrift fur Kristallographie* **1993**, *205*, 35-40.
(12)Xiao, Q.; Zhang, J.; Xiao, C.; Tan, X. *Catalysis Communications* **2008**, *9*, 1247-1253.
(13)Ge, L. *Journal of Molecular Catalysis A: Chemical* **2008**, *282*, 62-66.
(14)Drelinkiewicz, A.; Sobczak, J. W.; Sobczak, E.; Krawczyk, M.; Zieba, A.; Waksmundzka-Góra, A. *Materials Chemistry and Physics* **2009**, *114*, 763-773.
(15)V.I., N.; D., G.; Dzhurinskii, B. F.; N.P., S.; Y.V., S. *Zh. Neorg. Khimii* **1975**, *20*, 2307.
(16)Martensson, N.; Malmquist, P. A.; Svensson, S.; Basilier, E.; Pireaux, J. J.; Gelius, U.; Siegbahn, K. *Nouveau, J. Chim.* **1977**, *1*, 191
(17)Subramanian, V.; Wolf, E. E.; Kamat, P. V. Journal of the American Chemical Society 2004, 126, 4943-4950.

(374d) Facile Synthesis and Optical Characterization of Hybrid Upconverting and Plasmonic $NaGdF_4$: Yb^{3+}, Er^{3+}/Silica/Gold Nanoparticles

Sha Liu[1], Guanying Chen[2], Tymish Y. Ohulchanskyy[2], Paras N. Prasad[2] and Mark T. Swihart[1], (1) Chemical and Biological Engineering, The University at Buffalo (SUNY), Buffalo, NY, (2)Institute for Lasers, Photonics, and Biophotonics, The University at Buffalo (SUNY), Buffalo, NY

We present a simple process for preparing water dispersable hybrid $NaGdF_4$: Yb^{3+}, Er^{3+}/silica/gold nanoparticles. Monodispersed upconversion nanoparticles of $NaGdF_4$:2% Er, 20% Yb capped by oleic acid were first synthesized via a thermal-decomposition method. The hydrophobic UC NPs were then capped with silica via a reverse-micelle method to make them hydrophilic. Amino-propyltriethoxysilane (APTES) was then used to functionalize the silica-capped UC NPs with amine groups. Finally, gold NPs capped with PVP were mixed with the silica-coated UC NPs. Amine groups on the UC NPs displaced the PVP from the AuNP surface, yielding AuNP-decorated UC NPs that were dispersible in water or ethanol. The emission intensity and color of the upconverting cores are modulated by the plasmonic absorbance and field enhancement of the gold nanoparticles. Energy transfer happened in the hybrid nanoparticles from UC core to the decorating gold nanoparticles, especially at the wavelength of 530 nm at which gold nanoparticles have strong absorption. The lifetime analysis shows that the plasmon resonance effected the emission of upconversion nanoparticles. The applicability of hybrid NPs for multi-modal bioimaging probes was also illustrated with confocal microscopy of living cancer cells in vitro.

See more of this Session: Functional Nanoparticles and Nanocoatings on Particles I
See more of this Group/Topical: Particle Technology Forum

(755d) Mimicking the Myelin Sheath: Stable and Fluid Multilayer Phospholipid-Silica Thin Films:

Gautam Gupta[1], Srinivas Iyer[2], Plamen Atanassov[3], Gabriel Montano[1], Andrew M. Dattelbaum[1] and Gabriel P. Lopez[4], (1)Center for Integrated Nanotechnolgy, Los Alamos, Los Alamos, NM, (2) Biosciences Division, Los Alamos, Los Alamos, NM, (3)Chemical and Nuclear Engineering, University of New Mexico, Albuquerque, NM, (4)Biomedical Engineering and Mechanical Engineering and Materials Science, Duke University, Durham, NC

Phospholipids are a key component of cell membranes, which act as a physiochemical barrier to compartmentalize cells. The phospholipid molecules are dynamic and fluid in nature within the cell membrane. Transmembrane proteins (TMPs) perform various functions including solute transport, signal transduction, and ATP synthesis. There is enormous interest in developing various model systems including black lipid membrane, liposomes, supported lipid bilayer, and tethered bilayer lipid membrane to study TMPs. Although significant advancements have been made, very little has been achieved in developing multilamellar structures that are present in the nervous system e.g. myelin and in organelles e. g. mitochondria. Perturbations of myelin, including demyelination is responsible for at least a dozen diseases, most prominent of which is multiple sclerosis (MS). Despite its importance, minimal information on structure-function dynamics are available primarily due to lack of stable multi-lamellar lipid architectures that could be used as experimental models. We have developed two strategies to synthesize stabilized multilamellar lipid architectures: (a) one-step method in which multilamellar lipid template is obtained using spin-coating, which is followed by stabilization using vapors of silica (b) layer-by-layer method in which multilamellar lipid structures are obtained by alternating phospholipid layers, which is stabilized by biotin–avidin chemistry. Multi-lamellar thin films have enhanced stability and maintain fluidity in aqueous solutions. This robust platform will allow us to incorporate TMPs and will be utilized to synthesize rugged materials for device integration with a wide range of applications for mimicking myelin, sensing, and energy harvesting.

See more of this Session: Self-Assembled Biomaterials
See more of this Group/Topical: Nanoscale Science and Engineering Forum

(385b) Gas-Phase Synthesis of Gadolinium Nanoparticles for Magnetic Resonance Imaging Contrast Agents

Pooja Chakrabarty[1], William J. Scharmach[1], Folarin Erogbogbo[2], Raymond D. Buchner[1], Vasilis Papavassiliou[3] and Mark T. Swihart[1], (1)Chemical and Biological Engineering, University at Buffalo (SUNY), Buffalo, NY, (2)Institute for Lasers, Photonics, and Biophotonics, University at Buffalo (SUNY), Buffalo, NY, (3)Praxair Technology Center, Tonawanda, NY

Gadolinium ions are the paramagnetic component of the most common intravenous contrast agents for medical magnetic resonance imaging (MRI). Because gadolinium ions are toxic, these MRI contrast agents include chelating agents that bind the gadolinium ion strongly. Five such agents are FDA approved. Unfortunately, all of them are the subject of very recent FDA rulings that impose new labeling requirements, due to patients with severely impaired kidneys developing a rare but potentially fatal disease called Nephrogenic Systemic Fibrosis, which is linked to gadolinium exposure. Pure gadolinium or gadolinium oxide nanoparticles, with proper encapsulation, may provide a safer and equally or more effective alternative to chelated gadolinium ions in this application.

The High Temperature Reducing Jet (HTRJ) process is a novel flame based aerosol synthesis method that can produce metal nanoparticles with a thin carbon coating in a single step. Carbon-coated gadolinium and gadolinium oxide nanoparticles have been synthesized in this reactor system. The surface of these nanoparticles is modified with dextran and other biocompatible molecules to enable their dispersion in water and biological media. The resulting nanoconstructs were characterized by TEM, SEM/EDS, XRD, Raman spectroscopy, TGA, and DSC. Most importantly, their T1 relaxivity was measured and shown to be adequate for MRI contrast enhancement. Sensitive colorimetric chemical assays for gadolinium ions are used to monitor any release of gadolinium.

See more of this Session: Magnetic Nanoparticles In Biotechnology and Medicine
See more of this Group/Topical: Nanoscale Science and Engineering Forum

(693e) The Rheology of Double-Stranded DNA Stabilized Single-Walled Carbon Nanotube Dispersions

Geyou Ao, Department of Chemical Engineering, Auburn University, Auburn, AL and **Virginia A. Davis**, Chemical Engineering, Auburn University, Auburn, AL

Aqueous dispersions of single-walled carbon nanotubes (SWNTs) stabilized by double stranded DNA (dsDNA) can form isotropic, nematic or cholesteric microstructures depending on the relative concentrations and processing methodology. This enables the production of coatings, films and fibers with a broad range of properties. For example, dilute dispersions can be used to prepare coatings with heights as low as 1.6 nm and controlled orientation and nematic dispersions can be solution spun to produce fibers tens of microns in diameter. Cholesteric dispersions can be processed into films where the optical characteristics are dependent on the cholesteric pitch. We report rheological characterization of the initial dispersions and highlight how this rheological understanding can be used to better understand and control both the dispersion microstructure and the properties of the final assembled materials.

See more of this Session: Graphene and Carbon Nanotubes: Characterization, Functionalization, and Dispersion
See more of this Group/Topical: Nanoscale Science and Engineering Forum

(747a) Carbon-Iron Oxide Fluorescent-Magnetic Nanocomposites for In Vivo Imaging

C. Jenny Dorcna, Kristi Olesik and Jessica O. Winter, Chemical and Biomolecular Engineering, The Ohio State University, Columbus, OH

Here we present the design, synthesis, and evaluation of fluorescent-magnetic nanocomposites for *in vivo*, multimodal, brain tumor imaging. One of the primary limitations in brain tumor resection is difficulty distinguishing normal from cancerous tissue. Whereas this can be achieved to some extent via pre-operative MRI, there is a need for intra-operative tumor imaging technologies. In particular, it is desirable to have multifunctional imaging agents, which not only allow non-invasive pre-operative magnetic resonance imaging (MRI), but also direct tumor observation through fluorescence imaging during surgery. We are investigating carbon dots (C-dots) as potential, bright, photostable, and non-toxic fluorescent imaging agents in composites with superparamagnetic iron oxide nanoparticles (SPIONs), which, for some thirty years, have been utilized extensively as contrast agents for MRI. These multimodal nanocomposites, consisting of both fluorescent C-dots and SPIONs, are encapsulated in polymer nanosphere matrices made of poly (lactic-co-glycolic acid) (PLGA), yielding a dual functional imaging contrast agent that responds to both fluorescent and magnetic stimuli.

Fluorescent C-dots (diameter of 2 – 5 nm) whose quantum efficiency was ~ 2.6 %, were synthesized by acid reflux of non-fluorescent and commercially available carbon black. According to results obtained via Raman spectroscopy, this precursor material was graphitic in nature. Previous studies on the toxicity of other carbonaceous materials such as fullerenes and carbon nanotubes, have shown low levels of toxicity. Using preliminary MTT (3-(4, 5-dimethylthiazol-2-yl)-2, 5-diphenyltetrazolium bromide) evaluations, we detected low levels of toxicity in C-dots. An aqueous suspension of SPIONs was prepared from a mixture of iron (II) chloride tetrahydrate and iron (III) chloride hexahydrate, heated to 70 °C, and treated with a concentrated solution of sodium hydroxide. C-dots and SPIONs were incorporated into PLGA nanospheres by following a double emulsion procedure. This yielded PLGA nanocomposites ranging from 100 to 200 nm, which were not able to non-specifically bind to brain cancer cell lines. We also investigated the specific targeting of these cell lines by successfully conjugating PLGA-encpasulated C-dots-SPIONs nanocomposites to interleukin 13 receptor (IL-13R) antibodies using standard EDC (1-ethyl-3-(3-dimethylaminopropyl) carbodiimide) chemistry.

This research will broaden and improve *in vivo* imaging tools necessary for the location and surgical resection of brain tumors. These tumors, in spite of dramatic improvements in micro-neurosurgical techniques, neuroimaging, chemotherapy, and radiation therapy, remain very difficult to treat. Maximal surgical resection has been shown to be a critical factor in extending survival of patients suffering from these lesions. Our multifunctional nanocomposites will allow the selective targeting of brain tumor cells, as well as their detection by both diagnostic MRI and intraoperative fluorescence imaging. Thus, surgical resections are expected to improve, and patient survival will be extended.

(609b) Semiconductor Nanowire Fabric As a New Photovoltaic Material

Chet Steinhagen[1], Vahid Akhavan[1], Vince Holmberg[1] and **Brian A. Korgel**[2], (1)Chemical Engineering, University of Texas at Austin, Austin, TX, (2)Department of Chemical Engineering, University of Texas at Austin, Austin, TX

Significant quantities of semiconductor nanowires can be synthesized using solution-based methods, such as supercritical fluid-liquid-solid (SFLS) and solution-liquid-solid (SLS) growth. These nanowires are mechanically flexible because of their narrow (<50 nm) diameter and high aspect ratio, with lengths greater than millimeters in some cases. Because of the absence of crystallographic defects like grain boundaries, the nanowires are extremely strong, exhibiting nearly ideal mechanical strength. The nanowires can be formed into non-woven fabric, or paper, that is flexible and mechanically robust. We are in the process of trying to develop these nanowire fabrics as a new class of flexible photovoltaic material. This presentation will highlight the nanowire fabric itself--its properties and how it is made--as well as data from prototype photovoltaic silicon (Si) and CuInSe2 (CIS) nanowire fabric.

See more of this Session: Semiconductor Nanowires and Nanotubes for Next Generation Photovoltaics From Fundamentals to Devices
See more of this Group/Topical: Topical 5: Nanomaterials for Energy Applications

(439c) Anticancer Effect of Curcumin Liposomes Against Osteosarcoma and Synergistic Effect of Curcumin with Ceramide 6

Santosh S. Dhule[1], Patrice Penfornis[2], Radhika Pochampally[2] and Vijay T. John[1], (1)Chemical and Biomolecular Engineering, Tulane University, New Orleans, LA, (2)Center for Stem Cell Research and Regenerative Medicine, Tulane University Health Sciences Center, New Orleans, LA

Osteosarcoma (OS) accounts for approximately 20% of all bone cancers and it develops during the period of rapid growth in adolescence. Curcumin is being used as an anticancer drug against broad spectrum of tumors but its nanoformulation potential is not fully explored to treat OS. The drug is water insoluble that limits the bioavailability. In present work, an effective delivery route is through encapsulation in cyclodextrins followed by a second encapsulation in liposomes. Anticancer potential of curcumin liposomes in monolayer culture system (two dimensional) and spheroid culture system (three dimensional) has been explored. DMSO-curcumin (free drug) induced autophagic cell death in OS whereas liposomal curcumin initiates the caspase cascade that leads to apoptotic cell death *in vitro* as well as *in vivo*. Curcumin has already been shown to act by inducing ceramide generation in many cancer cell lines. Ceramide 6 (C6) is a sphingolipid with anticancer potential. We encapsulated curcumin and C6 in liposomes to estimate combinatorial effect against osteosarcoma cell lines. Curcumin and ceramide 6 show synergistic anticancer effect when encapsulated together. Major limitation in cancer drug delivery is inability of the delivery vehicle to reach center of the solid tumor due to lack of blood circulation and lymphatic system. To mimic the three dimensional tumor model, osteosarcoma spheroid culture system was optimized and the time dependent diffusion of curcumin liposomes was performed *ex vivo*. The *ex vivo* diffusion study shows promising results about penetration power of curcumin liposomes. The presentation will focus mainly on the formulation, characterization of the curcumin liposomes and its anticancer potential in monolayer versus spheroid culture system.

See more of this Session: Bionanotechnology for Gene and Drug Delivery I
See more of this Group/Topical: Nanoscale Science and Engineering Forum

(246a) Using Interfacial Manipulations to Control Ordering In Tapered Block Copolymers

Thomas H. Epps, Raghunath Roy, Jong Keun Park, Wei-Fan Kuan and Bin Wei, Chemical Engineering, University of Delaware, Newark, DE

Block copolymers permit the controlled fabrication of nanoscale materials for membrane and templating applications. To enable the increased utilization of block copolymer systems in commercial devices, additional understanding of the influence of interfacial manipulations on their bulk self-assembly is necessary. In this work, well-controlled tapered regions are inserted between blocks in a diblock copolymer to probe the effect of interfacial composition on block copolymer self-assembly. We systematically manipulate the interfacial region in tapered poly(isoprene-*b*-styrene) [P(I-S)] diblock copolymers to control ordering transitions. In addition to locating non-lamellar-forming materials, we find that normal and inverse tapered diblock copolymers show measurable decreases in the order-disorder transition temperature (T_{ODT}) relative to the corresponding non-tapered diblock copolymers; with the inverse tapered materials showing the greatest deviation in T_{ODT}. Additionally we have recently found the double-gyroid network morphology in several tapered diblock copolymers and show the ability to significantly reduce T_{ODT}'s while keeping mechanical properties nearly constant.

See more of this Session: Nanoscale Structure In Polymers I
See more of this Group/Topical: Materials Engineering and Sciences Division

(420f) The Chiral Self- Assembly: From Chiral CdTe Nanoparticles to Chiral CdTe Nanostructures

Yunlong Zhou Sr., Chemical Engineering, University of Michigan, Ann Arbor, MI

Abstract:

Self-assembly ofsemiconductor nanoparticles (NPs) have been proved a facile method for reproduciblyfabricating large-scale nanoelectronics or nanophotonics device arrays. Among others, chiral assembly by engineering metamterials,are demonstrated unprecedented photoelectric and electricmegneticproperties, which cannot produce by naturally materials. Taking advantage ofthe chirality of molecules in the assembly ofinorganic NPs, we can obtain novel and functional chiralnanostuctures. Current work are focusing onunderstanding the surface structure of nanoparticlesand assembly of chiral amino acids (cysteine) stabilized CdTe nanoparticles by adjusting the weak interaction forces,such as electrostatic interactions, hydrogen bonding, dipole-dipoleinteractions and hydrophobic force, between nanoparticles.Our research will facilitate the advance of semiconductor nanostructure andhave great potential application in development of new photoelectric devices.

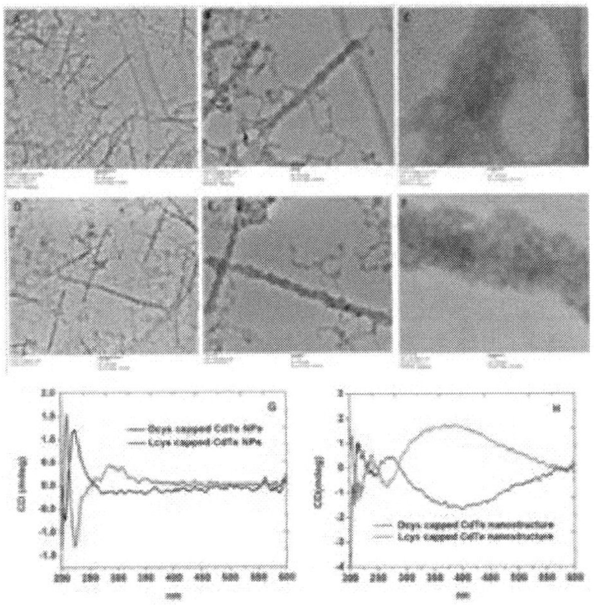

Figure 1 A), B), C) TEMof CdTe chiralnanostructure assembled by L-cysteine (cys) capped CdTe NPs. D), E), F)TEM of CdTe chiralnanostructure assembled by D-cys capped CdTe NPs. G) CD spectra of (D-, L-) cyscapped CdTe NPs. H) CD spectra of CdTechiral nanostructure assembled by (D-,L-) cys capped CdTe NPs

Lignin – A Renewable Precursor for Carbon Nanofibers

John F. Kadla,[1] Ian Dallmeyer,[1] and Frank Ko[2]
[1]Biomaterials Chemistry Lab, Dept. of Wood Science, Faculty of Forestry
[2]Advanced Fibrous Materials Lab, AMPEL, Dept. of Materials Engineering
[1,2]University of British Columbia, Vancouver, BC

Abstract

Here we report on our research efforts aimed at developing carbon nanofibers through electrospinning of Kraft lignin. The research described herein focuses on understanding the effect of polymeric additives in influencing viscoelastic fluid properties which control the fiber morphology and diameter of fibers electrospun from blend solutions of softwood Kraft lignin (SKL) and poly(ethylene oxide) (PEO). To characterize the relevant fluid properties we report on the solution shear properties as well as the extensional properties of spinning solutions. We demonstrate that viscoelasticity can be varied in order to control the spinning process in our system by changing the concentrations of SKL and PEO as well as the PEO molecular weight. We also report on the mechanical properties of CNF precursor fibers as a function of thermal processing parameters. The results of these studies indicate that electrospinning is a promising platform for the development of micro- and nanofibrous carbon materials from technical lignins.

Introduction

Lignin is a naturally abundant biopolymer which reinforces the carbohydrate matrix, facilitates water and nutrient transport, and protects against invasion by microorganisms in the stems of trees and other terrestrial plants.[1,2] In terms of its chemical structure, lignin is an aromatic heteropolymer consisting of an irregular sequence of phenylpropane units connected by a variety of interunit linkages.[1,2] The renewable nature, abundance, and availability of lignin from the pulp and paper industry[3] are driving forces for the development of value-added products from lignin. Carbon fibres (CFs) are one example of value-added products that can be derived from lignin.[4] CFs have numerous applications including reinforcement in advanced light weight composite materials,[5] catalyst substrates,[6] adsorbents,[7] electrodes,[8] and tissue engineering scaffolds.[9] The most common precursors for commercial CFs are petroleum pitch, synthetic copolymers based on polyacrylonitrile (PAN), and rayon.[5] While PAN-based CFs have superior tensile strength, the high cost of PAN and non-renewable nature of petroleum-derived precursors are disadvantages.

Lignin is a potential renewable precursor for CFs, but the complex, heterogeneous chemical structure, the variability in properties among different types of isolated lignins, and the comparably poorer mechanical properties of lignin-based CF are barriers to their application.[4,10,11] Nevertheless, significant research efforts are being devoted to the production of CFs based on lignin, for example, at Oak Ridge National Laboratory[12]. Researchers in our lab have reported on thermal extrusion of lignin and lignin-polymer blends and conversion of these precursors to CFs by thermal treatment during the last decade.[4,13,14] Polymer blending was employed as a strategy to improve the spinnability of lignin[4], increase the tensile strength[13], or modify the surface porosity[14], depending on the blending polymers and their extent of interactions with lignin.[15-18] While studies conducted to date have reported improvements in fibre properties, a deeper understanding of the relationship between the complex macromolecular structure of lignin, interactions with additives, processibility, and material properties is needed to expand into a wider range of applications for lignin-based CFs. In addition, novel processing strategies must be explored to improve properties for specific applications and overcome some of the difficulties associated with lignin mentioned above.

One strategy to improve the properties of lignin-based CFs is to reduce the fiber diameter, since the tensile strength of CFs is known to increase with decreasing fibre diameter,[4,19,20] and applications such as filtration and catalysis depend on high specific surface area, which increases with decreasing diameter. Electrospinning is a robust strategy to produce fibers with diameters in the range of a few hundred nanometers or less.[21-23] Electrospinning of CF precursors such as PAN,[24-26] phenolic resin,[27] and pitch[28] has been reported. Electrospinning of organosolv lignin[29,30] and blends of sulfonated alkali lignin with PAN[31] have also been reported in the literature. However, lignin in general has been observed to display a high tendency to electrospray into droplets instead of forming fibers. To overcome this limitation, we have reported on the electrospinning of blends of several technical lignins with small amounts of poly(ethylene oxide) (PEO).[32] The advantages of our system compared to other reported systems are applicability to different technical lignins, the ability to spin fibers with greater than 99% lignin content with a single spinneret, and the ability to control the fiber diameter by varying the viscoelasticity of the spinning solution. Our current research in the area of lignin electrospinning is aimed at two main objectives: 1) To understand the role of lignin structure and interactions with additives in determining the viscoelastic fluid properties which govern the spinning process, and 2) To thermally process the electrospun fabrics and characterize the material properties as a function of processing parameters. In the following sections we provide an overview of our research efforts in the areas of rheology and its effect on the electrospinning process and processing and characterization of nonwoven fabrics produced from fractionated softwood Kraft lignin.

Materials and Methods

Lignin and solution preparation:

Softwood Kraft Lignin (SKL, Indulin-AT, Meadwestvaco, Glen Allen, VA, USA) was washed with dilute HCl (pH 2) to remove Na^+ ions and fractionated by sequential extraction with organic solvents according to a procedure based on that of Morck et al.[33] The F_4 fraction (according to the nomenclature of Morck et al[33]) was selected for rheological and electrospinning experiments. Polyethylene oxide (PEO) with M_v of 1000 and 5000 kDa (denoted PEO1M and PEO5M) were obtained from Sigma-Aldrich. N,N'-dimethylformamide (DMF, ACS Reagent grade) was obtained from Fisher Scientific (Ottawa, ON) and used as received.

Solution preparation:

SKL and SKL:PEO solutions in DMF were prepared as follows: SKL solutions were first prepared by adding dry lignin powder to the solvent. In the case of SKL:PEO solutions, PEO was first dissolved by briefly heating the solution at 80°C for 10-15 minutes, then SKL was added to the PEO solution. The mixture was heated at 80°C for 2 hours with intermittent vortexing every 30 minutes for 1-2 minute intervals. Solutions without PEO were prepared at concentrations of 25, 30, 35, 40, 45, and 50% by mass. Solutions with PEO were prepared with 25-45% SKL and 0.1-0.4% PEO (M_v = 1 or 5 x 10^6 g/mol).

Rheology:

The shear viscosity η was obtained from small amplitude oscillatory shear as the limit of $G''(\omega)/\omega$ at low ω or from steady shear measurements increasing shear rate from 1-1000 s^{-1} using an AR2000 torsional rheometer (TA Instruments, Grimsby, ON) with 40 mm diameter cone geometry and cone angle of 2°. A capillary breakup extensional rheometer (Haake CaBER 1, Thermo Fisher Scientific, Waltham, MA) was used to characterize the extensional fluid properties. In CaBER experiments, plate diameters of 4 mm, initial plate separation of 2 mm, final plate separation of 12 mm, and plate separation time (strike time) of 50 ms were used. Where applicable (for elastic solutions), the characteristic time scale of viscoelastic stress growth in uniaxial extensional flow λ (herein referred to as the relaxation time) was obtained by fitting the portion of the thinning profile where exponential thinning was observed as described by Rodd et al[34] using the equation:

(1) $D_{mid}(t) = D_1((G^*D_1)/4\sigma)^{1/3}\exp(-t/3\lambda)$

Where $D_{mid}(t)$ is the midpoint filament diameter, D_1 is the filament midpoint diameter just after the upper plate stopped moving, G is the elastic modulus, and σ is surface tension.

Electrospinning:

Electrospinning experiments were conducted using two different apparatuses. One was a self-built apparatus previously described[32] with a flat-tip needle as spinneret (gauge 18, 21, 22, or 25), operating voltage of 15 kV, flow rate of 0.03 mL/min, and spinneret-to-collector distance of 15-20 cm. The second was a Kato Tech Nanofiber Electrospinning Unit (NEU) (Kato Tech Co., LTD, Kyoto, Japan) using a horizontal configuration, rotating drum collector, and the same operating parameters listed above.

Thermal processing of electrospun fabrics:

Electrospun fabrics were heated under different conditions in a Hewlett Packard 5890 Series II gas chromatography oven under air. The samples were mounted on a Teflon-wrapped glass sample holder and heated from 25°C at a specified heating rate up to a specified final temperature and holding at the maximum temperature for 1 hour. Heating rates of 0.5, 5, and ~40 °C/min and final temperatures of 150, 200, and 250°C were used. Each condition was repeated three times and tensile testing was carried out for each of the three runs using 20 samples per run. The results were reported as the average of the three runs +/- the standard deviation. Following thermostabilization, carbonization was carried out in a GSL 1100X2 tube furnace (MTI Corporation, Richmond, CA) under N_2 heating at a rate of 10°C/min to 1000°C and holding at 1000°C for 1 hour.

Characterization of fibers and fabrics:

Optical and scanning electron microscopy was used to characterize fiber appearance using and Olympus BX41 Microscope (Richmond Hill, ON) and Hitachi S3000N SEM using gold-coated samples and accelerating voltage of 5 kV. Fiber diameters were measured using ImageJ analysis software counting a minimum of N = 100 fiber diameters per sample. The tensile properties of electrospun nonwovens fabrics were measured on a KES-G1 Kawabata microtensile tester as described elsewhere.[35] A density of 1.3 g/mL was obtained from an MSDS for alkali lignin and used to convert tensile data from units of N/tex to units of MPa.

Results and Discussion

Shear rheometry:

Shear rheometry has been used to characterize the viscosity of electrospinning solutions in the literature and correlations between shear viscosity, electrospinnability, and fiber diameter have been reported for a variety of different systems.[36-38] We first reported on the shear properties of lignin solutions with and without PEO in DMF in a published study.[32] In general, the shear properties of lignin solutions was Newtonian, with a viscosity independent of shear rate. However, it was clear in previous work that different lignins have different viscosities.[32] The observed differences underscore the inherent variability in lignins obtained by different isolation procedures or from different botanical origins. To focus our study, we selected to characterize SKL in greater detail.[32] The viscosity of SKL solutions varied over roughly 3 orders of magnitude when increasing from 25-50%. The rather strong dependence of viscosity on SKL concentration $\eta \sim c^7$ indicated that intermolecular interactions between SKL molecules increased the viscosity. Interestingly, solutions with PEO displayed a similar dependency of viscosity with concentration, with slightly higher values of viscosity but overall showing the same trend. However, SKL:PEO solutions were electrospinnable at shear viscosities far lower than those of SKL solutions. This result indicated that shear viscosity was not the key fluid property influencing the electrospinning behavior, since typically higher viscosity promotes fiber formation.

CaBER:

While viscosity affects the electrospinning process, it is also known that the elastic properties of polymer solutions strongly affect the ability to form fibers and the fiber diameter.[39-43] Based on comparison of our results with reports from the literature, we hypothesized that elasticity resulting from PEO addition may be responsible for the improved electrospinnability of lignin:PEO solutions. To test our hypothesis we varied the elasticity by varying the PEO molecular weight and the concentrations of SKL and PEO and investigated the ability to form fibres and the fibre diameter. Furthermore, we characterized the extensional properties of our solutions using a capillary breakup extensional rheometer (CaBER).

Figure 1 shows the thinning profile of the SKL solutions without PEO for concentrations of 40, 45, and 50%. Time is on the abscissa and D_{mid} as recorded by a laser micrometer is shown on the ordinate normalized by the initial filament diameter D_1. In general SKL solutions at or below 35% concentration could not be measured using the CaBER due breakup occurring before the end of the initial step. In Figure 1 it can be observed that the filament thinning profiles were linear in time, consistent with the behavior of a Newtonian fluid, in agreement with shear measurements.[44]

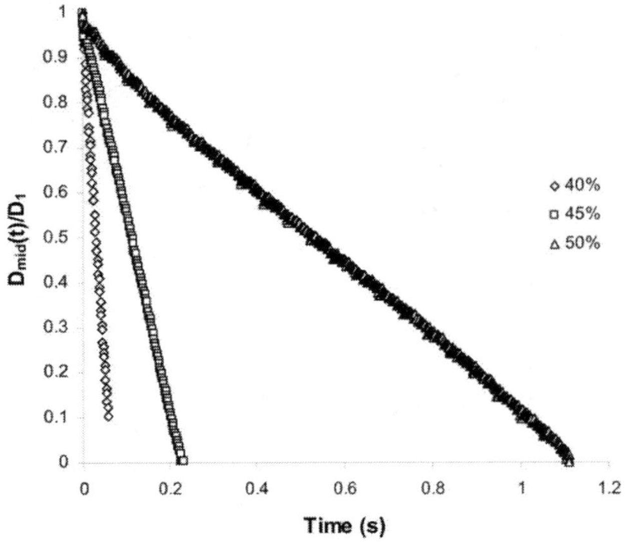

Figure 1: Thinning profiles of SKL solutions without PEO from CaBER experiments

In contrast, solutions with PEO displayed very different thinning behavior. Figure 2 shows the thinning profiles of several solutions with different SKL concentration, PEO concentration, and PEO molecular weight. Solutions with PEO display exponential thinning behavior, which indicates elasticity.[44] Elasticity implies that the extensional viscosity η_{ext} is not constant during thinning but increases with decreasing filament diameter (strain hardening). The relaxation time λ was determined from the CaBER data. Table 1 summarizes the compositions of the fluids and the values of λ obtained from the data. It can be seen in the table that λ increases with SKL concentration, PEO concentration, and PEO molecular weight, in agreement with published results.[45] Taken together these results show that the viscoelasticity of the SKL:PEO spinning solution can be varied by changing the relative amounts of SKL and PEO as well as the PEO molecular weight. The results also suggest that strain hardening occurring in extensional flow (which is absent in solutions without PEO) may provide the stability previously observed for lignin:PEO solutions in electrospinning.[32] We now discuss the effect of viscoelasticity on electrospinning.

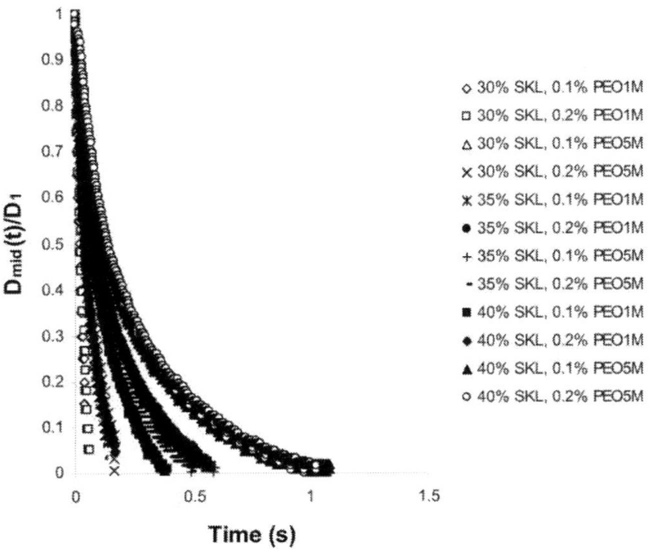

Figure 2: Thinning profiles of SKL:PEO solutions from CaBER experiments

Table 1: Fluid compositions and relaxation times

SKL concentration (%)	PEO concentration (%)	PEO mol. weight (g/mol)	λ (ms)
30	0.1	1×10^6	7.3
30	0.2	1×10^6	11.4
30	0.1	5×10^6	21.7
30	0.2	5×10^6	24.7
35	0.1	1×10^6	19.1
35	0.2	1×10^6	21.1
35	0.1	5×10^6	50.6
35	0.2	5×10^6	49.5
40	0.1	1×10^6	39.9
40	0.2	1×10^6	63.1
40	0.1	5×10^6	92.3
40	0.2	5×10^6	99.3
45	0.1	1×10^6	134.8
45	0.2	1×10^6	150.6
45	0.1	5×10^6	240.8
45	0.2	5×10^6	408.8

Electrospinning:

Example SEM images are shown in Figure 3 to illustrate the results of spinning experiments. Similar to our previous observations, only electrospray was formed without PEO below a concentration of 45-50%, with beaded fibers forming at 45%. In contrast, nearly all of the solutions with PEO were capable of forming fibers even though the amount of PEO in solution was less than 1% relative to SKL, producing fibers with SKL contents greater than 99%. The ability to achieve processibility with low PEO content is an advantage of the SKL:PEO system. Comparing rheology and electrospinning experiments, it can clearly be seen that fiber diameter increases with increasing λ. Figure 4 shows a plot of diameter vs. λ. The mean fiber diameter correlates well with λ, indicating that elasticity is correlated with the fiber diameter. Furthermore, the error bars show the standard deviations of the measured fiber diameters. It is also clear from the plot that the width of the diameter distribution tends to increase with increasing λ. This is partially due to the flattened, twisted morphology which tended to occur in solutions with higher viscoelasticity. The flattened morphology may be due to unevaporated solvent trapped in the core of the fiber followed by collapse of the fiber during the bending instability. The results presented here demonstrate that by controlling the viscoelasticity of the spinning solution, the mean diameter and width of the diameter distribution can be controlled.

Figure 3: SKL:PEO fibers produced from solutions with different viscoelasticities (Scale bar = 10 μm)

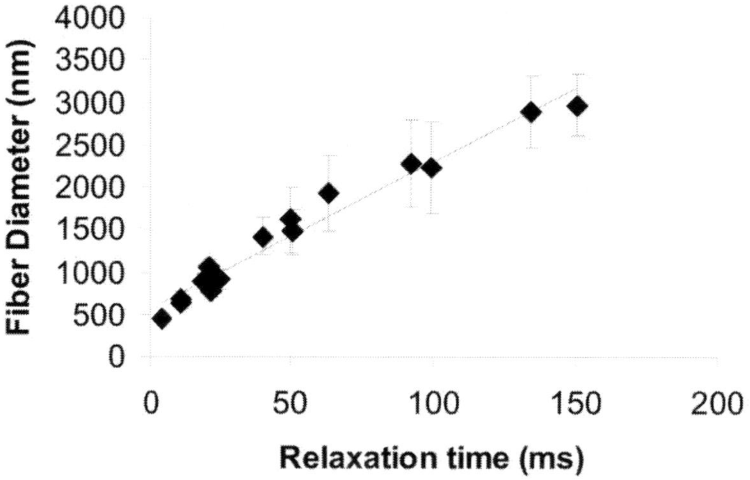

Figure 4: Fiber diameter vs. relaxation time for SKL:PEO solutions

Thermal Processing:

To prepare carbon fibers from lignin, an oxidative thermostabilization stage is necessary to prevent fibers from fusing during carbonization.[4] In previous work a slow heating rate was required to prevent fusion of melt-spun hardwood Kraft lignin (HKL) fibers.[4] The necessity of using a slow heating rate can be considered a disadvantage since it increases the time and energy required to generate carbon fibers. Due to its different chemical structure, SKL is expected to be more thermally stable.[4] However, previous attempts at melt spinning SKL were unsuccessful due to the tendency of SKL to char, so the thermostabilization of SKL fibers has not been reported in the literature. An SEM of a sample carbonized with and without thermostabilization is shown in Figure 5, showing fusion of fibers on the left and non-fused fibers on the right. To investigate thermostabilization of SKL fibers, we subjected our electrospun fabrics to different heating rates and final heating temperatures. It was observed that in contrast to melt-spun fibers or non-thermostabilized samples, none of the samples fused, indicating good stability. These samples were therefore suitable for conversion to carbon nanofibers. An SEM of carbonized 99% SKL fibers is also shown in Figure 5.

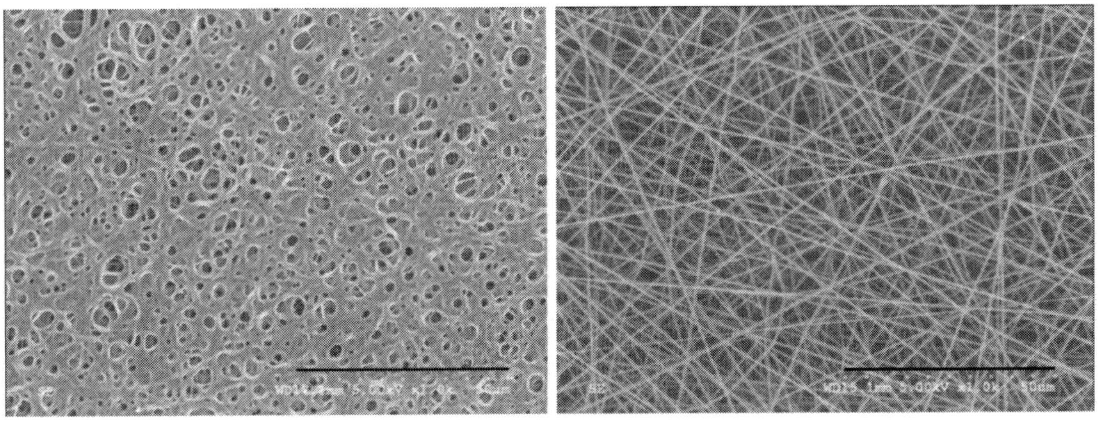

Figure 5: Carbonized 99% SKL fibers without thermostabilization (left), and after thermostabilization at 5°C/min to 250°C (right). Scale bar = 50 μm

Tensile testing:

To investigate the mechanical properties of SKL nonwoven mats, samples thermostabilized under different heating rates and temperatures but with similar diameters on the order of 1 μm were subjected to microtensile testing. The modulus and tensile strength values for different heating rates and thermostabilization temperatures are shown in Table 2. A steady increase in tensile strength was observed as the maximum treatment temperature increased from roughly 2-4 MPa for fibers heated to 150°C to values of more than 20 MPa for samples heated to 250°C. A similar trend was observed for elastic modulus, which increased from roughly 300 MPa to values of 1 GPa for fibers treated at 150 and 250°C, respectively. Interestingly, fibers heated at the lowest heating rate had a lower tensile strength, while comparable tensile strength was observed for samples heated at 5°C/min or faster. This could indicate that longer heating times lead to degradation of the mechanical properties of SKL fibers. The results presented here demonstrate the feasibility of generating precursors for carbon nanofibers from SKL by electrospinning.

Table 2: Mechanical properties of thermostabilized SKL:PEO nonwoven fabrics

Heating rate (°C/min)	T_{max} (°C)	Tensile strength (MPa)	Standard Deviation (MPa)	Elastic Modulus (MPa)	Standard Deviation (MPa)
As-spun	n/a	4.0	1.1	327.6	67.0
0.5	150	2.9	1.0	324.8	147.3
0.5	200	12.4	3.4	938.9	199.0
0.5	250	14.5	2.9	818.2	160.1
5.0	150	1.7	0.6	231.7	117.1
5.0	200	13.3	3.0	966.1	147.0
5.0	250	24.0	5.9	1055.9	269.5
47.0	150	3.7	1.0	322.6	121.6
41.0	200	10.7	1.9	849.4	101.6
39.0	250	22.1	3.7	1131.4	149.9

Summary:

The results from rheological experiments have demonstrated for the first time that while SKL solutions without PEO behave like Newtonian fluids while undergoing extensional flow, SKL:PEO solutions display non-Newtonian elastic behavior. We have demonstrated that SKL:PEO solution viscoelasticity strongly influences the fiber diameter and the width of the diameter distribution using shear rheometry and CaBER coupled with electrospinning experiments. By varying the SKL concentration, PEO concentration, and PEO molecular weight, we have demonstrated the ability to vary the viscoelastic properties over a wide range with less than 1% PEO added to the spinning solution (relative to SKL mass). With a sufficient understanding of SKL:PEO solution viscoelasticity, we then carried out thermostabilization experiments and measured the tensile properties of nonwoven SKL fabrics as a function of temperature and heating rate. In contrast with some previous reports on melt-spun fibers produced from HKL, electrospun SKL fibers did not fuse under any of the thermostabilization conditions tested, whereas carbonization performed without thermostabilization resulted in fiber fusion. In future work we hope to further optimize the process of lignin electrospinning to move toward toward carbon nanofiber precursors with properties suitable for real-world applications.

Acknowledgements

We would like to acknowledge North Carolina State University, the U.S. Department of Agriculture, and the Natural Sciences and Engineering Research Council of Canada (NSERC) for supporting this research.

References

1) Boerjan, W., Ralph, J., Baucher, M. (2003) Lignin Biosynthesis. *Annu. Rev. Plant Biol.* 54; 519-546
2) Ralph, J., Lundquist, K., Brunow, G., Lu, F., Kim, H., Schatz, P.F. Marita, J.M., Hatfield, R.D., Ralph, S.A., Christensen, J.H., Boerjan, W. (2004) Lignins: Natural polymers from oxidative coupling of 4-hydroxyphenylpropanoids. *Phytochemistry Reviews* 3; 29-60

3) Brodin, I., Sjoholm, E., Gellerstedt, G. (2009) Kraft lignin as feedstock for chemical products: The effects of membrane filtration. *Holzforschung*: 63, 290-297.
4) Kadla, J.F., Kubo, S., Venditti, R.A., Gilbert, R.D., Compere, A.L. (2002) Lignin-based carbon fibers for composite fiber applications. *Carbon* 40: 2913–2920
5) Morgan, P (2005). Carbon fibers and their composites Taylor and Francis, Boca Raton
6) Im, J.S., Park, S.J., Kim, T., Lee, Y-S. (2009) Hydrogen storage evaluation based on investigations of the catalytic properties of metal/metal oxides in electrospun carbon fibers International Journal of Hydrogen Energy; 34: 3382-3388
7) Osmond, N.M. (2000) Activated carbon fibre adsorbent materials. *Adsorption science and technology* 18, 6, 529-539
8) Sharma, C.S., Katepalli, H., Sharma, A., Madou, M. (2011) Fabrication and electrical conductivity of suspended carbon nanofiber arrays. *Carbon*; 49: 1727-1732
9) Aoki, K., Usui, Y., Narita, N., Ogiwara, N., Iashigaki, N., Nakamura, K., Kato, H., Sano, K., Ogiwara, N., Kametani, K., Kim, C., Taruta, S., Kim, Y.A., Endo, M., Saito, N. (2009) A thin carbon-fiber web as a scaffold for bone-tissue regeneration. *Small* 5: 13, 1540-1546.
10) Kubo S, Ishikawa N, Uraki Y, Sano Y. (1997) Preparation of lignin fibers from softwood acetic acid lignin—Relationship between fusibility and the chemical structure of lignin. *Mokuzai Gakkaishi* 43, 8, 655–662.
11) Dave,V., Prasad, A., Marand, H., Glasser, W. (1993) Molecular organization of lignin during carbonization. *Polymer* 34, 15, 3144-3154.
12) http://www.ornl.gov/ornlhome/news_items/news_100920.shtml
13) Kubo, S., Kadla, J.F. (2005a) Lignin-based Carbon Fibers: Effect of Synthetic Polymer Blending on Fiber Properties. *Journal of Polymers and the Environment*, Vol. 13, No. 2. 97-105.
14) Kubo, S., Yoshida, T., Kadla, J.F. (2007) Surface Porosity of Lignin/PP Blend Carbon Fibers. *J. Wood Chemistry and Technology* 27; 257-271.
15) Kubo, S., Kadla, J.F. (2005b) Kraft Lignin/Poly(ethylene oxide) Blends: Effect of Lignin Structure on Miscibility and Hydrogen Bonding. *Journal of Applied Polymer Science* 98; 1437–1444
16) Kadla, J.F., Kubo, S. (2003) Miscibility and Hydrogen Bonding in Blends of Poly(ethylene oxide) and Kraft Lignin. *Macromolecules 36:* 7803-7811
17) Kadla, J.F., Kubo, S. (2004) Lignin-based polymer blends: analysis of intermolecular interactions in lignin-synthetic polymer polymer blends. *Composites, Part A: Applied Science and Manufacturing* 35; 395-400
18) Kubo, S., Kadla, J.F. (2006) Effect of poly(ethylene oxide) molecular mass on miscibility and hydrogen bonding with lignin. *Holzforschung*. 60; 245–252.
19) Tagawa, T., Miyata, T. (1997). Size effect on tensile strength of carbon fibers. *Materials Science and Engineering* A238, 336–342
20) Tanaka, T., Nakayama, H., Sakaida, A., Horikawa, N., (1999) Estimation of tensile strength distribution for carbon fiber with diameter variation along fiber. *Materials Research Science International* 5, 2, 90-97.
21) Rutledge, G.C., Fridrikh, S.V. (2007) Formation of fibers by electrospinning. *Advanced Drug Delivery Reviews* 59, 1384-1391.
22) Reneker, D.H., Yarin, A.L. (2008) Electrospinning jets and polymer nanofibers. *Polymer* 49, 2387-2425

23) Greiner, A., Wendorff, J.H. (2007) Electrospinning: A fascinating method for the preparation of ultrathin fibers. *Angew. Chem. Int. Ed.* 46, 5670-5703.

24) Joshi, P., Zhang, L., Chen, Q., Galipeau, D., Fong, H., Qiao, Q. (2010) Electrospun Carbon Nanofibers as Low-Cost Counter Electrode for Dye-Sensitized Solar Cells. *ACS Applied Materials and Interfaces*; 2: 12, 3572-3577.

25) Arshad, S.N., Naraghi, M., Chasiostis, I. (2011) Strong carbon nanofibers from electrospun polyacrylonitrile. *Carbon*; 49: 1710-1719

26) Moon, S., Farris, R. (2009) Strong electrospun nanometer-diameter polyacrylonitrile carbon fiber yarns. *Carbon* 47: 2829-2839

27) Imaizumi, S., Matsumoto, H., Suzuki, K., Minagawa, M., Kimura, M., Tanioka, A. J. (2009) Phenolic Resin-Based Carbon Thin Fibers Prepared by Electrospinning: Additive Effects of Poly(vinyl butyral) and Electrolytes. *Polymer Journal* 41: 12, 1124-1128.

28) Park, S.H., Kim, C., Yang, K.S. (2004) Preparation of carbonized fiber web from electrospinning of isotropic pitch. *Synthetic Metals* 143: 175-179.

29) Lallave, M., Bedia, J., Ruiz-Rosas, R., Rodríguez-Mirasol, J., Cordero, T., Otero, J.C., Marquez, M., Barrero, A., and Loscertales, I.G. (2007) Filled and Hollow Carbon Nanofibers by Coaxial Electrospinning of Alcell Lignin without Binder Polymers. *Advanced Materials* 19; 4292-4296

30) Ruiz-Rosas, R., Bedia, J., Lallave, M., Loscertales, I.G., Barrero, A., Rodrı´guez-Mirasol, J., Cordero, T. The production of submicron diameter carbon fibers by the electrospinning of lignin. *Carbon* 48 (2010) 696–705

31) Seo, D.K., Jeun, J.P., Kim, H.B., Kang, P.H. (2011) Preparation and characterization of the carbon nanofiber mat produced from electrospun PAN/lignin precursors by electron beam irradiation. *Rev. Mater. Sci.*; 28: 31-34.

32) Dallmeyer, I., Ko, F., Kadla, J.F. (2010) Electrospinning of technical lignins for the production of fibrous networks. *Journal of Wood Chemistry and Technology* 30: 315-329.

33) Morck, R., Yoshida, H., Kringstad, K.P., Hatakeyama, H. (1986) Fractionation of Kraft lignin by successive extraction with organic solvents. *Holzforschung* 40(suppl.): 51-60.

34) Rodd, L.E., Scott, T.P. Cooper-White, J.J. McKinley, G.H. (2005) Capillary breakup rheometry of low viscosity elastic liquids. *Applied Rheology* 15: 12-27.

35) Gandhi,M., Yang,H., Shor, L., Ko, F. (2009) Post-spinning modification of electrospun nanofiber nanocomposite from Bombyx mori silk and carbon nanotubes. *Polymer* 50: 1918-1924.

36) Gupta, P., Elkins, C., Long, T.E., Wilkes, G.L. Electrospinning of linear homopolymers of poly(methyl methacrylate): exploring relationships between fiber formation, viscosity, molecular weight and concentration in a good solvent. *Polymer* 46 (2005) 4799–4810

37) McKee, M.G., Wilkes, G.L., Colby, R.H. Long, T.E. (2004) Correlations of solution rheology with electrospun fiber formation of linear and branched polyesters. *Macromolecules* 37, 1760-1767

38) Shenoy, S.L., Bates, W.D., Frisch, H.L., Wnek, G.E. (2005). Role of chain entanglements on fiber formation during electrospinning of polymer solutions: good solvent, non-specific polymer–polymer interaction limit. *Polymer* 46, 3372–3384

39) Feng, J.J. (2002) The stretching of an electrified non-Newtonian jet: A model for electrospinning. *Physics of Fluids* 14, 11, 3912-3926.

40) Feng, J.J. (2003) Stretching of a straight electrically charged viscoelastic jet. *J. Non-Newtonian Fluid Mechanics* 116, 55-70.

41) Carroll, C.P. Joo, Y.L. (2006) Electrospinning of viscoelastic Boger fluids: modeling and experiments. *Physics of Fluids* 18, 053102-1-14.

42) Yu, J.H., Fridrikh, S.V., Rutledge, G.C. (2006). The role of elasticity in the formation of electrospun fibers. *Polymer* 47, 4789–4797

43) Helgeson, M.E., Grammatikos, K.N., Deitzel, J.M. Wagner, N.J. (2008) Theory and kinematic measurements of the mechanics of stable electrospun polymer jets. *Polymer* 49, 2924-2936

44) Entov, V.M., Hinch, E.J. (1997) Effect of a spectrum of relaxation times on the capillary thinning of a filament of elastic liquid. *J. Non-Newtonian Fluid Mechanics* 72, 31-53.

45) Clasen, C., Plog, J.P., Kulicke, W.-M., Owens, M., Macosko, C., Scriven, L.E., Verani, M., McKinley, G.H. (2006) How dilute are dilute solutions in extensional flows? *Rheology* 50: 6, 849-881.

(462e) Application of Screening-Level Life Cycle Assessment to Emerging Nanoproducts: Nanosilver Textiles and CNT Electronics

David E. Meyer, National Risk Management Research Laboratory, U. S. Environmental Protection Agency, Cincinnati, OH and Venkata K. K. Upadhyayula, Oak Ridge Institute of Science and Education (ORISE), Oak Ridge, TN

As global concerns for environmental sustainability grow, emerging technologies must be scrutinized early in deployment to insure they represent the most environmentally conscious option. Screening-level Life Cycle Assessment (LCA) can provide a quick tool to identify the life cycle hot-spots and focus research efforts to help minimize the burdens of a technology while maximizing its benefits. This talk examines the application of LCA to emerging nanotechnologies using case studies of nanosilver socks and a field emission display device (FED) containing carbon nanotubes (CNTs). The use of nanoscale silver in consumer products has exploded in popularity because of antimicrobial effects, making it the largest nanocomponent in use. The use of CNTs in nanoproducts is expected to offer numerous environmental benefits such as energy efficiency and the waste reduction. Although these products are considered beneficial to society, some attention must be given to the potential environmental impacts arising throughout the life cycle as production demands escalate.

See more of this Session: Life Cycle Studies Associated with Nanomaterials and Nano-Enhanced Products
See more of this Group/Topical: Topical C: Environmental Aspects, Applications, and Implications of Nanomaterials and Nanotechnology

Nanoscale Dispersions of Polymers: Aramid Nanofibers

Ming Yang[1], Keqin Cao[2], Lang Sui[3], Ying Qi[4], Jian Zhu[1], Anthony M. Waas[5], Ellen M. Arruda[2], John Kieffer[3], M. D. Thouless[2] and Nicholas Kotov[1*], (1)Chemical Engineering, University of Michigan, Ann Arbor, MI, (2)Mechanical Engineering, University of Michigan, Ann Arbor, MI, (3)Materials Science & Engineering, University of Michigan, Ann Arbor, MI, (4)Electron Microbeam Analysis Laboratory, University of Michigan, Ann Arbor, MI, (5)Aerospace Engineering, University of Michigan, Ann Arbor, MI, *Corresponding author: kotov@umich.edu

Introduction: Kevlar is a well-known ultra-strong para-aramid synthetic macroscale fiber with a high tensile strength-to-weight ratio.[1,2] Typical threads and fibers of Kevlar consist of long molecular chains produced from PPTA (poly-paraphenylene terephthalamide) between which there are inter-chain bonds making the material extremely strong and stiff with a tensile strength 3.6 GPa and modulus of ~90 GPa,[3] comparable to the properties of CNT fibers.[4,5] Despite the fact that Kevlar processing might have involved the polymer in the nanoscale state, Kevlar fibers were never associated with nanoscale building blocks of composites. The formation of nanoscale versions of aramid fibers was never experimentally observed or theoretically predicted before.

Instead of the traditional synthesis from monomers used in many other methods of preparation of nanofibers, which can also be described as the bottom-to-top approach analogous to the terminology used in many nanoelectronic devices, alternative approaches must be explored for making polymeric structures that would satisfy the needs of both hybrid composites from INCs and multifunctional organic coatings. Importantly, the strength of polymeric components can come not only from the strength of individual polymeric chains but also from their assemblies, which, in turn, underscores the significance of the preparation method. Distribution of stress over multiple polymeric chains allows one to avoid stress concentration at particular bonds, which is the most daunting problem of molecular design of new ultrastrong polymers. Taking this further, we decided to use the so called top-down approach reminiscent to that being used for cellulose to prepare suitable polymeric components from macroscale fibers of synthetic polymers, which are known to have high mechanical strength but are processed in a form incompatible with INCs or other polymers.

Results and discussions: 1. Synthesis of ANFs: Inspired by Takayanagito's work on deprotonation of amide groups,[6,7] Burch *et al.* made aramid films by casting the PPTA solution followed by water immersion[8] or thermal decomposition.[9] In all these studies, the nature of the solution was not studied and it was assumed that the solution consisted of polyanions, which was probably only partially correct. Analyzing these works we realized that fine control over the deprotonation and dissolution process can lead to nanoscale fibers because the balance between electrostatic repulsion and attraction due to π-π and van der Waals interactions must lead to nanoscale version of this material. Notably in many ways the balance of forces and the process itself would be similar to the methods of preparation of graphite dispersions with intermediate step of charged graphene oxide. One could expect that the nanometer-scale forms of Kevlar fibers would possess equally high or better strength, stiffness, toughness and possibly other properties. Nanoscale fibers will allow the distribution of stress over many polymer chains and multiple bonds. This will avoid fast breakage of the weakest bond(s) taking place in non-assembled polymer chains.

High temperatures are known to induce the depolymerization of Kevlar,[8] so we decided to employ a different method of Kevlar dissolution than those used before, which allows for processing to be done at room temperature with minimal reduction of the molecular weight of aramid chains. It was found that the dissolution of Kevlar threads or fabric in KOH splits the bulk macroscale fibers into nanoscale fibers (Figure 1). This process is likely to occur by abstraction of mobile hydrogen from >NH groups and substantial reduction of the strength of hydrogen bonds between the polymer chains (Figures 1f,m), while increasing electrostatic repulsion. Importantly, the extent of this destruction is limited and does not appear to proceed down to the level of individual polymer chains. Complete disintegration of the fibers into individual chains is counteracted by hydrophobic attraction and π-π stacking in the polymer backbone. The diameters of ANFs after the dissolution of Kevlar fabric are 20-30 nm (Figure 1a,g). These diameters decrease with the addition of greater amounts of water (Figure 1b-1d, 1h-1j), reaching 5-10 nm when the volume ratio of water to DMSO is 1:40 (Figure 1d,j). From the materials perspective this is quite remarkable and very convenient for processing into composites although will require further studies. The length of these fibers appears to be in the range of 5 to 10 μm (Figure 1l). ANFs are thicker than single-walled CNTs, which are typically about 1 nm in diameter and several micrometers in length. ANFs are thinner and longer than the cellulose fibers, which are 5-15 nm in diameters and 200 nm to several micrometers in length.[10] Abundance

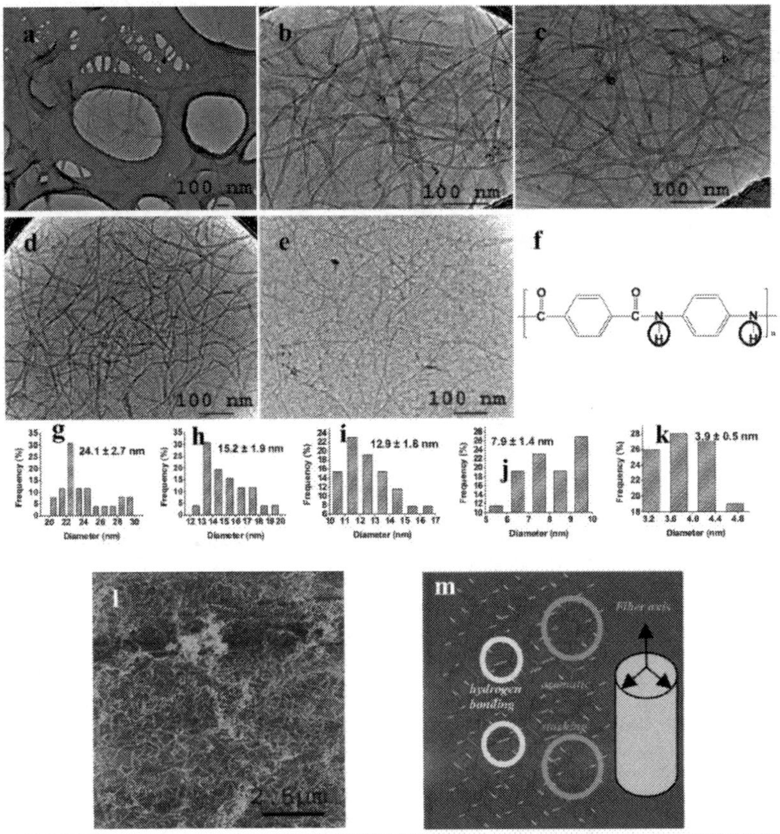

Figure 1. TEM images of ANFs in DMSO solution with different water to DMSO volume ratios: a) 0, b) 1/200, c) 1/100 and d) 1/40. e) TEM image of one bilayer PDDA/ANFs. f) The molecular structure of aramid; the H atoms removed by KOH are marked by circles. (g-k) Statistical analysis of diameter distribution for ANFs under same conditions as that shown in a-e, respectively. l) AFM image of one bilayer PDDA/ANFs. m) Molecular description of ANFs. Yellow circles mark the inter-molecular hydrogen bonds formed between the carbonyl groups and >NH groups; green circles mark aromatic stacking interactions between adjacent strands.

of polar functional groups on ANFs provides additional opportunities compared with CNTs to dramatically improve nanofiber-matrix stress transfer. Note that other potential nanofibers investigated so far, such as cellulose nanofibers, are quite rigid. While materials from them have remarkable properties, their successful utilization with INCs, for instance CNTs, to obtain effective stress transfer is impeded by insufficient flexibility as indicated by multiple recent studies.[11-14] Their sensitivity to humidity could also be problematic.[15]

*2. **ANF Multilayer Assembly:*** The availability of dispersions makes it possible to utilize a variety of different materials design techniques based on simple liquid processing. The layer-by-layer (LBL) assembly technique is one of the most versatile methods for the preparation of organic-inorganic composites with enhanced

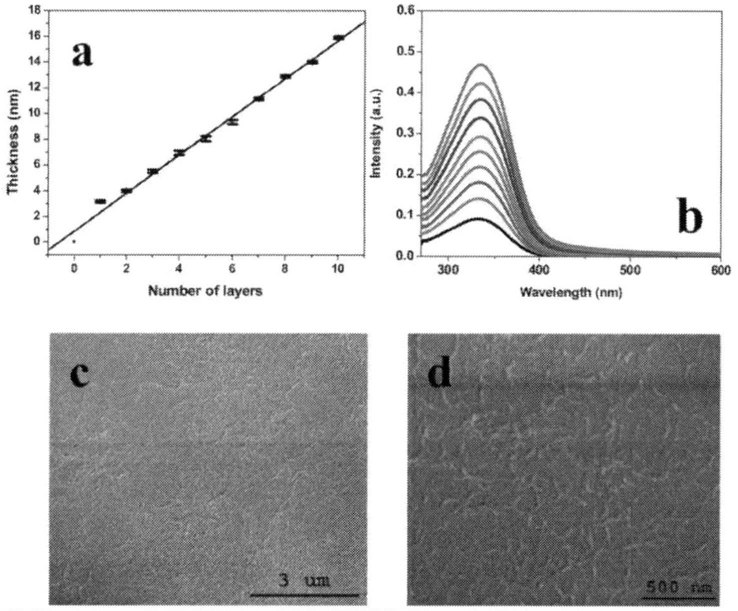

Figure 2. a) Ellipsometry results for LBL assembly of ANFs on silicon substrate. b) UV-Vis spectra of 1-10 layers of ANF films (from bottom to top: 1-10 layers of ANFs). c and d are SEM images of one bilayer PDDA/ANFs on a glass slide.

mechanical parameters. Among other methods of composite synthesis, LBL is advantageous due to intrinsically strong interactions between the components and great uniformity of the resulting materials. Successful examples of LBL composites with exceptional mechanical properties include several INCs.[16,17] It is essential to demonstrate that ANFs can be assembled in LBL systems. It was found that the ANF fibers were both dispersed and ionized in polar solutions and are therefore likely to be suitable for LBL processing. LBL assembly of ANFs was accomplished by using a traditional LBL dipping however only the first layer was PDDA and therefore will be denoted as PDDA/ANF$_n$ where n is the number of dipping cycles in ANF dispersion. The ANFs are fairly uniformly packed on the surface of glass slide during LBL assembly (Figure 2c,d). The gradual growth of ANF multilayers was monitored by ellipsometry and UV-Vis absorption spectra. The growth of PDDA/ANF$_n$ multilayers is nearly linear with the first 10 layers reaching a film thickness of about 16 nm (Figure 2a). This gives a thickness increment for each LBL bilayer of 1.6 nm, which corresponds to fairly loose two-dimensional packing of ANF in each adsorption cycle, as can be seen in the AFM image of Figure 1I.

As more LBL cycles were applied, the enhancement of an absorption peak at around 330 nm, which is characteristic of Kevlar, was observed in the UV-Vis spectrum (Figure 2b). Note that although the initial adsorption of negatively charged ANFs on positively charged PDDA can be attributed to electrostatic interactions, intermediate rinsing with water is expected to regenerate the H^+ on the aramid macromolecules removed by KOH (Figure 1f) and near-electroneutrality of the ANFs. At this moment, stability of ANF films outside of KOH is attributed primarily to

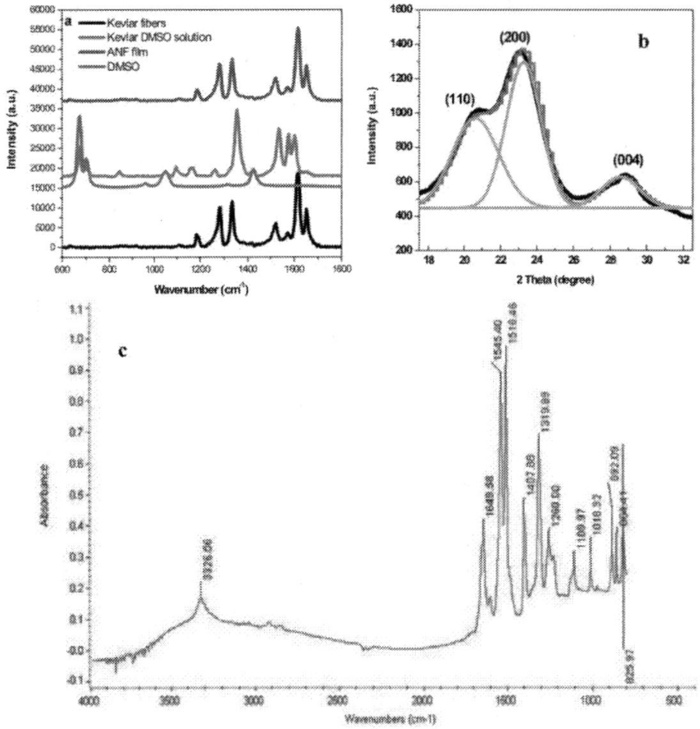

Figure 3. a) Raman scattering of Kevlar fibers (black line), Kevlar DMSO solution (red line) and ANF film (blue line). The Raman scattering of DMSO (green line) is also given near the spectra of Kevlar DMSO solution to clarify the solvent influence. b) XRD pattern of ANF film (black line) with Gaussian peak fitting (green and red lines). c) IR spectra of ANF film.

aromatic stacking interactions and hydrogen bonding between the multilayer components (Figure 1m), which are also believed to be the main interactions between individual nanofibers. Van der Waals interactions are also expected to contribute to the mechanical integrity of the fibrous network. The successful reconstruction of aramid films with the inclusion of ANFs during the LBL process can be confirmed by Raman scattering (Figure 3a). The Raman spectra of the original macroscale Kevlar material are similar to the results reported before,[18] but they are drastically changed after the dissolution in KOH DMSO solution (Figure 3a). These changes disappear after the construction of the ANF films and the Raman spectra of our final films are nearly the same as those of the original Kevlar fibers (Figure 3a). Identification of molecular structures of ANFs in the film was also confirmed by IR spectra (Figure 3c). Compared with Kevlar macrofiber,[19] the intensity and sharpness of IR peaks (for example, N-H stretching vibrations) is decreased due to broader distribution of bond lengths and surface states of the fibers. XRD measurements also show broadening of characteristic Kevlar peaks (Figure 3b), which can be assigned to (110),

(200) and (004) reflections, respectively.[20] Based on the Scherrer equation, $D = 0.9\lambda/\ \cos\theta$, where D is the size of crystallites, λ is the wavelength of X-ray beam used in diffraction experiments, is the line broadening at half the maximum intensity, θ is the Bragg angle, the diameter of the nanofiber can be calculated to be ca. 4.3 nm based on the broadening of (200) reflection obtained after Gaussian peak fitting. This size of crystalline domains matches

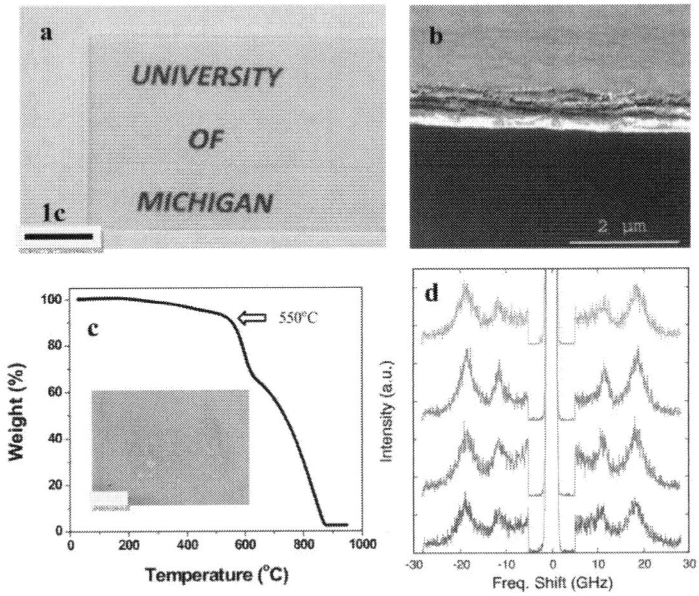

Figure 4. a) Optical image of 300 layers LBL ANF film on glass slide, b) cross-sectional SEM image of PDDA/ANF_{300} and c) thermogravimetric analysis of PDDA/ANF_{300}. Inset: free-standing PDDA/ANF_{300} film. d) Stokes and anti-Stokes peak of the Brillouin Spectra for PDDA/ANF_{300} taken at four different spots, the spectra are offset for the ease of viewing.

really well with the diameter of ANF from TEM observations (Figure 1e), and indicates that the nanoscale fibers retain substantial crystallinity despite reduction in diameter. This fact is also indicative of the retained mechanical properties characteristic of macroscale Kevlar fibers.

*3. **Physical Properties:*** The ANF LBL films are transparent allowing for potential use in ultrastrong surface coatings (Figure 4a). They can also be easily peeled from a surface by HF etching (Figure 4c inset). SEM images indicated that a PDDA/ANF_{300} film has a thickness of about 580 nm (Figure 4b), which is comparable to the results of ellipsometry measurements (Figure 2a) and suggests that the average ANF layer is about 1.9 nm thick. Thermogravimetric analysis (TGA) of PDDA/ANF_{300} (Figure 4c) is similar to the bulk Kevlar material.[1] This is very encouraging because heat-resistant, flexible, transparent films represent a bottleneck for many applications in flexible electronics and flexible photovoltaic cells based on semiconductors with high electron/hole mobility.

It is expected that PDDA/ANF_n films will be elastically orthotropic, with the in-plane and out-of-plane properties being different as is the case with clay and many other multilayers. Brillouin light scattering (BLS) is a convenient method to determine the orthotropic moduli. The in-plane and out-of-plane longitudinal moduli were determined to be 22 ± 1.2 GPa and 11.8 ± 0.6 GPa, respectively (Figure 4d , Table 1). A higher in-plane modulus is expected as ANFs are spread on the substrate with axes parallel to the plane. PDDA/ANF_n films outperform a well-known high-performance polymeric coating material, *i.e.*, polyimide thin film with 17.7 GPa in-plane longitudinal modulus and 7.7 GPa out-of-plane longitudinal modulus recently probed by a similar technique.[21] These mechanical properties of PDDA/ANF$_n$ films also exceed those for cellulose nanofibers with no effect of humidity that greatly reduces the performance of cellulose LBL films. Furthermore, the Young's modulus is 10-20 times higher than that of aligned cellulose fibers and CNTs.

The compression stiffness and hardness of the ANF LBL films were estimated using a Berkovich nanoindenter at various indentation depths. Care was taken to ensure that the depths of the indentations were shallow enough to avoid significant influence from the substrate, which was indeed detected at greater penetration depths (Figure 5). The tests revealed a hardness of 0.46 ± 0.04 GPa (Figure 5c, Table 1) for our film, which is higher than some high strength polymers, such as isotactic polypropylene (H = 0.125 GPa) and high-density polyethylene (H = 0.06 GPa). The Young's modulus was obtained using the continuous stiffness measurement technique of Oliver and Pharr during both the loading and unloading portions of the indentation. The film was found to have a

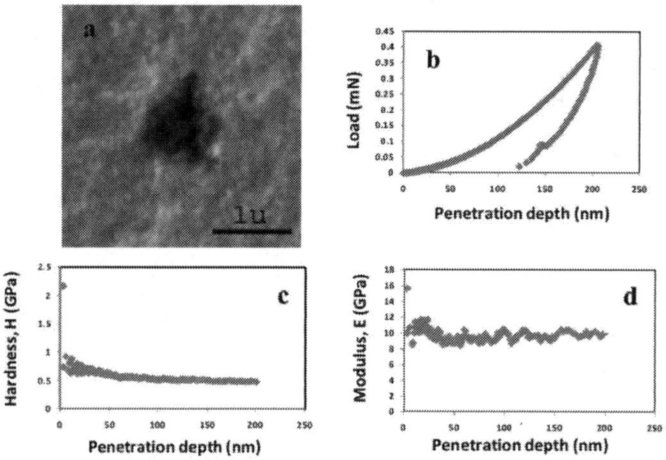

Figure 5. Typical results from nanoindentation experiments for PDDA/ANF$_{300}$ LBL film with maximum depth set at 200 nm. a) SEM images of an indent left by a Berkovich tip in PDDA/ANF$_{300}$ film. b) Represents load-displacement curve. c) Represents the hardness as a function of penetration depth and d) corresponding modulus. These results are from the loading part of the experiment.

modulus value of 9 ± 1 GPa (Table 1), and the value appeared not to be affected by the substrate for indentation depths of less than 200 nm (Figure 5d). The modulus of the films was higher than the 5.5 GPa reported for a cast film of pure Kevlar.[8] This indicates that the preparation of ANFs is indeed advantageous for preparation of high-performance materials. Solvent-based LBL process is particularly effective at producing reasonably dense and

A-nanoindentation	Loading	Unloading
Modulus (GPa)	9 ± 1	8 ± 1
Hardness (GPa)	0.46 ± 0.04	0.45 ± 0.05
B-Brillouin light scattering	In plane	Out of plane
Modulus (GPa)	22 ± 1.1	11.8 ± 0.6

Table 1. A) Summary of modulus and hardness derived from loading and unloading part. B) Summary of in-plane and out-of-plane modulus obtained from Brillouin light scattering.

better ordered arrays of the nanofibers, compared with the casting technique. The modulus is also higher than the reported elastic modulus obtained from nanoindentation of various uniaxial anisotropic polyimide films, which were in the range of 5 to 7 GPa.[22,23] Since BLS measures the longitudinal modulus, the corresponding Young's modulus can be extracted,[24] assuming a Poisson's ratio of 0.25,[25] resulting in a value of 9.8 ± 0.5 GPa, which compares remarkably well with the nanoindentation measurement of 9 ± 1 GPa.

Conclusions: Negatively-charged, uniformly-sized ANFs can be prepared in DMSO media by controlled deprotonation with KOH. The diameter of the fibers can be effectively controlled by addition of water. Such dispersions represent the first example of nanofiber dispersions of synthetic polymers. They are suitable for making nanoscale materials using many techniques and can be considered as new nanoscale building blocks for ultrastrong materials. They also reveal intriguing similarities to CNTs in respect to morphology and mechanical behavior in polymer composites. As a potential direction of future research, evaluation of the mechanical properties of individual ANF to compare those to carbon nanotubes would be quite interesting from both fundamental and practical perspectives. We believe that the method demonstrated here provides a way to prepare ANF-based high-performance films and to incorporate ANFs into multifunctional films lending its outstanding mechanical properties to the resulting hybrid material(s).

References:

1. Tanner, D.; Fitzgerald, J. A.; Phillips, B. R., The Kevlar Story - An Advanced Materials Case-Study. *Angew. Chem.-Int. Edit. Engl.* **1989**, *28*, 649-654.

2. Chae, H. G.; Kumar, S., Rigid-Rod Polymeric Fibers. *J. Appl. Polym. Sci.* **2006**, *100*, 791-802.

3. Vollrath, F.; Knight, D. P., Liquid Crystalline Spinning of Spider Silk. *Nature* **2001**, *410*, 541-548.

4. Koziol, K.; Vilatela, J.; Moisala, A.; Motta, M.; Cunniff, P.; Sennett, M.; Windle, A., High-Performance Carbon Nanotube Fiber. *Science* **2007**, *318*, 1892-1895.

5. O'Connor, I.; Hayden, H.; O'Connor, S.; Coleman, J. N.; Gun'ko, Y. K., Kevlar Coated Carbon Nanotubes for Reinforcement of Polyvinylchloride. *J. Mater. Chem.* **2008**, *18*, 5585-5588.

6. Takayanagi, M.; Katayose, T., N-Substituted Poly(p-phenylene terephthalamide). *J. Polym. Sci. Pol. Chem.* **1981**, *19*, 1133-1145.

7. Takayanagi, M.; Kajiyama, T.; Katayose, T., Surface-Modified Kevlar Fiber-Reinforced Polyethylene and Ionomer. *J. Appl. Polym. Sci.* **1982**, *27*, 3903-3917.

8. Burch, R. R.; Sweeny, W.; Schmidt, H. W.; Kim, Y. H., Preparation of Aromatic Polyamide Polyanions – A Novel Processing Strategy for Aromatic Polyamides. *Macromolecules* **1990**, *23*, 1065-1072.

9. Burch, R. R.; Manring, L. E., N-Alkylation and Hofmann Elimination from Thermal-Decomposition of R_4N^+ Salts of Aromatic Polyamide Polyanions – Synthesis and Stereochemistry of N-Alkylated Aromatic Polyamides. *Macromolecules* **1991**, *24*, 1731-1735.

10. Podsiadlo, P.; Choi, S. Y.; Shim, B.; Lee, J.; Cuddihy, M.; Kotov, N. A., Molecularly Engineered Nanocomposites: Layer-by-Layer Assembly of Cellulose Nanocrystals. *Biomacromolecules* **2005**, *6*, 2914-2918.

11. Chen, P.; Yun, Y. S.; Bak, H.; Cho, S. Y.; Jin, H.-J., Multiwalled Carbon Nanotubes-Embedded Electrospun Bacterial Cellulose Nanofibers. *Mol. Cryst. Liq. Cryst.* **2010**, *519*, 169-178.

12. Li, M.; Kim, I.-H.; Jeong, Y. G., Cellulose Acetate/Multiwalled Carbon Nanotube Nanocomposites with Improved Mechanical, Thermal, and Electrical Properties. *J. Appl. Polym. Sci.* **2010**, *118*, 2475-2481.

13. Miyauchi, M.; Miao, J.; Simmons, T. J.; Lee, J.-W.; Doherty, T. V.; Dordick, J. S.; Linhardt, R. J., Conductive Cable Fibers with Insulating Surface Prepared by Coaxial Electrospinning of Multiwalled Nanotubes and Cellulose. *Biomacromolecules* **2010**, *11*, 2440-2445.

14. Lu, P.; Hsieh, Y.-L., Multiwalled Carbon Nanotube (MWCNT) Reinforced Cellulose Fibers by Electrospinning. *ACS Appl. Mater. Interfaces* **2010**, *2*, 2413-2420.

15. Cranston, E. D.; Eita, M.; Johansson, E.; Netrval, J.; Salajkova, M.; Arwin, H.; Wagberg, L., Determination of Young's Modulus for Nanofibrillated Cellulose Multilayer Thin Films Using Buckling Mechanics. *Biomacromolecules* **2011**, *12*, 961-969.

16. Podsiadlo, P.; Kaushik, A. K.; Arruda, E. M.; Waas, A. M.; Shim, B. S.; Xu, J. D.; Nandivada, H.; Pumplin, B. G.; Lahann, J.; Ramamoorthy, A.; Kotov, N. A., Ultrastrong and Stiff Layered Polymer Nanocomposites. *Science* **2007**, *318*, 80-83.

17. Tang, Z. Y.; Kotov, N. A.; Magonov, S.; Ozturk, B., Nanostructured Artificial Nacre. *Nat. Mater.* **2003**, *2*, 413-418.

18. Washer, G.; Brooks, T.; Saulsberry, R., Investigating the Effects of Aging on the Raman Scattering of Kevlar Strands. *Res. Nondestruct. Eval.* **2008**, *19*, 144-163.

19. Crompton, T. R., *Polymer Reference Book* Smithers Rapra Technology, **2006**, p181-182.

20. Jain, A.; Vijayan, K., Kevlar 49 Fibres: Thermal Expansion Coefficients from High Temperature X-Ray Data. *Curr. Sci.* **2000,** *78*, 331-335.

21. Gomopoulos, N.; Saini, G.; Efremov, M.; Nealey, P. F.; Nelson, K.; Fytas, G., Nondestructive Probing of Mechanical Anisotropy in Polyimide Films on Nanoscale. *Macromolecules* **2010**, *43*, 1551-1555.

22. Lee, C.; Kwon, J.; Park, S.; Sundar, S.; Min, B.; Han, H., Nanoindentation Studies of Polyimide Thin Films with Various Internal Linkages in the Diamine Component. *J. Polym. Sci. Pt. B-Polym. Phys.* **2004**, *42*, 861-870.

23. Jang, W.; Seo, J.; Lee, C.; Paek, S. H.; Han, H., Residual Stress and Mechanical Properties of Polyimide Thin Films. *J. Appl. Polym. Sci.* **2009**, *113*, 976-983.

24. The longitudinal modulus, C_{11}, is related to the Young's modulus, E, and Poisson's ratio, ν, through the formula, $C_{11}=E(1-\nu)/(1+\nu)(1-2\nu)$.

25. Nakamae, K.; Nishino, T.; Xu, A. R., Poissons Ratio of the Crystal-Lattice of Poly(p-phenylene terephthalamide) by X-Ray-Diffraction. *Polymer* **1992,** *33*, 4898-4900.

(599f) Optimization of Synthetic Methods for Preparing Polypeptide-Gold Nanorods Plasmonic Matrices As Potential Therapeutic Systems

Huang-Chiao Huang, Alisha Nanda and Kaushal Rege, Chemical Engineering, Arizona State University, Tempe, AZ

Gold nanoparticles have been extensively investigated as potential theranostics due to their plasmonic properties (e.g. localized surface plasmon resonance or LSPR), biocompatibility, multiple imaging modalities and ease of surface functionalization. Interfacing gold nanorods (GNRs) with biological polymers can lead to smart materials with several applications in biosensing, regenerative medicine and therapeutic systems. In this study we describe "plasmonic matrices" generated using cross-linking of cysteine-containing elastin-like polypeptides (ELPs) and gold nanorods (GNRs) at physiological temperature (37 °C). Extensive cross-linking between the polypeptide and nanorods leads to the formation of plasmonic ELP-GNR matrices. Extensive investigations into matrix formation included elucidation of the role of nanoparticle concentration, electromagnetic radiation, and presence of cetyltrimethyl ammonium bromide (CTAB) surfactant, crucial to synthesis of gold nanorods. Biologically active molecules (e.g. chemotherapeutic drugs and DNA) were entrapped in the cross-linked GNR-polypeptide network. The photothermal property of the plasmonic matrix, combined with the thermal responsive property of the polypeptide, was employed in order to induce release of the entrapped molecules. Ongoing studies indicate the promise of these plasmonic matrices for combination treatment in cancer and for microarray applications in drug discovery.

See more of this Session: Nanostructured Biomimetic and Biohybrid Materials and Devices
See more of this Group/Topical: Nanoscale Science and Engineering Forum

(394d) Co-Products of Bioenergy System: Characterization of Saccharification Residuals

Han-Seung Yang, Shona Duncan, William T. Y. Tze and Jonathan Schilling, Department of Bioproducts and Biosystems Engineering, University of Minnesota, Saint Paul, MN

AIChE2011 Annual Meeting - Forest and PlantBioproducts Division

Co-Products of Bioenergy System: Characterizationof Saccharification Residuals

Han-Seung Yang[1], WilliamT. Y. Tze[1], Jonathan Schilling[1] and Shona Duncan[1]

[1]Departmentof Bioproducts and Biosystems Engineering

University of Minnesota

Saint Paul, MN 55108-6130 USA

Tel: 612-625-5200 Fax: 612-625-6286

e-mail: hsyang@umn.edu

The objective of this study is to investigate the effects ofdilute acid pretreatment and enzymatic saccharification on the reactivity,porosity, and crystallinity of solid residuals of biomass. Aspen wood flour was pre-treated with adilute acid treatment followed by enzymatic saccharification. The solid phase of the saccharificationprocess was analyzed using standard NREL protocols for amounts of lignin, pentosans,and hexosans. The functional groupsof solid residuals were analyzed using mid-infrared spectroscopy and hydroxylnumbers. The porosity andcrystallinity of the solids were also evaluated usingrespectively, nitrogen adsorption and X-ray diffraction. Cellulose nanocrystals were extractedfrom the solid residuals and verified using atomic force microscopy. The results will be discussed in thelight of reactivity, porosity, and crystallinity of saccharification residues topave the way for down-stream conversion into co-products of bioenergy(bio-alcohol) systems.

See more of this Session: Nanocellulose-Based Materials and Composites
See more of this Group/Topical: Forest and Plant Bioproducts Division

(755e) A Cell-Free Approach to Optimized Production and Self-Assembly of Novel Monodisperse Virus-Based Nanoparticles

Bradley C. Bundy, Mark T. Smith and Chad T. Varner, Chemical Engineering, Brigham Young University, Provo, UT

Virus-like particles (VLPs) are composed of viral proteins that assemble into highly symmetrical nanoparticles similar to viruses but do not contain the infectious genome. A wide variety of VLPs have been discovered and engineered for diverse applications such as drug/gene delivery, vaccine development, hydrogen storage, MRI imaging, and biosensing. A particularly attractive VLP for the above applications is the Qbeta VLP which is composed of 180 copies of the coat protein. The Qbeta VLP is very stable under typically bio-incompatible conditions due to disulfide bonds which covalently cross-link the protein monomers of the VLP together. Herein, we present the use of a cell-free protein synthesis system to directly control Qbeta VLP protein production for optimal VLP production and self-assembly. Cell-free protein synthesis is an in vitro expression system wherein cytotoxic proteins can be produced at high yields and the DNA expression template concentrations can be directly controlled to optimize the relative expression rates of multiple proteins. Using this system we expressed at high yields the cytotoxic A2 protein which is incorporated into the Qbeta virus such that only a single protein is inserted into each virus. Due to its cytotoxic nature, the crystal structure for the A2 protein has not been solved and the mechanism and interactions of its insertion into the Qbeta virus is not well understood. However, by coexpressing the Qbeta coat protein and A2 protein a novel VLP which incorporated the A2 protein was found to self-assemble. We also found that the relative expression rate and concentrations of the A2 protein relative to the coat protein had a strong impact on A2 incorporation efficiency and VLP self-assembly efficiency. This novel VLP would be useful for applications where a single unique attachment site on a nanoparticle is desired.

See more of this Session: Self-Assembled Biomaterials
See more of this Group/Topical: Nanoscale Science and Engineering Forum

(395c) Template-Directed Synthesis of Micro and Nano-Structures of Functional Conducting Copolymers by Oxidative Chemical Vapor Deposition

Dhiman Bhattacharyya and Karen K. Gleason, Chemical Engineering, Massachusetts Institute of Technology, Cambridge, MA

Template-directed synthesis of micro and nano-structures of functional conducting copolymers by oxidative chemical vapor deposition

Dhiman Bhattacharyya, and Karen K. Gleason*

Department of Chemical Engineering, Massachusetts Institute of Technology, Cambridge, MA 02139

Abstract:

Template-directed syntheses of inorganic and organic materials are becoming very popular because it is easy to control the desired structures in the products. Micro and nanostructures are important because they provide very high surface area to volume ratio. It increases the interfacial area of the surfaces. Thus synthesized increased surface areas have many applications, for example, to increase the sensitivities of sensors, performances of the photovoltaic cells. Previously, nano and micro-structures made of conductive polymers have been reported. However, they were free floating nanostructures of conductive polymers. In this work, we present a single step preparation of micro-tubules and nanoforests of functional conductive copolymers by oxidative chemical vapor deposition (oCVD) process at low temperature. Here, 3,4-ethylenedioxythiophene (EDOT) and 3-substituted thiophene monomers were employed to synthesize these structures. Iron chloride was used as an oxidant during the vapor phase polymerization step, as well as a dopant in the polymeric system in order to make it conductive. The copolymeric films were characterized by FT-IR, UV-Vis and X-ray photoelectron spectroscopies and 4-point probe measurements and SEM. The available functional groups in the copolymers were used for further derivatization reactions.

See more of this Session: Nanoscale Structure In Polymers III
See more of this Group/Topical: Materials Engineering and Sciences Division

(543b) Photothermally Triggered Drug Release From Temperature Sensitive Liposomes

Natalie Forbes, Chemical Engineering, University of California, Santa Barbara, Santa Barbara, CA and Joseph A. Zasadzinski, Department of Chemical Engineering and Materials Science, University of Minnesota, Minneapolis, MN

A drug delivery system with rapid contents release under continuous-wave near-infrared (NIR) light will enable precise spatial and temporal control of drug release. The drug is encapsulated within temperature sensitive liposomes and rapid drug release is triggered by the photothermal heating of hollow gold nanoshells coupled to the liposome surface. The nanoshells locally heat the liposome membrane to the phase transition temperature, allowing the drug to passively diffuse across the membrane. The phospholipid composition of the liposome membrane was tuned to control the temperature at which release was initiated and to enhance the rate of release. Rapid release of 60-70% of the encapsulated contents was observed at physiological temperatures within minutes of irradiation. This system could be used for cancer therapy to deliver anticancer agents directly to a tumor site.

See more of this Session: Nanotechnology for Biotechnology and Pharmaceuticals II
See more of this Group/Topical: Nanoscale Science and Engineering Forum

(744e) Enzyme Coatings On Magnetic Nanoparticles for Rapid Protein Digestion In Proteomic Analysis

Byoungsoo Lee[1], Daniel Lpez-Ferrer[2], Richard D. Smith[2] and Jungbae Kim[1], (1)Department of Chemical and Biological Engineering, Korea University, Seoul, South Korea, (2)Pacific Northwest National Laboratory, Richland, WA

Protein digestion is a critical step in the bottom-up proteomic analysis, but takes a long time due to the poor stability of a digestive enzyme called trypsin. We prepared trypsin coatings on magnetic nanoparticles (EC-TR/NPs) via a simple multilayer random crosslinking of the trypsin molecules onto magnetic nanoparticles. The resulting EC-TR/NPs were highly stable and could be easily captured using a magnet after the digestion was complete. EC-TR/NPs showed a negligible loss of trypsin activity after multiple uses and continuous shaking, while the conventional immobilization of covalently-attached trypsin on magnetic nanoparticles resulted in a rapid inactivation under the same conditions due to the denaturation and autolysis of trypsin. A single model protein, a five protein mixture, and a whole mouse brain proteome were digested at traditional digestion condition or in combination with pressure cycling technology (PCT) at room temperature for 1 min. Highly stable and magnetically separable EC-TR/NPs were successfully employed in protein digestion for proteomic analysis. In addition, the concomitant use of EC-TR/NPs and PCT resulted in very rapid (~1 min) and efficient digestions with more reproducible digestion results.

See more of this Session: Nanoscale Science and Engineering In Biomolecular Catalysis II
See more of this Group/Topical: Nanoscale Science and Engineering Forum

(703a) Scalable Nanomanufacturing of Millimeter Length 2D Nanosheets of Thermoelectric $Na_{0.7}CoO_2$

Mahmut Aksit, David P. Toledo and Richard D. Robinson, Material Science and Engineering, Cornell University, Ithaca, NY

A scalable nanomanufacturing technique is reported for batch fabrication of electrically-conducting 2D metal-oxide nanosheets of $Na_{0.7}CoO_2$. We report a sol-gel based, high temperature bottom-up synthesis which is a cost-effective route capable of producing tens of thousands of nanosheet layers packed into a macro-scale pellet. The synthetic procedure consists of sol-gel coordination of metal ions, auto-combustion, pressurized pellet formation, kinetic demixing, and calcination. The nanosheets are uniform in length and shape with highly anisotropic dimensions of nanometer sheet thickness and millimeter lateral lengths (10^{-5}:1:1), and are readily delaminated into free-standing nanosheets. According to GIXRD, the nanosheet stacking has turbostratic arrangement with nanosheets misaligned rotationally with respect to stacking axis. $Na_{0.7}CoO_2$, in bulk form, is already a surprisingly good thermoelectric material with a thermoelectric (TE) power factor as high as the industry standard Bi_2Te_3 (50 vs. 40 μW/ K^2cm respectively). Also, it has been reported in the literature that 2D nanosheets are observed to reduce thermal conductivity of materials with a factor ~10^2 due to phonon confinement, which provides a significant increase in the thermoelectric efficiency. Therefore, in conjunction with low fabrication cost, the reported millimeter length $Na_{0.7}CoO_2$ nanosheets are expected to provide high thermoelectric efficiency with practical importance. Preliminary thermoelectric measurements have been obtained for turbostratically aranged millimeter-length nanosheet stackings.

See more of this Session: Nanomanufacturing
See more of this Group/Topical: Materials Engineering and Sciences Division

(246c) Oligosaccharide/Silicon-Containing Block Copolymers for Lithography Applications

Julia D. Cushen[1], Issei Otsuka[2], Sami Halila[2], Sbastien Fort[2], Redouane Borsali[2], Erica Rausch[1], C. Grant Willson[3] and **Christopher J. Ellison**[1], (1)Department of Chemical Engineering, The University of Texas at Austin, Austin, TX, (2)Centre de Recherche sur les Macromolcules Vgtales, CERMAV, CNRS UPR 5301-ICMG affiliated with Joseph Fourier University, Grenoble, France, (3)Departments of Chemical Engineering and Chemistry, The University of Texas at Austin, Austin, TX

The need to overcome feature-size limitations in conventional lithography has led to the development of new patterning techniques using block copolymer templates. Ideal block copolymer systems for these applications have high etch contrast between blocks to promote good feature resolution and high chi-parameters to achieve small features. An additional desireable attribute is polymers with high silicon content such that they form a robust oxide mask during reactive ion etching with oxygen. To achieve etch contrast, these silicon-containing polymers can be incorporated into a block copolymer where the adjacent block is organic and etches easily. It is also observed that, in some cases, incorporating silicon into one of the blocks increases chi compared to similar silicon-deficient block copolymers. Here we will present three new block copolymer systems exhibiting cylindrical morphologies that incorporate fast-etching oxygen-rich oligosaccharides coupled to a silicon-containing polymer. The silicon-containing block provides sufficient etch resistance to achieve robust patterns in addition to promoting high chi parameters which allows access to cylinder diameters between 2 and 5nm.

See more of this Session: Nanoscale Structure In Polymers I
See more of this Group/Topical: Materials Engineering and Sciences Division

(462c) Mobility and Deposition of Silver Nanoparticles On Silica Surfaces Under Environmentally Relevant Conditions

B. Reginald Thio, Engineering Product and Design, Singapore University of Technology and Design, Singapore, Singapore, **Milka O. Montes**, Bren School of Environ. Sci & Mgmt, University of California, Santa Barbara, Santa Barbara, CA, Mahmoud A. Mahmoud, School of Chemistry and Biochemistry, Georgia Institute of Technology, Atlanta, GA and Arturo A. Keller, Bren School Environ. Sci & Mgmt, University of California, Santa Barbara, Santa Barbara, CA

Increasing use of nanomaterials in commercial products has led to widespread concerns about their potential environmental and health impacts. To better understand the transport, fate and behavior of nanoparticles in aquatic systems, it is essential to understand their interactions with different components of natural waters including natural organic matter (NOM) over a broad range of physicochemical conditions. Ag nanoparticles are widely used in many consumer applications such as antibacterial socks, and pigments. Deposition and mobility rates and the effects of NOM on particle deposition, as well as transport behavior of Ag nanoparticles onto a silica surface were evaluated using thin layer chromatography (TLC), and quartz crystal microbalance with dissipation (QCM-D) over a broad range of solution conditions.

See more of this Session: Life Cycle Studies Associated with Nanomaterials and Nano-Enhanced Products
See more of this Group/Topical: Topical C: Environmental Aspects, Applications, and Implications of Nanomaterials and Nanotechnology

(353d) Award Submission: Fabrication of Nanocarbon Fibers for Neural Tissue Engineering

John M. Landers[1], Dan Y. Lewitus[2], Jonathan R. Branch[2], Gerardo Callegari[3], Karen L. Smith[4], Joachim Kohn[5] and Alexander V. Neimark[6], (1)Department of Chemical and Biochemical Engineering, Rutgers University, Piscataway, NJ, (2)The New Jersey Center for Biomaterials, Rutgers, the State University of New Jersey, Piscataway, NJ, (3)TRI/Princeton, Princeton, (4)Wadsworth Center, New York State Department of Health, Albany, NY, (5)Chemistry and Chemical Biology, Rutgers University and The New Jersey Center for Biomaterials, Piscataway, NJ, (6)Chemical and Biochemical Engineering, Rutgers, The State University of New Jersey, Piscataway, NJ

We report a novel approach for fabricating nanocarbon fibers (NCF) as a potential candidate material for neural prosthetic devices. The current crop of neural probes, most notably metal or silicone electrodes, suffer from the lack of long term performance due to scarring and issues with biocompatibility. However, carbon allotrope structures such as single walled carbon nanotubes and graphene, not only circumvent issues of biocompatibility by being comprised only of carbon, but also provide an advantage of being conductive, that is the critical attribute for electrode performance. Fabrication of NCF are aided with the incorporation of polysaccharide agarose whose inherent gelation properties can be manipulated to produce flexible, durable and conductive fibers. Furthermore, agarose can be chemically functionalized to express biological moieties through free hydroxly groups on the agarose chain. Therefore, by combining the conductive properties of nanocarbon materials with the gelling and functionalization properties of agarose, we succeeded to produce fibers that are conductive and nontoxic in nature, which in turn enables cell attachment and response,as shown by both in vitro and in vivo experiments.

See more of this Session: Bionanotechnology Graduate Student Award Session
See more of this Group/Topical: Nanoscale Science and Engineering Forum

(747d) Gold Nanoclusters with Strong near Infrared Absorbance for Biomedical Imaging

Ameya U. Borwankar[1], Brian W. Willsey[2], Tianyi Wang[3], Veronika Sapozhnikova[4], April Twu[1], Marc D. Feldman[5], Thomas E. Milner[3] and Keith P. Johnston[1], (1)Chemical Engineering, University of Texas at Austin, Austin, TX, (2)Dept. Materials Science, University of Texas at Austin, Austin, TX, (3) Biomedical Engineering, University of Texas at Austin, Austin, TX, (4)South Texas Veterans Affairs Hospital System, San Antonio, TX, (5)Division of Cardiology, Department of Medicine, University of Texas Health Science Center at San Antonio, San Antonio, TX

We have developed small 40-50 nm gold nanocluster NIR imaging agents, stabilized by carboxymethyldextran (CMD) The shape asymmetries and the closely spaced primary particles within the clusters produce a strong near infra red (>800 nm) absorbance via a surface plasmon resonance (SPR), as a result of multipole interactions. The cluster size was characterized by dynamic light scattering(DLS) and TEM, whereas the quantity of polymer was determined by Thermogravimetric analysis. We develop a theory to explain the synthesis of these particles based on nucleation, clustering and growth upon reduction of a Au precursor in the presence of CMD. Molecular imaging of plaque in coronary arteries before rupture is a crucial step in treating atherosclerosis. These particles are preferentially engulfed by macrophages associated with vulnerable plaque due to the CMD associated with the particles. The particles also exhibit multi-photon luminescence and hence the plaque can be imaged by multi-photon microscopy.

See more of this Session: Nanotechnology for In Vivo and In Vitro Imaging
See more of this Group/Topical: Nanoscale Science and Engineering Forum

(342e) Novel Porous Ceramic Materials Via Magnetically Driven Self-Assembly of Non-Magnetic Nanoparticles

Marco Furlan, Institute for Chemical and Bioengineering, Department of Chemistry and Applied Biosciences, ETH Zurich, Zurich, Switzerland and **Marco Lattuada**, Department of Chemistry and Applied Biosciences , Institute for Chemical and Bioengineering, ETH Zurich, Zurich, Switzerland

The preparation of porous materials via self-assembly of nano- and microparticles is a topic of great scientific and technological relevance. Applications for such materials can be found in many areas, such as chromatography, membrane production, catalyst supports, bio-inspired materials, and scaffolds for tissue engineering.

Conventional materials, which can be prepared from colloidal suspension of polymers or ceramics, have usually a random porous structure. The preparation of these materials with better organized pores is of paramount interest for many applications.

In this work we introduce a new technique for the preparation of porous materials with anisotropic structure. This new method takes advantage of the alignment in the presence of a magnetic field of non-magnetic colloidal nanoparticles, either polymeric or ceramic (e.g. silica, alumina), dispersed in an aqueous highly stable ferrofluid. Once the magnetic field is applied, the non-magnetic nanoparticles act as magnetic holes, i.e., they acquire a magnetic moment in the opposite direction to that of the external field. These moments generate dipolar interactions capable of aligning non-magnetic particles in the direction of the applied field.[1]

The obtained structures can be frozen by adding a water-soluble monomer, a crosslinker and an initiator to the aqueous solution, so that once the structure is formed a free radical polymerization process can be utilized to produces a hydrogel that locks it. Once the obtained anisotropic structure is blocked inside the hydrogel, the ferrofluid can be removed by immersing the hydrogel in a concentrated hydrochloric acid solution. The anisotropic structure can then be hardened by depositing via a sol gel process either the same material of the nanoparticles or a different one, by simply immersing the monolith in a solution of the sol-gel precursor. After removing of the hydrogel by thermal treatment a hard porous monolith can be obtained. The resulting materials have a complex and organized structure, which was studied and characterized using SEM microscopy and mercury porosimetry.

[1] A. T. Skjeltorp, *Phys. Rev. Lett.* **1983**, *51*, 2306.

See more of this Session: Templated Assembly of Inorganic Nanomaterials II

(385c) Peptide-Functionalized Superparamagnetic Iron Oxide Nanoparticles for Diagnosis and Treatment of Atherosclerosis

Laurie B. Drews, Chemical Engineering, University of California, Berkeley, Berkeley, CA and Matthew Tirrell, Bioengineering, University of California, Berkeley, Berkeley, CA

Atherosclerosis is one of the major causes of cardiovascular disease, which remains the leading cause of death in the United States. The development and progression of atherosclerosis varies on a case-by-case basis, but certain universal markers exist which provide a means for a targeted approach to diagnosing and treating the disease. Current methods of detection of atherosclerosis are both expensive and lack portability. Magnetic particle imaging (MPI), however, offers the promise of increased resolution and sensitivity as well as portability as compared to alternative imaging techniques.

To this end, we are functionalizing iron oxide particles with peptide targeting agents via 2-pyridyl disulfide. Functionalization of the cysteine residue in the peptide CREKA with 2-pyridyl disulfide provides a means of quantifying the average number of peptides per particle [1]. Preliminary work has shown the ability of CREKA to selectively target fibrin at the areas most prone to rupture, namely the plaque corners [2]. Iron oxide nanoparticles functionalized with CREKA can be used to image the extent of occlusion of the plaque in a given artery using magnetic particle imaging. Infrared spectroscopy confirmed successful conjugation of the peptide to iron oxide nanoparticles. Transmission electron microscopy images and dynamic light scattering show that spheres smaller than 40 nm in diameter are formed. To provide proof of principle *in vitro*, conjugation of a cyclic RGD peptide, which has previously been used for its cell adhesion properties [3], is currently being explored to test the ability of magnetic particle imaging to detect peptide-functionalized iron oxide particles attached to the cell exterior. Successful detection of RGD-conjugated iron oxide nanoparticles will pave the way for future *in vivo* studies using CREKA.

[1] Josephson, L, et al. High-Efficiency Intracellular Magnetic Labeling with Novel Superparamagnetic-Tat Peptide Conjugates. *Bioconjugate Chem.* 1999; **10**: 510-520.

[2] Peters, D, et al. Targeting Atherosclerosis by Using Modular Multifunctional Micelles. *PNAS*. 2009; **106**: 9815-9819.

[3] Sugahara, K, et al. Tissue-Penetrating Delivery of Compounds and Nanoparticles into Tumors. *Cancer Cell*. 2009; **16**: 186-191.

See more of this Session: Magnetic Nanoparticles In Biotechnology and Medicine
See more of this Group/Topical: Nanoscale Science and Engineering Forum

(121b) A Coarse-Grained Molecular Dynamics Study On the Effect of Nanoparticles On Cylindrical Confined Assembly of Symmetric and Asymmetric Block Copolymers

Jay Hoon Park, Chemical and Biomolecular Engineering, Cornell University, Ithaca, NY, Vibha Kalra, Chemical and Biological Engineering, Drexel University, Philadelphia, PA and Yong L. Joo, School of Chemical and Biomolecular Engineering, Cornell University, Ithaca, NY

Our recent experimental study of electrospinning the block copolymer (BCP)/nanoparticle (NP) nanocomposites has revealed that various self-assembling structures can be formed in submicron scale fibers. [1,2] In an attempt to understand and control the self-assembly of BCP/NP nanocomposites in nanofibers, model symmetrical and asymmetrical BCPs with and without NPs under cylindrical confinement were studied using a coarse-grained molecular dynamics simulation. First, the effect of important parameters such as the ratio of the confinement radius to the length of BCP domain, *d/l*, polymer chain lengths, and the polymer-wall interactions on the confined assembly was thoroughly investigated. By examining both cross-section and assembled structures in the core along the cylinder axis, we have constructed a phase diagram for both symmetrical and asymmetrical BCP systems where structures were categorized by three features: the number of layers of domains, the number of axially connected domains, and the number of radially connected domains. Secondly, various concentrations (10% - 30%) of NPs with three different interactions with BCP, i) *A* domain selective, ii) *B* domain selective and iii) non-selective (neutral), were incorporated into BCPs. It is predicted that the swelling of either domain caused by the inclusion of selective NPs gives rise to different morphologies when they are compared with the corresponding BCP systems with the same domain ratio. Finally, the predicted BCP/NP structures will thoroughly be validated by those observed in electrospun BCP/NP nanofibers. The current study demonstrates that coarse-grained MD simulation can offer a useful guidance in the formation of self assembled structures in electrospun BCP/NP nanofibers.

[1] V. Kalra, J. H. Lee, J. H. Park, M. Marquez and Y. L. Joo, Small, 2009, **5**, 2323-2332.

[2] V. Kalra, J. H. Lee, S. G. Lee, M. Marquez, U. Wiesner, and Y. L. Joo, *Small,* 2008, **4**, 2067-2073.

See more of this Session: Self and Directed Assembly at the Nanoscale
See more of this Group/Topical: Nanoscale Science and Engineering Forum

(746d) Single-Walled Carbon Nanotube Films (SWNTs) for Biosensing Applications

Jie Chen, Bioengineering, University of Oklahoma, Norman, OK, Tu O. Tran, University of Oklahoma, Norman, OK and **David W. Schmidtke**, School of Chemical, Biological, and Materials Engineering, University of Oklahoma, Norman, OK

Amperometric biosensors were constructed by a two step procedure. In the fist step, a thin SWNT film was formed on a glassy carbon electrode surface by solution casting of a dispersed SWNT solution. The SWNT film was then coated with a redox polymer/enzyme hydrogel. The influences of (i) the SWNT solution dispersed by different surfactants (water, Triton X-100, Sodium cholate); (ii) different redox polymers (PVP-OS, Fc-C6-LPEI, Fc-C3-LPEI), and (iii) enzymes (GOx, HRP) on the electrochemical and enzymatic response were investigated by cyclic voltammetry, constant potential amperometry, and electrochemical impedance spectroscopy. The results showed that, for the three redox polymer/GOx based biosensors, the type of surfactant used to form the initial SWNT film was a critical factor in determining whether the sensors produced a significant increase (~ 2.5 mA/cm^2) or decrease (~ 0.1 mA/cm^2) of the enzymatic response to glucose.

See more of this Session: Nanotechnology and Nanobiotechnology for Sensors
See more of this Group/Topical: Nanoscale Science and Engineering Forum

(532d) Evaluating the Dynamics of An Integrated Electrokinetic and Zero-Valent Iron Nanoparticle System for Treatment of Hexavalent Chromium In Groundwater

Ryan Thacher and Massoud Pirbazari, Sonny Astani Department of Civil and Environmental Engineering, University of Southern California, Los Angeles, CA

Electrokinetic technology is a promising groundwater treatment strategy capable of transporting ionic contaminants through groundwater by application of an electric potential across a contaminated area. While this is effective in the transport and concentration of contaminants at an electrode, complimentary remedial action may still be required to enhance treatment of the contaminants. Zero-valent iron nanoparticles (nZVI) may potentially offer an ideal solution for this genre of *in-situ* environmental remediation. Their small size coupled with their high surface area to volume ratio as well as reductive capacity, make nZVI a seemingly ideal medium for subsurface remediation where other classical remediation strategies such as excavation are not possible. This may be in deep aquifers, urban areas, or under existing infrastructure.

A comprehensive understanding of the fate and transport of nZVI in the subsurface is just as critical to *in-situ* remediation as the high reactivity. Upon introduction to an aquifer environment, nZVI must be readily dispersible and remain stable so that they reach the targeted contamination zone. Several researchers have investigated transport of nZVI through porous media, however most of the published work is conducted in idealized batch studies. These investigations have indeed provided a good knowledge base, but limited information exists concerning the fate, transport, and toxicity of nZVI in the natural environment. The challenging issues at hand include: nanoparticle agglomeration, nZVI storage and age, components of soil matrices, aquifer hydraulic properties, presence of natural organic matter, pH, and redox potential, among others.

This study investigates the treatment of simulated groundwater containing hexavalent chromium (Cr^{VI}), a known carcinogen, in a dynamic system using combined electrokinetic and zero-valent iron nanoparticle technology. Continuous flow columns are charged with simulated natural soils, followed by saturation with solutions containing Cr^{VI}, variable humic acid concentration, and at multiple pH values. Upon saturation, electric potential is applied, and dilute solutions of nZVI are added at regular intervals for the entirety of the experiment.

This presentation will discuss the influence of natural aquifer characteristics such as pH, natural buffer, and natural organic matter content on the transport of nZVI through subsurface environments, and how this impacts treatment of contaminants. Furthermore, this research will evaluate the dynamics of coupling electrokinetic and nZVI technologies in decontamination of the subsurface environment.

See more of this Session: Green and Renewable Energy Aspects of Nanotechnology
See more of this Group/Topical: Topical C: Environmental Aspects, Applications, and Implications of Nanomaterials and Nanotechnology

BACTERIAL COLONIZATION OF NANOMODIFIED ENDOTRACHEAL TUBES IN A BENCH TOP AIRWAY MODEL

MARY C. MACHADO[1], KEIKO M. TARQUINIO[2], AND THOMAS J. WEBSTER[1]
[1]BROWN UNIVERSITY, PROVIDENCE RI; [2]RHODE ISLAND HOSPITAL, PROVIDENCE RI

ABSTRACT

VENTILATOR ASSOCIATED PNEUMONIA (VAP) IS A SERIOUS AND COSTLY CLINICAL PROBLEM. SPECIFICALLY, RECEIVING MECHANICAL VENTILATION OVER A 48 HOUR TIME PERIOD INCREASES THE RISK OF VAP AND IS ASSOCIATED WITH HIGH MORBIDITY, MORTALITY AND MEDICAL COSTS. ENDOTRACHEAL TUBES (ETTS) ARE ESSENTIAL TO THE PROCESS OF MECHANICAL VENTILATION. ENDOTRACHEAL TUBES PRESENT A SPECIAL CONCERN TO CLINICIANS BECAUSE THEY ARE OFTEN COLONIZED BY OROPHARYNGEAL BACTERIA DURING LONG-TERM MECHANICAL VENTILATION. COST EFFECTIVE ETTS THAT ARE RESISTANT TO BACTERIAL INFECTION WOULD BE ESSENTIAL TOOLS IN THE PREVENTION OF VAP. THE OBJECTIVE OF THIS STUDY WAS TWO FOLD, FIRST TO DEVELOP STRATEGIES TO DECREASE BACTERIAL ADHESION ON ETTS USING NANOTECHNOLOGY AND SECONDLY TO DEVELOP BETTER METHODS TO ASSESS *IN VITRO* BACTERIAL ADHESION AND BIOFILM FORMATION ON ETTS USING A BENCH TOP EXPERIMENTAL MODEL AND COMPUTER SIMULATIONS OF AIR FLOW. NANOROUGHENED ETTS WERE CREATED BY EXPOSING POLYVINYL CHLORIDE (PVC) ETTS (SHERIDAN® TO A 0.1% MASS SOLUTION OF *R. ARRHIZUS* LIPASE DISSOLVED IN A POTASSIUM PHOSPHATE BUFFER. NANOMODIFIED ETTS WERE TESTED IN A CUSTOM MADE BENCH TOP MODEL AIRWAY. THESE TUBES WERE EITHER STAINED WITH CRYSTAL VIOLET AND ANALYZED WITH OPTICAL DENSITY OR PROCESSED

USING A VORTEXING PROTOCOL TO DETERMINE THE NUMBER OF BACTERIA ON THE SURFACE OF THE TUBE. FLUID ANALYSIS OF THE TUBES CONSISTED OF A NUMERICAL ANALYSIS OF THE CURVATURE OF THE TUBES AND THE CREATION OF A 3D FINITE ELEMENT MODEL TO SIMULATE FLOW WITHIN THE SYSTEM. BOTH STATIC AND DYNAMIC STUDIES SHOWED LIPASE ETCHING CAN CREATE NANO-ROUGH SURFACE FEATURES ON PVC ETTS THAT SUPPRESSES *S. AUREUS* GROWTH. ADDITIONALLY, FLOW CONDITIONS WITHIN THE ETT INFLUENCED BOTH THE LOCATION AND CONCENTRATION OF BACTERIAL GROWTH ON THE ETTS. THE DIFFERENCES IN BOTH BACTERIAL NUMBER AND OPTICAL DENSITY RECORDED ALONG THE LENGTH OF THE TUBE SUGGESTED THAT WALL SHEAR STRESS PLAYS A SIGNIFICANT ROLE IN BACTERIAL COLONIZATION. THUS, NANOMODIFIED ETTS COULD PROVIDE CLINICIANS WITH AN EFFECTIVE AND INEXPENSIVE TOOL TO COMBAT HOSPITAL ACQUIRED INFECTIONS LIKE VAP, AND SHOULD BE STUDIED IN GREATER DEPTH.

INTRODUCTION

THE PREVENTION OF VENTILATOR ASSOCIATED PNEUMONIA (VAP) IS OF GREAT IMPORTANCE TO MODERN MEDICINE. HOSPITAL ACQUIRED INFECTIONS (HAIS), SUCH AS VAP, POSE A SIGNIFICANT PROBLEM WORLDWIDE AND ARE ASSOCIATED WITH INCREASES IN MORTALITY, MORBIDITY, DURATION AND COST OF HOSPITALIZATION [1]. IN THE UNITED STATES, THE ESTIMATED YEARLY COST OF THESE DELETERIOUS CONDITIONS CAN BE UP TO 6.7 BILLION DOLLARS [1]. DEVICE RELATED INFECTIONS, SUCH AS VAP AFFECT 28% OF THE PATIENTS WHO USE MECHANICAL VENTILATION. WITHIN PEDIATRIC INTENSIVE CARE UNITS (ICUS), VAP IS THE SECOND MOST COMMON HAI (15% TO 23% OF ALL HAIS, RESPECTIVELY) [1]. THE TWO PATHOGENS MOST COMMONLY ASSOCIATED WITH VAP ARE *PSEUDOMONAS AERUGINOSA* (*P. AERUGINOSA*) AND *STAPHYLOCOCCUS AUREUS* (*S. AUREUS*) [3].

VAP PRESENTS A PARTICULAR INTRACTABLE PROBLEM TO THE PEDIATRIC ICU BECAUSE IT IS OFTEN DIFFICULT TO DISTINGUISH PNEUMONIA FROM OTHER RESPIRATORY CONDITIONS COMMON IN MECHANICALLY VENTILATED PATIENTS [4]. THIS IS ESPECIALLY TRUE IN VENTILATED NEONATES. RADIOGRAPHIC AND CLINICAL CRITERIA FOR THE DIAGNOSES OF VAP ARE OFTEN UNSPECIFIC, WHILE MICROBIOLOGICAL TESTS OFTEN YIELD MISLEADING RESULTS. SPECIFICALLY, BACTERIAL COLONIZATION IN THE TRACHEA MUST BE DIFFERENTIATED FROM LOWER RESPIRATORY INFECTION. FALSE POSITIVES CAN OCCUR WHEN TESTS DETECT BENIGN BACTERIAL COLONIZATION. FALSE NEGATIVES OCCUR WHEN TESTS MISS THE ACTIVE AREA OF BACTERIA INFECTION. MORE THAN HALF OF PATIENTS DIAGNOSED WITH CLINICAL VAP HAVE NEGATIVE CULTURES [5]. SUCH UNCERTAINTY CAN CAUSE DELAYS IN TARGETED TREATMENT OR PROLONGED USE OF A BROAD- SPECTRUM OF ANTIBIOTICS, SUCH AS AMPICILLIN, LEADING TO SIGNIFICANT RISKS SUCH AS THE DEVELOPMENT OF MULTIDRUG RESISTANT ORGANISMS [1].

ONE OF THE MAIN SOURCES OF BACTERIAL COLONIZATION WITHIN THE AIRWAY IS THE ENDOTRACHEAL TUBE (ETT). AS WITH ANY OTHER MEDICAL DEVICE IMPLANTED WITHIN THE BODY, THE ETT IS EXPOSED TO NUMEROUS PATHOGENS. ETTS ARE OF SPECIAL CONCERN BECAUSE THESE TUBES PROVIDE A DIRECT CONDUIT FROM THE OUTSIDE ENVIRONMENT TO THE LUNGS BY IMPAIRING SOME OF THE BODY'S NATURAL DEFENSES LIKE COUGHING AND MUSCOCILIARY MOTION OF THE TRACHEA [6].

ONE OF THE KEY CHALLENGES TO THE INHIBITION OF BACTERIAL GROWTH ON ETTS OCCURS WHEN CERTAIN BACTERIA EXUDE AN EXOPOLYSACCHARIDE, WHICH ADHERE BACTERIA TOGETHER. BACTERIA IN THIS TYPE OF EXTRACELLULAR MATRIX (KNOWN AS A BIOFILM) ARE ESPECIALLY RESISTANT TO BOTH ANTIBIOTICS AND THE IMMUNE SYSTEM OF THE PATIENT. IN SUCH BIOFILMS, ANTIBIOTICS CANNOT PENETRATE THROUGH THE BIOFILM, EFFECTIVELY SHIELDING THE INTERIOR OF THE COLONY. BACTERIA ALSO MUST PROTECT THEMSELVES FROM THE INNATE IMMUNE SYSTEM OF THE HOST ORGANISM. THIS IS ACHIEVED THROUGH THE PREVENTION OF PHAGOCYTIC ACTION WITHIN THE BIOFILM

ACCOMPLISHED BY EITHER ELECTROSTATIC REPULSION OR PROTEOLYSIS OF THE ANTIMICROBIAL PEPTIDES OF HOST CELLS. VENTILATION THROUGH SUCH INFECTED ETTS, OR EVEN CONDENSATION OF HUMIDIFIED AIR, CAN BREAK OFF PORTIONS OF THE BIOFILM BRINGING BACTERIA DEEPER INTO THE LUNGS AND SPURRING GROWTH ON OTHER AREAS OF THE TUBE [7].

THE UNIQUE PROPERTIES OF NANOMODIFIED SURFACES (OR SURFACES WITH FEATURES <100 NM IN ONE DIRECTION) COULD PROVIDE A SOLUTION TO THE PERSISTENT PROBLEMS OF VAP [8]. MICRO- AND NANO-SCALE BUILDING BLOCKS FORM THE FOUNDATION FOR CELLS AND TISSUES WITHIN THE HUMAN BODY. IT IS THOUGHT THAT THE DIFFERENCE IN THE ACTIVITY OF SOME CELLS ON NANOMODIFIED SURFACES IS DUE TO THE ABILITY OF THESE MATERIALS TO MIMIC THE NATURAL DIMENSIONS OF CONSTITUENTS OF BIOLOGICAL TISSUES [9]. NANOMATERIALS HAVE UNIQUE SURFACE ENERGETICS DUE TO THEIR SIGNIFICANTLY GREATER SURFACE AREA COMPARED TO CONVENTIONAL, NANO-SMOOTH, MATERIALS. SUCH CHANGES IN SURFACE ENERGY UNDOUBTEDLY INFLUENCE INITIAL PROTEIN INTERACTIONS IMPORTANT FOR MEDIATING BACTERIAL ADHESION [10]. SPECIFICALLY, ONE OF THE FIRST STEPS WITHIN THE PROCESS OF CELL ADHESION IS THE ASSOCIATION BETWEEN PROTEINS ADSORBED ON IMPLANT SURFACES AND CELL MEMBRANE RECEPTORS. THE LARGE SURFACE TO VOLUME RATIO CHARACTERISTICS OF NANOMATERIALS HAS BEEN SHOWN TO AFFECT THIS ASSOCIATION TO INHIBIT BACTERIA ATTACHMENT AND PROMOTE NON-BACTERIAL CELL (SUCH AS OSTEOBLASTS, SMOOTH MUSCLE CELLS, ENDOTHELIAL CELLS, CHONDROCYTES, ETC.) ADHESION [11]. THE TYPE, CONCENTRATION, CONFORMATION, AND BIOACTIVITY OF PROTEINS ADSORBED ONTO A MATERIAL DEPENDS ON ITS SURFACE CHEMISTRY, HYDROPHILICITY, CHARGE, TOPOGRAPHY, ROUGHNESS, AND ENERGY: ALL OF WHICH CAN BE INFLUENCED BY NANOTECHNOLOGY [10]. SURFACE PROPERTIES INCLUDING ROUGHNESS, STIFFNESS AND TOPOGRAPHY ALSO EFFECT BACTERIAL ADHESION AND MAY PLAY AN IMPORTANT ROLE IN THE INITIAL STAGES OF BIOFILM FORMATION [12], [13]. REDUCING BACTERIAL ADHESION

ON THE SURFACE OF AN ETT COULD REDUCE BOTH BIOFILM FORMATION AND BACTERIAL CONTAMINATION WITHIN THE TUBE.

MOREOVER, PREVIOUS STUDIES ON VAP HAVE CONCENTRATED ON BIOFILM FORMATION IN STRAIGHT, ADULT, ETTS UNDER STATIC CONDITIONS. HOWEVER, DYNAMIC MEASUREMENTS ARE AN IMPORTANT PART OF THE UNDERSTANDING OF BOTH BIOFILM FORMATION AND MAINTENANCE IN THE ETT AND THE PATHOGENESIS OF VAP. FLUID WITHIN CURVED TUBES, LIKE ETTS, DISPLAYS UNIQUE SECONDARY FLOWS. THESE FLOWS ARE VORTICAL IN NATURE AND ARE THOUGHT TO CREATE LARGE BOUNDARY LAYERS ALONG THE WALLS OF THE CURVE, ALLOWING FOR BACTERIAL ATTACHMENT AND BIOFILM FORMATION [14].

BOUNDARY LAYER THICKNESS SUBSTANTIALLY IMPACTS CELL SURFACE INTERACTIONS IN LAMINAR OR MINIMALLY TURBULENT FLOWS [15]. IT HAS BECOME INCREASINGLY APPARENT THAT BIOFILMS CAN ALTER THEIR STRUCTURAL PROPERTIES IN RESPONSE TO FLUID EFFECTS (SUCH AS SHEAR STRESS) BOTH RESISTING DETACHMENT AND FACILITATING FURTHER COLONIZATION [16], [17]. NANOMODIFIED SURFACES CAN ALSO CAUSE SIGNIFICANT CHANGES IN THE STRUCTURE OF BIOFILMS, INCLUDING INCREASED LEVELS OF EXTRACELLULAR POLYMERIC SUBSTANCES AS WELL AS CHANGES IN COLONY SHAPE AND SIZE [13]. THEREFORE, DYNAMIC TESTS OF NANOMODIFIED SURFACES ARE NECESSARY TO CHARACTERIZE THE FULL ANTIMICROBIAL POTENTIAL OF THESE NOVEL NANOMODIFIED MATERIALS, ESPECIALLY FOR ETT APPLICATIONS.

DUE TO THE ABOVE, THE OBJECTIVE OF THIS *IN VITRO* STUDY WAS TO FABRICATE AND TEST NANOMODIFIED ETTS (WITHOUT THE USE OF ANTI-BACTERIALS), UNDER SIMULATED DYNAMIC AIRFLOW CONDITIONS OF THE RESPIRATORY SYSTEM, IN ORDER TO DETERMINE THEIR ANTIMICROBIAL EFFECTIVENESS. TO ACHIEVE THIS, BOTH EXPERIMENTAL AND COMPUTATIONAL ANALYSES WERE PERFORMED.

MATERIALS AND METHODS

NANOMODIFICATION OF ETTS

THE FIRST TYPE OF ANTIBACTERIAL ETT PREPARED FOR ANALYSIS WITHIN OUR SYSTEM WAS A POLYVINYL CHLORIDE (PVC) ETT (SHERIDAN®) WITH NANO-ROUGHENED SURFACES. THESE TUBES WERE TRANSFORMED TO HAVE NANOFEATURES USING LIPASES FROM THE FUNGI *RHIZOPUS ARRHISUS* (SIGMA ALDRICH) WHICH ENZYMATICALLY DEGRADED THE PVC MATERIAL. THE PROCEDURE INCLUDED EXPOSING THE PVC TUBING TO A 0.1% MASS SOLUTION OF *R. ARRHISUS* LIPASE DISSOLVED IN A POTASSIUM PHOSPHATE BUFFER AT 37° C. THE SAMPLES WERE GENTLY AGITATED FOR A TOTAL OF 48 HOURS AND THE LIPASE MEDIA WAS CHANGED EVERY 24 HRS.

THE ACTIVITY OF THE *R. ARRHISUS* USED IN THE EXPERIMENT WAS 10.5 U/G, WHERE ONE UNIT WAS DEFINED AS THE AMOUNT OF ENZYME THAT CATALYZED THE RELEASE OF 1 μMOL OF OLEIC ACID PER MINUTE AT PH 7.4 AND 40°C. ETTS PLACED WITHIN THE SYSTEM WERE MODIFIED ON BOTH THE INNER AND OUTER SURFACES. ETTS WERE STERILIZED USING ETHYLENE OXIDE EXPOSURE IN A 16 HOUR STERILIZATION CYCLE.

MATERIAL CHARACTERIZATION

EACH OF THE SAMPLES WAS COATED WITH A GOLD PALLADIUM MIXTURE TO INCREASE CONDUCTIVITY FOR SCANNING ELECTRON MICROSCOPY (LEO 1530VP FE-4800 FIELD-EMISSION SEM, CARL ZEISS SMT, PEABODY, MA) IN ORDER TO VISUALLY ASSESS THE NANOFEATURES. NANOROUGHNESS WAS ALSO ASSESSED USING AN ATOMIC FORCE MICROSCOPE (AFM: XE-100, PARK SYSTEMS, SANTA CLARA, CA) UNDER TAPPING MODE USING A 10 NM AFM TIP (PPP-NCHR, PARK SYSTEMS, SANTA CLARA, CA) WITH A SCAN RATE OF 0.5 HZ. AVERAGE ROOT MEAN SQUARE ROUGHNESS DATA WAS COLLECTED FOR ALL SURFACES. CONTACT ANGLE DATA WAS ASSESSED TO DETERMINE SURFACE HYDROPHOBICITY (EASYDROP DSA20S, KRUSS, HAMBURG, GERMANY). CHARACTERIZATION OF THE ELEMENTAL COMPOSITION OF THE NANOMODIFIED AND

CONTROL ETTS WAS ALSO CONDUCTED BY XPS (5500 MULTITECHNIQUE SURFACE ANALYZER, PERKIN ELMER, WALTHAM, MA).

BACTERIA DYNAMIC TESTING IN A BENCH TOP AIRWAY MODEL

THE ULTIMATE TEST OF THE EFFECTIVENESS OF NANOMODIFIED ETTS WAS PERFORMED UNDER THE AIRFLOW CONDITIONS PRESENT WITHIN THE AIRWAY. NANOMODIFIED ETTS WERE TESTED IN A CUSTOM MADE BENCH TOP MODEL AIRWAY ADAPTED FROM THE GENERAL DESIGN OF HARTMANN ET AL. (1999) [18]. THIS MODEL AIRWAY CAN BE SEEN BELOW IN FIGURE 1.

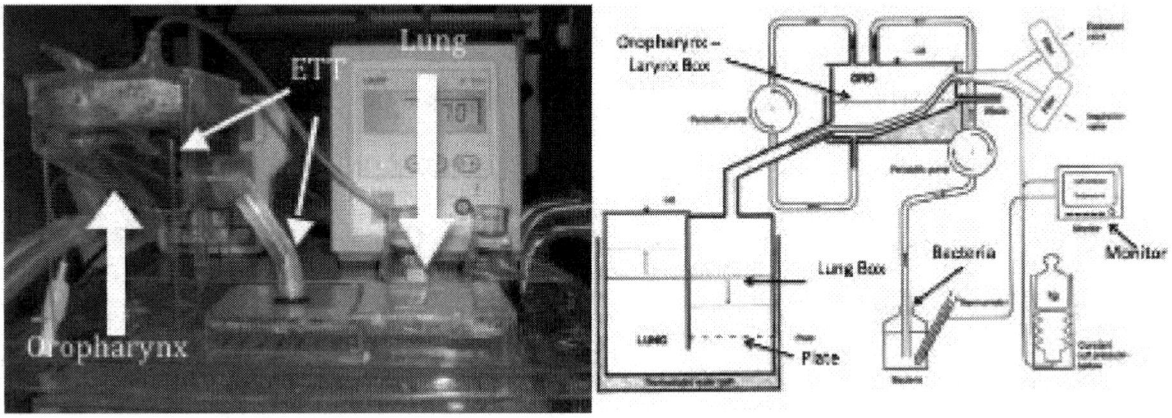

FIGURE 1: MODEL OF THE ETT SYSTEM USED IN THIS STUDY [18].

ETTS WITH A 3.5 MM INTERNAL DIAMETER (ID) WERE PLACED IN THE SYSTEM. THE BOXES WERE THEN FILLED WITH STERILE TRYPTICASE SOY BROTH MEDIA (TSB) AND PLACED IN A WATER BATH THAT WAS MAINTAINED AT 37° C. THIS ETT WAS THEN CONNECTED TO AN INFANT STAR 950 VENTILATOR (PURITAN BENNETT, COVIDIEN, MANSFIED, MA) WITH POSITIVE END-EXPIRATORY PRESSURE (PEEP) OF 1 CMH_2O AND FRACTION OF INSPIRED OXYGEN (FIO_2) OF 0.5. 480 ML OF TSB INOCULATED WITH 10^3 COLONY FORMING UNIT/ MILLILITER (CFU/ ML) *S. AUREUS* (ATCC #25923) WAS CIRCULATED INTO THE OROPHARYNX BOX OVER THE DURATION OF THE 24 HOUR TEST.

AT THE END OF EACH TRIAL, ETTS WERE CUT INTO TEN 1.5 CM PIECES. THESE TUBES WERE EITHER STAINED WITH CRYSTAL VIOLET AND ANALYZED WITH OPTICAL DENSITY USING A SPECTROPHOTOMOTER (SPECTRAMAX 340PC384, MOLECULAR DEVICES, SUNNYVALE, CA) OR PROCESSED USING A VORTEX PROTOCOL TO DETERMINE THE NUMBER OF COLONY FORMING UNITS. THE VORTEX PROTOCOL CONSISTED OF ADDING THE ETT PIECES TO 2 ML OF TSB MEDIA AND VORTEXING THE ETTS FOR 1 MIN AT 3000 RPM (DIGITAL VORTEX MIXER, FISHER SCIENTIFIC, PITTSBURGH, PA). THE SUPERNATANT WAS THEN SERIALLY DILUTED AND PLATED. COLONY COUNTS WERE PERFORMED TO DETERMINE THE NUMBER OF BACTERIA ON EACH AREA OF THE TUBE.

AIR FLOW ANALYSIS

TO FULLY UNDERSTAND THE DYNAMIC AIR FLOW CONDITIONS WITHIN THIS SYSTEM, FURTHER ANALYSIS OF THE FLUID EFFECTS OF ETT ORIENTATION WAS CONDUCTED BY GRAPHING THE CURVED TUBES BY HAND AND USING COMPUTER ANALYSIS OF THE THREE DIMENSIONAL MODEL IN SOLIDWORKS (2010, DASSAULT SYST MES). JARREAU ET AL. REPORTED THAT THE MEAN RANGE OF ETT CURVATURE DIAMETERS FROM MRI SCANS WERE 8-18 CM WITH A <16% VARIATION IN PRESSURE BETWEEN THESE EXTREMES [19]. BOTH OF THE PRESENTLY STUDIED CURVES IN THE ETT, WERE WITHIN THESE PARAMETERS. THE RADIUS OF CURVATURE (R_C) OF FIRST CURVE MEASURED ABOUT 18 CM WHILE THE R_C OF THE SECOND CURVE WAS 9 CM.

JARREAU ET AL. [19] SHOWED EXPERIMENTALLY THAT THE PRESSURE DROP WITHIN PURELY LAMINAR ETT FLOW COULD BE APPROXIMATED USING THE ITO FORMULA (SHOWN BELOW), WHICH IS APPLICABLE WHEN 30< DE< 2000:

$$\frac{\Delta P_{vd}}{0.5\rho \bar{u}^2} \bullet \frac{D}{L} = 64 \bullet \mathrm{Re}^1 \bullet [0.103 \bullet De^{0.5}(1+3.945 \bullet De^{0.5} + 7.782 \bullet De^1 + 9.097 \bullet De^{1.5} + 5.608 \bullet De^2 + ...)] \quad (1)$$

IN THE ABOVE EQUATION, D IS TUBE DIAMETER, L IS TUBE LENGTH, ρ IS THE DENSITY OF THE FLUID, RE IS THE REYNOLDS NUMBER, DE IS THE DEAN NUMBER, ΔP_{VD} IS THE PRESSURE DROP DUE TO VISCOUS DISSIPATION, AND $\bar{u}$ IS THE AVERAGE VELOCITY. FOR THE ETTS STUDIED IN THE BENCH TOP INVESTIGATION, THE PRESSURE DROP DUE TO VISCOUS DISSIPATION ESTIMATED USING THE ITO FORMULA WAS 1.24 CMH_2O INSIDE OF THE BOXES.

SHEAR STRESS AT THE WALL OF THE TUBES WAS ALSO CALCULATED USING THE FOLLOWING EQUATION WHERE w_m IS THE AVERAGE VELOCITY, K_2 IS THE CURVATURE RATIO (R_C/D), DE IS THE DEAN NUMBER AND τ_w IS THE SHEAR STRESS AT THE WALL. FOR THE ETTS USED IN THIS STUDY THE VALUE OF THE SHEAR STRESS DERIVED FROM THE ITO EQUATION WAS DETERMINED TO BE 0.3557 PA [20].

$$\frac{\tau_w}{(\frac{1}{2}\rho w_m^2)} k_2^{1/2} = 0.00513 + \frac{0.0760}{De} \quad (2)$$

THE FINAL PART OF THE PRESENT AIR FLOW ANALYSIS OF THE ETT CONSISTED OF THE CREATION OF A THREE DIMENSIONAL COMPUTATIONAL MODEL. A FINITE ELEMENT MODEL OF THE EXACT CURVATURE OF THE ETTS WAS CREATED USING COMSOL MULTIPHYSICS (V4.0A, COMSOL GROUP, BURLINGTON MA). AIR FLOW THROUGH THE ETT WAS CONSIDERED TO BE LAMINAR, NEWTONIAN AND INCOMPRESSIBLE. THE UNSTEADY STATE, THREE DIMENSIONAL FORM OF THE NAVIER STOKES EQUATIONS WERE THEN SOLVED USING THE FINITE ELEMENT METHOD. BOUNDARY CONDITIONS PRESCRIBED OVER THE DOMAIN WERE AS FOLLOWS: I) ON THE SOLID WALL OF THE TUBE A NO SLIP CONDITION WAS APPLIED; II) AN ABSOLUTE PRESSURE OF 1 ATM WAS ASSUMED AT THE OUTLET, AS IS COMMON WITHIN THE CLINICAL SETTING; III) THE INITIAL INLET FLOW RATE WAS SET AT 9.63 M/S; AND IV) FINALLY INLET PRESSURE (P) WAS CALCULATED USING THE FOLLOWING EQUATION FOR PRESSURE LIMITED, CONSTANT FLOW VENTILATION. IN THIS

EQUATION "C" IS THE POSITIVE END EXPIRATORY PRESSURE (PEEP) OF 1 CMH_2O. "A" EQUALS THE FLOW RATE DIVIDED BY COMPLIANCE OF OUR MODEL: 18.77 CMH_2O/SEC. "T" IS EQUAL TO THE TIME AND "B" IS EQUAL TO ONE.

$$P = a \times t^b + c \quad (3)$$

RESULTS

MATERIAL CHARACTERIZATION

PVC ETTS CUT AND SUBMERSED IN THE LIPASE SOLUTION *R. ARRHISUS* SHOWED VISIBLE NANOFEATURES (FIGURE 2). IN CONTRAST, CONTROL UNTREATED PVC ETT DID NOT SHOW ANY NANOFEATURES.

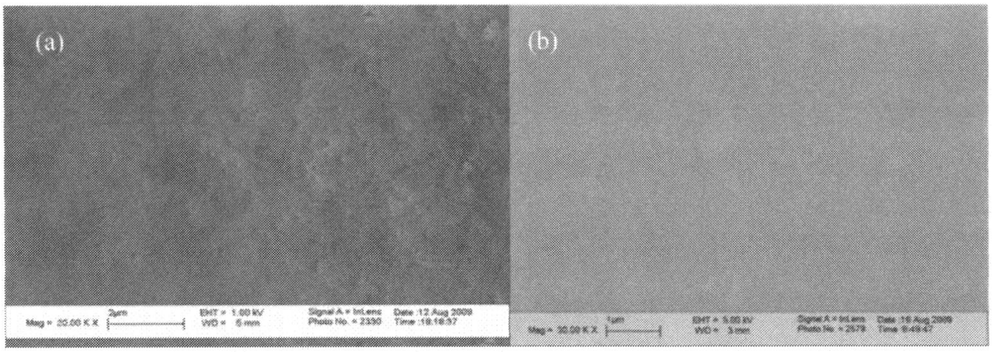

FIGURE 2: SEM IMAGES OF NANOMODIFIED PVC ETTS: (A) NANO-R: MAG. X 20K (B) UNTREATED: MAG. X 30K. NANO-R WAS MODIFIED WITH *R. ARRHISUS*. SCALE BARS = 2 μM.

AFM ANALYSIS OF THE NANOMODIFIED ETTS CONFIRMED SEM ANALYSIS AND CAN BE SEEN IN FIGURE 3. THE AVERAGE ROOT MEAN SQUARE (RMS) ROUGHNESS OF THE ETTS NANOMODIFIED WITH *R. ARRHISUS* WAS 12.5 NANOMETERS COMPARED TO A VALUE OF 2.2 NANOMETERS FOR THE UNTREATED PVC TUBES.

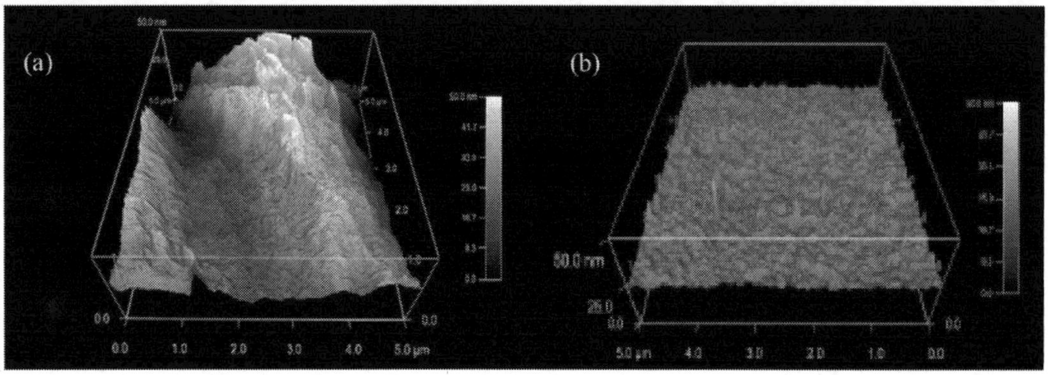

FIGURE 3: AFM IMAGES OF PVC ETTS TREATED WITH (A) 0.1% *R. ARRHISUS* SOLUTION (NANO-R) FOR 48 HOURS COMPARED TO (B) UNTREATED PVC SAMPLES.

SURFACE ENERGY ANALYSIS FOR THE NANOMODIFIED PVC WAS ALSO PERFORMED. RESULTS SHOWED A WATER CONTACT ANGLE OF 84.4° FOR UNMODIFIED PVC AND AN ANGLE OF 68.9° FOR *R. ARRHISUS* TREATED NANO-R SAMPLES. THUS, NANO-R HAD INCREASED THE HYDROPHILICITY COMPARED TO UNTREATED PVC.

DYNAMIC TESTING IN BENCH TOP MODEL

FOR THE PRESENT ANALYSIS, TWENTY-FOUR HOUR STUDIES PERFORMED IN A DYNAMIC AIR FLOW CHAMBER SHOWED A MARKED DIFFERENCE IN BIOFILM FORMATION ON DIFFERENT AREAS OF UNMODIFIED ETTS (FIGURES 4 AND 5).

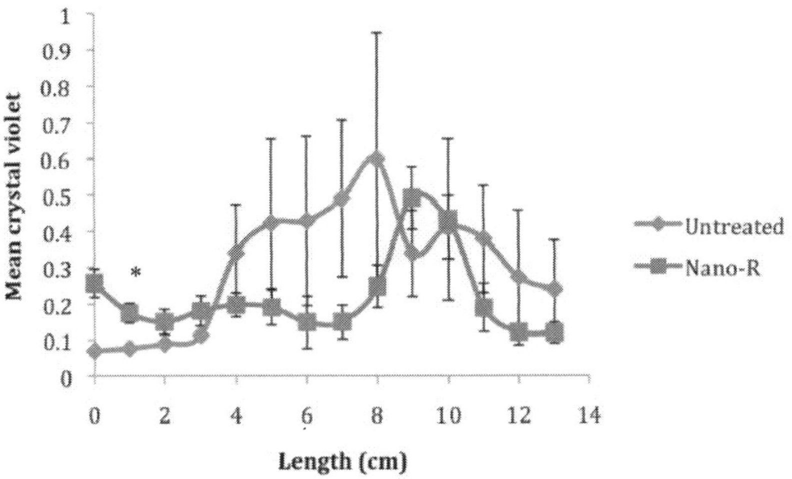

FIGURE 4: DYNAMIC LUNG SYSTEM RESULTS FOR UNTREATED TUBES VERSUS NANOMODIFIED PVC ETTS: CRYSTAL VIOLET STAIN; N=3 (X-AXIS LENGTH= LONGITUDINAL ETT LENGTH). ERROR BARS +/- 1 SE. *P< 0.05 COMPARED TO UNTREATED ETTS AT THE SAME TIME POINTS.

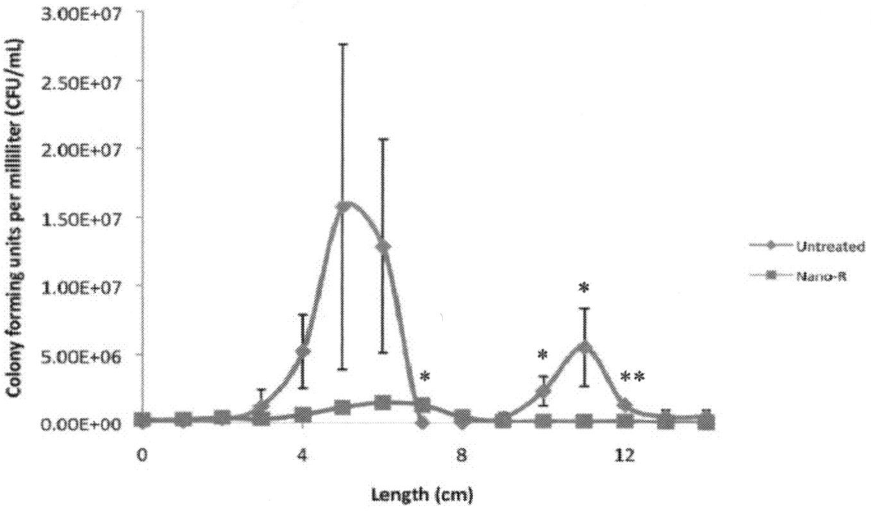

FIGURE 5: DYNAMIC LUNG SYSTEM RESULTS FOR UNTREATED VERSUS NANOMODIFIED PVC ETTS: COLONY COUNTS OF *S. AUREUS*, N=3 (X-AXIS LENGTH= LONGITUDINAL ETT LENGTH). ERROR BARS +/- 1 SE. *P< 0.05 AND **P< 0.01 COMPARED TO UNTREATED ETTS AT THE SAME TIME POINTS.

AREAS OF ETT CURVATURE, SUCH AS AT THE ENTRANCE TO THE MOUTH AND THE CONNECTION BETWEEN THE OROPHARYNX AND THE LARYNX, WERE CORRELATED WITH LARGER AMOUNTS OF BIOFILM FORMATION. CRYSTAL VIOLET STAINING REVEALED ROUND PATCHES IN THESE REGIONS. COLONY FORMING UNITS WERE SIGNIFICANTLY SMALLER (UP TO 1.5 LOG REDUCTIONS) ON THE NANOMODIFIED ETT THAN ON CONTROL ETTS.

FLUID ANALYSIS

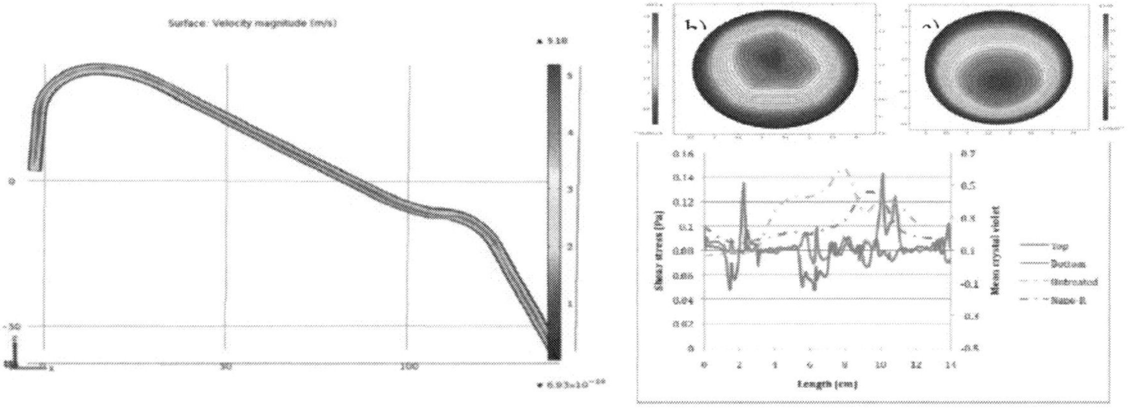

FIGURE 6: FEM MODEL: A) VELOCITY MAGNITUDE, B) FIRST CURVE, C) SECOND CURVE, D) SHEAR STRESS VERSUS BACTERIAL DENSITY

COMPUTATIONAL MODEL RESULTS SEEN IN FIGURE 6 SHOWED SKEWING OF THE AIR VELOCITY PROFILE AT BOTH CURVES IN THE ETTS (B AND C), WITH CONSEQUENT VARIATIONS IN THE WALL SHEAR RATES ALONG THE ETT.

DISCUSSION

DYNAMIC TESTING IN BENCH TOP MODEL

VAP POSES SIGNIFICANT PROBLEMS IN THE PEDIATRIC ICU LEADING TO INCREASED MEDICAL COSTS AND MORTALITY AMONG A CRITICALLY ILL POPULATION. THE PROLONGED USE OF BROAD SPECTRUM ANTIBIOTICS BROUGHT ON BY THE UNSPECIFIC SYMPTOMS OF THESE CONDITIONS IS ESPECIALLY TROUBLING. NANOTECHNOLOGY COULD PROVIDE A UNIQUE SOLUTION TO VAP BY PREVENTING BACTERIAL COLONIZATION OF THE ETT, ONE OF THE MAIN SOURCES OF CONTAMINATION OF THE AIRWAY. THE OBJECTIVE OF THIS STUDY WAS TO FABRICATE AND TEST NANOMODIFIED TUBES IN A DYNAMIC AIRWAY SYSTEM.

PREVIOUS STATIC STUDIES SHOWED SIGNIFICANT DIFFERENCES, (FROM 50-74%) IN *S. AUREUS* BIOFILM FORMATION BETWEEN THE PRESENTLY UNTREATED ETTS AND THE NANOMODIFIED PVC ETTS AFTER 4,12, 24 HOURS [21], [8]. THE MECHANISM FOR THIS REDUCTION ON THE NANOMODIFED TUBES IS UNCLEAR BUT IS THOUGHT TO BE ASSOCIATED WITH THE REDUCED ADHESION OF THE BACTERIA TO THE SURFACE OF THE NANOMODIFIED ETT. THIS COULD BE DUE TO ALTERED INITIAL PROTEIN INTERACTIONS, WHICH DICTATE BACTERIA ATTACHMENT DUE TO THE GREATER SURFACE ENERGY OF THE NANOMODIFIED ETTS.

SIMILAR TO PREVIOUS STATIC STUDIES, RESULTS FROM THIS STUDY SHOWED SIGNIFICANT REDUCTIONS IN THE NUMBER OF COLONY FORMING UNITS ON THE NANOMODIFIED ETT. THE DIFFERENCE BETWEEN CRYSTAL VIOLET AND COLONY COUNTS WAS THOUGHT TO BE A RESULT OF A COMBINATION OF INCREASED EXTRACELLULAR POLYMERIC MATERIAL ON THE NANOMODIFIED SURFACES AND THE REACTION OF THE NANO-MODIFIED SURFACES WITH THE CRYSTAL VIOLET DYE. TO MITIGATE THESE EFFECTS, OPTICAL DENSITY VALUES FROM ETTS MODIFIED WITH NANO-R, STRERILIZED, AND STAINED WITH CRYSTAL VIOLET WERE SUBTRACTED FROM THOSE OF THE BACTERIA CULTURED ETTS.

NOTABLY, THE DYNAMIC VALUES FOR BIOFILM DENSITY ALONG MANY AREAS OF THE ETT WERE SUBSTANTIALLY LESS THAN THE VALUES OBTAINED IN PREVIOUS STATIC STUDIES ON THE SAME SAMPLES ILLUSTRATING THE IMPORTANCE OF DYNAMIC MEDIA ANALYSIS OF ETTS.

FLUID ANALYSIS

RESULTS OF THE COMPUTATIONAL MODEL SHOWED AREAS OF LOW WALL SHEAR (0.0800 PA), CREATED BY SECONDARY MOTION IN THE CURVED PORTION OF THE TUBE, WHICH, CORRESPONDED TO EXPERIMENTAL AREAS OF HIGH BACTERIAL DENSITY. THESE VALUES WERE COMPARED TO LITERATURE VALUES FOR SHEAR STRESS IN THE HEALTHY LUNG, WHICH CAN VARY WIDELY BASED UPON DIAMETER AND FLOW RATE, BUT ARE BETWEEN 0.0294 AND 0.1961 PA IN THE TRACHEA [22].

CONCLUSIONS

BOTH STATIC [8] AND DYNAMIC STUDIES SHOW LIPASE ETCHING CAN CREATE NANO-ROUGH SURFACE FEATURES ON PVC ETTS THAT SUPPRESS *S. AUREUS* GROWTH. IMPORTANTLY, THE NANOMODIFIED TUBES DECREASED *S. AUREUS* GROWTH BY ONLY CHANGING THE DEGREE OF NANOFEATURES NOT BY ALTERING CHEMISTRY OR USING ANTIBIOTICS. DYNAMIC AIR FLOW CONDITIONS HAD AN EFFECT ON BOTH THE LOCATION AND CONCENTRATION OF BACTERIAL GROWTH ON THE ETT. PARTICULARLY, AREAS OF INCREASED SHEAR STRESS ALONG THE CURVES OF THE TUBE SAW DECREASED INITIAL BACTERIAL COLONIZATION. THE DIFFERENCES RECORDED ALONG THE LENGTH OF THE TUBE SUGGEST THAT FLOW AND CONTINUOUS CONTAMINATION PLAY A SIGNIFICANT ROLE IN BACTERIAL COLONIZATION. THIS PROVIDES FURTHER EVIDENCE THAT STATIC TESTING PROTOCOLS MAY NOT REVEAL THE IMPORTANT INTERACTIONS AT THE SURFACE OF SUCH DEVICES *IN VIVO*. RESULTS OF BOTH THE STATIC AND DYNAMIC MODELS SUGGEST THAT

NANOMODIFIED TUBES COULD PROVIDE CLINICIANS WITH AN EFFECTIVE AND INEXPENSIVE TOOL TO COMBAT HOSPITAL ACQUIRED INFECTIONS LIKE VAP, AND SHOULD BE STUDIED IN GREATER DEPTH.

ACKNOWLEDGEMENTS

THE AUTHORS WOULD LIKE TO THANK THE HERMANN FOUNDATION, THE RI SCIENCE AND TECHNOLOGY ADVISORY COUNCIL AND NSF GK-12 FOR FUNDING.

REFERENCES

1. VENKATACHALAM V., HENDLEY J.O., WILLSON D.F. THE DIAGNOSTIC DILEMMA OF VENTILATOR-ASSOCIATED PNEUMONIA IN CRITICALLY ILL CHILDREN. PEDIATR CRIT CARE.12: 286–296, 2011.
2. HALEY R.W. "INCIDENCE AND NATURE OF ENDEMIC AND EPIDEMIC NOSOCOMIAL INFECTIONS", IN: BENNETT AND BRACHMAN'S HOSPITAL INFECTIONS, EDITED BY J.V. BENNETT, P.S. BRACHMAN AND W.R. JARVIS. BOSTON: LIPPINCOTT WILLIAMS & WILKINS, 2007, 359–374.
3. GAYNES R. AND J.R. EDWARDS. OVERVIEW OF NOSOCOMIAL INFECTIONS CAUSED BY GRAM-NEGATIVE BACILLI. CLIN. INFECT. DIS. 41:848-854, 2005.
4. BALTIMORE R.S. THE DIFFICULTY OF DIAGNOSING VAP. PEDIATRICS 112: 1420-1421, 2003.
5. BAHRANI-MOUGEOT F.K., PASTER B.J., AND S. COLEMAN. MOLECULAR ANALYSIS OF ORAL AND RESPIRATORY BACTERIAL SPECIES ASSOCIATED WITH VENTILATOR-ASSOCIATED PNEUMONIA. J. CLIN. MICROBIOL. 45: 1588-1593, 2007.

6. KOERNER R.J. CONTRIBUTION OF ENDOTRACHEAL TUBES TO THE PATHOGENESIS OF VENTILATOR-ASSOCIATED PNEUMONIA. J. HOSP. INFECT. 35: 83-89, 1997.
7. CARDINAL P., JESSAMINE P., CARTER-SNELL C., S. MORRISON, AND G. JONES. CONTRIBUTION OF WATER CONDENSATION IN ENDOTRACHEAL TUBES TO CONTAMINATION OF THE LUNGS. CHEST 127-129, 1993.
8. MACHADO M.C., CHENG D., TARQUINIO K.M., AND T.J. WEBSTER. NANOTECHNOLOGY: PEDIATRIC APPLICATIONS. PEDIATR. RES., 67(5): 500–504, 2010.
9. SANVICENS, N.; MARCO, M. P. MULTIFUNCTIONAL NANOPARTICLESO PROPERTIES AND PROSPECTS FOR THEIR USE IN HUMAN MEDICINE. TRENDS BIOTECHNOL. 2008, 26, 425–433.
10. KLEIN J. PROBING THE INTERACTIONS OF PROTEINS AND NANOPARTICLES. PROC. NATL. ACAD. SCI. U.S.A. 104: 2029-2030, 2007.
11. LIU H. AND T.J. WEBSTER. NANOMEDICINE FOR IMPLANTS: A REVIEW OF STUDIES AND NECESSARY EXPERIMENTAL TOOLS. BIOMATERIALS 28: 354-369, 2006.
12. LICHTER J.A, THOMPSON M.T., DELGADILLO M., NISHIKAWA T., RUBNER M.F., AND K. J. VAN VLIET. SUBSTRATA MECHANICAL STIFFNESS CAN REGULATE ADHESION OF VIABLE BACTERIA. BIOMACROMOLECULES 9(6):1571-1578, 2008.
13. DIAZ C., CORTIZO M.C, SCHILARDI P.L., SARAVIA S.G.G., AND M.A.F.L. MELE. INFLUENCE OF THE NANO-MICRO STRUCTURE OF THE SURFACE ON BACTERIAL ADHESION. MAT. RES. 10(1): 11-14, 2007.
14. BERGER SA, AND L TALBOT. FLOW IN CURVED PIPES. ANNU. REV. FLUID MECH. 15:461–512, 1983.
15. RUSCONI R., LECUYER S., GUGLIELMINI, L., AND H.A. STONE. LAMINAR FLOW AROUND CORNERS TRIGGERS THE FORMATION OF BIOFILM STREAMERS. J. R. SOC. INTERFACE 7:1293-1299, 2010.

16. RUPP C.J., FUX C.A., AND P. STOODLEY. VISCOELASTICITY OF *STAPHYLOCOCCUS AUREUS* BIOFILMS IN RESPONSE TO FLUID SHEAR ALLOWS RESISTANCE TO DETACHMENT AND FACILITATES ROLLING MIGRATION. APPL. ENVIRON. MICROBIOL. 71(4): 2175–2178, 2005.
17. DONLAN R.M. BIOFILMS: MICROBIAL LIFE ON SURFACES. EMERG. INFECT. DIS. 8: 881-890 2002.
18. HARTMANN, M., GUTTMANN, J., MULLER, B., HALLMANN, T., AND K. GEIGER. REDUCTION OF THE BACTERIAL LOAD BY THE SILVER-COATED ENDOTRACHEAL TUBE (SCET) A LABORATORY INVESTIGATION. TECHNOL. HEALTH CARE 7: 359-370, 1999.
19. JARREAU P.H., LOUIS B., DASSIEU G., DESFRERE L., BLANCHARD P.W., MORIETTE G., ISABEY D., AND A. HARF. ESTIMATION OF INSPIRATORY PRESSURE DROP IN NEONATAL AND PEDIATRIC ENDOTRACHEAL TUBES. J. APPL. PHYSIOL. 87: 36 – 46, 1999.
20. MOULIN P., VEYRET D., AND F. CHARBIT. DEAN VORTICES: COMPARISON OF NUMERICAL SIMULATION OF SHEAR STRESS AND IMPROVEMENT OF MASS TRANSFER IN MEMBRANE PROCESSES AT LOW PERMEATION FLUXES. J. MEMBR. SCI. 183: 149–162, 2001.
21. SEIL J.T., RUBIEN N.M., WEBSTER T.J., AND K.M. TARQUINIO. COMPARISON OF QUANTIFICATION METHODS ILLUSTRATES REDUCED PSEUDOMONAS AERUGINOSA ACTIVITY ON NANOROUGH POLYVINYL CHLORIDE. J. BIOMED. MATER. RES. B, 98(1): 1-7, 2011..
22. NUCCI G., SUKI B., AND K. LUTCHEN. MODELING AIRFLOW-RELATED SHEAR STRESS DURING HETEROGENEOUS CONSTRICTION AND MECHANICAL VENTILATION. J. APPL. PHYSIOL. 95: 348-356, 2003.

(121c) Computationally Derived Rules for the Persistence of C60 Nanowires On Recumbent Pentacene Bilayers

Rebecca A. Cantrell[1], Christine N. James[2] and **Paulette Clancy**[1], (1)Chemical and Biomolecular Engineering, Cornell University, Ithaca, NY, (2)Chemical Engineering Department, University of Michigan, Ann Arbor, MI

Research in organic electronic devices has seen an increasing focus on designing molecularly scaled materials. This has been driven by the exploitation of quantum size confinement effects of nanoparticles, nanotubes and nanowires that lead to different electronic, optical and other properties. In this study, we have considered the structural properties of all-organic heterojunctions, in particular, the C_{60} /pentacene interface, as a prototypical planar *p-n* junction. Achieving one-dimensional C_{60} structures on recumbent pentacene layers is the goal of this study, and we have discovered ways to significantly improve the likelihood of C_{60} nanowire persistence using fully atomistic Molecular Dynamics (MD) for explicit all-atom models of the two components. We report three main effects that affect C_{60} nanowire persistence sensitively: the angle that the recumbent pentacene molecules make with the surface normal, the amount of initial tensile strain applied to the nanowire, and the presence of surface step edges. At room temperature, C_{60} nanowires oriented along the pentacene short axes are stable within the inherent timescale constraints of MD and are more likely to occur if they reside between, or within, pentacene rows for tilt angles, $\phi \leq \sim 60°$, with increasing likelihood of persistence the smaller the off-normal tilt angle. Nanowires oriented along the long axes of pentacene molecules (denoted N-S), are unlikely to form. The stability of nanowires to withstand thermal fluctuations was tested by raising the temperature from room temperature to 400 K, just below the temperature at which pentacene desorption occurs. Nanowires located *between* pentacene rows withstood this thermal inducement to disorder. However, C_{60} nanowires located initially *within* pentacene rows are only stable in the range of tilt angles, ϕ_1=30–50° . Increasing temperatures leads to a higher likelihood of losing 1D order, or at least creating indecisive switching of the location of the nanowire from between to within pentacene rows. Flatter pentacene surfaces, characterized by tilt angles above about 60° , induce C_{60} molecules to "burrow" into the pentacene surface creating considerable disorder. This result is also suggested by experimental STM observations by Dougherty *et al.* [2009]. An initial strain of 5% applied to the C_{60} nanowires significantly *decreases* the likelihood of finding nanowires. In contrast, any appreciable surface roughness, even half a monolayer high, strongly enhances the likelihood of nanowire formation due to the strong binding energy of C_{60} molecules to step edges.

(703d) The Structural Evolution and Diffusion During the Chemical Transformation From Cobalt to Cobalt Phosphide Nanocrystals

Don-Hyung Ha, Liane M. Moreau, Clive R. Bealing, Haitao Zhang, Richard G. Hennig and Richard D. Robinson, Materials Science and Engineering, Cornell University, Ithaca, NY

Nanoscalesystems can display interesting and unpredictable transformation kinetics that increase the structural complexity of the original material. Characterization of thetransformation routes to form the complex final structure is one of the majorchallenges in nanoscience. A more completeunderstanding of the transformation pathways would provides a means to improve synthesis techniques aswell as an insight into the control of chemical and physical properties, in orderto optimize the nanocrystals (NCs) for use in applications. We report the structural evolution and the diffusionprocesses which occur during the phase transformation of nanocrystal ε-Co to $Co_2P$to CoP, from a reaction with tri-*n*-octylphosphine (TOP). Extended X-ray absorption fine structure (EXAFS) investigations wereused to elucidate the changes in the local structure of cobalt atoms whichoccur as the chemical transformation progresses. Results from EXAFS show boththe Co_2P and CoP phases contain excess Co. Results from EXAFS, transmission electron microscopy, X-ray diffraction, and density functionaltheory calculations reveal that the inward diffusion of phosphorus is morefavorable at the beginning of the transformation from ε-Co to $Co_2P$by forming of an amorphous Co-P shell, while retaining a crystalline cobaltcore. When the major phase of the sample turns to Co_2P, thediffusion processes reverse and cobalt atom out-diffusion is favored, leaving ahollow void, characteristic of the nanoscale Kirkendalleffect. For the transformation from Co_2P to CoP theory predicts anoutward diffusion of cobalt while the anion lattice remains intact. In realsamples, however, the Co-rich nanocrystals continue Kirkendallhollowing.

See more of this Session: Nanomanufacturing
See more of this Group/Topical: Materials Engineering and Sciences Division

(395d) Polymeric Nanofiber Braid Manufacturing and Characterization

Ji Wang, Engineering Science and Mechanics, Virginia Tech, Blacksburg, VA and **Amrinder S. Nain**, Mechanical Engineering, Virginia Tech, Blacksburg, VA

Using our previously reported Spinneret based tunable Engineering Parameters (STEP) technique, we demonstrate aligned deposition of high aspect ratio uniform diameter Polystyrene (PS) fibers (diameter: sub 50nm to 800nm, length: mm-cm). PS fibers made from STEP technique were deposited in well aligned configuration with parallelism ≤ 2.5 degrees which makes it possible to prepare unidirectional PS fiber bands composed of thousands of individual fibers. PS fiber bands having 600nm diameter uniform fibers of 22 mm in length were made from 10% wt PS(M_w=2000K) dissolved in xylene. PS fiber braids were made by twisting two parallel PS fiber bands using home-designed braiding device and the braiding angle was controlled by number of twists added to the braids. For 3 , 6, 9 and 12 twists braids, the braiding angles were measured to be 6°, 15°, 21° and 32 ° respectively. The break load of PS fiber braids increases as braiding angle becomes larger and peaks (around 700mN)at 21°. Further increase of braiding angle leads to decrease of break load. Compared to non braided PS fiber bands, PS fiber braids demonstrate higher ultimate elongation, higher yield strength and lower modulus. The ultimate elongation increases from 15% to 36% as braiding angle increases from 0° to 15°while even larger braiding angle do not contribute to elongation. SEM images of braids fracture reveals that sheath-core structure forms during the elongation of densely packed fiber braids, which causes fibers at inner portion of braids to fuse together. To further enhance mechanical properties of PS fiber braids, 2%wt multi wall carbon nanotubes(MWNTs) were embedded into polystyrene fiber, which causes the fiber diameter to increase from 600nm to 1000nm. The break load for non fiber bands and 9 twists braids of PS-CNT fibers are 431 mN and 1000mN respectively, both of which are much larger than corresponding pure PS fiber bands and braids. Besides, 9twists braids of PS-CNT fiber demonstrates ultimate elongation up to 82% which is twice of 9 twists braids of pure PS fibers. Ongoing efforts are devoted towards detecting reinforcing effect of CNTs with different aspect ratio and understanding the influence on CNTs on polymer chain orientation in PS fibers.

See more of this Session: Nanoscale Structure In Polymers III
See more of this Group/Topical: Materials Engineering and Sciences Division

(394e) Bio-Nano Reinforcement of Polylactic Acid with Surface Modified Cellulose Nanocrystals

Jos Luis Orellana, Esteban E. Urea-Benavides and Christopher L. Kitchens, Department of Chemical and Biomolecular Engineering, Clemson University, Clemson, SC

Cellulose nanocrystals (CNC) have the potential to serve as a nanoscale reinforcement for polylactic acid (PLA), which has become a popular polymer but some of the deficiencies still need to be overcome. The main goal of this work is to improve the mechanical properties of the PLA films – especially toughness – by the addition of surface-modified CNC in the polymer matrix without sacrificing the thermal stability and processability.

CNC were isolated via acid hydrolysis from cotton and surface modified with an esterification process. Transmission Electron Microscopy (TEM) images demonstrated the isolation of the nanocrystals although aggregation can be evidently observed. CNC was further functionalized with different surfactants including decylamine (DA) and cetyltrimethylammonium bromide (CTAB). The dispersibility of the CNC was studied in different solvents such as THF, chloroform and ethyl acetate. Surface modified CNC were added in different concentrations to PLA. Tensile properties of the films were studied using an Instron universal testing machine.

It was found that the stability of the surface-modified CNC improved in organic solvents. DA greatly improved the stability of CNC in both THF and ethyl acetate; while CTAB did in chloroform. No settling was observed in these suspensions for at least three weeks. The toughness and elongation at break of PLA-CNC nanocomposites were improved by the addition of CNC by at least two-fold increase compared to plain PLA. Tensile strength and tensile modulus were not significantly affected by the addition of the crystals. The mechanical properties of the composites were compared and related to the CNC dispersion quality in the different solvents.

See more of this Session: Nanocellulose-Based Materials and Composites
See more of this Group/Topical: Forest and Plant Bioproducts Division

(472e) Nanoparticles Masquerade As "self" to Inhibit Phagocytosis

Pia L. Rodriguez[1], Takamasa Harada[1] and Dennis E. Discher[2], (1)Chemical and Biomolecular Engineering, University of Pennsylvania, Philadelphia, PA, (2)Chemical & Biomolecular Engineering, University of Pennsylvania, Philadelphia, PA

A major challenge for injecting particles or implanting biomaterials into the body is that they activate immune cells such as macrophages, the cells that normally function to clear invading pathogens. Interestingly, macrophages have a surface receptor mechanism which prevents them from phagocytosing our own "self" cells. During initial macrophage engulfment, macrophages recognize foreign and self targets because they both have antibodies or plasma complement proteins on their surface. However, before the macrophage engulfs the target, self cells are checked for the presence of the surface protein CD47 which will bind to the macrophage receptor SIRPa (CD172), and we show that CD47-SIRPa interactions in cell-cell adhesion, with human macrophages in sparse culture, produce phagocytosis inhibition (Tsai & Discher, JCB 2008). Whether the CD47 interaction is functional with small targets of phagocytosis is unclear and relevant perhaps to nano-sized targets. We show that CD47 coupled to a series of synthetic beads can inhibit uptake by macrophages. However, we need to test this interaction and observe how these results are consistent with in vivo systems. We are currently testing whether nanoparticles that have surface immobilized hCD47 or a portion of it will not be phagocytosed in nobese diabetic (NOD) mice, we believe that hCD47 will bind to this mSirpa strain and that could bring a better understanding of the interaction at a nanoscale.

See more of this Session: Nanotechnology for Biotechnology and Pharmaceuticals I
See more of this Group/Topical: Nanoscale Science and Engineering Forum

(175a) Mechanical Characterization of Polymeric Nanofibers Using An Integrated Approach

Mohammad Khan, Department of Mechanical Engineering, Virginia Tech, Blacksburg, VA and **Amrinder S. Nain**, Mechanical Engineering, Virginia Tech, Blacksburg, VA

Recent advances in polymeric fiber manufacturing platforms enable integration of polymeric micro/ nanofibers as 1D building blocks in a bottom-up assembly environment. These building blocks are attributed to possess superior mechanical, thermal and electrical properties compared with larger bulk counterparts, which potentially allows using them in applications of sensors, tissue scaffolds, fiber reinforced composites, drug delivery carriers, airborne structures, etc. The superior physical properties are attributed to the effect of surface tension, defect free one dimensional anistropic materials and ordered molecular chain orientation. Characterization of those fibers is essential to achive improved design and reliable performace for targetted application. The primary challengest associated with material characterization involve precise deposition of nanofibers with near exact boundary conditions in aligned configurations, availability of large number of samples and finally appropriate nanoscale characterization platform. Current techniques for repeated fabrication of polymeric nanofibers having appropriate boundary conditions for characterization involve multiple processing steps which often lead to inconsistent boundary conditions and makes experiments very tedious. Here we report an ***Integrated Approach*** for the fabrication and characterization of polymeric nanofibers with near-exact clamped-clamped boundary conditions at each end of fiber. In the proposed approach, polymer dissolved in a solvent is extruded through a glass micropipette, which in turn is moved over a commercially available Transmission Electron Microscopy (TEM) grid. Fiber formation on the individual grid patterns occurs by solvent evaporation and this technique provides flexibility in depositing aligned fibers having diameters ranging from sub-50 nm to sub- microns. The deposited fibers on the grid have clamped-clamped boundary conditions, which allows for fiber material characterization under the TEM (polymer molecular chain arrangement) and also provides a pathway for mechanical characterization using force deflection spectroscopy under the Atomic Force Microscope (AFM). This simple, straightforward and integrated approach is demonstrated for mechanical characterization of polymeric nanofibers: polystyrene (PS), poly methyl methacralate (PMMA), and carbon nanotube (CNT) embedded PS fibers. At first the fibers are deposited across commercially available TEM grids and a map of the fiber location across the TEM grid is generated under TEM, thus, providing information on the alignment of the deposited fibers and location of the fibers in the grids. The grids are then mounted under the AFM and using the lateral force microscopy (LFM), fibers are deformed till breakage. Specifically, the probe tip is then moved in the lateral direction and it makes contact with the fiber, at which point fiber starts to deflect and undergoes plastic deformation at high strains and finally breaks. This process causes the probe tip to deflect, thereby changing the voltage signal, which drops to the original signal after the fiber breaks. The magnitude of the sharp drop in voltage signal at the point of fiber breakage is converted to a force value. Finally, accurate fiber dimensions are determined using SEM.

(499d) Development of Nylon Biocomposites Through the Torrefaction of Waste Stream Agricultural by-Products

Jessica L. Lattimer and Chad A. Ulven, Mechanical Engineering, North Dakota State University, Fargo, ND

In the last several years the use of biobased materials for fillers in thermoplastics has seen a remarkable increase. The desire to produce lighter, stronger, and more ecologically friendly materials being key driving forces behind the movement. While the focus has thus far been on producing biobased composites out of commodity polyolefins or bio-derived resins, little work has been done in the realm of engineering thermoplastics. For engineering thermoplastics, the increased processing temperatures are unfavorable for biocomposite production, causing significant damage to the biobased fillers. As the automotive industry is one of the largest consumers of nylon and the ever increasing push to make cars green, there is a natural fit for research into producing biobased nylon composites. Finding a way to drive off the constituents within the biomass that degrade above the traditional processing temperatures of polyolefins would provide a new biomass that could withstand the processing of engineering thermoplastics.

From previous studies into the use of biobased materials for energy production it is known that torrefaction, conducted between 220-300 °C, will drive off the hemicellulose, fats, waxes, and other constituents yielding a biomass consisting of mostly cellulose, degrading between 300-375 °C, and lignin, degrading slowly over 250-500 °C. As cellulose is the constituent in biomass that acts as a reinforcing agent in the biocomposite, this is a potential method to prepare biobased fillers for introduction in to engineering thermoplastics. This study describes the development of nylon based biocomposites using torrefied sunflower hulls (TSFH), oat hulls (TOH), and flax shive (TFS).

To prepare the biocomposites, nylon was blended with each of the torrefied biomasses analyzed based on weight percentage. This study looks at three grades of the biocomposite; neat nylon-6, 16 wt% filler in nylon-6, and 20 wt% filler in nylon-6. The biocomposite samples prepared were tested for static and dynamic properties in addition to characterizing their resistance to moisture uptake as a function of torrefied biomass loading. Correlations between biomass input and loading are discussed.

See more of this Session: Biobased Materials II: Lignin-Based Materials
See more of this Group/Topical: Forest and Plant Bioproducts Division

(499e) Lignin Based BioFoam Composites From Functionalized Soy Oil Based Biopolyurethane

Manju Misra, School of Engineering and the Department of Plant Agriculture, University of Guelph, Guelph, ON, Canada

uploading in progress

See more of this Session: Biobased Materials II: Lignin-Based Materials
See more of this Group/Topical: Forest and Plant Bioproducts Division

(543f) Controlled Delivery of Functional Antibody From Amphiphilic Polyanhydride Nanoparticles

Brenda R. Carrillo-Conde[1], Steven J. Seiler[1], Amanda Ramer-Tait[2], Michael J. Wannemuehler[2] and Balaji Narasimhan[1], (1)Chemical and Biological Engineering, Iowa State University, Ames, IA, (2) Veterinary Microbiology and Preventive Medicine, Iowa State University, Ames, IA

The most widely used type of passive immunotherapy involves the administration of antibodies (e.g., monoclonal antibodies) to patients to provide immunity against disease, or help fight off infections. For example, it has been suggested that antibodies could be used in early, passive immune treatments for HIV and the passive transfer of IgG has also been shown to protect against Ebola in mice. As with the development of protein pharmaceutics, one of the most challenging tasks in passive immunotherapies is to deal with the physical and the chemical instabilities of the protein (i.e., antibodies), which invariably leads to loss of biological activity. In order to avoid these problems, it is necessary to design delivery vehicles that will minimize the degradation, maximize the *in vivo* activity, and provide controlled release of the encapsulated protein.

Biodegradable polyanhydrides have been shown to stabilize and provide sustained release of proteins to be used in vaccine formulations. However, the mechanisms of protein alteration are protein-specific, so it is necessary to select polymer formulations that can stabilize the specific protein of interest. In this study, the use of polyanhydride nanoparticles is suggested for the delivery of two antibodies in order to show their ability to release stable therapeutic payloads. Tetanus anti-toxin and anti-TNF-α monoclonal antibody were encapsulated and released from polyanhydride nanoparticles fabricated by an anti-solvent nanoencapsulation method. The polyanhydride systems based on copolymers of 1,6-bis(*p*-carboxyphenoxy)hexane (CPH), sebacic acid (SA), and 1,8-bis(*p*-carboxyphenoxy)-3,6-dioxacotane (CPTEG) monomers were used in this work. *In vitro* release kinetics were evaluated at physiological conditions and quantified using the micro-bicinchonic acid assay. Enzyme-linked immunosorbent assay (ELISA) was used to quantify the antigenicity of both the full tetanus anti-toxin and the monoclonal anti-TNF-α after release from polyanhydride nanoparticles. A biological assay was used to determine the biological function of the released anti- TNF-α by measuring their neutralization ability to the killing effect of TNF- α in L929 cells. Experiments were also performed to determine the effect of nanoparticle chemistry on the primary, secondary, and tertiary structure of the released protein. Results indicated that the amphiphilic environment provided by CPTEG-containing polyanhydrides preserved protein antigenicity, suggesting that this platform can be effectively used as a delivery system for therapeutic proteins.

See more of this Session: Nanotechnology for Biotechnology and Pharmaceuticals II
See more of this Group/Topical: Nanoscale Science and Engineering Forum

(649f) Graphene As a Support for Nanocrystals

Matthew G. Panthani, Colin M. Hessel, Aaron M. Chockla, Justin T. Harris, Dariya K. Reid and Brian A. Korgel, Department of Chemical Engineering, The University of Texas at Austin, Austin, TX

Graphene, a two-dimensional carbon network, provides an intriguing platform for anchoring inorganic nanostructures due to its extremely high surface area and conductivity. We demonstrate the application of graphene-supported nanocrystals in two applications: 1. Transmission electron microscopy (TEM) supports for enhanced imaging contrast and 2. Lithium-ion battery electrodes.

We use graphene as a TEM support for silicon nanocrystals in place of conventional amorphous carbon. Because of its low electron scattering, graphene enables silicon nanocrystals to be imaged at high resolution with higher contrast compared to conventional supports. We have demonstrated the application of graphene-supported nanocrystals as a high capacity anode material for lithium ion batteries. Graphene/nanocrystal composites exhibit enhanced charge capacity compared to non-supported nanocrystals.

See more of this Session: Graphene and Carbon Nanotubes: Applications
See more of this Group/Topical: Nanoscale Science and Engineering Forum

(599b) Hybrid Material Systems for Controlling Lipid Bilayer Assembly

Noah Malmstadt, Mork Family Department of Chemical Engineering and Materials Science, University of Southern California, Los Angeles, CA

We have developed a set of tools and techniques for controlling the assembly of lipid bilayers to produce biomimetic membrane structures. These structures can serve as platforms for fundamental studies of membrane biophysics or as biocompatible materials for the integration of biomolecules with synthetic structures.

Using a two-phase microfluidic flow system, lipid bilayers can be constructed in an "artificial cell" format that mimics the scale and geometry of eukaryotic cell membranes as well as their leaflet-to-leaflet compositional asymmetry. These artificial cells provide a unique experimental format for probing the role that compositional asymmetry plays in physiological processes. We have deployed this technique in an investigation of the passive transport of small molecular and ionic species across lipid bilayers--a phenomenon that controls the bioavailability and tissue distribution of drugs. Early results show that asymmetric membranes support directional passive transport: the permeability of the membrane depends on how its two distinct leaflets are oriented with respect to the concentration gradient of a transported species. This result has profound implications for our understanding of the role that lipid bilayers play in generating and supporting physiological concentration gradients.

Asymmetric artificial cells are also ideal platforms for investigating the mechanics of lipid bilayers. Membrane mechanics can be probed both with direct measurements based on membrane fluctuation amplitude and by analysis of membrane deformation under osmotic stress.

We have also constructed a series of hybrid lipid materials for orienting biomolecules on nanoengineered surfaces. This includes a technique for covalently attaching lipid molecules to the surfaces of carbon nanotubes that leads to the formation of well-organized molecular lipid bilayers on nanotube surfaces. These membrane-mimetic bilayers support the integration of membrane proteins with synthetic carbon nanomaterials. We have also developed a class of polymer nanogels in which the surface of a nanoscale polymer hydrogel sphere is covalently decorated with a lipid monolayer. These nanogels support the spontaneous formation of membrane-like bilayers and are potential drug delivery vehicles.

See more of this Session: Nanostructured Biomimetic and Biohybrid Materials and Devices
See more of this Group/Topical: Nanoscale Science and Engineering Forum

(420g) Effects of DNA Methylation On the Self-Assembly of a Chromatin Fiber

Chongli Yuan, School of Chemical Engineering, Purdue University, West Lafayette, IN and Isabel Jimenez-useche, Chemical Engineering, Purdue University, West Lafayette, IN

DNA in eukaryotic cells is folded into an organized structure facilitated by a protein complex, i.e., a histone octamer. This protein-DNA complex, known as a chromatin, regulates gene expression in eukaryotic cells by limiting the accessibility of genomic DNA sequences. DNA methylation, i.e., the addition of a methyl group to a nucleotide, is known as an epigenetic mark for regulating gene expression in eukaryotic cells. Besides being a targeted site for many chromatin-binding proteins, DNA methylation has been recently postulated to directly mediate gene expression by affecting the structure of a chromatin. To verify this hypothesis and to reveal the detailed roles of DNA methylation in chromatin structure, this study examines the effects of DNA methylation on the assembling and stability of a nucleosome, i.e., the structural repeating unit of a chromatin, using various biophysical approaches. This knowledge will shed light on the mechanism of DNA packaging in eukaryotic cells.

See more of this Session: Poster Session: Nanoscale Science and Engineering
See more of this Group/Topical: Nanoscale Science and Engineering Forum

(385a) Magnetic Block Ionomer Complexes for Imaging and Therapeutics

Nikorn Pothayee[1], Nipon Pothayee[1], Neeta Jain[2], Lindsay M. Johnson[1], Sharavanan Balasubramaniam[1], Nammalwar Sriranganathan[2], Alexander V. Kabanov[3], **Richey M. Davis**[4] and Judy S. Riffle[5], (1) Macromolecules and Interfaces Institute, Virginia Tech, Blacksburg, VA, (2)VA-MD Regional School of Veterinary Medicine, Virginia Tech, Blacksburg, VA, (3)College of Pharmacy, University of Nebraska Medical Center, Omaha, NE, (4)Chemical Engineering, Virginia Tech, Blacksburg, VA, (5) Chemistry, Virginia Tech, Blacksburg, VA

Multifunctional nanocarriers comprising a combination of therapeutic and imaging agents are of great interest for delivering drugs and tracking their biodistribution. With an aim to develop nanocarriers with high drug loadings that integrate magnetic nanoparticles for imaging into one system, we report antibiotic-laden magnetic block ionomer complexes (MBICs) based on assembly of block ionomers with nanomagnetite and cationic drugs. The magnetite core was synthesized via reduction of an Fe(III) organometallic precursor. Nonionic-ionic block copolymers were synthesized through controlled free-radical polymerization (ATRP). The polymer was bound to the magnetic nanoparticle surfaces via ligand adsorption of the PAA, thereby creating a double corona structure with a nonionic PEO shell and an ionic layer of PAA. The portion of carboxylates that were not attached to the magnetite provided binding sites for drug loading via ionic complexation. The multi-cationic drugs, gentamicin and streptomycin, were employed as models for encapsulation. Results suggest that the MBICs are sufficiently stable in physiological conditions to be suitable as drug carriers. Dispersions of the MBICs were stable in PBS containing 0.14 M NaCl for up to 10 days, and were stable in PBS containing fetal bovine serum (FBS). Physicochemical properties of the MBICs were significantly altered upon gentamicin encapsulation (31 wt % drug). A near zero-order release of gentamicin (pH 7.4 in PBS) that reached ~35 wt% of the initial gentamicin within 10 h was observed, and this was followed by slower release of another 7% by 18 h. We will show how the transverse relaxivities (r_2) of the complexes correlate with cluster size as it is affected by drug complexation.

See more of this Session: Magnetic Nanoparticles In Biotechnology and Medicine
See more of this Group/Topical: Nanoscale Science and Engineering Forum

(747e) Tailoring Surface Charge and Hydrophicity In Colloidal Quantum Dot Biosensors

Yanjie Zhang[1], Amanda Riddle[1], Elizabeth Whitley[2], Ian C. Schneider[3] and Aaron R. Clapp[1], (1)Dept. of Chemical & Biological Engineering, Iowa State University, Ames, IA, (2)Veterinary Pathology, Iowa State University, Ames, IA, (3)Iowa State University, Ames, IA

Broad and improved use of quantum dots as biosensors requires precise control over the surface chemistry to improve colloidal stability, retain desirable optical properties, and target extra- and sub-celluar structures. We have explored a facile procedure for producing water-soluble CdSe-ZnS quantum dots having a tunable mixture of ligands bound to their surface. This biphasic synthetic route results in quantum dots with compact hydrodynamic size which is critical for many biosensing applications. The affinity of quantum dots for live cells and fixed tissues was investigated using fluorescence microscopy and flow cytometry. We found that the addition of lipid molecules to the quantum dot surface resulted improved partitioning in the lipid bilayer as well as lipid-rich or hydrophobic sections of tissue. Additionally, the net charge was seen to strongly influence the binding characteristics of quantum dots. Overall, the interplay between these characteristics results in a complex interaction mechanism suggesting that quantum dots can be engineered to target specific regions of cells and tissues by varying fundamental surface parameters.

See more of this Session: Nanotechnology for In Vivo and In Vitro Imaging
See more of this Group/Topical: Nanoscale Science and Engineering Forum

(472c) Evaluation of a Nanoparticle Delivery Vehicle with Bacterial Targeting Ligand for Respiratory Treatment

Timothy Brenza[1], Mai Tu[1], Michael A. Apicella[2] and Jennifer Fiegel[3], (1)Pharmaceutics and Translational Therapeutics, The University of Iowa, Iowa City, IA, (2)Microbiology, The University of Iowa, Iowa City, IA, (3)Pharmaceutics and Translational Theraputics; Chemical and Biochemical Engineering, The University of Iowa, Iowa City, IA

A variety of phylogeneticallydistinct bacterial pathogens, such as non-typeable *Haemophilus influenzae (NTHi)*, invade host cells in the upper airwaysby binding the platelet-activating factor (PAF) receptor. Lipooligosaccharide(LOS) glycoforms naturally expressed on the bacterialcell surface facilitate bacteria-epithelial interactions. We are utilizing LOS ligands isolated from the bacteria surface to coat thesurface of nanosized delivery vehicles enablingtargeting of drug molecules to the respiratory epithelium. We have observed that modification of thesurface of nanoparticles with absorbed LOS glycoforms from the 2019 NTHi strainincreased cellular adherence and penetration when applied to bronchial epithelialcell cultures *in vitro*.

Commercially available fluorescent nanoparticles were modified with various NTHi LOS glycoforms (2019 and3198). The efficacy of the targeted nanoparticles to attach and be internalized by therespiratory epithelium was evaluating using a cell line to model the upperrespiratory tract. Calu-3 bronchialepithelial cells were grown to confluent monolayersunder air-interfaced culture conditions on semi-permeable Transwellmembranes. The adherence and uptake of nanoparticles intothe epithelial cells was determined under two apical fluid conditions: removedapical fluid and natural secretions produced by the cells. Flow cytometry and confocal microscopy were used to evaluate the subpopulationof the cell monolayer with particles present on the cell membrane orinternalized within the cell at different time periods. We observe that when LOS modified nanoparticles were applied to a more physiologicallyrelevant model of the respiratory system their diffusion through therespiratory fluid barrier is hindered.

Cellulartoxicity of the was measured through the use of a MTS (tetrazoliumsalt) assay to quantify metabolically active cells, NR (neutral red) assay toexamine membrane permeability and lysosomalintegrity, and a Live/Dead Fixable dye from Invitrogento evaluate cell membrane integrity. TheMTS and NR assays are designed for 96 well plates so the seeding density, cellincubation time, amount of MTS and NR dyes, and dye incubation time wereoptimized for the 96 well plates. TheLive/ Dead Fixable dye with the Calu-3 cells grown to confluence underair-interface conditions on semi-permeable Transwellmembranes was utilized to evaluate cellular health in a more relevant model ofthe respiratory tract. For the Calu-3cell line the presence of the natural secretions on the apical surface reducedthe toxicity observed in the Live/Dead stain when compared to particles directlyapplied to a washed surface of Calu-3 cell monolayer. This suggests that for the Calu-3 cell linethe MTS and NR assays in the 96 well plates overestimated the amount of celltoxicity for particles when compared to the more physiologically relevant confluentmonolayer grown under air-interface culture conditions.

See more of this Session: Nanotechnology for Biotechnology and Pharmaceuticals I
See more of this Group/Topical: Nanoscale Science and Engineering Forum

(176h) Tunable Mirrors Made From Gold Nanoparticle Assembly At the Oil-Water Interface

Mingxiang Luo, Gloria Olivier and Joelle Frechette, Chemical and Biomolecular Engineering, Johns Hopkins University, Baltimore, MD

Nanoparticle self-assembly has opened the door to the design of materials with unique optical, electronic, and chemical properties. The interface between two immiscible liquids is a promising platform for nanoparticle assembly due to the rapid correction of defects afforded by the dynamic fluid interface, and optical applications afforded by the ability of light to access both sides of the interface. In particular, assembly of a two-dimensional array of metal nanoparticles at the oil-water interface enables the optical reflectance of the interface to be tuned according to the diameter of and interparticle spacing between the particles. Here we demonstrate the reversible or switchable assembly of reflective array of nanoparticles at a liquid interface. The films described herein possess both reversibility and mirror-like reflectance. Specifically, we show that 5 and 10 nm diameter gold nanoparticles, functionalized with an ion-pair, assemble spontaneously into pH-sensitive reflective sheets at the oil-water interface. The assembly occurs across a wide range of pH values (pH $<$ 12) and more importantly, can be switched off and on reversibly by shifting the pH above and below pH 12, respectively. We address the role of the surface functionality on the reversible assembly and optical properties of the film at the oil-water interface. We also demonstrate control over the interfacial assembly based on an applied electrical potential.

See more of this Session: Nanostructured Materials for Electronic and Photonic Applications
See more of this Group/Topical: Materials Engineering and Sciences Division

(531b) Inkjet Patterned Carbon Nanotube Multilayer Devices

Christine M. Andres and Nicholas A. Kotov, Chemical Engineering, University of Michigan, Ann Arbor, MI

Layer-by-layer assembly (LBL) of carbon nanotubes serves as a method for the creation of unique combinations of electrical, mechanical and optical material properties allowing for flexible electronic applications with integrated strength and conductivities. Inkjet printing serves as a continuous direct-write approach to patterning LBL films for the simple, low-cost integration into advanced functional devices. The strictly additive approach allows for the incorporation of a variety of different materials and substrates for a versatile, multicomponent and noncontact patterning technique. Here we will demonstrate the use of this technique for the development of a wide range of application from electrical actuators as building blocks for complex micromechanical devices to low cost paper sensors for rapid detection and diagnostics.

See more of this Session: Graphene and Carbon Nanotube Based Devices
See more of this Group/Topical: Materials Engineering and Sciences Division

(747f) Luminescent CuIn(SeS)2 Nanocrystals for Diagnostic Imaging

Matthew G. Panthani[1], Tarik A. Khan[1], Michael R. Rasch[1], Dariya K. Reid[1], Daniel J. Hellebusch[2] and Brian A. Korgel[1], (1)Department of Chemical Engineering, The University of Texas at Austin, Austin, TX, (2)Department of Chemical and Biomolecular Engineering, University of California, Berkeley, Berkeley, CA

Semiconductor quantum dots could potentially be used as near-infrared fluorescent probes that are brighter and more photostable than organic dyes. However, the toxicity of the components of near-infrared (NIR) emitting II–VI and IV–VI nanocrystals (Cd, Pb, Hg, *etc.*) is an obstacle to clinical due to their toxicity. We report a method to synthesize copper indium selenium sulfide ($CuIn(SeS)_2$) nanocrystals with strong photoluminescence in the visible to NIR. The hydrophobic nanocrystals are encapsulated with an amphiphilic polymer, which enables dispersion in aqueous media while retaining high quantum yield. We demonstrate the potential of these nanocrystals for use as a noninvasive imaging probe by performing in-vivo fluorescence imaging in mice.

See more of this Session: Nanotechnology for In Vivo and In Vitro Imaging
See more of this Group/Topical: Nanoscale Science and Engineering Forum

(472a) Targeting C-Type Lectin Receptors On Alveolar Macrophages: A Novel Strategy In the Design of Intranasal Vaccines

Ana V. Chavez-Santoscoy[1], Rajarshi Roychoudhury[2], Amanda Ramer-Tait[3], Nichola Pohl[2], Michael J. Wannemuehler[4] and Balaji Narasimhan[1], (1)Department of Chemical and Biological Engineering, Iowa State University, Ames, IA, (2)Department of Chemistry, Iowa State University, Ames, IA, (3) Department of Veterinary Microbiology and Preventive Medicine, Iowa State University, Ames, IA, (4) Veterinary Microbiology and Preventive Medicine, Iowa State University, Ames, IA

The goal of this work is to target C-type Lectin receptor (CLRs) on antigen presenting cells (APCs) using "pathogen-mimicking" carbohydrate-functionalized polyanhydride nanoparticles. Polyanhydride nanoparticles represent a promising platform in the development of effective single dose vaccines, providing an appropriate environment for protein stabilization, sustained release, and modulation of immune response. There is a critical need for the development of efficacious vaccines against respiratory pathogens capable of inducing protective mucosal immune responses. In this context, alveolar macrophages (AMs) are the first phagocytic cells to encounter antigens in the respiratory track and transport them to the lymph nodes draining the lung. Thus, their activation is required for the design of effective vaccines against respiratory pathogens. Our approach is to target CLRs on AMs as a means to enhance antigen presentation and modulate cytokine production. Polyanhydride nanoparticles based on a 50:50 ratio of 1,8-bis(p-carboxyphenoxy)-3,6-dioxaoctane (CPTEG) and 1,6-bis(p-carboxyphenoxy) hexane (CPH) were fabricated by an anti-solvent nano-precipitation method and the particle surface was modified by attaching di-mannose or galactose residues by an amine-carboxylic acid coupling reaction. AMs were harvested from C57BL/6 mice by bronchoalveolar lavage (BAL) and stimulated for 48 h with the nanoparticles. Cell surface marker expression (i.e., MHC I, MHC II, CD40, CD86, CD206, CD209, CD301a/b) was assessed by flow cytometry and cytokine production (i.e., IL-1β, TNF-α, IL-6, IL-12) was evaluated using a multiplexed bead assay. The expression of activation makers was enhanced by di-mannose as well as galactose modified nanoparticles in comparison with the non-modified particles. The data also showed that di-mannose modified nanoparticles selectively upregulated the expression of the macrophage mannose receptor (MMR) and CIRE (also termed as SIGNR5 or mouse DC-SIGN) receptors, while galactose-modified nanoparticles upregulated the expression of macrophage galactose lectin (MGL). In addition, di-mannose functionalization enhanced the production of pro-inflammatory cytokines. In summary, the "pathogen-like" mannose functionalized polyanhydride nanoparticles were shown to possess superior adjuvant properties based on their effective activation of AMs, rendering them as promising components of intranasal vaccines against respiratory infectious diseases.

See more of this Session: Nanotechnology for Biotechnology and Pharmaceuticals I
See more of this Group/Topical: Nanoscale Science and Engineering Forum

(105f) Controllable Assembly of Magnetic Janus Particles and Their Electric Properties

Bin Ren[1], Alex Ruditskiy[2] and Ilona Kretzschmar[2], (1)Chemical Engineering, The City College of New York, The City University of New York, New York, NY, (2)Chemical Engineering, City College of New York, New York, NY

Magnetic nanoparticles have potential applications in position sensing, ultrahigh density magnetic storage media, clinical diagnosis and treatment, biological labeling, and separation applications. However, their use is hampered by non-specific aggregation due to the interaction of their magnetic moments. Janus particles with magnetic caps have a distinct advantage over symmetric particles. They exhibit an asymmetric magnetization that makes them suitable building blocks for directed three-dimensional assembly under the exposure of external magnetic field. If the Janus particles have also metallic properties, the addition of an electric field provides an additional experimental parameter.

Here, we present a study of the preparation and assembly behavior of iron oxide capped Janus particles, in presence of both magnetic and electric fields with the orientation of the two fields either being parallel or perpendicular to each other. Iron oxide capped Janus particles are prepared via deposition of iron onto a polystyrene particle monolayer using physical vapor deposition in an oxygen/argon atmosphere. Subsequently, the modified particles are assembled in a magnetic field for 30 min. Three assembly behaviors are observed: (i) staggered chains, (ii) double chains, or (iii) no assembly. Subsequently, samples are exposed to parallel and perpendicular AC electric and magnetic fields.

We will report on the changes observed in the assembly behavior of iron-oxide capped Janus particles as a function of deposition rate and type of the applied field. A diagram will be presented relating deposition rate and assembly behavior. In addition, we will report on the behavior of double and staggered chains in parallel and crossed magnetic and electric fields.

See more of this Session: Nanofabrication and Nanoscale Processing
See more of this Group/Topical: Nanoscale Science and Engineering Forum

(599g) Synthetic Tannins for Self-Assembled Nanocomplexes

Omar Z. Fisher[1], Robert Langer[2] and Daniel G. Anderson[2], (1)Koch Institute for Integrative Cancer Research, Massachusetts Institute of Technology, Cambridge, MA, (2)Chemical Engineering, Massachusetts Institute of Technology, Cambridge, MA

Polyphenols have been extensively researched as therapeutics but primarily as plant derived, low molecular weight compounds. We have designed synthetic polyphenols that mimic the structure of those that occur naturally. Specifically, we have been able to attach phenolic residues to polysaccharide scaffolds through the careful tuning of acyl transfer reaction conditions. By controlling the number of pendant phenolic groups we can design synthetic polymers with antioxidant power greater than what can be derived from natural sources. As water soluble polymers these can also be potentially targeted to sites of disease within the human body in ways unavailable to small molecules. We are investigating the unusually strong hydrogen bonding between polyphenols, poly(ethylene glycol) (PEG), and various protiens as a way to form self-assembled, nanoscale drug carriers. By synthesizing synthetic polyphenols with controlled size and shape, better control over the PEG-polyphenol self-assembly process can be achieved. We are developing a combinatorial library of these structures that achieve the following goals: diameters < 200 nm, neutral surface charge, and the ability to be package small and large molecular weight chemotherapeutic agents. Preliminary results showed that the spontaneous complexation between the two polymers and resulting particle size could be controlled through polymer chain length and chain configuration. Our long term goal is to select compounds capable of targeting chemotherapeutics to cancer cells

See more of this Session: Nanostructured Biomimetic and Biohybrid Materials and Devices
See more of this Group/Topical: Nanoscale Science and Engineering Forum

(662b) Connected but Confined: Surface Functionalization of Semiconducting Nanocrystals

Haitao Zhang[1], Bo Hu[1], Tobias Hanrath[2] and **Richard D. Robinson**[1], (1)Materials Science and Engineering, Cornell University, Ithaca, NY, (2)Chemical and Biomolecular Engineering, Cornell University, Ithaca, NY

Connected but Confined: Surface Functionalization of Semiconducting Nanocrystals

Semiconductor colloidal nanocrystals (NCs) have attracted much attention in the past decade due to their unique size and shape dependent properties. In most of the successful synthetic routes to colloidal NCs, the use of bulky hydrocarbon (C_8-C_{18}) molecules with coordinating functional groups (such as -COOH, -NH_2, etc) as surfactant ligands is crucial for stabilization, as well as for size and shape control of NCs. The presence of these large organic molecules, however, creates highly insulating barriers which block the electronic communication between NCs, limiting the applications of assemblies of colloidal NCs. We have developed a novel chemical method to efficiently remove these surfactant ligands. This unique surface modification process produces bare nanocrystals without any surfactant ligands, which has been confirmed by FTIR and elemental analysis. GISAX study also reveals a significant decrease on inter-particle spacing. The bare nanocrystals are attached but maintain quantum confinement. Such "connected but confined" nanocrystal assemblies exhibit much stronger inter-particle coupling, and are promising new materials for fabricating electronic and optoelectronic devices.

See more of this Session: Nanocrystal Science and Technologies
See more of this Group/Topical: Materials Engineering and Sciences Division

(395e) Nanofiber Structure Influenced by Air Gap In a Collector Plate of Electrospinning

Jong Kyu Hong[1], Guan Xu[2], Daqing Piao[2] and Sundararajan. V. Madihally[1], (1)School of Chemical Engineering, Oklahoma State University, Stillwater, OK, (2)Department of Electrical and Computer Engineering, Oklahoma State University, Stillwater

Electrospinning has recently emerged as a technique due to the possibility of fabricating nanofibers. Electrospraying apparatus consisted of a syringe pump, syringe, needle tip, high voltage power supply, earth grounding, and an air gap having collector plate wrapped with aluminum foil. In this study, we address the influence of the shape and size of air gap in a collector plate on fiber density, fiber configuration, and pore size of fibers using synthetic and natural polymers.

SEM morphology showed that the shape of air gap affects the pattern of fibers. For the circular gap in the collector plate, the fibers were randomly deposited throughout the entire void of the collector plate. For the triangular gap, the fibers aligned perpendicular to the bisector of the vertices at the edge of the vertices. For the rectangular void, the fibers aligned from one short side to the other side perpendicular to the long sides of the rectangle in large scale. Quantification study revealed that when the size of air gap was increased from 0.9 cm via 1.4 cm to 1.9 cm, the average pore size of fibers increased from 14.71, 22.83, and 57.77 μ m since fiber density on air gap decreased when the gap size increased.

Simulation study using electrostatics module of COMSOL with passion equation revealed that the equi-potential surface collapses into the air gap. Since the dragging force generated by the electric field will be perpendicular to the equi-potential surfaces and the polymer fiber with be directed along the steepest potential decent direction, that is, to take the shorter path with respect to its initial launching direction and accumulate at the aluminum surfaces. Otherwise, fiber density is low at air gap due to the collapsed equi-potential surface. Also, the strong electric forces at the aluminum gap edges will stretch the polymer fiber and form aligned fiber at the edge of triangular gap.

See more of this Session: Nanoscale Structure In Polymers III
See more of this Group/Topical: Materials Engineering and Sciences Division

(420z) Nanostructured Electrode of Multi-Walled Carbon Nanotubes and Polyaniline Nanofibers

MD Nasim Hyder[1], Fevzi Cebeci[1], Seung Woo Lee[2], Yang Shao-Horn[2] and Paula T. Hammond[1], (1) Chemical Engineering, Massachusetts Institute of Technology, Cambridge, MA, (2)Mechanical Engineering, Massachusetts Institute of Technology, Cambridge, MA

Polyaniline (PANi), an electroactive polymer, is known for its high electrical conductivity, excellent thermal stability and interesting redox behavior. On the other hand, multi-walled carbon nanotubes (MWNT) are known for their high aspect ratio, rich functionalization chemistry and high conductivity. Nanostructured films of PANi and MWNTs could serve as potential electrode materials for supercapacitors, batteries, sensors, and photovoltaic devices.

We are investigating the physicochemical and electrical properties of ultrathin electrodes of polyaniline nanofibers and acid functionalized multiwalled nanotubes fabricated using layer-by-layer (LbL) assembly.The resulting elecreodes consist of a randomly oriented nanoscale interpenetrating network structure with well developed nanopores indicates that yield excellent electrochemical performance that could potentially be applied in supercapacitors, fuel cells, and battery electrodes.

See more of this Session: Poster Session: Nanoscale Science and Engineering
See more of this Group/Topical: Nanoscale Science and Engineering Forum

(420aa) Electromagnetic Stimuli-Responsive Hybrid Nanoparticle-Biopolymeric Materials

Huang-Chiao Huang, Alisha Nanda and Kaushal Rege, Chemical Engineering, Arizona State University, Tempe, AZ

Thermally responsive elastin-like polypeptide (ELP) mimics mammalian elastin by characterizing the sequence VPGXG (V=valine, P=proline, G=glycine, and X=any amino acid except proline). ELPs exhibit an entropy-dominated phase transition at the inverse transition temperature (T_t). Upon heating, ELPs undergo a reversible intramolecular contraction leading to phase separation, commonly known as coacervation. Further heating of ELP leads to coalescence of coacervates into larger droplets. The thermal transition behavior of ELPs is reversible and has been exploited in a variety of applications including, bioseparation, drug delivery, sensors, and tissue engineering. This poster focuses on the development of ELP with a several nanoparticles (e.g. gold, iron oxide, polyplexes and polymers). Nanoparticles (e.g. gold, iron oxide) were first interfaced with ELP for sensing of external electromagnetic (thermal, photothermal, magnetic) and chemical changes. Visible responses (e.g. cross linked-phase separation, release of indicator molecules) resulting from structural changes in the polypeptide following exposure to radiation or chemicals will be demonstrated. A focus on the irreversible formation of gold nanorod (GNR) hybrid elastin–like polypeptide (ELP) matrices upon external thermal changes will be discussed in detail. This allowed better understanding of GNR-ELP matrices formation which further facilitated the development of GNR-ELP matrices as therapeutic platform for chemotherapy and non-viral gene deliver for potential high throughput screening of cellular response.

See more of this Session: Poster Session: Nanoscale Science and Engineering
See more of this Group/Topical: Nanoscale Science and Engineering Forum

(342a) Development of Double-Gyroid Nanowire Arrays for Photovoltaics

Hugh W. Hillhouse, School of Chemical Engineering, Purdue University, West Lafayette, IN

Multiple exciton generation (MEG) has been observed and studied in PbS and PbSe quantum dots (QDs). However, despite many advances in the understanding of the photoconversion physics and electron transport in quantum dot arrays, the grand challenge of developing photovoltaic devices that exhibit IQE greater than 100% has yet to be realized. Quantum dots are zero dimensional objects and must be electronically coupled in order to extract any photocurrent. A broader class of structures may be sought in pursuit of achieving this goal that have 0D photophysics but electronic transport more similar to 3D structures [1]. Here, we present experimental and theoretical results on solar cells based on highly ordered "double-gyroid" arrays of PbS quantum wires. The wire segments that compose the ordered array are 4 nm in diameter and intersect at y-junctions every 12 nm. These unique nanomaterials are fabricated using self-assembled double-gyroid nanoporous films as the nanowire template [2,3]. 2D grazing incidence small-angle x-ray scattering (GISAXS) patterns exhibit 96 diffraction peaks from the nanoscale order. The patterns are indexed with a (211) oriented cubic unit cell with Ia-3d symmetry and lattice constant equal to 18 nm. Thus, the PbS wire segments are interconnected with crystallographic periodicity. We have used an envelope formalism with parameters fitted from PbS QD data to calculate the electronic structure of the PbS double-gyroid wire array. The electronic density of states shows a significant blue-shift in the band gap and additional small gaps in the conduction and valence band [4]. All-inorganic Schottky junction and interpenetrated p-n junction photovoltaic devices have been fabricated with a variety of contacts and device architectures. Currently, the best devices show poor power conversion efficiencies, but show unusually high open circuit voltages (as high as 0.45 eV). The presentation will focus on the electronic structure calculations, photovoltaic device fabrication, and device performance.

References:

[1] Hillhouse H.W. & Beard M.C., "Solar Cells from Colloidal Nanocrystals: Fundamentals, Materials, Devices, and Economics," Current Opinion in Colloid & Interface Science, 14, 245-259 (2009).

[2] Urade VN, Wei TC, Tate MP, Kowalski JD, & Hillhouse HW, "Nanofabrication of double-gyroid thin films," Chem. Mater. 19 (4), 768 (2007).

[3] Tate, M.P., Urade, V.N., Gaik, S.J., Muzzillo, C.P., Hillhouse,* H.W., "How to Dip-Coat or Spin-Coat Nanoporous Double-Gyroid Silica Films with EO19-PO43-EO19 Surfactant (Pluronic P84) and Know it Using a Powder X-ray Diffractometer," Langmuir 26 (6), 4357-4367 (2010).

[4] Khlebnikov S & Hillhouse HW, "Electronic Structure of Double-Gyroid Nanostructured

Semiconductors: Perspectives for Carrier Multiplication Solar Cells," Phys. Rev. B 80, 115316 (2009).

See more of this Session: Templated Assembly of Inorganic Nanomaterials II
See more of this Group/Topical: Nanoscale Science and Engineering Forum

(369d) Adsorption of Gold Nanoparticles and Humic Acid On Activated Carbon Used In Drinking Water Treatment

Holly A. Stretz, Department of Chemical Engineering, Tennessee Technological University, Cookeville, TN, Vasanta Pallem, Department of Chemical Engineering and Center for the Management, Utilization and Protection of Water Resources, Tennessee Technological University, Cookeville, TN and Martha J. M. Wells, Department of Pharmaceutical Sciences, Midway College School of Pharmacy, Paintsville, KY

Granular activated carbon (GAC) is primarily used in drinking water treatment processes for removing dissolved organic matter comprising of humic and fulvic acids (DOM), and other pollutants. The increasing applications of gold nanoparticles (GNPs) in diagnostic imaging, chemical and biological sensing, also present potential concerns for adsorption and removal of GNPs in water treatment units, following post-clinical discharge. The current study employs spectroscopic separation to evaluate the extent of adsorption of citrate coated GNPs with core size 50 nm (GNPs-50), and HA separately and in mixtures. Modified Freundlich isotherms were used to describe the adsorption behavior of humic acid and gold in mixtures. The novelty of the approach was to examine the adsorption behavior for synergistic or antagonistic effects when all three materials (gold nanoparticles, HA and activated carbon) were present, and to carefully account for the fluorescence of the controls in determining fractional fluorescence prior to establishing the isotherms. The GNPs-50 were completely adsorbed on activated carbon when present individually, however in the presence of HA the adsorption of GNPs-50 decreased significantly. The adsorption of humic acid also decreased in the presence of GNPs-50 showing an antagonistic effect. Therefore, the mixtures of HA/GNPs-50 present significant implications for the fate and removal of GNPs and HA in water processing units.

See more of this Session: Environmental Implications of Nanomaterials: Fate and Transport
See more of this Group/Topical: Topical C: Environmental Aspects, Applications, and Implications of Nanomaterials and Nanotechnology

(175c) High Strength Graphene Nanocomposites with Tunable Architectures Through Layer-by-Layer Assembly

Jian Zhu, Department of Chemical Engineering, University of Michigan, Ann Arbor, MI and Nicholas A. Kotov, Chemical Engineering, University of Michigan, Ann Arbor, MI

Despite the tremendous efforts in utilizing the high strength of graphene sheet to produce high quality engineering composites, the results are not satisfying. Common techniques including filtration or solution casting have limited controllability over the fabrication process, thus are poor in manipulating the interactions between graphene and various polymers and often result in non-uniform structures and inadequate mechanical properties. In addition, these techniques are difficult in controlling thickness with nanoscale accuracy and are incompatible with current micro fabrication procedures. Here, we demonstrate that high strength and toughness graphene nanocomposite can be made through layer-by-layer assembly. It was discovered that sonication time can influence the thickness of multilayer graphene building blocks, which could be taken advantage of to manipulate the structures, thus the properties of final material. In the end, it was shown that micro features can be made on these composites using photolithography.

See more of this Session: Nanoscale Structure In Polymers I: Nanostructured Polymeric Materials
See more of this Group/Topical: Nanoscale Science and Engineering Forum

(649e) Phase Behavior and Fiber Spinning of Lysozyme-Single Walled Carbon Nanotube Dispersions

Daniel W. Horn[1], **Geyou Ao**[2], Maryse Maugey[3], Cecile Zakri[3], Philippe Poulin[3] and Virginia A. Davis[4], (1)Department of Chemical Engineering, Auburn University, Auburn, AL, (2)Department of Chemical Engineering, Auburn University, Auburn University, AL, (3)Centre de Recherche Paul Pascal, CNRS Bordeaux, Pessac, France, (4)Chemical Engineering, Auburn University, Auburn, AL

The phase behavior of lysozyme-single-walled carbon nanotube (LSZ-SWNT) dispersions was characterized using a combination of techniques including cross-polarized optical microscopy and rheology. Due to the antibacterial properties of LSZ, the strength of SWNT, and the strong associations, the dispersions can be processed into antibacterial films and fibers. The phase behavior of dispersions using HiPCo and CoMoCat SWNT were compared. The effects of [LSZ]:[SWNT], processing conditions, and the addition of additives were also explored. Films and fibers were produced from the most promising dispersions; the microstructure and antibacterial activity of these bulk materials were characterized.

See more of this Session: Graphene and Carbon Nanotubes: Applications
See more of this Group/Topical: Nanoscale Science and Engineering Forum

(342d) Synthesis of Oxide Nanobowls' and Armor-Coated' Active Sites by Templated ALD: a New Paradigm In Heterogeneous Catalyst Synthesis

Christian P. Canlas[1], Natalie Ray[2], Junling Lu[3], Sungsik Lee[4], Randall Winans[4], Jeffrey Elam[3], Peter C. Stair[2] and **Justin M. Notestein**[1], (1)Chemical and Biological Engineering, Northwestern University, Evanston, IL, (2)Chemistry, Northwestern University, Evanston, IL, (3)Energy Systems Division, Argonne National Laboratory, Argonne, IL, (4)X-ray Science Division, Advanced Photonic Source, Argonne National Laboratory, Argonne, IL

Oneof the most important goals in heterogeneous catalyst development is toincrease selectivity and stability. Zeolites have long been used to impartselectivity based on size discrimination, and they are efficient at maintainingcatalyst sites dispersed and active. Here, we utilize atomic layer deposition (ALD), which has been used for years in microelectronics fabrication but onlyrecently applied to heterogeneous catalysts synthesis, to grow thin oxidelayers perforated with cavities over an existing oxide catalyst. We call thesenanoscale cavities (<2 nm deep, <2 nm in diameter) nanobowls'. Theirpresence introduces a size-sieving feature to an existing catalyst surface, andthey can armor-coat' individual active sites on a surface. In this study, thenanobowls are generated by using a molecular template immobilized on thecatalyst surface before ALD. Bulky template molecules such asp-tertbutylcalix[4]arene and adamantanecarboxylic acid are ideal for thispurpose. They are deposited on oxide supports and a wall of oxide is grown aroundthe immobilized template molecule by ALD. Template removal generates thenanobowl. For the case of armor-coated' catalyst sites, the template is boththe metal-containing catalyst precursor and the structure directing agent. Nanobowlsynthesis is monitored by DRUV-vis, TGA and N_2 physisorption, whilethe existence of the nanobowls is elucidated by *in-situ* QCM studies,small angle x-ray scattering (SAXS) and transmission electron microscopy. Sizeselectivity of the catalysts is demonstrated through selective photooxidationof benzyl alcohol, 1-hexanol and 2-adamantanol over a reference, thoughtraditionally non-selective titania photocatalyst. For ALD films more than afew tenths of a nanometer thick, bulky secondary alcohols like 2-adamantanol areable to access fewer active sites on the titania surface than do terminalalcohols like benzyl alcohol and 1-hexanol. Thus, the relative rate of alcoholoxidation, or conversion after a set time, is a strong function of the nanobowldepth. This selective photocatalytic oxidation of alcohols demonstrates hownanobowls can discriminate size on an arbitrary catalyst surface and introducesa new paradigm in heterogeneous catalyst synthesis.

See more of this Session: Templated Assembly of Inorganic Nanomaterials II
See more of this Group/Topical: Nanoscale Science and Engineering Forum

(420h) Nanomaterials Engineering with Sequential Infiltration Synthesis In Block Copolymers

Qing Peng[1], Yuchih Tseng[2], Seth Darling[2] and Jeffrey Elam[1], (1)Energy Systems Division, Argonne National Laboratory, Argonne, IL, (2)Center for Nanoscale Materials, Argonne National Laboratory, Argonne, IL

Nanomaterials Engineering with Sequential Infiltration Synthesis in Block copolymers

Qing Peng,[1] Yu-Chih Tseng,[2] Seth B. Darling,[2] and Jeffrey W. Elam[1]

[1]Energy Systems Division and [2]Center for Nanoscale Materials at Argonne National Laboratory, Argonne, 9700 South Cass Ave., IL, 60439

Corresponding address: jelam@anl.gov

Abstract for AICHE 2011

Nanoscale patterning of materials with controllable characteristic feature sizes, shapes, and compositions is of considerable interest in a wide range of fields. Various lithography processes have been used for nanosclae patterning including photolithography, Ebeam lithography, and block copolymer lithography. For a majority of these processes, polymers are used for transferring a pattern onto the desired material. On the other hand, atomic layer deposition (ALD) has been used to modify the physical and chemical properties of polymers.

In this work, we introduce a new process based on ALD, sequential infiltration process (SIS), capable of chemically modifying the polymers used in lithography to achieve more sophisticated control in nanomaterials engineering. Specifically, we will present applications of SIS process to self-assembled block copolymers (BCPs) thin films which form ordered periodic nanostructures (*e.g.* spheres, cylinders, lamellar and bicontinuous structures) due to microphase separation of the incompatible blocks. In this way we generate ordered nanoscale patterns of different materials with tunable dimensions. Mechanism studies will be shown to illustrate the unique reaction and diffusion phenomena involved when applying SIS to different polymer systems. Extended applications of the process will also be included.

Through rational design of the polymer chemistries and selection of SIS parameters, patterned designer materials with controlled size, shapes, spacing, symmetry, and composition can be synthesized on various substrates for various applications.

(346e) Long-Circulating Magnetic Nanoparticles As Platforms for Multifunctional Nanomedicine

Allan E. David[1], Adam J. Cole[2] and Victor C. Yang[1], (1)Department of Pharmaceutical Sciences, The University of Michigan, Ann Arbor, MI, (2)Pharmaceutical Sciences, University of Michigan, Ann Arbor, MI

Interest in utilizing magnetic nanoparticles (MNP) for biomedical applications has increased considerably over the past two decades. This excitement has been driven in large part by the success of MNPs as contrast agents in magnetic resonance imaging. The recent investigative trend has also turned toward concurrent therapy, giving rise to the MNPs as potential ''theranostics''. Recently we reported the synthesis of long-circulating polyethylene glycol-modified MNPs (PEG-MNP) suitable for magnetic targeting. Using a rat model, this work explored the biodistribution of PEG-MNPs in organs of elimination and showed that enhanced magnetic targeting to brain tumors can be achieved due to their improved circulation lifetime. Reductions in liver (~12 fold) and spleen (~2.5 fold) concentrations at 1 hour compared to parent unmodified-MNPs confirm plasma pharmacokinetics observed previously. Time-course biodistribution analysis revealed that concentrations of PEG-MNP in the liver remained considerably lower than those observed for unmodified-MNP throughout their plasma clearance, while it was observed to continually increase in the spleen for up to 60 hrs – a trend visualized with histology. Limited to no uptake of PEG-MNPs was observed in lungs or kidneys. Additionally, with magnetic targeting, the accumulation of PEG-MNPs selectively in brain tumors (1.0% injected dose/g tissue) was 15-fold greater than that of unmodified-MNPs (0.07% injected dose/g tissue). This enhanced delivery of PEG-MNP to tumors was further confirmed by MRI and histological analyses, which also suggested that the contribution of passive targeting to tissue retention of nanoparticles was limited. Our exciting results justified the further development of PEG-MNP as a multifunctional platform for nanomedicine.

See more of this Session: Applications of Magnetic Nanoparticles
See more of this Group/Topical: Nanoscale Science and Engineering Forum

Vapor-Based Reactive Polymer Coatings: A Robust Tool for Biointerface Engineering

Hsien-Yeh Chen[1], **Wei-Chieh Liang**[1], Meng-Yu Tsai[1] and Joerg Lahann[2], (1)Department of Chemical Engineering, National Taiwan University, Taipei, Taiwan, (2)Department of Chemical Engineering, Macromolecular Science and Engineering, Biomedical Engineering, and Materials Science and Engineering, University of Michigan, Ann Arbor, MI

Multifunctional reactive polymers can be synthesized via double-site chemical vapor deposition (CVD) co-polymerization. The resulting ultra-thin coatings are pinhole-free and can be conformally co-deposited to a wide range of substrates and materials. More importantly, the equipped multifunctional groups can serve as anchoring sites for tailoring the surface properties, making these reactive coatings a robust platform to deal with sophisticated challenges faced in biointerfaces. These advanced coatings can improve the interfacial biocompatibility of biomedical surfaces or can be compatible with complex biological features as they represent a designable interlayer; stable under the conditions of the biological environments. On the other hand, accessibility of reactive coatings within confined microgeometries was systematically studied, and the preparation of homogeneous polymer thin films within the inner surface of microchannels was demonstrated. Moreover, surface microstructures and surface patterns using reactive coatings via microcontact printing (μCP), vapor-assisted micropatterning in replica structures (VAMPIR), and dip-pen nanolithography (DPN) are introduced. These patterning techniques can be complimentarily used and provide access to precisely confined microenvironments on flat and curved geometries regardless of the underlying material used. Finally, the recently developed alkyne-containing and maleimide-containing coatings are also introduced. These coatings are applicable to a wide range of substrates and can be used to conjugate with desired biomolecules through subsequent click chemistry, copper-free click chemistry, or thiol-maleimide chemistry. Reactive coatings provide a technology platform that creates active, long-term control and may lead to improved mimicry of biological systems for effective bio-functional modifications.

(121g) The Importance of Curvature At the Nano-Scale: How and Why Curvature Influences Self-Assembly

David A. Walker and Bartosz Grzybowski, Chemical and Biological Engineering, Northwestern University, Evanston, IL

While there has been significant progress in controlling the size, shape and properties of nanomaterials, their controlled assembly into nanostrucutres remains a significant challenge. To date, many nanoparticles (NPs) of unusual shapes have been reported - but the role which curvature plays on the surface chemistry of these particles remains a mystery. We recently found that the pK_a of carboxylic acids bound to the surface of NPs deviates drastically compared to values for free ligands in solution. Interestingly, simply by controlling the size/shape of a NP, one can control the degree of charge on charged NP surface, shift the redox potentials of ligands on a NP surface and more. A mixture of experimental and theoretical results will be presented.

See more of this Session: Self and Directed Assembly at the Nanoscale
See more of this Group/Topical: Nanoscale Science and Engineering Forum

(172h) Electrical Properties of Metallic Nanoparticle Films

David A. Walker, Yong Yan, Hideyuki Nakanishi and Bartosz Grzybowski, Chemical and Biological Engineering, Northwestern University, Evanston, IL

Nanoparticles (NPs) display a wide range of electrical behavior depending on the particle material, shape, and local environment. Particularly, the electronic properties of metallic NP films can be greatly influenced by the self-assembled monolayers (SAMs) which coat the particles' surfaces. This talk will focus on a class of nanoparticle-based materials whose conductivity can change drastically based on how the SAM is influenced by various external stimuli (i.e. irradiation with light causes photoconductance and inverse photoconductance). Such materials can be used to create elementary circuitry made entirely out of metallic NPs.

See more of this Session: Nanoelectronic Materials
See more of this Group/Topical: Nanoscale Science and Engineering Forum

(420ab) Electrostatics At the Nanoscale: Anisotropy

David A. Walker and Bartosz Grzybowski, Chemical and Biological Engineering, Northwestern University, Evanston, IL

The field of electrostatics is thought, by many, to be 'fully understood' - however at the nano-scale we continue to discover intriguing and counter-intuitive phenomena. Here we present the results from a series of recent fundamental studies investigating the electrostatic interactions of non-spherically symmetric nanoparticles (NPs). We show that electrostatic interactions can be used to control the assemblies of charged nano-objects (of both like and oppositely charged polarities), the important role which curvature can play in patterning charge density on NP surfaces, and how electrostatic interactions can be used to selectively assemble reactive NPs into more complex geometries. A mixture of both experimental and theoretical evidence will be presented.

See more of this Session: Poster Session: Nanoscale Science and Engineering
See more of this Group/Topical: Nanoscale Science and Engineering Forum

(374f) Phase Separation of Mixed Monolayers On Silica Nanoparticles Induced by Hydrogen Bonding

Daniel Sunday, Chemical Engineering, University of Virginia, Charlottesville, VA and **David L. Green**, Department of Chemical Engineering, University of Virginia, Charlottesville, VA

Controlling the self-assembly of mixed monolayers on silica nanoparticles is challenging due to irreversible ligand attachment. In the course of attempting to control the graft density of polystyrene brushes on silica nanoparticles, we synthesized both a surface atom transfer radical polymerization (ATRP) initiator along with an inert analog containing an amide group. The initiator and analog were reacted to the nanoparticles in methylisobutyl ketone to form a mixed monolayer from which polymers could grow. The inert analog reacted preferentially to the nanoparticle surface, which affected the polymerization of styrene with ATRP. In particular, analysis of the initiation efficiency, or the ratio between the ATRP and graft polymer surface coverage provides evidence for phase separation between the ATRP initiator and inert analog in the mixed monolayer. This is most likely due to hydrogen bonding primarily between the inert analog in solution and on the particle surface. Small-angle x-ray scattering (SAXS) and NOESY NMR were used to show that in solution, the analog and initiator form disc-shaped molecular aggregates about 1 nm thick and 1.6 nm in diameter through hydrogen bonding between amide groups. Based on kinetic deposition experiments and FTIR measurements, the molecular aggregates, composed primarily of the inert analog, adsorb strongly to the surface. This adsorption leads to phase separation between the inert analog the ATRP initiator in the monolayer as inferred from FTIR by tracking the non-linear shift in the methylene stretches of the carbons in the ligands, which was a function of the amount of the initiator and analog used to functionalize the nanoparticles. To our knowledge, this reaction mechanism in which the ligand solution behavior affects mixed monolayer structure on silica nanoparticles has not been previously observed.

See more of this Session: Functional Nanoparticles and Nanocoatings on Particles I
See more of this Group/Topical: Particle Technology Forum

(439e) Enhancement of Polymer-Mediated Transgene Expression Using Histone Deacetylase Inhibitors (HDACi)

Sutapa Barua, Chemical Engineering, Arizona State University, Tempe, AZ, Jennifer Lehrman, Harrington Biomedical Engineering, Arizona State University, Tempe, AZ and Kaushal Rege, Chemical Engineering, Biomedical Engineering, Arizona State University, Tempe, AZ

Polymeric vectors are promising agents for transgene delivery because of their flexibility of structural design, capacities for delivering transgenes in the size of a few hundred nanometers, reduced immunogenicities, and low production costs. However, polymer-mediated gene delivery suffers from low transgene expression levels. Here we demonstrate that chemotherapeutic inhibitors of intracellular histone deacetylases (HDACs) enhance polymer-mediated transgene expression levels in cancer cell lines. Chemotherapeutic inhibitors of both cytoplasmic and nuclear HDACs were investigated for enhancing polymer-mediated transgene expression of reporter proteins (e.g. luciferase). The potential role of nuclear vs. cytoplasmic HDACs, and the HDAC-interacting proteins on enhancing the efficacy of polymer gene delivery was also investigated. Our results demonstrate that inhibiting HDACs can be employed as a synergistic strategy for enhancing polymer-mediated transgene delivery to cancer cells.

See more of this Session: Bionanotechnology for Gene and Drug Delivery I
See more of this Group/Topical: Nanoscale Science and Engineering Forum

(600i) A Finite-Element Based Global Model for Multiphase Flows In a Convective Assembly System

Gaurab Samanta[1], Andrew Yeckel[1], Satish Kumar[2] and Jeffrey J. Derby[1], (1)Chemical Engineering and Materials Science, University of Minnesota, Minneapolis, MN, (2)Department of Chemical Engineering and Materials Science, University of Minnesota, Minneapolis, MN

Convective assembly is a promising method for manufacturing self-assembled, colloidal crystals. In this process, a moving meniscus and fluid flow is employed to achieve self-assembly of particles from a suspension --- mono-disperse spheres crystallize near the moving contact line on a stationary substrate immersed in an evaporating medium. In this process, a thin fcc crystal is deposited on the substrate in a matter of few hours. In order to understand the mechanism behind such fast deposition of ordered arrays of particles, Norris et al. put forth the convective steering hypothesis, which posits that solvent flow through pore spaces of close-packed spheres could preferentially direct advancing spheres into "clear or fcc niches" rather than into the "obstructed niches". Although calculations based on porous-media network theory suggested that flow is up to 33% higher into the clear niches than into obstructed niches, subsequent calculations based on elaborate three-dimensional finite element analysis of flows into an fcc structure showed that inertial effects of the global flow may overwhelm the local steering effects induced by the particle arrangement.

In this presentation, we focus on the character on the global flows in convective assembly system by employing a two-dimensional, finite element analysis of the temporal flows in the system. We account for time-dependent flows in the solvent as well as the gas phase above it, including a rigorous accounting for evaporation, capillary, and Marangoni forces at the liquid-gas interface. We study the effect of varying gas inflow over the liquid surface, substrate and growth front inclinations, and solute concentration in the suspension on the flow behavior of the system. We validate our model against prior experimental results.

See more of this Session: Nanostructured Thin Films
See more of this Group/Topical: Materials Engineering and Sciences Division

(385e) Enhanced Transport of PEG-Based Magnetic Nanocomposites In Artificial Mucus Barriers

Nathanael Stocke[1], Heidi Mansour[2] and J. Zach Hilt[1], (1)Chemical and Materials Engineering, University of Kentucky, Lexington, KY, (2)Department of Pharmaceutical Sciences, University of Kentucky, Lexington, KY

Chemotherapy causes many adverse side effects when administered systemically through oral or IV delivery, and therefore, the localization of this therapy will result in improved therapeutic outcomes. Pulmonary administration of chemotherapy has shown enhanced therapeutic response while reducing side effects [1,2], and pulmonary delivery of multifunctional nanoparticles could serve as a multimodal treatment with enhanced patient response. However, the delivery of nanoparticles to mucosal tissue such as the lungs is greatly inhibited by a layer of viscoelastic mucus, and diseased states can enhance the mucus barrier. In this work, iron oxide core-shell nanocomposites were synthesized, characterized, and investigated in order to determine the effects of heating and surface chemistry on the rate of transport through a PEG-based artificial mucus. Atom transfer radical polymerization was used to grow PEG-based polymer coatings from iron oxide (Fe_3O_4) cores. These magnetic cores possess the ability to heat upon exposure to an alternating magnetic field due to Brownian and Neel relaxation processes. Fourier transform infrared spectroscopy was used to confirm the presence of PEG on the surface of the nanoparticles by examining characteristic peaks in the spectra, and thermogravimetric analysis was used to quantify the amount of PEG attached by calculating mass loss. Dynamic light scattering was used to determine the average hydrodynamic radius of the nanoparticles. A PEG-based artificial mucus was synthesized with 95:5 water:macromer ratio by weight, and the transport of nanoparticles into this mesh was studied at room temperature and at an elevated temperature. Preliminary results indicate that the transport of PEG-coated iron oxide nanoparticles can be increased by heating these particles.

References:

1. Otterson, G.A., et al., *Phase I/II Study of Inhaled Doxorubicin Combined with Platinum-Based Therapy for Advanced Non-Small Cell Lung Cancer.* Clinical Cancer Research, 2010. **16**(8): p. 2466-2473.

2. Otterson, G.A., et al., *Phase I study of inhaled doxorubicin for patients with metastatic tumors to the lungs.* Clinical Cancer Research, 2007. **13**(4): p. 1246-1252.

(543a) Imipramine Blue-Doxorubicin Co-Loaded Nanoparticles Increase Survival In Glioblastoma Over Doxorubicin Alone In a Single Treatment

Jennifer M. Munson[1], Ravi V. Bellamkonda[2], Jack L. Arbiser[3] and Rania Khan[2], (1)School of Chemical & Biomolecular Engineering, Georgia Institute of Technoloy, Atlanta, GA, (2)Biomedical Engineering, Georgia Institute of Technoloy, Atlanta, GA, (3)Dermatology, Emory Medical School, Atlanta, GA

Introduction: Though only accounting for 2% of new cancer cases, brain tumors are a large clinical problem with poor survival times. Glioblastoma (GBM), the most common brain tumor, is a highly invasive tumor which infiltrates the brain. Invasion causes failure of common treatments as cells move away from the site or are missed by the treatment due to surgical or diffusional limitations. In this study we aim to incorporate a novel anti-invasive compound, Imipramine Blue (IB), with a common chemotherapeutic, Doxorubicin (DXR), in a single nanoparticle to enhance response to treatment in a rat model of glioblastoma. In this way, we test the hypothesis that inhibition of invasion will assist in the treatment paradigm of GBM.

Materials and Methods: *Liposomal encapsulation of IB and Doxorubicin* Liposomal nanoparticles were made from 10mol% cholesterol (Sigma), 85mol% DSPC (Avanti), 5mol% mPEG-DSPE (Avanti), dissolved in ethanol with 0.9 mg/ml IB, and hydrated with ammonium sulfate to create a gradient. DXR was remotely loaded and liposomes were concentrated to yield a final IB concentration of 1.6 mg/ml. Control liposomes were made without one or both components (IB, DXR, or control). *In vitro treatment* RT2 glioma cells were treated with liposomes (Control, IB, DXR, IB-DXR) in the same concentrations of IB and DXR, then measured after 72 hours by CCK8 assay (Dojindo). Viability was normalized to control. *In vivo survival studies* Male Fisher 344 rats were inoculated with RT2 glioma (250,000 cells) into the cortex (5-7 animals per group based on MRI confirmation of tumor at time of treatment). The tumor established for 7 days at which point MRI was performed using a Bruker 7T microMRI with T_2 weighted measurements. Animals were then injected with liposomes via tail vein (6.8 mg/kg DXR and 3 mg/kg IB). MRI was used to monitor progression of tumor and animals were euthanized upon display of signs of morbidity and histology performed for tumor morphology. *Statistical Analysis* One-way ANOVA was used for *in vitro* study analysis. Survival data was analyzed using Gehan-Breslow-Wilcoxon analysis.

Results and Discussion: Liposomes were approximately equivalent in size and lipid:drug ratio (average size of liposomes 160 nm). *In vitro* studies showed no synergistic effect between IB and DXR on the RT2 glioma. This is a desirable effect since our central hypothesis is that the anti-invasive effect of IB will assist the efficacy of DXR and not some other effect of the combination. This was also expected from previous microarray pathway analysis data of the effects of the drug. Survival studies revealed a significant increase in survival in the IB-DXR group over the DXR alone group to survive after treatment (X^2 (1) = 4.033, $p<0.05$) with a median survival time increase from 44.5 days to >210 days. Interestingly, at the end of the study (210 days), MRI revealed in the DXR animals evidence of tumor including blood brain barrier breach and inflammation whereas the co-treated animals had no evidence of tumor existence. This was confirmed by histology showing that the original tumor site in the co-treated animals was mostly scar tissue and remnant blood products. The use of this combination nanoparticle enhances survival significantly in just a single treatment which is unprecedented in the literature. Further, use of a single nanoparticle yields easier delivery to patient and reduced toxicity from possible accumulation of multiple treatments. Histological analysis confirming absence of tumor at the end of study further enhances the possibility of this compound combination for future trials and studies.

Conclusions: Integration of a novel anti-invasive Imipramine Blue significantly enhances the efficacy of Doxorubicin in malignant glioma after a single treatment.

Acknowledgments: Johannes Leisen, Alexander Ortiz, Safkat Alkindi, Daniel Brat.

See more of this Session: Nanotechnology for Biotechnology and Pharmaceuticals II
See more of this Group/Topical: Nanoscale Science and Engineering Forum

(125d) Sustainability Implications of Integrating Renewable Power Sources and Electric Vehicles Into the Grid

Gintaras V. Reklaitis, Chemical Engineering, Purdue University, West Lafayette, IN

Sustainability Implications of Integrating Renewable Power Sources and Electric Vehicles into the Grid.

See more of this Session: Sustainability Plenary
See more of this Group/Topical: Sustainable Engineering Forum

(395g) Impact Resistance and Adhesive Properties of a Functionally Graded Polymer Composite Interlayer

Michael K. Opoku[1], Robb M. Winter[1] and David R. Salem[2], (1)Department of Chemical and Biological Engineering, South Dakota School of Mines and Technology, Rapid City, SD, (2)Composites and Polymer Engineering Laboratory, South Dakota School of Mines and Technology, Rapid City, SD

There is growing interest in lightweight materials with high impact resistance for various applications. In the development of these materials, researchers have started to explore polymer composites containing very low concentrations of ceramics and/or metals with at least one dimension ranging from nanometers to a few microns. In this research, we attempt to synthesize an interlayer nanocomposite polymer which provides a gradual transition of elastic modulus between two laminates of dissimilar modulus. This is expected to minimize the stress concentration gradient at the interlayer boundaries and improve impact resistance.

Formulation of polyurea and polyurethane from disocynate and diamine and/or polyol prepolymers was investigated, and promising candidates were mixed with nanoparticles, nanowhiskers or nanofibers of alumina or silicon carbide at different loading levels and size/loading distributions. These nanoparticle/ nanofiber resin systems were designed to serve as a bonding layer between materials of dissimilar modulus (such as ceramics, metals and polymers) while creating a functionally graded composite with minimal interfacial stress and high energy absorbing properties. We will present details of the synthesis of the polyurethane/polyurea resins and the preparation of the nanocomposite interlayer material; characterization of the nanocomposite layer and interface regions; and results from mechanical property testing.

See more of this Session: Nanoscale Structure In Polymers III
See more of this Group/Topical: Materials Engineering and Sciences Division

(369e) High Throughput Collection and Detection of Environmental Nanoparticles

Fanxu Meng[1], Serdar Ozturk[1], Maria D. King[2], Yassin A. Hassan[3] and Victor M. Ugaz[1], (1)Artie McFerrin Department of Chemical Engineering, Texas A&M University, College Station, TX, (2) Department of Mechanical Engineering, Texas A&M University, College Station, TX, (3)Departments of Nuclear and Mechanical Engineering, Texas A&M University, College Station, TX

The increasing prevalence of manufactured nanoparticles in the environment poses new risks to human health, pointing to the need for advanced nanoparticle sampling and analysis approaches capable of continuously monitoring room-sized volumes so that dynamic changes in exposure levels can be rapidly detected. Here we describe our efforts to address this need through development of a new technique that enables nanoparticles to be collected and concentrated at very high flow rates, so that they can be directly detected using a continuous-flow microfluidic approach coupled with particle sizing analysis. Nanoparticle detection is accomplished by harnessing interfacial fluorescence phenomena that uniquely emerge in a microfluidic environment. Fundamental parameters including sensitivity and specificity are characterized to establish detection limits for comparison with current-generation instrumentation. We then explore feasibility of coupling this component with an autonomous Wetted Wall Cyclone (WWC) collector developed in our laboratory and fabricated at Northrop Grumman Inc. This system is capable of performing continuous air sampling at 400 L/min with a cutpoint of 500 nm. Performance was characterized by sampling colloidal (anatase, 5-30 nm) and amorphous powder (5-30 nm) TiO_2 nanoparticles aerosolized and collected in a laboratory flowcell, and comparing the results against a standard glass fiber filter sampler. Environmental monitoring capability was assessed by sampling air in a freshly painted classroom for 24 hours at 100 L/min. Each hydrosol sample was weighed and quantitated alongside suspensions extracted from the reference filter using inductively coupled plasma mass spectrometry. This analysis indicated that the WWC system achieved collection efficiencies of 83.5% and 56.5% for aerosolized colloidal and amorphous power TiO_2 nanoparticles, respectively. The level of TiO_2 in the painted room air was around 0.2 ppt, however, SiO_2 was also detected at 1 ppb. Spectral analysis of the interfacial fluorescence offers potential to enable chemical fingerprinting to be rapidly performed in order to obtain a detailed picture of nanoparticle exposure levels while simultaneously providing the ability to respond to sudden environmental changes

See more of this Session: Environmental Implications of Nanomaterials: Fate and Transport
See more of this Group/Topical: Topical C: Environmental Aspects, Applications, and Implications of Nanomaterials and Nanotechnology

(385f) Thermal Behavior of Electroless CNT-FeCo Composite In Simulated Body Fluid In Applied RF Magnetic Field

Egwu E. Kalu[1], Lauren Martin[1], Lauren Wilson[1] and Michael E. McHenry[2], (1)Department of Chemical & Biomedical Engineering, Florida A& M Univ. and Florida State Univ., Tallahassee, FL, (2) Materials Science and Engineering, Carnegie Mellon University, Pittsburg, PA

Thermal Behavior of Electroless CNT-FeCo Composite in Simulated Body Fluid in Applied RFMagnetic Field

E. E Kalu[1].Lauren Martin[1], Lauren Wilson[1], M.E McHenry[2]

[1]ChemicalEngineering, FAMU-FSU College of Engineering, Tallahassee, Florida

[2]MaterialScience and Engineering, Carnegie Mellon University, Pittsburgh, Pennsylvania

The use of electrolessdeposition of magnetic alloy on carbon nanotubes isexplored for use in thermotherapy treatment of cancer cells. Carbon nanotubes were plated with Iron-Cobalt (FeCo)using varying amounts of methanol, palladium and electrolessbath in the process. The samples with the highest and lowest saturationmagnetizations were examined in simulated body fluid (SBF) using an alternatingradio frequency coil for their temperature response with time. Results will bepresented on the response of the electroless metallized-CNT in an alternating magnetic field and theirbehavior when suspended in the SBF.

See more of this Session: Magnetic Nanoparticles In Biotechnology and Medicine
See more of this Group/Topical: Nanoscale Science and Engineering Forum

(649a) Graphene Field-Effect Transistors with Boron-Nitride Dielectric Interfaces for RF Applications

Kenneth Shepard, Electrical Engineering, Columbia University, New York, NY

There has been growing interest in graphene as a replacement for III-V materials in MMIC applications because of its high mobility, its potential for high saturation velocity, and its nearly perfect two-dimensional electrostatics. We summarize ongoing efforts to design and characterize graphene field-effect transistors (GFETs). We demonstrate the importance of high quality dielectric interfaces to graphene, and the advantages of boron nitride as a dielectric interface material. We present detailed measurement and analysis of velocity saturation in GFETs, demonstrating the potential for velocities approaching 10^8 cm/sec and the effect of an ambipolar channel on current-voltage characteristics. We find that the saturation velocity is sheet-carrier dependent and limited by phonon emission. We describe on-going challenges in achieving low contact resistances and making scalable device structures.

See more of this Session: Graphene and Carbon Nanotubes: Applications
See more of this Group/Topical: Nanoscale Science and Engineering Forum

Enhanced Cardiomyocyte Function on Poly-Lactic-co-Glycolic Acid: Carbon Nanofiber Composites under Electrical Stimulation

David A. Stout, Brown University, Providence, RI
Dr. Thomas J. Webster, Brown University, Providence, RI

Introduction

Cardiomyocytes (a specialized contractile muscle cell that generates part of the myocardium tissue of the heart) depends on continuous conductivity to function [1, 2]. However, such conductivity may break down during heart disease or malfunction. For instance, a myocardial infarction, also known as a heart attack, usually occurs because a major blood vessel supplying the heart's left ventricle is suddenly blocked by an obstruction, such as a blood clot [3-5]. During myocardial infarction, part of the cardiac muscle, or myocardium, is deprived of blood and therefore oxygen, which destroys cardiomyocytes leaving dead tissue [3-5] as well as denervation of the myocardium [6-8]. In particular, nerve damage to cardiac tissue can result in nerve sprouting in the left ventricle [6, 9] and development of arrhythmias [10]. Scarred cardiac muscle results in heart failure for millions of heart attack survivors worldwide. In 2009, an estimated 785,000 Americans had a new coronary attack and about 470,000 had a recurrent heart attack leading to a coronary event [11].

In recent years, various techniques have been developed to promote cardiomyocyte growth around dead tissue after a myocardial infarction. Each approach has their own advantages [12], as well as disadvantages [12], but in general, all of the above can be divided into two groups: a) conductive cardiovascular patches (using scaffolding and 3D printing techniques usually with polypyrrole) and b) non-conductive cardiovascular patches (mostly involving direct cell injection, scaffolding, and injectable scaffolds). One can also observe that cardiovascular biomaterials can be either based on biodegradable or on non-biodegradable materials. Within this matrix of conductive vs. non-conductive and biodegradable vs. non-biodegradable material lie the most commonly studied materials and techniques used for promoting heart health. However, one area that has been largely omitted to date is the exploration of nanotechnology (or materials with one dimension less than 200nm) in cardiovascular applications. Numerous articles suggest that using nanotechnology can specifically promote cell functions on a variety of materials, ranging from titanium to silicon [13-15] due to optimal surface chemistry, wettability which control protein adsorption onto the surface. While some degree of nanostructured features may promote tissue growth, it is also known that certain topographies can hinder cell activity [16-18]. For example, when comparing nano to micro diamond features, Yang et al. showed that osteoblast (or bone forming cell) adhesion and proliferation increased on the nano-rough topography [16]. On the contrary, using a 1718 gene microarray Dalby et al. were able to suggest that nano-rough topography greater than 40 nm decreased human fibroblast cell adhesion [17, 18].

In recent research, it was demonstrated that the use of nano-materials can promote the growth of cardiomyocytes when compared to the use of conventional or micro-structured materials. One such nano-material consisted of poly-lactic-co-glycolic acid (PLGA) with carbon nanofibers (CNFs) where it was shown that cardiomyocytes proliferated faster on the composites compared to a pure PLGA film after 5 days.

The object of this study was to better understand the cytocompatibility properties of this composite material for myocardial applications by using electrical stimulation which mimics that of the heart. For this reason, an *in vitro* continuous electrical stimulation model was used to determine cardiomyocyte functions on PLGA: CNF composite materials.

Experiments

Fabrication

Purified carbon nanofibers (CNFs) (99.9% by weight %, Catalytic Materials, MA) with diameters of 60, 100, and 200 nanometers were sonicated in 20 ml of chloroform at 20W for 30 minutes. Two pellets of poly-lactic-co-glycolic acid or PLGA (50:50 PLA:PGA wt.) (Polysciences Cat #23986) were diluted in a 50 ml flask with 30 ml of tetrahydrofuran and were sonicated in a water bath below 30 ^{0}C for thirty minutes.

After the PLGA and CNF solutions were prepared, various PLGA:CNF weight percent ratios were created (100:0, 75:25, 50:50, 25:75, 0:100) by adding the appropriate amount of CNF to PLGA in 20 ml disposable scintillation vials. When the appropriate ratios were reached, each composite material was sonicated at 10W for 20 minutes. 1 ml of the appropriate PLGA:CNF solution was placed onto a glass substrate and then placed into an oven below 50 ^{0}C for 15 minutes. Each composite film was then vacuum dried at -20 inches of Hg gauge pressure for 48 hours to allow the THF and chloroform to evaporate.

Composite Characteristics

A Hitachi 2700 scanning electron microscope was used to characterize the surface of the PLGA:CNF samples (Fig. 1). The electrical resistance of the PLGA:CNF samples was determined using a multi-meter (HP 34401A) by connecting two probes, via alligator clip connectors, of the meter to opposite ends of the sample. The sample was tested dry at room temperature and at opposite ends of the sample, which were 22mm apart. Three measurements were taken for each sample.

X-Ray diffraction (Bunker AXS: D8 Focus) was used with settings at 0.5° per 2 minutes between $2\Theta = 10\text{-}35°$ to characterize the crystallinity of the composites. Also, Raman spectroscopy results were obtained and recorded using an argon ion laser attached with a spectrophotometer (Acton Research Corporation, model AM505F). A low laser power of 7mW was applied to avoid any local surface heating. To acquire the spectrum, an incident light with a wavelength of 514.5 nm was used in a wavenumber region of 200–3700 cm^{-1}. The scattered radiation was collected at 180° (backscattered geometry) to the incoming beam and detected using a CCD cooled to -120°C. The spectral resolution of the Raman experimental set up was better than 1 cm^{-1} and the total integration time was 3 minutes.

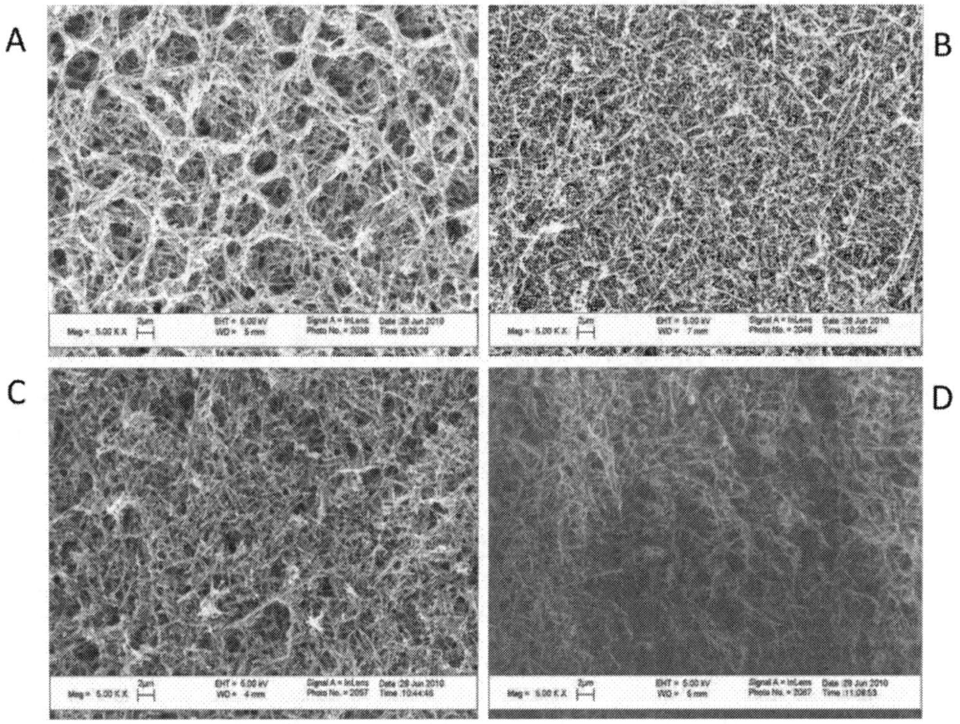

Figure 1: SEM images at 5K magnification showing the different PLGA:CNF (wt:wt) material ratios of 75:25 (A), 50:50 (B), 25:75 (C), 0:100 (D). Scale bar = 2 µm. SEM image analysis showed that more CNFs were observed at higher CNF ratio samples and the material width, with glass substrate, was 15.01 ± 1.22 µm (n = 4, mean ± s.d.).

Cytocompatibility

All samples and controls were sterilized using ultraviolet light for 24 hours prior to cell seeding. Human cardiomyocytes (Celprogen, Cat #36044-15) were seeded in human cardiomyocyte complete stem cell culture growth media with serum (Celprogen, Cat #M36044-15S) at a cell concentration of 10 x 10^4 cells/cm^2 for the cell adhesion assay and 1.5 x 10^4 cell/cm^2 for the cell proliferation assay. Cells were seeded into 8-well rectangle nunc cell plates on PLGA:CNF samples, and 22 mm diameter microscope cover slips as controls. Samples were incubated and were continuously stimulated (rectangular, 2 nm, $5\,\frac{V}{cm}$, 1 Hz) for 24, 72, and 120 hours under standard incubation conditions (at 5% CO_2 95% humidified air and 37°C, changing the media every other day) (Fig. 2). MTT assay was used and compared to determine cell proliferation.

Troponin-I ELISA biomarker assays were completed in conjunction with the proliferation assay. This was done to make sure that the cardiomyocytes had the appropriate protein complexes that are specific to cardiomyocyte cell characteristics and did not differentiate throughout the experiment or loose phenotypes.

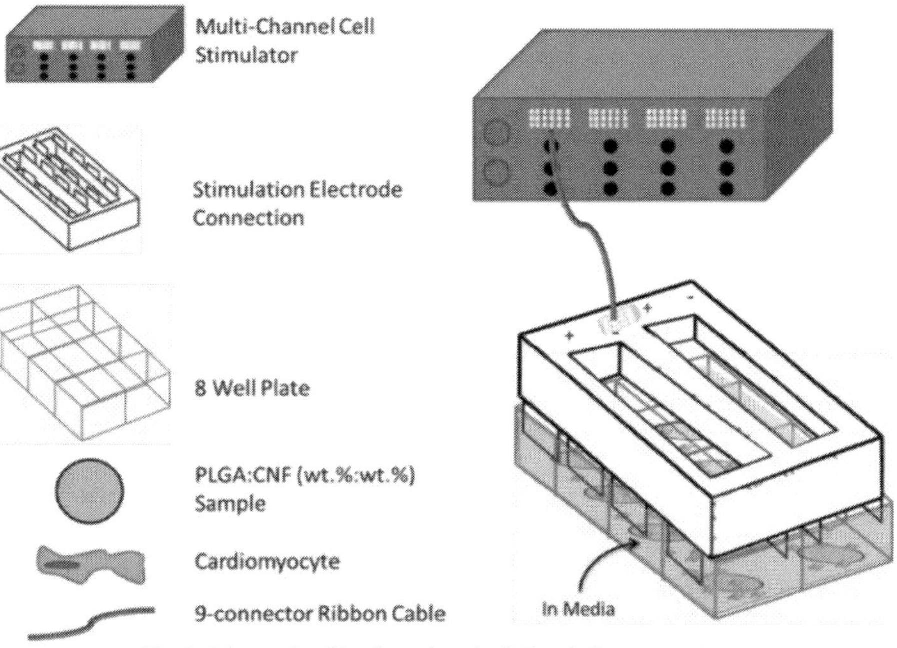

Fig. 2: Schematic of in vitro electrical stimulation apparatus.

Statistical Analysis

All proliferation assays were performed at least in triplicate and results were compared to a controlled glass surface at 24, 72, and 120 hours. Data were plotted as mean ± standard error of the mean and statistical analyses were performed. When data were compared, ANOVA software and a student T-test were used. A p-value of < 0.05 was considered to be significant.

Results

Material Properties

Scanning electron microscopy showed that the CNFs were uniformly dispersed within the PLGA material and more CNFs were observed at higher CNF ratio samples. It was shown that increasing the CNF weight ratio in PLGA increased conductivity of the sample. Comparing the same sample weight ratio with different CNF sizes, it was shown that decreasing the CNF size increased sample conductivity (Fig. 3). Also, XRD spectra obtained from the as-synthesized PLGA:CNF composites are shown in Fig. 2. As expected an extremely broad and flat peak was recorded for the PLGA matrix, confirming its amorphous nature. However, with addition of CNF in PLGA, the characteristic X-ray peak at $2\Theta = 26.5°$ evolved into a rather strong and sharp peak in PLGA:CNF composites for the 25:75 [PLGA:CNF (wt:wt)] ratio (Fig 4). The curves were fitted to the characteristic peak by Gaussian distribution. Clearly, the intensity increased as full width at half maximum (FWHM) decreased as CNFs were added to.

Complimentary evidence of the confirmation of the crystallinity was obtained using Raman spectroscopy and the results are provided in. Two sets of Raman bands are clearly visible in the composites with one set of Raman shift at 173 and 278 cm^{-1} recorded in the presence of CNF. These Raman bands are characteristic of C-C bonds [19]. The Raman bands at 1345 cm^{-1} and 1570 cm^{-1} arose particularly due to C=C bending or bonding [19]. Another observation was the slight shift of these Raman bands in higher CNF containing PLGA composites or with 100% CNF.

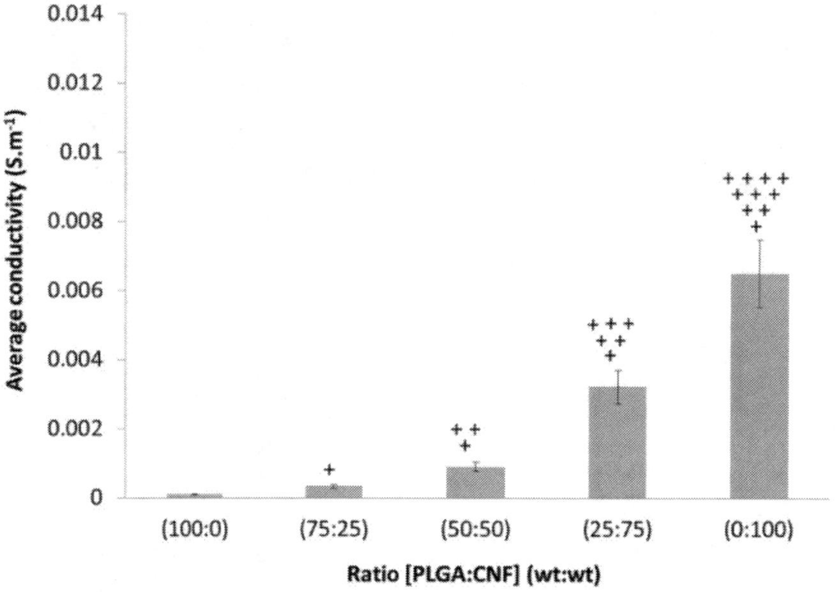

Figure 3: Comparative conductivity analysis of different PLGA:CNF (wt:wt) composite ratios with d. Data = mean ± SEM; N =3 + $p < 0.05$ compared to the 100 nm diameter CNF 100:0 (PLGA:CNF) ratio. ++ $p < 0.05$ compared to the 100 nm diameter CNF 75:25 (PLGA:CNF) ratio. +++ $p < 0.05$ compared to the 200 nm diameter CNF 50:50 (PLGA:CNF) ratio. ++++ $p < 0.05$ compared to the 100 nm diameter CNF 25:75 (PLGA:CNF) ratio.

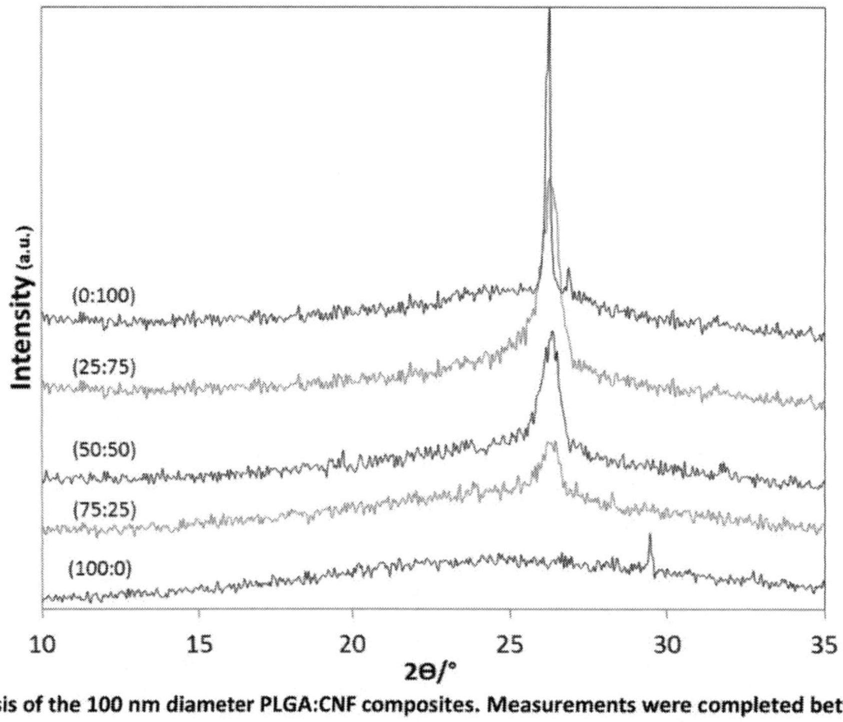

Fig. 4: XRD analysis of the 100 nm diameter PLGA:CNF composites. Measurements were completed between 10-35 degrees at a speed of 0.5 degrees per 2 minutes. Increasing the CNF weight ratio, Increased the intensity and slenderness of the peak. This is due to more crystallinity of the composites, therefore the 0:100 ratio [PLGA:CNF (wt:wt)] had the most slender and intense peak.

Cytocompatibility

Results indicated that PLGA:CNF composites promoted human cardiomyocyte proliferation better when electrically stimulated. General trends were seen between each ratio

(Fig. 5) and support published results [20]. For 100nm diameter CNFs, it was determined that the 50:50 ratio [PLGA:CNF; (wt:wt)] had the highest cardiomyocyte density all three time points, whereas the lowest density was on the 100:0 ratio [PLGA:CNF; (wt:wt)].

Troponin-I assay indicated that cardiomyocytes had a functional Troponin Complex throughout the cytocompatible assays. All CNF experiments for all time points showed a slight increase of Troponin-I indicating more cell proliferation and production. It is also good to note that Troponin-I assays presented the same trends seen in the proliferation cytocompatibility assays.

Discussion

Many materials have been developed to promote cardiomyocyte function. Recently, it has been shown that composite materials composed of different matrices (conductive vs. non-conductive and biodegradable vs. non-biodegradable) can stabilize heart tissue to decrease the consequences of a myocardial infarction and such approaches include but are not limited to: culturing patient tissue [21-30], direct cell injection [31-36], biomaterial scaffolds (materials ranging from elastin to 3D PLGA fibers) [37-43], 3D printing [44-46], and injectable scaffolds using fibrin based materials [47-49]. This present work assessed a brand new material, PLGA:CNF composites, for enhancing conductivity and cytocompatibility properties necessary for cardiovascular applications.

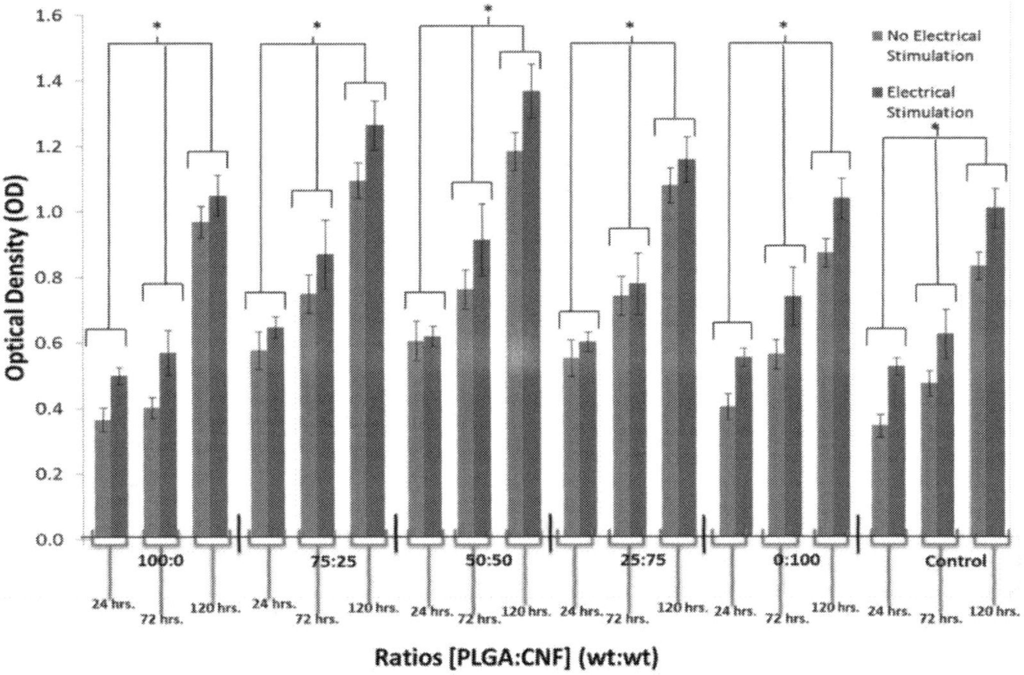

Fig. 5: Comparative electrical stimulation analysis of different PLGA:CNF (wt:wt) composite ratios with d. Data = mean ± SEM; N =3. * $p < 0.05$ compared to its non-electrically stimulated counterpart as well to other PLGA:CNF (wt:wt) ratios.

Results showed improved cardiomyocyte functions on electro-stimulated composites than those samples that were not electro-stimulated.

It is worth speculating why greater cardiomyocyte density on electro-stimulated composited were observed. Pedrotty et al. showed that numerous cardiac cell functions (including adhesion, proliferation, and migration) may be modulated by electrical stimulation [50], hence requiring the use of a conductive material in cardiac applications. Also, Mihardjo et al. demonstrated that enhanced myocardial repair following an ischemic injury could be achieved using conductive polymers, such as polypyrrole [51]. The conductivity values measured in the present work were lower than that of the heart tissue (ranging from 0.16 S/m longitudinally to 0.005 S/m transversely) [52], but future techniques (such as CNF alignment) may increase the anisotropic conductivity [53] to match that of heart tissue.

Conclusion

In this study, it was shown that adding CNFs to PLGA improved that conductivity and compatibility properties of PLGA for myocardial tissue engineering applications. Different PLGA:CNF ratios were created to improve conductivity and cytocompatibility properties with human cardiomyocytes. It was further shown that increasing CNF weight ratio increased composite conductivity. Also, when then the samples were in vitro electrically stimulated, increased cell proliferation were observed. In this manner, the results of this study suggest that CNFs should be further studied in cardiac tissue engineering applications.

Acknowledgements

The authors would like to thank the National Science Foundation Graduate Student Fellowship Program (NSF #069063) and the Hermann Foundation for funding.

References

1. M. Radisic, M.V. Sefton, Cardiac Tissue, in: Principles of Regenerative Medicine (Second edition), Academic Press, San Diego, 2011, pp. 877-909.
2. J.A. Armour, Autonomic Nervous System: Cardiovascular Control, in: Larry R. Squire (Ed.), Encyclopedia of Neuroscience, Academic Press, Oxford, 2009, pp. 849-854.
3. D. Locca, C. Bucciarelli-Ducci, G. Ferrante, A. La Manna, N.G. Keenan, A. Grasso, P. Barlis, F. Del Furia, S.K. Prasad, J.C. Kaski, D.J. Pennell, C. Di Mario, New Universal Definition of Myocardial Infarction: Applicable After Complex Percutaneous Coronary Interventions? 3 (2010) 950-958.
4. K. Thygesen, J.S. Alpert, H.D. White, Universal Definition of Myocardial Infarction 50 (2007) 2173-2195.
5. A.K. Jain, C. Knight, Myocardial infarction - new definitions and implications 38 (2010) 421-423.
6. M. Barber, T. Mueller, B. Davies, R. Gill, D. Zipes, Interruption of sympathetic and vagal-mediated afferent responses by transmural myocardial infarction, Circulation 72 (1985) 623-631.
7. E.F. Hirsch, M. Jellinik, T. Cooper, G.C. Kaiser, H.C. Barner, A.M. Borghard-Erdle, The Innervation of the Vertebrate Heart (1970).
8. Y. Oh, A.Y. Jong, D.T. Kim, H. Li, C. Wang, A. Zemljic-Harpf, R.S. Ross, M.C. Fishbein, P. Chen, L.S. Chen, Spatial distribution of nerve sprouting after myocardial infarction in mice, Heart Rhythm 3 (2006) 728-736.
9. R. Vracko, D. Thorning, R. Frederickson, Fate of nerve fibers in necrotic, healing, and healed rat marycardium, Laboratory Investigation 63 (1990) 490-501.

10. W.C. Randall, M.P. Kaye, G.R. Hageman, H.K. Jacobs, D.E. Euler, W. Wehrmacher, Cardiac dysrhythmias in the conscious dog after surgically induced autonomic imbalance, Am. J. Cardiol. 38 (1976) 178-183.
11. WRITING GROUP MEMBERS, Committee,for the American Heart Association Statistics, Stroke Statistics Subcommittee, Heart Disease and Stroke Statistics--2009 Update: A Report From the American Heart Association Statistics Committee and Stroke Statistics Subcommittee 119 (2009) e21-181.
12. S. Cohen, J. Leor, REBUILDING BROKEN HEARTS, Scientific American 291 (2004) 45-51.
13. D. Khang, J. Lu, C. Yao, K.M. Haberstroh, T.J. Webster, The role of nanometer and sub-micron surface features on vascular and bone cell adhesion on titanium 29 (2008) 970-983.
14. R.L. Price, M.C. Waid, K.M. Haberstroh, T.J. Webster, Selective bone cell adhesion on formulations containing carbon nanofibers 24 (2003) 1877-1887.
15. A. Ranella, M. Barberoglou, S. Bakogianni, C. Fotakis, E. Stratakis, Tuning cell adhesion by controlling the roughness and wettability of 3D micro/nano silicon structures 6 (2010) 2711-2720.
16. L. Yang, B.W. Sheldon, T.J. Webster, The impact of diamond nanocrystallinity on osteoblast functions 30 (2009) 3458-3465.
17. M.J. Dalby, D. Pasqui, S. Affrossman, Cell response to nano-islands produced by polymer demixing: a brief review, Nanobiotechnology, IEE Proceedings - 151 (2004) 53-61.
18. M.J. Dalby, D. Giannaras, M.O. Riehle, N. Gadegaard, S. Affrossman, A.S.G. Curtis, Rapid fibroblast adhesion to 27 nm high polymer demixed nano-topography, Biomaterials 25 (2004) 77-83.
19. S. Aslan, C.Z. Loebick, S. Kang, M. Elimelech, L.D. Pfefferle, P.R. Van Tassel, Antimicrobial biomaterials based on carbon nanotubes dispersed in poly(lactic-co-glycolic acid), Nanoscale 2010.
20. D.A. Stout, B. Basu, T.J. Webster, Cytocompatible Poly-Lactic-co-Glycolic Acid: Carbon Nanofiber Composite Analysis for Cardiovascular Applications, IEEE 2011 Northeast Bioengineering Conference 11 (2011).
21. L. Bengtsson, K. Radegran, A. Haegerstrand, In vitro endothelialization of commercially available heart valve bioprostheses with cultured adult human cells 7 (1993) 393-398.
22. C.R. Horres, J.F. Aiton, M. Lieberman, E.A. Johnson, Electrogenic transport in tissue cultured heart cells 11 (1979) 1201-1202, IN31, 1203-1205.
23. H.J. Jongsma, L. Tsjernina, J. de Bruijne, The establishment of regular beating in populations of pacemaker heart cells. A study with tissue-cultured rat heart cells 15 (1983) 123-133.
24. E.C. Martinez, T. Kofidis, Adult stem cells for cardiac tissue engineering In Press, Corrected Proof.
25. S. Ramaswamy, D. Gottlieb, G.C. Engelmayr Jr, E. Aikawa, D.E. Schmidt, D.M. Gaitan-Leon, V.L. Sales, J.E. Mayer Jr, M.S. Sacks, The role of organ level conditioning on the promotion of engineered heart valve tissue development in-vitro using mesenchymal stem cells 31 (2010) 1114-1125.
26. D. Schmidt, A. Mol, S. Neuenschwander, C. Breymann, M. GÃ¶ssi, G. Zund, M. Turina, S.P. Hoerstrup, Living patches engineered from human umbilical cord derived fibroblasts and endothelial progenitor cells 27 (2005) 795-800.
27. A. Wada, H. Yoneda, N. Shibata, Y. Inui, H. Fushimi, K. Takemura, S. Onishi, Tissue-cultured heart cells from the cardiomyopathic hamster 8 (1976) 619-620.

28. H. Wei, C. Chen, W. Lee, I. Chiu, S. Hwang, W. Lin, C. Huang, Y. Yeh, Y. Chang, H. Sung, Bioengineered cardiac patch constructed from multilayered mesenchymal stem cells for myocardial repair 29 (2008) 3547-3556.
29. C. Yang, R. Sodian, P. Fu, C. LÃ¼ders, T. Lemke, J. Du, M. HÃ¼bler, Y. Weng, R. Meyer, R. Hetzer, In Vitro Fabrication of a Tissue Engineered Human Cardiovascular Patch for Future Use in Cardiovascular Surgery 81 (2006) 57-63.
30. K. Zitta, B. Brandt, A. Wuensch, P. Meybohm, B. Bein, M. Steinfath, J. Scholz, M. Albrecht, Interleukin-1[beta] regulates cell proliferation and activity of extracellular matrix remodelling enzymes in cultured primary pig heart cells 399 (2010) 542-547.
31. N. Borenstein, V. Chetboul, P. Bruneval, M. Hekmati, R. Tissier, L. Behr, G. Derumeaux, D. Montarras, Non-cultured cell transplantation in an ovine model of non-ischemic heart failure 31 (2007) 444-451.
32. C. Chen, Y. Chang, C. Wang, C. Huang, C. Huang, Y. Yeh, S. Hwang, H. Sung, Construction and characterization of fragmented mesenchymal-stem-cell sheets for intramuscular injection 28 (2007) 4643-4651.
33. S.C. Li, L. Wang, H. Jiang, J. Acevedo, A.C. Chang, W.G. Loudon, Stem cell engineering for treatment of heart diseases: Potentials and challenges 33 (2009) 255-267.
34. E.C. Perin, H.F. Dohmann, R. Borejevic, S.A. Silva, A.L. Sousa, G.V. Silva, J.A. Assad, L. Belem, A.C. Carvalho, R. Esporcatte, W.K. Vaughn, H.J. Dohmann, J.T. Willerson, 814-2 Time course of improvement following stem cell injections in humans with heart failure 43 (2004) A187-A187.
35. Y. Shintani, S. Fukushima, A. Varela-Carver, J. Lee, S.R. Coppen, K. Takahashi, S.W. Brouilette, K. Yashiro, C.M.N. Terracciano, M.H. Yacoub, K. Suzuki, Donor cell-type specific paracrine effects of cell transplantation for post-infarction heart failure 47 (2009) 288-295.
36. L. Zhang, T.J. Webster, Nanotechnology and nanomaterials: Promises for improved tissue regeneration, Nano Today 4 (2009) 66-80.
37. A. Callegari, S. Bollini, L. Iop, A. Chiavegato, G. Torregrossa, M. Pozzobon, G. Gerosa, P. De Coppi, N. Elvassore, S. Sartore, Neovascularization induced by porous collagen scaffold implanted on intact and cryoinjured rat hearts 28 (2007) 5449-5461.
38. J. Hu, X. Sun, H. Ma, C. Xie, Y.E. Chen, P.X. Ma, Porous nanofibrous PLLA scaffolds for vascular tissue engineering 31 (2010) 7971-7977.
39. Q. Lu, K. Ganesan, D.T. Simionescu, N.R. Vyavahare, Novel porous aortic elastin and collagen scaffolds for tissue engineering 25 (2004) 5227-5237.
40. M. Siepe, M. Giraud, M. Pavlovic, C. Receputo, F. Beyersdorf, P. MenaschÃ©, T. Carrel, H.T. Tevaearai, Myoblast-seeded biodegradable scaffolds to prevent post-myocardial infarction evolution toward heart failure 132 (2006) 124-131.
41. P.M. Taylor, A.E.G. Cass, M.H. Yacoub, Extracellular matrix scaffolds for tissue engineering heart valves 21 (2006) 219-225.
42. Y. Wang, L.B. Wong, H. Mao, Creation of a long-lifespan ciliated epithelial tissue structure using a 3D collagen scaffold 31 (2010) 848-853.
43. W.Y. Yeong, N. Sudarmadji, H.Y. Yu, C.K. Chua, K.F. Leong, S.S. Venkatraman, Y.C.F. Boey, L.P. Tan, Porous polycaprolactone scaffold for cardiac tissue engineering fabricated by selective laser sintering 6 (2010) 2028-2034.
44. X. Cui, T. Boland, Human microvasculature fabrication using thermal inkjet printing technology 30 (2009) 6221-6227.

45. V. Mironov, T. Boland, T. Trusk, G. Forgacs, R.R. Markwald, Organ printing: computer-aided jet-based 3D tissue engineering 21 (2003) 157-161.
46. T. Xu, C.A. Gregory, P. Molnar, X. Cui, S. Jalota, S.B. Bhaduri, T. Boland, Viability and electrophysiology of neural cell structures generated by the inkjet printing method 27 (2006) 3580-3588.
47. G. Morales, M.I. Barrena, J.M.G.d. Salazar, C. Merino, D. RodrÃguez, Conductive CNF-reinforced hybrid composites by injection moulding 92 (2010) 1416-1422.
48. K.L. Christman, A.J. Vardanian, Q. Fang, R.E. Sievers, H.H. Fok, R.J. Lee, Injectable Fibrin Scaffold Improves Cell Transplant Survival, Reduces Infarct Expansion, and Induces Neovasculature Formation in Ischemic Myocardium 44 (2004) 654-660.
49. H. Wang, J. Zhou, Z. Liu, C. Wang, Injectable cardiac tissue engineering for the treatment of myocardial infarction, Journal of Cellular and Molecular Medicine 14 (2010) 1044-1055.
50. D.M. Pedrotty, J. Koh, B.H. Davis, D.A. Taylor, P. Wolf, L.E. Niklason, Engineering skeletal myoblasts: roles of three-dimensional culture and electrical stimulation 288 (2005) H1620-1626.
51. S.S. Mihardja, R.E. Sievers, R.J. Lee, The effect of polypyrrole on arteriogenesis in an acute rat infarct model, Biomaterials 29 (2008) 4205-4210.
52. K.C. Roberts-Thomson, P.M. Kistler, P. Sanders, J.B. Morton, H.M. Haqqani, I. Stevenson, J.K. Vohra, P.B. Sparks, J.M. Kalman, Fractionated atrial electrograms during sinus rhythm: Relationship to age, voltage, and conduction velocity 6 (2009) 587-591.
53. S. Bal, Experimental study of mechanical and electrical properties of carbon nanofiber/epoxy composites 31 (2010) 2406-2413.

(746c) Sensitive Detection of Circulating Tumor Cells by Graphene Oxide Nanoassembly

Hyeun Joong Yoon, Trinh M. Pham, Kilho Lee and Sunitha Nagrath, Department of Chemical Engineering, University of Michigan, Ann Arbor, MI

Technological obstacles in isolation and capture of circulating tumor cells (CTCS) from blood continue to stymie progress in quantitating CTCs for early diagnosis and effective monitoring of therapeutic response in patients. CTCs are extremely rare cells (about 1 tumor cell per 1 billion blood cells) that are known to circulate in the peripheral blood of patients with metastatic cancer. Nagrath et al. demonstrated the efficient and reproducible microfluidic chip, 'the CTC chip' to isolate CTCs from the blood of metastatic patients with common epithelial tumors in previous work. Here we have fabricated functionalized graphene oxide nanoassemblies on the patterned gold layer through a microfluidic silicon device as an effective tool to isolate, capture, identify and characterize extremely rare CTCs aimed at the early diagnosis of metastatic disease. Multifunctional biocompatible nanographene oxides with polyethyleneglycol-amine (PEG-NH_2) of various physical sizes have been used to selectively capture cancer cells with high sensitivity. Graphene itself has potential applications in advanced nanoelectronics, sensors, membranes and composites. The presented application of graphene materials will be a promising new tool in biomedical translational research.

See more of this Session: Nanotechnology and Nanobiotechnology for Sensors
See more of this Group/Topical: Nanoscale Science and Engineering Forum

(665d) Homo- and Hetero-Epitaxial Growth On Nanowire Substrates

Chandrashekhar Pendyala[1], Jeong H. Kim[2], Jacek Jasinski[3] and Mahendra K. Sunkara[1], (1)Chemical Engineering, University of Louisville, Louisville, KY, (2)Chemical Engineeering, University of Louisville, Louisville, KY, (3)Conn Center for Renewable Energy Research, University of Louisville, Louisville, KY

Epitaxial growth on planar substrates has been pursued for creating single crystal layers of new materials. In many circumstances, hetero-epitaxy is employed due to the limited availability of native substrates. In many important materials systems, such heteroepitaxy onto planar substrates leads to phase segregation and misfit dislocations due to lattice mismatch-induced stresses and strain. On the other hand, it has been proposed that the epitaxial growth on nanowire substrates could alleviate problems associated with planar substrates due to the lower interfacial contact area and stress relaxation within nanowire substrates.

In this presentation, we will present the fundamental studies conducted to follow the growth dynamics on nanowire substrates. In particular, the case of ternary alloy ($In_xGa_{1-x}N$) synthesis on to GaN and ZnO nanowire substrates will be discussed. Ternary $In_xGa_{1-x}N$ alloys are promising materials for a variety of opto-electronic devices, photovoltaics and photoelectrochemical systems owing to composition dependent tunability of the band gap from ultraviolet to near infrared. Specifically, the $In_xGa_{1-x}N$ alloys with indium composition from 45-65 % can have the right band gap between 1.7-2.2 eV necessary for photoelectrochemical water splitting applications. We also present a discussion on the mechanism of the unique growth modes observed for both ternary alloys ($In_xGa_{1-x}N$) and GaN growth on "a" and "c" plane oriented GaN nanowires.

See more of this Session: Nanowires I: Synthesis and Modeling
See more of this Group/Topical: Nanoscale Science and Engineering Forum

(665c) Post-Synthesis Decomposition of III-Nitride Nanowires Into Quantum Wires

Lance Brockway[1], Chandrashekhar Pendyala[2], Mahendra K. Sunkara[2] and Sreeram Vaddiraju[1], (1) Department of Chemical Engineering, Texas A&M University, College Station, TX, (2)Chemical Engineering, University of Louisville, Louisville, KY

A post-synthesis strategy is employed for shrinking the diameters of group III-nitride nanowires. This simple strategy involved the decomposition of pre-synthesized III-nitride nanowires at elevated temperatures in the presence of ammonia. Experiments performed using the compound semiconductors gallium nitride (GaN) and indium nitride (InN) as examples indicated that the decomposition is dependent on the initial diameter of the nanowires. The decomposition of GaN nanowires with diameters greater than 100 nm resulted in polycrystalline powder, while GaN nanowires with diameters of 20 nm underwent layer-by-layer decomposition to result in quantum wires with average diameters less than 5 nm. It is expected that this layer-by-layer decomposition in thin GaN nanowires is a direct result of the ballistic diffusion of gallium adatoms along the length of the nanowires. Similar results were also observed in the case of InN nanowires. These diameter- and growth direction-dependant decomposition phenomena along with the kinetics of decomposition will be discussed.

See more of this Session: Nanowires I: Synthesis and Modeling
See more of this Group/Topical: Nanoscale Science and Engineering Forum

SELF-ASSEMBLED NANOSTRUCTURES FOR BIOMEDICAL APPLICATIONS

Yupeng Chen
Department of Orthopaedics and School of Engineering, Brown University, Providence, RI

Introduction

Biomimetic strategy has been widely studied to build advanced biological materials for tissue engineering and related biomedical applications. Such strategy usually includes two aspects: structure and process. As many have recognized, natural tissues in a biological system are not simplex. Structurally, they are composed of highly-ordered hierarchical units (down to nanometer scale). In biological process, basic units self-assemble to form functional tissue structures. Therefore, to better mimic natural tissues for improved biological functions, self-assembled nanostructures are developed nowadays.

With the ordered self-assembly processes to generate bio-functional structures, these nanomaterials have unique advantages in biological uses. For example, they can form biomimetic scaffolds under physiological conditions for tissue engineering. Moreover, self-assembly of molecules often leads to an increase of viscosity or solidification of such materials. Therefore, they are suitable for injection applications in tissue healing. Additionally, drugs can be encapsulated with self-assembly of materials, so they are excellent candidates for controlled drug delivery. Here, we will review a variety of self-assembled nanomaterials including inorganic materials, polymers and biological-inspired molecules. Especially, we will demonstrate and emphasize the properties and biomedical applications of a type of most recently investigated nanostructures, self-assembled rosette nanotubes.

Self-Assembled Rosette Nanotubes (RNTs) for Biomedicine

Properties of RNTs

Based on nanotechnology, a promising biomaterial emerged recently is bio-inspired, self-assembled RNTs. Such materials have compatible biological properties to collagens, but are easy to synthesize and engineer for specific tissue engineering needs. RNTs (Figure 1) are novel biomimetic self-assembled supramolecular structures, whose basic building blocks are guanine (G) and cytosine (C) DNA base-pairs which can solidify into a viscous gel [1~3]. The G^C heteroaromatic bicyclic base possesses the Waston-Crick donor-donor-acceptor properties of guanine and the acceptor-acceptor-donor of cytosine. G^C undergoes a hierarchical self-assembly process under physiological conditions to form a six-membered supermacrocycle by the formation of 18 hydrogen bonds. Because of electrostatic forces, base stacking interactions and hydrophobic effects, the rosettes form a stable stack with an inner channel 11Å in diameter. An amino acid side chain (lysine) was chosen to impose chirality and surface chemistry on the RNTs. Importantly, the lysine side chain (with an amine group and a carboxyl group) provide the possibility to functionalize a variety of drugs onto RNTs (for example, to modify peptide growth factors through peptide bonds). In addition, some small bioactive molecules could be trapped in the inner channel of RNTs by hydrogen bonds. This is another choice for prolonged drug delivery using RNTs. In addition, two drug carrier sites of RNTs could have different release kinetics to satisfy specific requirements.

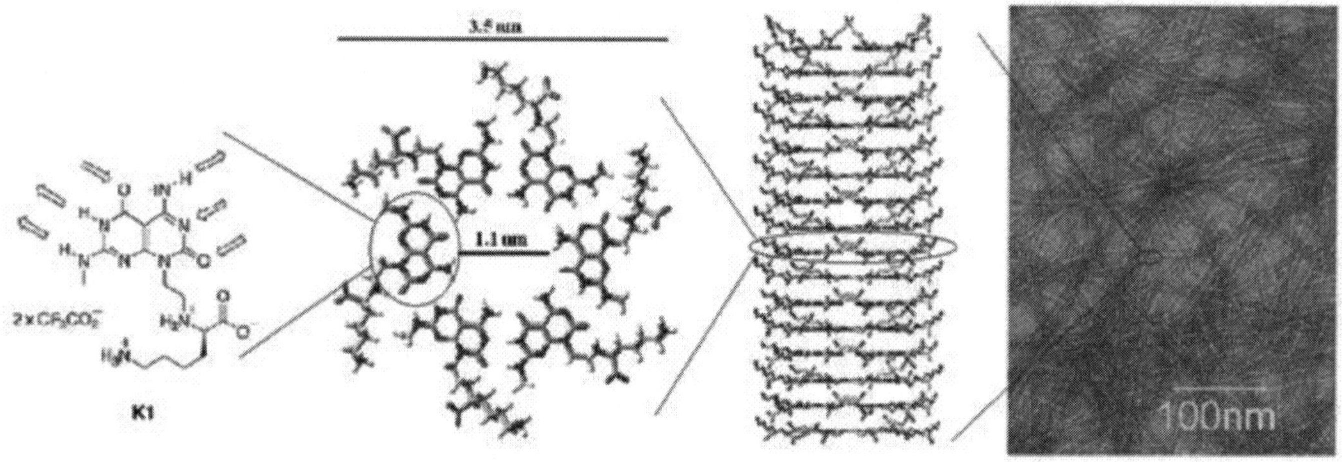

Figure 1. RNTs (with a lysine side chain) undergo spontaneous self-assembly under physiological conditions. (Adapted and redrawn from [1~2])

Thus, because RNTs can: 1) increase viscosity when heated to low temperatures (40~60°C); 2) be easily functionalized with a high-density of peptides and drugs; 3) mimic the nanostructure of natural cartilage, bone and the transitional zone between them; 4) create a surface environment which improves protein (like collagen and fibronectin) adsorption as well as enhance cell adhesion and subsequent functions; 5) be injected into cartilage defect sites and in situ repair damaged tissue; and 6) biodegrade with the recovery of natural tissue, RNTs are predicted to serve as a novel, multifunctional implant material to regenerate bone and cartilage. Previous studies have found that RNTs increased various cell (i.e., osteoblasts and endothelial cells) attachment and subsequent functions [4~5].

In addition, *in vitro* and *in vivo* toxicity studies have showed that RNTs leaded lower inflammatory response than carbon nanotubes at the same dose [6]. Compared with other nanotubes, like carbon nanotubes, RNTs showed low cyto-toxicity. Specifically, 25μg dose of RNTs had low or no inflammatory markers in lung of mice compared with significant inflammation of 10μg dose of single wall carbon nanotubes in lung of mice. Importantly, the inflammatory responses induced by a 50μg of RNTs resolved after 7 days. However, the inflammatory responses caused by single wall carbon nanotubes were still observed after 60 days. The reason may because carbon nanotubes mostly formed by carbon-carbon covalent bonds, which are difficult to degrade, but RNTs formed by hydrogel bonds and super-molecular interactions, which are relatively easy to dis-assemble. In addition, RNTs presented low-toxicity, because of the metal free synthesis procedure and water-soluble property, since trace metal catalyst residues (like Fe, Ni, Y, or Co) can generate iron-catalyzed free-radicals which contribute to oxidative stress and cyto-toxicity [6].

Previous in vivo studies have allowed mice to breath in RNTs into the lung and found inflammatory responses with RNT concentration equal or higher than 0.05 mg/mL [7]. A corresponding in vitro study confirmed such results to show no or low inflammatory activation on epithelial cells [8]. Especially, a lysine control was added in those experiments. Results demonstrated that lysine is not the key factor for the promotion of inflammatory. Thus, the physical structure of materials and bolus nature of the exposure could be the main causation of aggressive inflammatory responses.

Applications of RNTs

Because of the self-assembly of RNTs mimics the natural biological process of some important biomolecules (like collagens), one of biomedical uses of RNTs is their bioactivity to enhance cell adhesion and functions. For instance, since bone is formed with the assembly of hydroxyapatite (HA) and collagens, RNTs have the potential to serve as a biomimetic template for HA deposition in natural bone tissue [9]. Figure 2 (A) showed that HA crystals rapidly formed on RNTs when mixed with $CaCl_2$ and Na_2HPO_4 solutions for 2 minutes and were regularly aligned on RNTs in a pattern similar to the HA/collagen pattern in bone. Figure 2 (B) shows randomly agglomerated HA on grids without RNTs. Because the deposition, nucleation and growth of HA crystals in pores of an extracellular collagen matrix are important steps during bone remodeling *in vivo*, it can be anticipated that the RNT/HA composites formulated in this study not only mimic the nanostructure of bone but also resemble the HA/collagen self-assembly process to promote natural mineralization. Studies have found that such nanoHA/RNT structures increased osteoblast adhesion and subsequent functions [9]. Therefore, RNTs were proposed as promising scaffolds for tissue engineering.

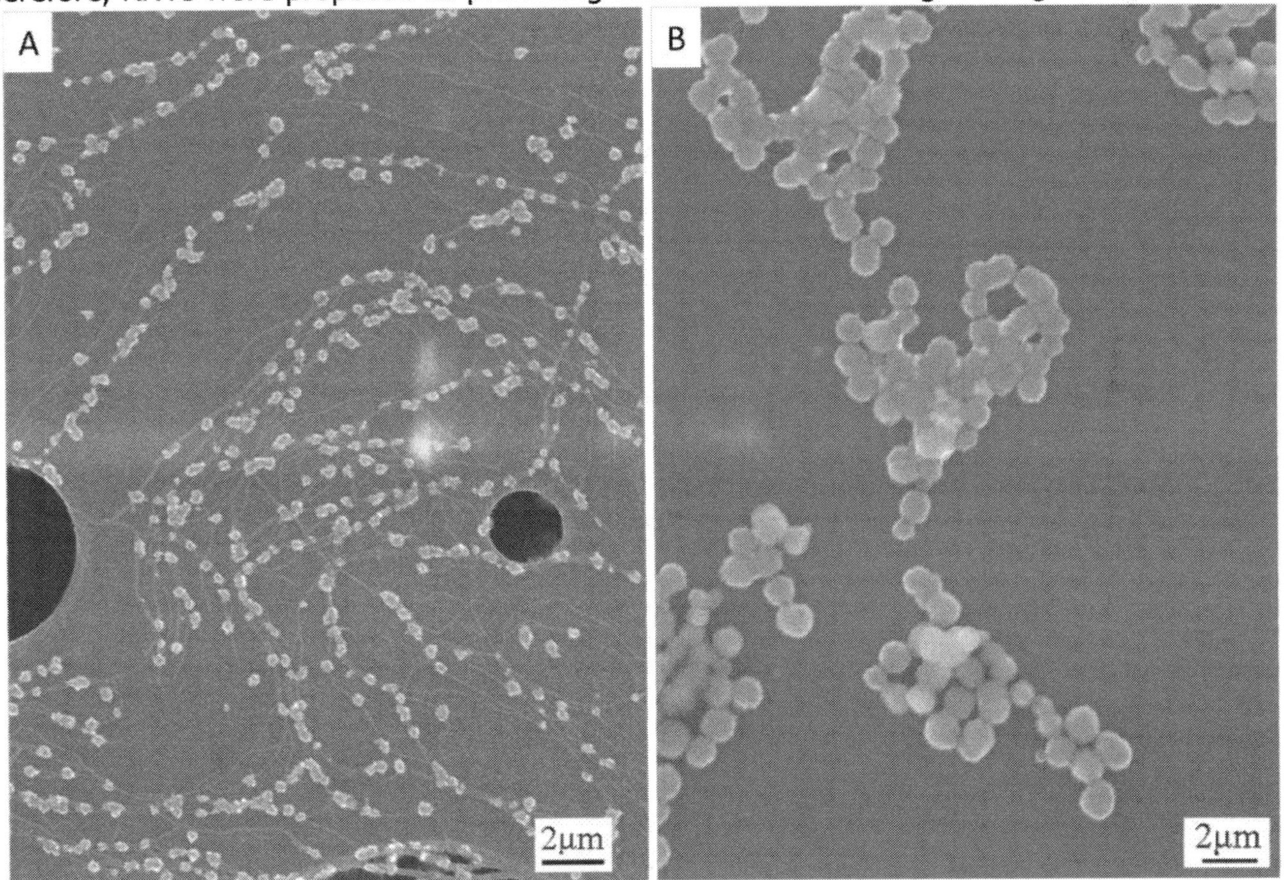

Figure 2. SEM (scanning electron microscopy) images of (A) biomimetic HA with 0.1 mg/ml RNT; and (B) only HA.

Other than biological interactions, one critical property for tissue engineering applications of RNTs is their ability to bond to the host tissue, potentially serving as an adhesive between severed tissues to initiate a healing process. Although RNTs are not mechanically as strong as nature bone, it was reasoned that in combination with other materials (for example, polymers) their "stickiness" may be improved. Previous studies provided the evidence that PEG (polyethylene glycol)/RNT composites dissolved in serum supplemented media resulted in higher adhesive strength to collagen casings compared to PEG alone placed in serum supplemented medium or RNTs alone placed in serum free medium. Such results suggested that RNTs may bond strongly to some proteins in serum to aid in bridging severed tissue which is in agreement with an earlier report showing that RNTs enhanced osteoblast (bone forming cell) adhesion and long-term functions [10] by improving initial fibronectin adsorption from serum. To further understand the properties of RNTs that may have improved bonding strength to collagen casings, contact angle studies were conducted. As suggested by the decreased contact angles of the PEG/RNT composite on collagen casing surfaces, this improvement in adhesive strength may be due to an improved infiltration into the collagen casings of the hydrogel in the presence of the RNTs. Coupled with data from previous studies [10], RNTs may have promoted collagen (a major constituent in cartilage) adsorption.

As shown in Figure 3, RNT nanotubular structure is quite similar to collagen. Since collagen is the major organic component in connective tissues (like bone and cartilage), RNTs, combined with the chemical affinity with collagen and structural similarity to collagen, could be a good substitute for collagens in engineering bone and cartilage for tissue bonding, cell attachment, physical support, nutrition exchange and tissue regeneration.

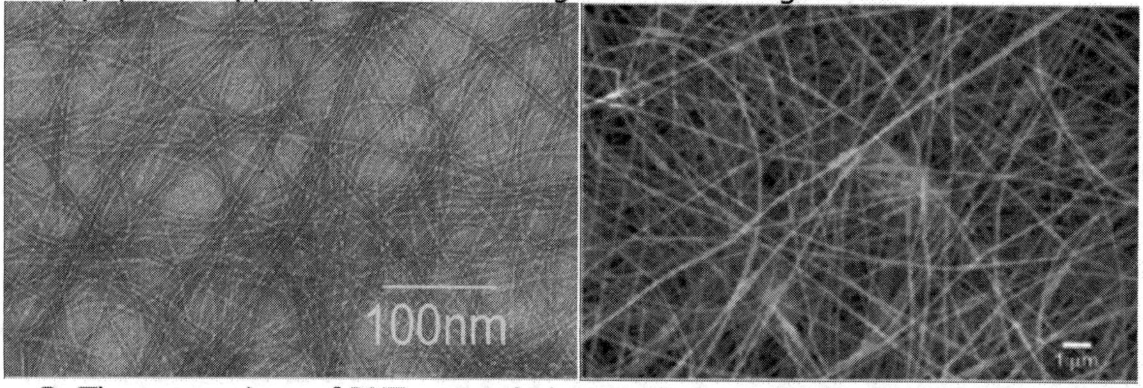

Figure 3. The comparison of RNT nanotubular structure and collagen type I. (The collagen SEM scan adapted from http://www.people.vcu.edu/~glbowlin/research.html)

Another important application of RNTs is for advanced drug delivery. According to the unique structure of RNTs, there were two strategies for drug delivery: on one hand, RNTs were consisting of guanine (G) and cytosine (C) base pairs. Each RNT unit has a lysine side chain to maintain their stability and chirality, so peptide based drugs (like BMP-7 short peptides [11]) can be linked with such lysine side chains by peptide bonds. For example, hydrogels coated with RGDSK (Arg-Gly-Asp-Ser-Lys) modified RNTs have been showed to improve osteoblast functions compared to unmodified RNT hydrogels [5].

On the other hand, RNTs could also be incorporated with drugs through super-molecular interactions (by physical adsorption without chemical modification). The self-assembly of RNTs is governed by hydrophobic interactions, π-π electron interactions and base stacking effects (an effect caused by the conjugated and high density π electrons in their aromatic rings) between base pairs which creates a hydrophobic core and hydrophilic outer surface. Importantly, because of such unique properties, RNTs are able to incorporate small molecular water-insoluble drugs rich of п electrons (like DEX (dexamethasone)) into their tubular structures by hydrophobic interactions, п-п electron interactions or base stacking effects. Thus, unlike other drug carriers, the hydrophilic outer surface of RNTs can suspend such hydrophobic drugs in physiological environment for tissue regeneration. For example, DEX is a water-insoluble drug, so DEX had no UV-vis absorbance in water. However, after DEX incorporated with RNTs, DEX-RNTs were able to suspend in water and present a different absorbance curve compared with only RNTs (Figure 4) [12]. Such results demonstrated that RNTs can be successfully incorporate dwith DEX and suspend hydrophobic DEX in a water environment. Not only DEX, a similar study showed that hydrophobic tamoxifen can also be incorporated with RNTs in physiological environment [13]. More importantly, drug activity experiments demonstrated that drugs released from RNTs were still bioactive [12].

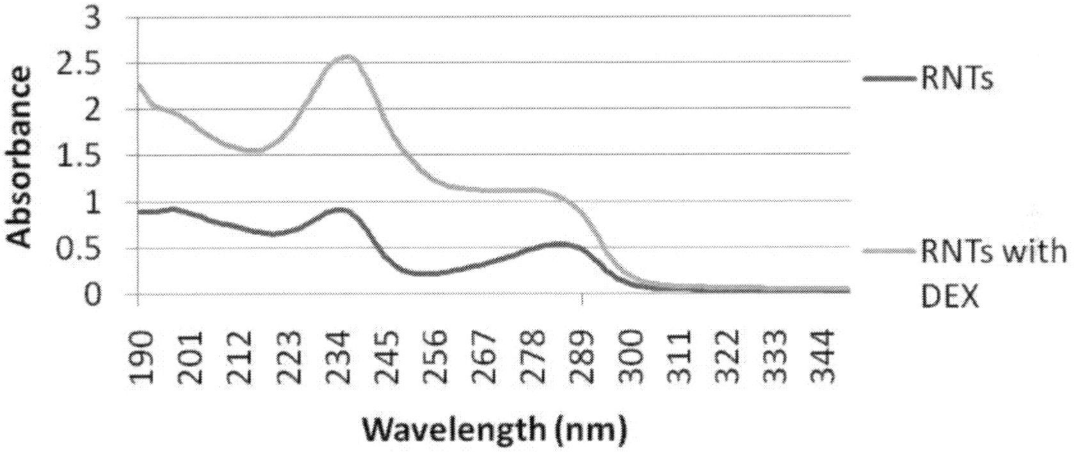

Figure 4. Comparison of UV-vis absorbance of RNTs and DEX-RNTs in water. (Adapted from [12])

Other Self-Assembled Structures for Biomedicine

Self-Assembled Peptides/Proteins

Similar to the self-assembly of DNA, peptides/proteins can also self-assemble in biological process. Therefore, another most common studied self-assembled nanostructure has also been inspired from proteins and peptides. These nanostructured biomaterials are genetically selected/designed peptides with specific binding to functional solids, tailoring their binding and self-assembly characteristics, developing bifunctional peptide/protein genetic constructs with both material binding and biological activity, and using these as molecular synthesizers, erectors and assemblers [14]. PuraMatrix™ hydrogel peptide is a new synthetic 16-amino-acid peptide developed by 3DM (Cambridge, Massachusetts, USA). When it is exposed to physiological salt conditions, the peptide forms a hydrogel because of ionic bonds and hydrophobic interactions. It was stated that the nano-fiber structure resembled natural collagen with fiber diameters of 5-10nm and promoted cell (like osteoblasts and fibroblasts) proliferation greater than any other synthetic scaffold, such as PGA (polyglycolic acid).

Self-Assembled Polymers

Since polymers are relatively easy to be synthesized and modified, many polymers (especially engineered co-polymers) were developed with self-assembly properties. For example, a double hydrophilic copolymer with one PEG block and one beta-cyclodextrin (beta-CD) flanking block (PEG-b-PCDs) were synthesized through the post-modification of macromolecules. The self-assembly of PEG-b-PCDs in aqueous solutions enabled the encapsulation of hydrophobic drugs for delivery purpose [15].

Self-Assembled Inorganic Materials

Self-assembly of inorganic materials were discovered for many years. However, in recent studies, more and more researchers were focusing the biological applications of self-assembled inorganic materials, like quantum dots (QDs). A study reported the potential of luminescent semiconductor QDs to enable development of hybrid inorganic-bioreceptor sensing materials and build biosensors [16].

Conclusions

With the understanding and development of biomimetic self-assembly process, more and more nanostructures will be investigated. They may have unique biological, chemical or physical properties. Especially, their self-assembly property similar to natural biomolecules and tissues will enlighten numerous applications in biomedical field.

References:

1. Chun AL, Moralez JG, Fenniri H, Webster TJ. Helical rosette nanotubes: a biomimetic coating for orthopedics? Biomaterials, 26: 7304-7309, 2005
2. Chun AL, Moralez JG, Fenniri H, Webster TJ. Helical rosette nanotubes: A more effective. orthopaedic implant material. Nanotechnology 15: s234-s239, 2004.
3. Fenniri H, mathivanan P, Vidale KL, Sherman DM, Hallenga K, wood KV and Stowell JG. Helical rosette nanotubes: design, self-assembly and characterization. J. Am. Chem. Soc. 123, 3854-3855, 2001.
4. Fine E, Zhang L, Fenniri h, Webster TJ. Enhanced endothelial cell functions on rosette nanotube-coated titanium vascular stents. Int J Nanomedicine. 4, 91~97, 2009.
5. Zhang L, Rakotondradany F, Myles AJ, Hicham Fenniri, Thomas Webster. 2009. Arginine-glycine-aspartic acid modified rosette nanotube-hydrogel composites for bone tissue engineering. Biomaterials, 30(7):1309~1320.
6. Journeay WS, Suri SS, Moralez JG, Fenniri H, Singh B. Rosette nanotubes show low acute pulmonary toxicity in vivo. Int J Nanomedicine. 3(3):373-383, 2008.
7. Lam CW, James JT, McCluskey R, Arepalli S, Hunter RL. A review of carbon nanotube toxicity and assessment of potential occupational and environmental health risks. Crit Rev Toxicol. 36(3):189-217, 2006.
8. Journeay WS, Suri SS, Moralez JG, Fenniri H, Singh B. Low inflammatory activation by self-assembling Rosette nanotubes in human Calu-3 pulmonary epithelial cells. Small. 4(6):817-823, 2008.

9. Zhang L, Chen Y, Fenniri H and Webster TJ, "Biomimetic Helical Rosette Nanotube and Nanocrystalline Hydroxyapatite Coatings on Titanium for Improving Orthopaedic Implants", International Journal of Nanomedicine, 2008;3(3):323-333.
10. Chen Y, Bilgen B, Pareta RA, Fenniri H, Ciombor DM, Aaron R and Webster TJ, "Self-assembled Rosette Nanotube/Hydrogel Composites for Cartilage Tissue Engineering", Tissue Engineering, Tissue Eng Part C Methods. 2010 Dec;16(6):1233-43.
11. Chen Y and Webster TJ, "Increased Osteoblast Functions in the Presence of BMP-7 Short Peptides for Nanostructured Biomaterial Applications", Journal of Biomedical Materials Research: Part A, 2009 91(1):296-304.
12. Chen Y, Song S, Yan Z, Fenniri H, Webster TJ., "Self-assembled Rosette Nanotubes Encapsulate and Slowly Release Dexamethasone", Int J Nanomedicine. 2011;6:1035-44.
13. Song S, Chen Y, Yan Z, Fenniri H, Webster TJ.," Self-assembled rosette nanotubes for incorporating hydrophobic drugs in physiological environments", Int J Nanomedicine. 2011 Jan 10;6:101-7.
14. Tamerler C, Sarikaya M. Molecular biomimetics: nanotechnology and bionanotechnology using genetically engineered peptides. Philos Transact A Math Phys Eng Sci. 13;367(1894):1705~1726, 2009.
15. Zhang J, Ellsworth K, Ma PX. Hydrophobic pharmaceuticals mediated self-assembly of beta-cyclodextrin containing hydrophilic copolymers: novel chemical responsive nano-vehicles for drug delivery. J Control Release. 2010 Jul 14;145(2):116-23.
16. Medintz IL, Clapp AR, Mattoussi H, Goldman ER, Fisher B, Mauro JM. Self-assembled nanoscale biosensors based on quantum dot FRET donors. Nat Mater. 2003 Sep;2(9):630-8.

(248b) Direct Stem Cell Differentiation Using Nanomaterials

Huinan Liu, Bioengineering, University of California at Riverside, Riverside, CA

Nanomaterials provide unique advantages in controlling stem cell functions due to their biomimetic characteristics and special biological and mechanical properties. Controlling adhesion and differentiation of stem cells is critical for tissue regeneration. This *in vitro* study investigated the effects of nanophase hydroxyapatite (nano-HA), nano-HA/polymer composites and a bone morphogenetic protein (BMP-7) derived short peptide on osteogenic differentiation of human mesenchymal stem cells (hMSCs). Poly (lactide-co-glycolide) (PLGA) was used as a model polymer in this study. The peptide was attached onto nano-HA, incorporated into nano-HA/PLGA composite and PLGA control, or directly injected into the culture media. The results showed that the nano-HA/PLGA composites promoted the hMSC adhesion as compared to the PLGA controls. Importantly, nano-HA and nano-HA/PLGA composites promoted osteogenic differentiation of hMSCs, comparable to the direct injection of the peptide into the culture media. The applications of nano-HA and nano-HA in polymer composites provide promising alternatives in controlling the adhesion and differentiation of hMSCs. Nanocomposites and nanophase ceramics should be further studied to understand their effects on mesenchymal stem cell functions and bone remodeling *in vivo*, eventually translating to clinical applications.

See more of this Session: Nanostructured Scaffolds for Tissue Engineering
See more of this Group/Topical: Nanoscale Science and Engineering Forum

Development of a Biomimetic 3D Nanostructured Tissue Engineered Bone Construct

Lijie Grace Zhang

Department of Mechanical and Aerospace Engineering and Institute for Biomedical Engineering, The George Washington University, 801 22nd Street, NW, 726 Phillips Hall, Washington DC 20052
Email: lgzhang@gwu.edu; Phone:202-994-2479

The objective of this study is to design a novel 3D nanostructured bone construct based on magnetically treated single-walled carbon nanotubes (SWCNTs), biomimetic nanocrystalline hydroxyapatites (nHA), and a biocompatible natural hydrogel (chitosan). Both of the SWCNTs and nHA have biomimetic nanostructure, excellent osteoconductivity and high potential to improve load-bearing capacity of hydrogels. Specifically, in this study the SWCNTs were produced via an arc plasma method under an applied magnetic field. These magnetically treated SWCNTs have a much smaller diameter and are twice as long as their non-magnetically prepared counterparts. Furthermore, 3D porous chitosan scaffolds with different concentrations of nHA and SWCNTs prepared with and without the magnetic field were created to support the growth of human osteoblasts (bone forming cells) and mesenchymal stem cells. Our study provided the first evidence that the chitosan nanocomposite with both magnetically treated SWCNTs and 20% nHA can achieve the highest osteoblast density when compared to all of the other experimental groups, thus making them promising for further exploration.

(125c) Meeting Sustainability Challenges In the Chemical Industry

Carol English, SHE Services, Cytec Industries, Inc, West Paterson, NJ

A variety of drivers have been identified for integrating sustainability into the chemical industry including demand by customers, new legislative initiatives, continuous focus on cost reduction and increased resource efficiency, enhanced reputation and alignment of company core values to name a few. While sustainability is regarded as an important issue, in many organizations it is slow to reach the core business strategy and to be implemented across the entire organization. The challenges evolve around making this integration visible and achieving the necessary engagement and implementation within the organization to obtain the business value from following sustainable practices.

Cytec Industries has been focusing on integrating sustainability into business practices and R&D and operational activities for some time. The emphasis has been on developing organizational support across all functions while developing tools that bring awareness and practice to key users in the development and manufacture of products. The discussion will cover the progress of this effort to integrate sustainability into the Cytec business process, and the development of key tools and in-house support for these initiatives.

See more of this Session: Sustainability Plenary
See more of this Group/Topical: Sustainable Engineering Forum

(693a) Graphene Oxide Composites

Christopher W. Macosko, Department of Chemical Engineering and Materials Science, University of Minnesota, Minneapolis, MN

Graphene has emerged as a subject of enormous scientific interest due to its exceptional electron transport, mechanical properties and high surface area. When incorporated appropriately, these atomically thin carbon sheets can significantly improve physical properties of host polymers at extremely small loading. We have made graphene by thermal reduction from graphite oxide and incorporated it into several polymers by both solvent- and melt-based strategies. We characterize dispersion of the graphene by TEM, melt rheology and conductivity percolation as well as measure electrical, thermal, mechanical and gas barrier properties of the graphene/polymer nanocomposites. We will discuss current challenges associated with processing and scalability of graphene composites, and future perspectives for this new class of nanocomposites.

See more of this Session: Graphene and Carbon Nanotubes: Characterization, Functionalization, and Dispersion
See more of this Group/Topical: Nanoscale Science and Engineering Forum

NANOMATERIALS: THE LATEST EMERGING ENVIRONMENTAL CONTAMINANT IDENTIFICATION AND MEASUREMENT

Presented by:
Katrina E. Varner
US Environmental Protection Agency/Office of Research & Development
Las Vegas, NV, USA

BIOGRAPHY:

My primary research focus as an environmental research scientist is method development and evaluation of parameters monitoring for contaminants and hazardous waste. Method and monitoring development involve the detection, speciation, and identification of nanomaterials in various media. This includes the development of screening techniques for groundwater and soil. My goal is to advance analytical techniques for the characterization and remediation of nanomaterials in various media including water, soil, and tissue to estimate the daily exposure of animals and humans to these materials.

Introduction

The basic research presented is a novel tool for the detection, characterization, quantification, and monitoring of nanomaterials in the environment. EPA is becoming more proactive concerning environmental issues. Emerging contaminants is an area EPA is investigating to determine if new products introduced by man are harmful to people or the environment. Nanomaterials are the latest emerging chemical group. These materials are being incorporated into commercial products at a faster rate than the development of knowledge and regulations to mitigate potential environmental impacts. Product technology is rapidly advancing, but are we, scientists and engineers, keeping pace with understanding and communicating information about potential health risks and effect to the public from these products? EPA's Office of Research and Development charged its laboratories to perform research on nanomaterials to gain understanding of the transformation, transportation, and longevity of these materials. The National Exposure Research Laboratory is specifically responsible for conducting nanotechnology studies.

EPA's National Exposure Research Laboratory-Environmental Sciences Division is researching the detection, characterization, quantification, and monitoring of nanomaterials in the environment. Before the possible risks of these materials in the environment can be determined, methods are needed for their collection, separation, detection, identification and quantification. The areas of research encompass the sources, fate and transport, exposure, and mitigation or prevention of the risks associated with nanomaterials. This research requires technology be applied to detect, identify, quantify and minimize exposure to engineered nanomaterials in the environment. Currently, nano-scale metallic and carbon based particles are not detectable in water or soil.

EPA is concerned with the transport of nanomaterials in the environment which requires understanding transfer mechanisms within environmental media, and measurement of various environmental parameters and nanomaterials parameters. Nanomaterial parameters include particle size and charge, chemical or elemental composition and surface modifications. Environmental media parameters include pH, ionic strength, flow rate, composition, and the presence of naturally occurring (such as dissolved organic carbon) as well as anthropogenic contaminant. Understanding interactions between nanomaterials and environmental media is also critically important.

The goals of this research are to develop and test novel monitoring and characterization methods for engineered nanomaterials (ENMs) with the initial focus on carbon nanoparticles (e.g. fullerenes) and metal nanoparticles (silver and gold) using electrochemical analyses. Increasing ENM content in consumer products may have a negative environmental impact. Many nano-enabled products including fullerene lotions and cosmetics contain reactive oxygen species that are readily introduced as non-point source pollution into the environment. Application onto the skin might pose more problems that may include immediate health concerns. Measurement of these materials is challenging because of their low concentrations and the presence of natural and anthropogenic nanomaterials containing the same elements. We are investigating the use of electrochemical probes with Quartz Crystal Microbalance (QCM) to quantitatively analyze for C_{60} versus C_{70}. The initial environmental matrix of water was utilized in the laboratory and further environmental matrices analyzed will include ground and surface water, soil and sediment.

Approach

A QCM gold electrode is modified with gold nanoparticles (AuNPs) prior to fictionalization using N-dicyclohexylcarbodiimide/N-hydroxysulfosuccinimide (DCC/NHS) chemistry (Figure 1). Decrease in frequency or mass increase is used to measure the adsorption/enumeration of C_{60} by the beta-cyclodextrin amine (ß–CD-NH_2) cavity attached on the QCM.

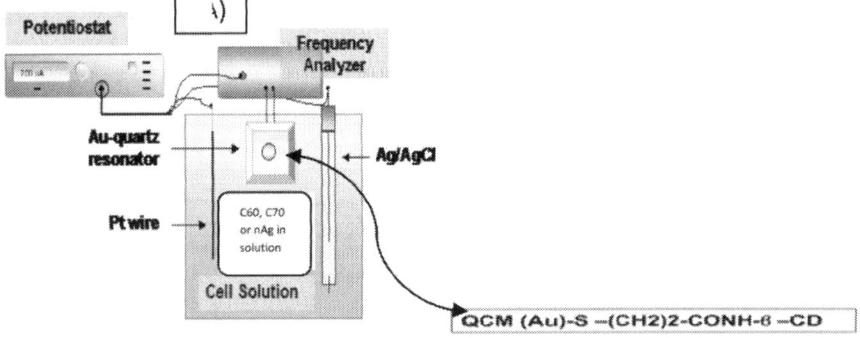

(B)

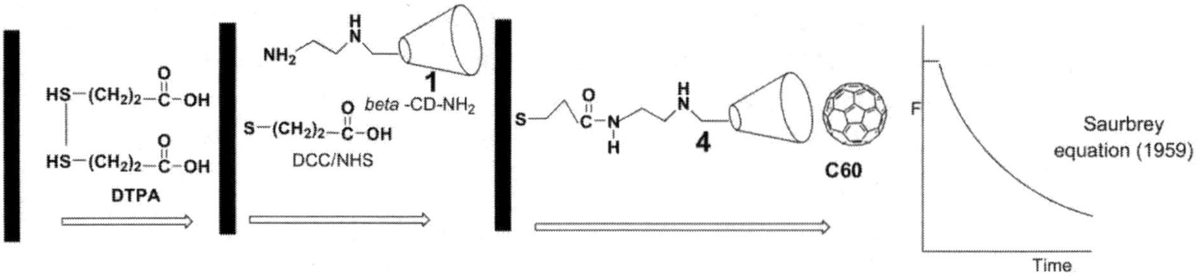

Δf is the change in frequency to coating (Hz), f0 is the frequency of quartz plate (MHz), Δm is the mass of engineered nanomaterial (in pico-grams), A is the area of quartz plate or electrode

Figure 1: *Schematic of modified QCM (A); Sensor design and surface modification using N-DCC/NHS chemistry (B).*

Discussion

The QCM is used to measure mass changes as well as the in-situ measurements of charge, current, potential and the resistance changes following the uptake of the particles. The unique sugar complex bucket-like structure is used to capture the spherical soccer ball-like structured C_{60}. Using amino modified ß–CD (a water-soluble polymer with synthetic receptor), the interactions of selected nanostructured materials (e.g. gold, silver, and fullerenes) are directly monitored by measuring the time-dependent changes in series resonant frequency by the QCM sensor. QCM provides a relationship between the resonance oscillation frequency of piezoelectric crystal and the interfacial mass change. The data are used to determine the quantity of the nanomaterial and comparisons are made with conventional methods such as Ultra Visible Spectroscopy (UV-Vis), Transmission Electron Microscopy (TEM) and Field Emission Scanning Electron Microscopy (FESEM). When the QCM is saturated with the nanomaterial, the attached nanomaterials are quantitatively released from the probe during a regeneration step washing with acid or alkaline hydrogen peroxide due to its selective association and dissociation behavior with C_{60}. And measure the concentration levels of C_{60} in water using simple QCM mass changes Application of a voltage between electrodes results in a shear deformation of the crystal

Results

Reproducible results are obtained using AuNPs in the immobilization chemistry. Figure 2A shows a plot of frequency versus time for the control, the QCM (AuNPs)-S–$(CH_2)_2$-CONH-ß–CD-NH_2 electrode in toluene. The figure demonstrates that no adsorption was noted on the modified QCM. In contrast, the frequency decreased (evidence of mass increase according to the Saurbrey equation) in the presence of 0.14 mg of C_{60} ml toluene (Figure 2B). These results suggest that the sensor can detect C_{60} in solution.

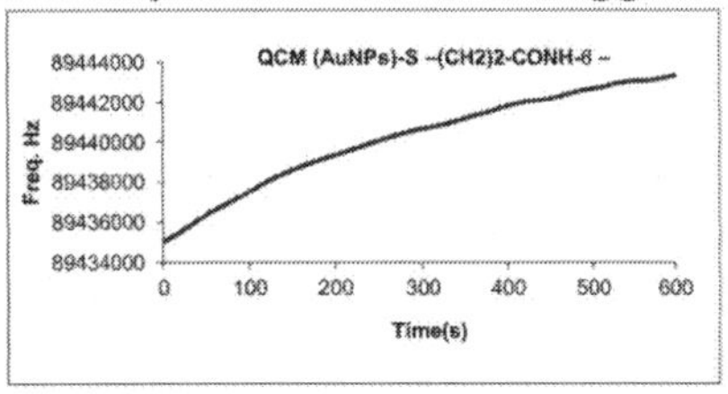

(A)

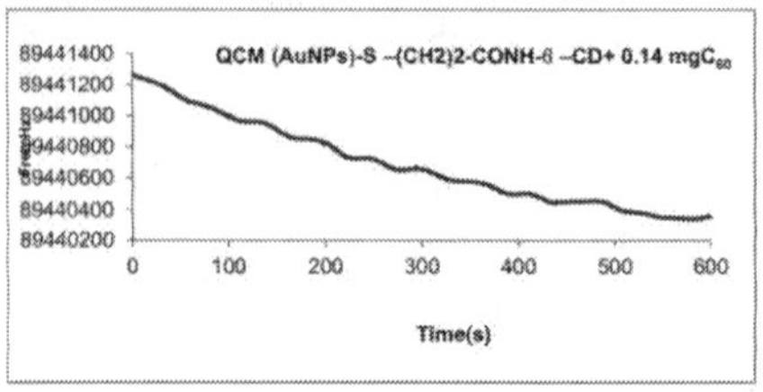

(B)

Figure 2: *Frequency vs. time for QCM (Au)-S-$(CH_2)_2$-CONH-ß-CD in presence of (A) toluene only (B) 0.14 mg C 60 + toluene (run time = 600 seconds)*

Figure 2B shows that the decrease in frequency continued up to 600 seconds. Additional frequency vs time change data showed that at 700s the maximum adsorption of C_{60} from 0.14 mg C_{60} /ml toluene was reached. This is observation indicated that all available active sites on the sensor were saturated. Sensor calibration data also showed that at much lower concentrations, the sensor can be calibrated up to 700 seconds. The QCM sensor has been tested for varying C_{60} concentrations.

Characterization

SEM images show the expected gold SEM micrograph. SEM micrographs of a clean modified QCM electrode (Figure 3A) and after exposure to C_{60} in toluene are shown. In the presence of C_{60} the SEM micrograph showed evidence of near spherical nanoparticles attributed to C_{60} clusters (Figure 3B), which were not observed on the pre-cleaned bare QCM.[1, 2] The presence of fullerene clusters of about 200 nm on the sensor surface suggests the aggregation of fullerenes since they exist as fractal systems which is consistent with previous studies.[3]

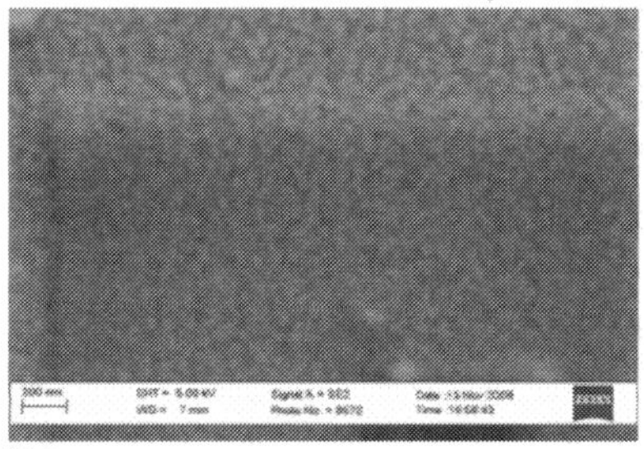

(A)

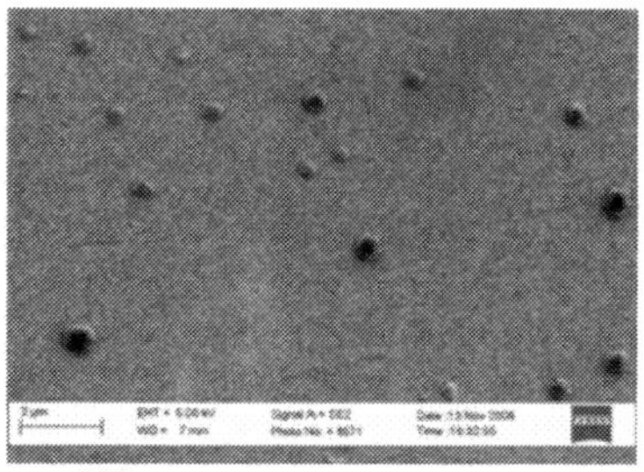

(B)

Figure 3: *Pre-cleaned QCM, image @ 200nm (A) and QCM after evaporating drop of 2.8 mg C_{60}/toluene solution, image@ 200nm (B)*

Conclusions

The outcome of this research is to provide quantitative data to exposure risk assessors that make possible predicative estimates of compounds most likely to be of concern for the protection of human health and the environment. The ultimate goal, to discover and define previously unknown or poorly understood vulnerabilities of nanomaterials, is being accomplished. The successful demonstration using the synthesis of ß-CD-NH_2, its attachment and modification of QCM transducer via DCC/NHS chemistry is a promising new tool to address that goal. The QCM sensor showed specificity for C_{60} extraction when compared to the control solution by measuring the mass changes as well as the in-situ measurements of charge, current, potential and the resistance changes following the absorption and isolation of the particles to determine the presence of nanomaterials. A new practical measurement method for engineered, C_{60}, fullerenes is demonstrated as a new analytical tool; the first nanosensor to measure ENMs utilizing a size exclusive approach for the quantitative measurement of fullerenes.[4]

References:

1. Sadik, O.: Kikandi, S.; Wang, Q.; Varner, K. "Sensors as Tools for Quantitation, Nanotoxicity and Nanomonitoring Assessment of Engineered Nanaomaterials", *J. Environ. Monit.* **2009**, 11, 1782-1800.

2. Liu, Y.; Yang, W.; Chen, Y. Chemical Communications, **2005**, 4208-4210

3. Kikandi, S; Okello, V.; Wang, Q.; Sadik, O.; Varner, K.; Burns, S. "Size-Exclusive Nanosensor for Quantitative Analysis of Fullerene C_{60}", *Environ Science and Tech.* Accepted **2011**.

4. Bulavin, L.; Adamenko, I.; Yashchuk, V.; Ogul'chansky, T.; Prylutskyy, Y.; Durov, S.; Scharff, P. "Self-Organization C60 Nanoparticles in Toluene Solution", J. *of Molecular Liquids* **2001**, 93, (1-3), 187-191.

Work performed by:

Samuel Kikandi (Student Services Contract #RFQ-RT-08-00130), Qiong Wang, Veronic Okello under guidance and mentorship of Omowunmi Sadik, Director, Center for Advanced Sensors & Environmental System, State University of New York-Binghamton, Department of Chemistry, POB 6000, Binghamton, NY 13902-6000

(744a) Salt Induced Irreversible Protein Adsorption with Extremely High Loadings On Electrospun Nanofibers

Ping Wang[1], Chunxia Liu[2], Song-Ping Zhang[3], Guanghui Ma[4] and Zhiguo Su[3], (1)Department of Bioproducts and Biosystems Engineering, University of Minnesota, St Paul, MN, (2)Institute of Process Engineering, Bijing, China, (3)Institute of Process Engineering, Chinese Academy of Sciences, Beijing, China, (4)Institute of Process Engineering, National Key Laboratory of Biochemical Engineering, Beijing, China

LiCl is a kosmotrope that generally promotes protein salvation in aqueous solutions. Herein we report that LiCl embedded in electrospun polymeric nanofibers interestingly induced an abnormal protein adsorption and substantially augmented the adsorption capacity of the fibers. As a result, equilibrium protein loadings reached over 64% (wt/wt) of the dry mass of fibers, 9-fold higher than that observed in the absence of the salt. The adsorption appeared to be irreversible such that little protein loss was observed even after washing the fibers vigorously with fresh buffer solutions. We further examined the application of such intensified protein adsorption for enzyme immobilization. Proteins including bovine serum albumin (BSA) and protamine were first adsorbed, followed by covalent attachment of an outer layer of an enzyme, a-chymotrypsin. Such a multilayer-structured nanofibrous enzyme exhibited extremely high stability with no obvious activity loss even after being incubated for 8 months at 4°C in aqueous buffer solution. The LiCl induced irreversible protein adsorption, which has been largely ignored in previous studies with electrospun materials, rendering an interesting scenario of interfacial protein-material interactions. It also reveals a new mechanism in controlling and fabricating molecular interactions at interfaces for development of a broad range of biomaterials.

See more of this Session: Nanoscale Science and Engineering In Biomolecular Catalysis II
See more of this Group/Topical: Nanoscale Science and Engineering Forum

(744d) Nanomagnetic Multiplex Biology Assay and High THROUGHPUT Biosensing Instrument

Jian-Ping Wang, Electrical Engineering and Computer Science, University of Minnesota, Minneapolis, MN

Bioassays that detect and quantify biomolecules at ultra-low quantity, are of great need in biomedical engineering. Spintronic sensors, such as giant magnetoresistive (GMR) sensor, have emerged as one promising platform for biomolecule detection because of its sensitivity, matrix-insensitive, low-cost and ease-to use. We herein report the development of a novel detection system based on highly sensitive GMR sensor and sub-13nm high-magnetic-moment FeCo nanoparticles, and the usefulness of this system for quantificating protein markers with extremely low concentration down to 150 copies per sensor surface (~10 fM sensitivity) within 5 mins. Magnetic label with high magnetic moment is preferred for the sake of generating larger magnetic signal, thereby higher sensitivity for the entire detection system. Furthermore, micro and sub-micro sized magnetic labels are not favorable for biomolecule detection due to the low diffusivity and poor binding selectivity. We have successfully demonstrated the application of sub-13 nm high moment FeCo magnetic nanoparticles with GMR biosensor for the first time. Furthermore, we will report a bench-top biosensing instrument and the developed nanomagnetic multipex protein-assay, 64-sensor per reaction well (5 mm in diameter), to directly quantify multiple proteins simultanuously and also to monitor and characterize the binding dynamic of the protein with different binding surfaces/sites.

See more of this Session: Nanoscale Science and Engineering In Biomolecular Catalysis II
See more of this Group/Topical: Nanoscale Science and Engineering Forum

(125a) Sustainability: Real World Issues and Real World Solutions

John Leazer, U.S. Environmental Protection Agency, Cincinnati, OH

EPA scientists are not only charged with solving real-world problems related to human health and the environment, but they are also challenged to solve these problems in a manner consistent with the principles of sustainability. Green chemistry addresses the fundamentals of most of the work done within the EPA's chemistry research programs. Green chemistry, first developed within EPA by Dr. Paul Anastas has taken a prominent position within EPA and the entire chemical research community. Advances in green chemistry have reached new levels in the scientific community and society, thereby resulting in unprecedented innovations in chemistry, materials, economics, pollution prevention, and human health.

EPA researchers have taken a leadership role in the development of numerous catalytic systems that have fundamentally changed chemical research in some areas (*e.g.* catalysis) as well as business practices in many sectors (*e.g.* waste remediation and water decontamination). Not only have EPA scientists developed novel methodologies to make new classes of catalysts, but they have also developed these processes in a sustainable manner using green techniques. The work presented herein will focus primarily on recent activities in the arena of catalysis, as well as the human health and environmental impacts resulting thereof.

See more of this Session: Sustainability Plenary
See more of this Group/Topical: Sustainable Engineering Forum

(353e) Award Submission: Photo-Activated Antibacterial and Antiviral Activity of Porphyrin-Conjugated Multiwalled Carbon Nanotubes

Indrani Banerjee, Marc P. Douaisi, Dhananjoy Mondal and Ravi S. Kane, Chemical and Biological Engineering, Rensselaer Polytechnic Institute, Troy, NY

We describe the design of porphyrin-conjugated multi-walled carbon nanotubes (MWNT-P) that show significant activity against *S. aureus* and influenza viruses. *S. aureus* infections can lead to life-threatening situations, especially when caused by antibiotic-resistant strains. Influenza spreads around the world in seasonal epidemics, resulting in thousands of deaths every year, up to millions in some pandemic years. To design antibacterial and antiviral conjugates, we attached Protoporphyrin IX (PPIX) to acid-functionalized multiwalled carbon nanotubes (MWNTs). We selected carbon nanotubes as scaffolds because of their ease of recovery from solution via simple filtration, and their ability to form large area coatings. In the presence of visible light, MWNT-P deactivated *S. aureus* and also significantly reduced the ability of Influenza A virus to infect mammalian cells. Films made of MWNT-P also exhibited potent biocidal activity against *S. aureus*. This work thus provides a facile approach to incorporate porphyrins into coatings and makes them re-usable which may lead to their more widespread use as antibacterial and antiviral agents/coatings.

See more of this Session: Bionanotechnology Graduate Student Award Session
See more of this Group/Topical: Nanoscale Science and Engineering Forum

(420a) Solution Spinning of Neat Carbon Nanotube Fiber, Effect of Carbon Nanotube Length and Processing

N. Behabtu[1], C. Young[2], D. Tsentalovich[2], A. Ma[2], F. Matteni[2], A. Bengio[2], O. Kleinerman[3], J. Schmidt[3], E. Kesselman[3], Yachin Cohen[3], Yeshayahu Talmon[4], R. Hoogerwerf[5], R. Waarbeek[5], J. deJong[5], Marcin Otto[5] and Matteo Pasquali[1], (1)Chemical and Biomolecular Engineering, Rice University, Houston, TX, (2)Rice University, Houston, TX, (3)Technion-Israel Institute of Technology, Haifa 32000, Israel, (4)Department of Chemical Engineering, Technion-Israel Institute of Technology, Haifa 32000, Israel, (5)Teijin Aramid BV, Arnhem, Netherlands

Transferring the exceptional single molecule properties of carbon nanotubes (CNTs) into macroscopic articles such as fibers and films has been a major challenge in the past two decades. Two approaches have merged. The first is solid state processing, where fibers and films are directly spun from the synthesis reactor or nanotube forest. The second is wet spinning, where carbon nanotubes are first dissolved and spun in fibers or cast into films.

Here we show recent progress on super-acid solution spinning. Nanotube length and quality have a significant effect on processability, mechanical and electrical properties. We also show the impact of processing variables on fiber alignment and properties and demonstrate process scalability by demonstrating high throughput spinning, which is orders-of-magnitude higher than solid state processing.

See more of this Session: Poster Session: Nanoscale Science and Engineering
See more of this Group/Topical: Nanoscale Science and Engineering Forum

(204b) Directed Co-Assembly of Live Cells and Colloidal Particles Into Biocomposites with Engineered Structure and Functionality

Orlin D. Velev, Department of Chemical and Biomolecular Engineering, North Carolina State University, Raleigh, NC

We will overview strategies for engineered biocolloidal assembly of live cells and synthetic micro- and nanoparticles. Such directed co-assembly could result in biomaterials, where the functionality of the cells is supplemented by the one of the particles. In the first part of the talk we will discuss how controlled drying of liquid suspensions in thin menisci driven on surfaces can be used for efficient deposition of thin particle-cell coatings. The organization of the particles and/or cells is a result of the combined action of convective transport and capillary forces at the liquid/air interface. The live cell and cell-particle films deposited by the convective assembly process can find applications in hydrogen generating devices, self-cleaning coatings and test beds for toxicity evaluation. The second part of the talk will be focused on more complex and precise methods for biocomposite assembly by electric fields. The process is based on dielectrophoresis (DEP), mobility and interaction of particles in AC fields. We will demonstrate how on-chip dielectrophoresis can be used to assemble chains and membranes of latex spheres and cells such as Baker's yeast, or mouse fibroblasts. Experimental observations and electrostatic simulations of the DEP co-assembly dynamics showed that particles smaller in size than the cells were drawn and captured into the cell junctions by the electric field. The arrays can be bound into permanent biocomposites by using lectin-conjugated spheres as biocolloidal binders. Binding by particles with magnetic cores yielded live cell membranes and chains that could be manipulated by magnetic field and interfaced with on-chip electrodes. Such cell-particle assemblies may find applications in sensors, microassays, microsurgery, or in responsive biomaterials.

See more of this Session: Bionanotechnology: Plenary Session I
See more of this Group/Topical: Nanoscale Science and Engineering Forum

(204c) Exploiting Nanotechnology to Visualize Viral Infection and Deliver Protein Therapeutics

Pin Wang, Chemical Engineering and Materials Science, University of Southern California, Los Angeles, CA

Nanotechnology has many promising applications for biomedicine and biomedical research. We have been exploring nanotechnology-enabled means for elucidating intracellular trafficking features of individual viral particles within target cells. Elucidating molecular mechanisms of viral infection can reveal novel therapeutic opportunities for controlling virus pandemics and pathogenesis. We have recently developed methods to label both enveloped and non-enveloped viruses with semiconductor quantum dots (QDs). It was demonstrated that such labeling schemes can enable us to visualize the detailed steps of viral infection routes and the dynamic interactions between viruses and target cells. The enveloped viruses can be reliably tagged with QDs through the insertion of a short acceptor peptide onto the viral membrane that is further site-specifically biotinylated and attached with streptavidin-conjugated QDs. We have also identified a mild condition to directly conjugate QDs to capsid proteins of non-enveloped viruses. We are able to monitor the trafficking of these QD-tagged viral particles within cells using the live cell imaging technique and reveal some previously unappreciated interactions between viruses and cellular structures. In collaboration with colleagues from other laboratories, we have developed a nanoparticle-based method for intracellular delivery of therapeutic proteins. The method involves the encapsulation of proteins into a nanosized cage that is responsible to the cellular environment for cargo release. Currently we are exploring this delivery system for therapeutic applications against cancer.

See more of this Session: Bionanotechnology: Plenary Session I
See more of this Group/Topical: Nanoscale Science and Engineering Forum

(204a) Self-Folding Materials and Devices for Biomedical Applications

David H. Gracias, Department of Chemical and Biomolecular Engineering, Johns Hopkins University, Baltimore, MD

The human body is a 3D structure and elaborately patterned from the nano to the macroscale. Hence, there is a need to extend planar micro and nanopatterning techniques into the third dimension to enable precisely structured diagnostic and therapeutic devices. Self-folding refers to self-assembly processes wherein planar structures fold up spontaneously, typically when released from a substrate or exposed to specific stimuli. The processes are versatile and can be used across length scales and with diverse materials. I will describe the use of self-folding methods to fabricate precisely patterned hollow polyhedral containers, physical and chemical scaffolds and bio-chemically actuated wireless surgical tools for less-invasive surgical diagnostics. These examples highlight the use of self-folding to achieve unprecedented precision and machine-based autonomous functionality on a hierarchy of length scales for micro and nanostructured biomedical implants and devices.

See more of this Session: Bionanotechnology: Plenary Session I
See more of this Group/Topical: Nanoscale Science and Engineering Forum

(277a) Multifunctional Nanocomposites for Single Cell and Molecule Manipulation

Jessica O. Winter, Chemical and Biomolecular Engineering and Biomedical Engineering, the Ohio State University, Columbus, OH

As the promise of individual nanoparticles begins to be realized, researchers are attempting to combine the properties of different nanomaterials into a single multifunctional construct. Here, I will describe our recent efforts to create fluorescent-magnetic nanocomposites using direct synthesis, bifunctional crosslinking, and encapsulation strategies, addressing the specific advantages and disadvantages of each approach. These composites, which combine quantum dots and superparamagnetic iron oxide nanoparticles, display both the fluorescence and magnetism properties of their constituent components. As such, they are particularly suited to applications in single cell/molecule tracking and separations. I will also discuss our recent successes in these areas including the development of alternating color quantum dots with near-continuous fluorescence and magnetic quantum dots, which permit simultaneous magnetic manipulation and fluorescent tracking of individual particles, molecules, and cells. Taken together, these technologies herald new possibilities in nanoscale manipulation, assembly, and chromatography.

See more of this Session: Bionanotechnology: Plenary Session II
See more of this Group/Topical: Nanoscale Science and Engineering Forum

(277c) Understanding the Properties of Biomolecular Motifs and Their Applications

Rajesh R. Naik, Materials and Manufacturing Directorate, Air Force Research Laboratory, Wright-Patterson-AFB, OH

Biological systems exemplify the utilization of highly specific recognition processes with a diverse set of building blocks for the synthesis and assembly of precisely defined nanomaterials. Consequently, these specialized biological components are appealing for materials synthesis and have included the use of oligonucleotides, proteins, viruses, and antibodies as the primary organizational element. Researchers exploit the knowledge gained by understanding how biological materials are designed to inspire the design, fabrication and assembly of functional materials with tailored properties. Our work is aimed at understanding how to utilize the attractive characteristics of biological components to develop hybrid functional materials and devices. In addition, we will describe some our recent work on understanding biomolecular interactions with materials using both experimental and modeling tools. The fundamental understanding gained from studying the unique properties of biological processes and materials will lead to identifying strengths (as well as limitations) of biomaterials, aimed towards addressing the material needs for a variety of applications including structural materials, sensing and catalysis.

See more of this Session: Bionanotechnology: Plenary Session II
See more of this Group/Topical: Nanoscale Science and Engineering Forum

(277b) Development of Virus/Polymer Chimeras As Gene Delivery Vectors

Daniel W. Pack, Chemical & Biomolecular Engineering, University of Illinois Urbana-Champaign, Urbana, IL

Human gene therapy holds the potential to revolutionize treatment of diseases ranging from cystic fibrosis to cardiovascular disease to cancer. The primary bottleneck hindering the technology, however, is delivery of the therapeutic genetic material. Recombinant viruses are highly efficient and are used in the majority of gene therapy clinical trials, but face serious safety concerns, are difficult to target to specific cells and tissues, and are expensive to produce. Polymer-based vectors are generally safer, less expensive and more versatile, but lack the efficiency needed for clinical relevance. The success of human gene therapy awaits development of safe and efficient methods for gene delivery. We have recently pioneered a top-down approach to construction of novel vectors assembled from viral and synthetic components that are designed to exhibit advantages of both types of vectors. Thus, we have prepared hybrid vectors through electrostatic complexation of retro- or lentivirus-like particles with polycations including polylysine, polyethylenimine and chitosan. These hybrid vectors are internalized by cells via macropinocytosis, the polymer provides escape from macropinosomes into the cytosol, and the viral components provide highly efficient trafficking of their genetic cargo into the nucleus. Further, targeting moieties may be conjugated to the polymer to provide cell-specific delivery. This new class of vectors provides gene delivery efficiency similar to native viruses, provides genomic integration for sustained transgene expression, and is capable of transfecting quiescent cells. Hybrid vectors, therefore, represent a new and promising approach for safe and efficiency gene delivery.

See more of this Session: Bionanotechnology: Plenary Session II
See more of this Group/Topical: Nanoscale Science and Engineering Forum

AUTHOR INDEX

AUTHOR INDEX

AUTHOR INDEX

AUTHOR INDEX

AUTHOR INDEX

AUTHOR INDEX

AUTHOR INDEX

AUTHOR INDEX

AUTHOR INDEX